D1191455

ORGANIC CHARGE-TRANSFER COMPLEXES

ORGANIC CHEMISTRY
A SERIES OF MONOGRAPHS

Edited by

ALFRED T. BLOMQUIST

Department of Chemistry, Cornell University, Ithaca, New York

ORGANIC CHARGE-TRANSFER COMPLEXES

R. FOSTER

Chemistry Department
University of Dundee, Scotland

1969
ACADEMIC PRESS
London and New York

ACADEMIC PRESS INC. (LONDON) LTD.
Berkeley Square House
Berkeley Square
London, W1X 6BA

U.S. Edition published by
ACADEMIC PRESS INC.
111 Fifth Avenue
New York, New York 10003

Copyright © 1969 by ACADEMIC PRESS INC. (LONDON) LTD.

Library of Congress Catalog Card Number: 70-85459

SBN: 12–262650–8

Printed in Great Britain by
Spottiswoode, Ballantyne & Co. Ltd.
London and Colchester

Preface

This book describes various aspects of interactions commonly termed "charge-transfer complexes". They comprise electron-donor–electron-acceptor associations for which an intermolecular electronic-charge-transfer transition (or transitions) is usually observed.

The expression "charge-transfer", used to describe these complexes, has been criticized on the grounds that in many systems the contribution of charge-transfer forces to the stabilization of the ground state is negligible, and that in any case the term presupposes a valence bond description of the complex. However, the term is widely used in this sense; other terminology tends to be either too general or too cumbersome. This problem of nomenclature is discussed further in Chapter 1 and elsewhere in the book.

The large amount of work which has been published in this field has meant that an effectively comprehensive study is impracticable. Although, therefore, some selection has been unavoidable, an attempt has been made to make the coverage reasonably wide within the field of *organic* charge-transfer complexes. Nevertheless, I am aware that the more detailed treatment of certain aspects reflects my particular interests.

There has been a general coverage of the literature up to the beginning of 1968, although a few more recent references are also included. The present rapid development of this field is indicated by the fact that at least 400 more papers relevant to this book have been published within the past year.

I am indebted to many colleagues and friends for their help and encouragement. My interest in charge-transfer complexes was initiated by the late Dr. D. Ll. Hammick, F.R.S., who was one of the pioneers in the study of these systems. My thanks are particularly due to the research students who have collaborated with me and have helped so much to maintain the interest in this work. Individual references to their published and unpublished contributions are made throughout the book.

Finally, I should like to thank Academic Press for their helpfulness, consideration and patience at all stages of publication.

Dundee
May, 1969   R. FOSTER

Acknowledgements

Grateful acknowledgements are due to the following for permission to use published data. References to the authors concerned are made in the text.

The Executive Secretary, Acta Chemica Scandinavica: Figs. 8.7, 8.14, 8.15, 8.16, 8.17.

The American Chemical Society (*Journal of the American Chemical Society*): Figs. 3.7, 3.12, 3.15, 3.17, 6.4, 7.3, 7.4, 7.9, 11.2; (*Journal of Physical Chemistry*): Fig. 4.1; (*Inorganic Chemistry*): Fig. 8.13.

The American Institute of Physics (*Journal of Chemical Physics*): Figs. 3.22, 8.20, 9.2, 9.4, 9.5, 9.6, 9.7, 9.9, 12.2.

Annual Reviews, Inc. (*Annual Reviews of Physical Chemistry*): Fig. 3.6.

The Chemical Society (*Journal of the Chemical Society*): Figs. 8.27, 8.28, 8.30, 8.31, 8.32, 8.33.

The Chemical Society of Japan (*Bulletin of the Chemical Society of Japan*): Figs. 3.3, 3.20, 8.21, 8.22, 8.29, 9.1, 9.3.

Elsevier Publishing Company (*Biochimica et Biophysica Acta*): Fig. 12.1.

The Faraday Society (*Transactions of the Faraday Society*): Fig. 3.2.

International Union of Crystallography (*Acta Crystallographica*): Figs. 8.8, 8.9, 8.10, 8.23, 8.24, 8.26, 8.34.

Koninklijke Nederlandse Chemische Vereniging (*Recueil des Travaux chimiques des Pays-bas*): Fig. 3.16.

National Research Council of Canada (*Canadian Journal of Chemistry*): Fig. 9.8.

Pergamon Press Ltd. (*Tetrahedron*): Fig. 3.9; (*Spectrochimica Acta*): Fig. 6.10; (Progress in Nuclear Magnetic Resonance Spectroscopy Vol. 4 ed. J. W. Emsley, J. Feeney and L. H. Sutcliffe): Fig. 5.1.

The Photoelectric Spectrometry Group (*Bulletin of the Photoelectric Spectrometry Group*): Fig. 11.1.

The Royal Society (*Proceedings of the Royal Society*): Fig. 4.2.

Springer-Verlag (Elektronen-Donator-Acceptor-Komplexe by G. Briegleb): Fig. 7.11.

Taylor and Francis, Ltd. (*Molecular Physics*): Figs. 9.10, 9.11.

Contents

List of Commonly Used Symbols

A = electron acceptor

D = electron donor

AD = 1:1 charge-transfer complex

ν_{CT} = frequency of intermolecular charge-transfer band

E_{CT} = energy of intermolecular charge-transfer band

I^D = ionization potential of an electron donor

E^A = electron affinity of an electron acceptor

A = absorbance [optical density]

ϵ = molar absorptivity [molar (decadic) extinction coefficient]

δ = chemical shift with reference to some non-interacting species

Δ = population-averaged chemical shift involving a component in the free and complexed states with reference to the same nucleus in the uncomplexed component

Δ_0 = chemical shift in a completely complexed component with reference to the same nucleus in the uncomplexed state

K = association constant

K^{AD} = association constant for the equilibrium $A + D \rightleftharpoons AD$

$K_c^{AD} = K^{AD}$ (*molar* concentration units)

$K_x^{AD} = K^{AD}$ (*mole-fraction* concentration units)

$K_m^{AD} = K^{AD}$ (*molal* concentration units)

$K_r^{AD} = K^{AD}$ (mol/kg solution concentration units)

Chapter I

General

I.A. Introduction

The historical development of chemistry has led to the concept of the molecule. Although this idea has been vital, not all associations of atoms conform to the concept. The anomalous "unsaturated valencies" or "residual affinities" of an earlier phase in organic chemistry are a reflection of our inadequate description, rather than the implied peculiarities of chemical interaction. All molecules interact with other molecules, but for the most part the intermolecular forces involved are small compared with the inter-atomic forces within molecules, and there is no problem in defining the molecule and concluding that intermolecular forces are "non-chemical" in the sense that they do not require a description in terms of classical covalent or ionic bonding. Nevertheless, while not being immediately amenable to description in terms of classical chemical bonds, interactions between molecules may be sufficiently strong or show such features as to make it impossible to provide an explanation in terms of dispersion, dipole–dipole, dipole–induced-dipole and suchlike van der Waals forces. The recognition of the usefulness of the concept of the molecule, and subsequently of chemical bonding within molecules, has led to a reluctance to recognize the possibility of graded interaction between the extremes of the classical covalent or ionic bond and the weakest van der Waals interactions.

Even when we do recognize this possibility, there is a danger of further compartmentalizing the systems of intermediate interaction into various distinct types, members of each group being expected to conform to type. There may be many apparently obvious representations of a particular type—for example hydrogen-bonded complexes. However, this does not mean that an interaction which only marginally conforms necessarily needs a separate special explanation.

I.B. Definition of the Term "Complex" in Organic Chemistry

It is difficult to find an adequate definition of the term "complex" because its usage has been so varied. On the one hand it has been used to describe the product of the weak reversible interaction between two or more components to account for small deviations from ideality. It has also been used as the description of the product of strong interactions where new covalent (σ) bonds are formed. The terminology is further confused by the use of the expression "molecular compound" by some authors as a synonym of (molecular) complex.

In this book the term "complex" will be taken to mean, experimentally, a substance formed by the interaction of two or more component molecules (and/or ions), which may have an independent crystal structure and which will reversibly dissociate into its components, at least partially, in the vapour phase and on dissolution. This definition suggests that there is no, or only a very small, contribution from covalent binding in the ground state. However, it must be recognized that there is a gradation from these weaker interactions to classical covalent (σ) bond formation.

I.C. Charge-Transfer (Electron Donor–Acceptor) Complexes

The complexes presently discussed are formed by the weak interaction of electron donors with electron acceptors. Using valence-bond theory, a rational explanation of many of the ground-state properties has been given in terms of a structure which involves mainly dispersion, dipole and similar forces together with a usually small contribution from a covalent dative structure in which one electron has been transferred from the donor to the acceptor component of the complex (see Chapter 2). The observed electronic absorption spectra, characteristic of the complex as a whole, are accountable as intermolecular charge-transfer transitions (Chapter 3).

There is not complete agreement concerning a general term to describe these interactions. Mulliken[1-3] has called all electron donor–electron acceptor interactions (*electron*) *donor–acceptor complexes*, and has emphasized the range of component sub-types by a detailed terminology. Interactions in which covalent (σ) bonds are formed are automatically included. In other contexts such products would simply be considered as new covalent compounds.

A criticism of the term "charge transfer" is that it reflects the particular valence-bond description.[4] In many so-called "charge-transfer" complexes, charge-transfer forces do not provide the major contribution to the binding forces in the ground state. This point has been made strongly by Dewar and his co-workers.[5, 6] Similar conclusions have been drawn from recent considerations of the measured dipoles of "charge-transfer" complexes in the ground state.[7] Measurements of halogen pure-electrical quadrupole resonances for complexes of p-xylene with carbon tetrabromide and with carbon tetrachloride, and for the benzene–bromine complex suggest that there is little or no transfer of charge in the ground states of these complexes.[8] Nevertheless, the term has a wide common usage to describe the weak interactions between electron donors and electron acceptors which are the subject of this book. This terminology will be retained, although it must be emphasized that the term cannot be taken to imply that transfer of charge is the major mode of binding in these complexes.

Charge-transfer complexes usually involve simple integral ratios of the components,* the enthalpy of formation is usually of the order of a few kcal/mol or less, and the rates of formation and decomposition into the components are so high that the reactions appear to be instantaneous by normal techniques. An electronic absorption extra to the absorption of the components is often observed. This is the result of an intermolecular charge-transfer transition involving electron transfer from the donor to the acceptor (see Chapters 2 and 3). Consequently, for many so-called charge-transfer complexes it would be semantically more correct to speak of "complexes showing charge-transfer absorption". However, this expression lacks the facileness of "charge-transfer complexes". Furthermore, there are many cases of complexes so classified in which it is experimentally difficult to demonstrate the presence of the supposed charge-transfer absorption.

It should be emphasized that charge-transfer interaction in the ground state is not an exclusive process. The description of a particular complex as, for example, a hydrogen-bonded or a charge-transfer complex is, on occasion, only one of differing terminology.

I.D. Electron Donors and Electron Acceptors

The components of the complex have been described as electron donors and electron acceptors. This description is preferred to *Lewis* bases and acids, since Lewis's concept originally applied to the donation or

---

* Exceptions are known: for example, the iodine complexes of perylene, violanthrene and p-phenylenediamine (see Chapter 9).

acceptance of *pairs* of electrons,[9] whereas we shall not wish to make such a restriction.

In a wider context Mulliken[2] has divided even-electron donor and acceptor species into increvalent and sacrificial types. Increvalent donors are lone-pair (n)-donors such as aliphatic amines, amine oxides, ethers, phosphines, sulphoxides, alcohols, iodides and the like. Some of these interact strongly with electron acceptors. Sacrificial donors are compounds which donate an electron from a bonding orbital. They include σ-donors such as hydrocarbons, especially small cyclic hydrocarbons. These are very weak electron donors. By contrast, sacrificial π-donors such as aromatics, particularly polycyclic systems and systems containing electron-releasing groups, may be strong donors. Some compounds such as aza-aromatics and aromatic amines may behave as n-donors towards some acceptors and π-donors towards others (see Chapters 3 and 7). Relatively little work has been reported on aliphatic π-donor hydrocarbons. It is doubtful whether the colours obtained by the addition of tetranitromethane to olefins are the result of charge-transfer complex formation (see Chapter 10).

Increvalent acceptors are of the vacant orbital (v) type. Many of these, for example B(hal)$_3$, Al(hal)$_3$, SnCl$_4$ are outside the scope of the present discussion. The binding of electron donors (particularly n-donors) to such acceptors is often so strong that covalent binding results. However, with other v-acceptors such as Ag$^+$, weak charge-transfer complexing may occur. Sacrificial acceptors may be of the σ- or π-type. σ-Acceptors include the hydrogen halides, the weakly accepting halo-substituted paraffins and the more strongly accepting halogens and pseudo-halogens, e.g. I$_2$, Br$_2$, ICl, ICN. π-Acceptors are the most common organic acceptors. They include aromatic systems containing electron-withdrawing substituents such as nitro, cyano and halo, also acid anhydrides, acid chlorides and quinones,[10] as well as planar aromatic cations (see Chapter 10). Some π-acceptors such as 1,3,5-trinitrobenzene, p-benzoquinone, chloranil (tetrachloro-p-benzoquinone) and picric acid (2,4,6-trinitrophenol) are well known as complexing agents. A list of some more novel π-acceptors is given in Table 1.1. Apart from these species, compounds with unpaired electrons may behave as electron donors or acceptors, e.g. I, O$_2$, diphenylpicrylhydrazyl (DPPH).

Although this division into electron donor and electron acceptor species may be made, it must be emphasized that the terms are only relative. A single species may, for example, behave both as a donor and as an acceptor. This may possibly occur in self-complexes of benzene.[11] Similarly, it has been suggested that the larger condensed polycyclic aromatic hydrocarbons, which are good donors, should also be good

TABLE 1.1. Some π-electron acceptors

Formula	Name*	Reference†
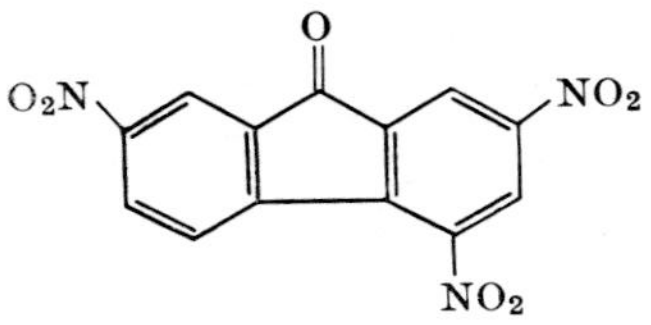	2,4,7-Trinitrofluorenone	M. Orchin and E. O. Woolfolk, *J. Am. chem. Soc.* **68**, 1727 (1946)
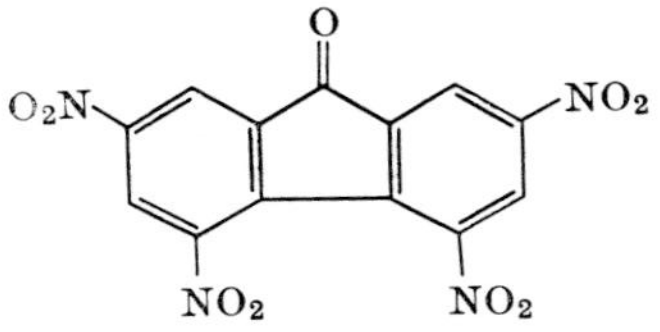	2,4,5,7-Tetranitrofluorenone	M. S. Newman and H. Boden, *Org. Synth.* **42**, 95 (1962)
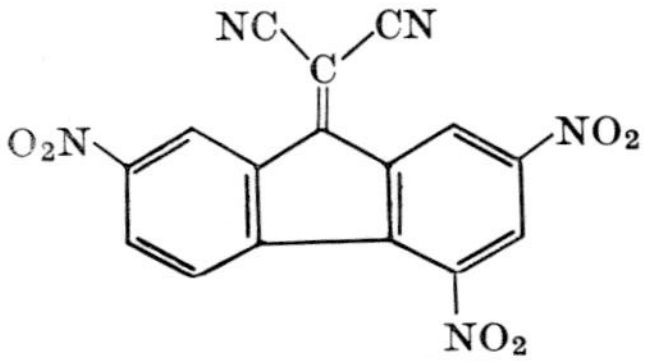	9-Dicyanomethylene-2,4,7-trinitrofluorenone	T. K. Mukherjee and L. A. Levasseur, *J. org. Chem.* **30**, 644 (1965)
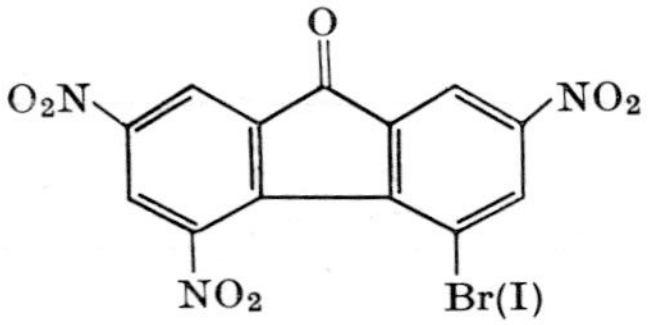	4-Bromo- and 4-iodo-2,5,7-trinitrofluorenone	M. S. Newman and J. Blum, *J. Am. chem. Soc.* **86**, 5600 (1964)

TABLE 1.1.—*continued*

Formula	Name*	Reference†
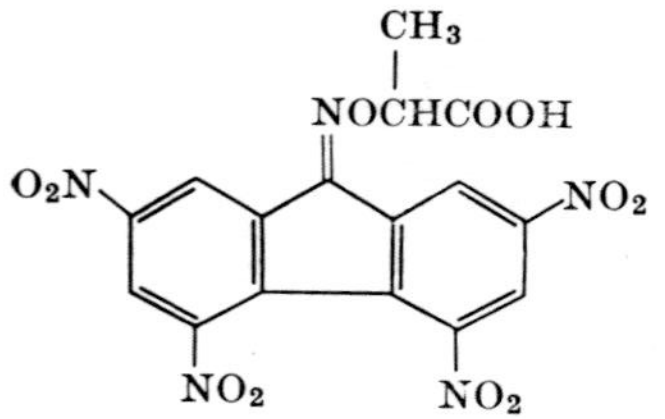	2-(2,4,5,7-Tetranitro-9-fluorenylideneaminoöxy)-propionic acid	M. S. Newman and W. B. Lutz, *J. Am. chem. Soc.* **78**, 2469 (1956)
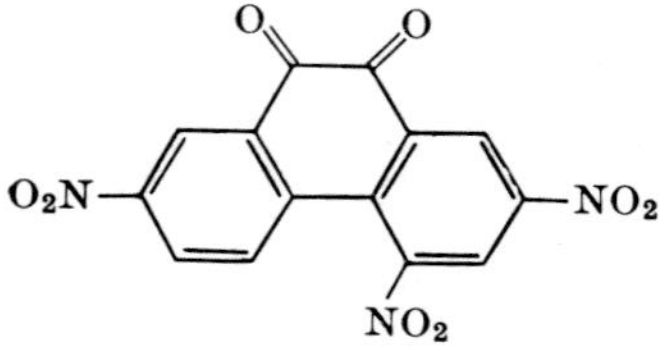	2,4,7-Trinitrophenanthraquinone (also 2,7-; 3,6- and 2,5-dinitro)	T. K. Mukherjee, *J. phys. Chem., Ithaca* **71**, 2277 (1967)
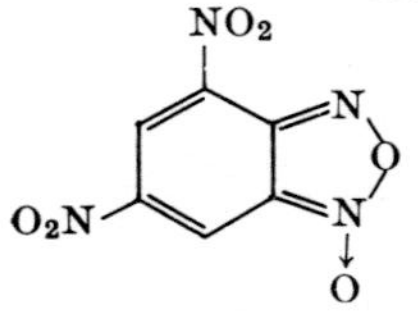	4,6-Dinitrobenzofuroxan	P. Drost, *Justus Liebigs Annln Chem.* **307**, 49 (1899); A. S. Bailey and J. R. Case, *Proc. chem. Soc.* 176, 211 (1957); *Tetrahedron* **3**, 113 (1958)
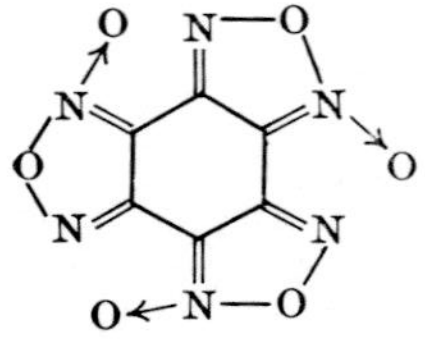	Benzotrifuroxan (sometimes described as hexanitrosobenzene)	A. S. Bailey and J. R. Case, *Proc. chem. Soc.* 176 (1957); *Tetrahedron* **3**, 113 (1958); A. S. Bailey, *J. chem. Soc.* 4710 (1960)

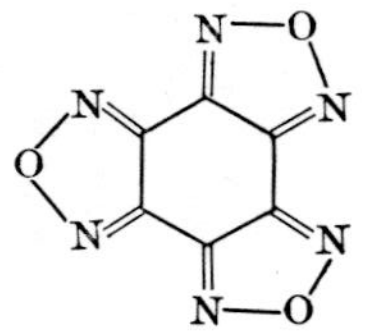

Benzotrifurazan

A. S. Bailey and J. M. Evans, *Chemy Ind.* 1424 (1964); *J. chem. Soc.* (C) 2105 (1967)

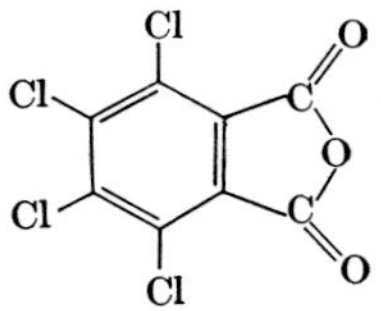

Tetrachlorophthalic anhydride

J. Czekalla, G. Briegleb, W. Herre and R. Glier, *Z. Elektrochem.* **61**, 537 (1957)

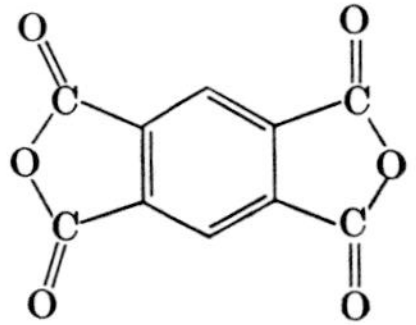

Pyromellitic dianhydride

L. L. Ferstandig, W. G. Toland and C. D. Heaton, *J. Am. chem. Soc.* **83**, 1151 (1961)

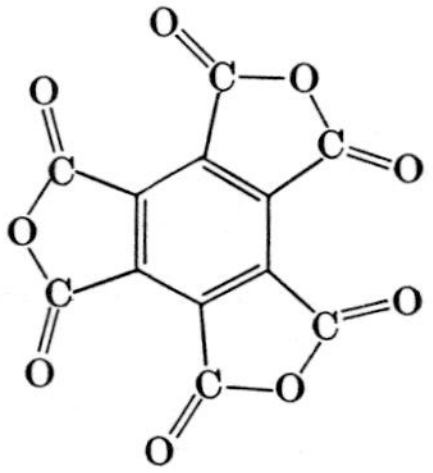

Mellitic trianhydride

I. S. Mustafin, *J. gen. Chem. U.S.S.R.* **17**, 560 (1947); H. M. Rosenberg, E. Eimutis and D. Dale, *J. phys. Chem., Ithaca* **70**, 4096 (1966)

TABLE 1.1.—*continued*

Formula	Name*	Reference†
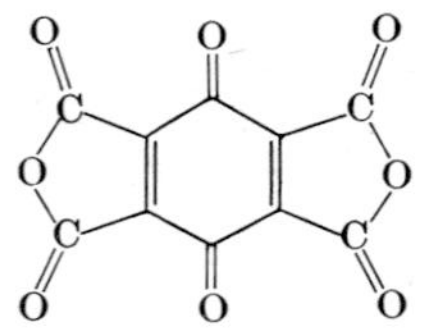	*p*-Benzoquinone tetracarboxylic dianhydride	P. R. Hammond, *Science, N.Y.* **142**, 502 (1963)
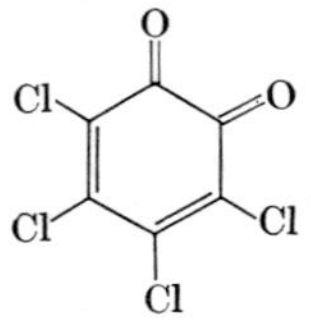	*o*-Chloranil(tetrachloro-*o*-benzoquinone)	D. R. Kearns and M. Calvin, *J. Am. chem. Soc.* **83**, 2110 (1961)
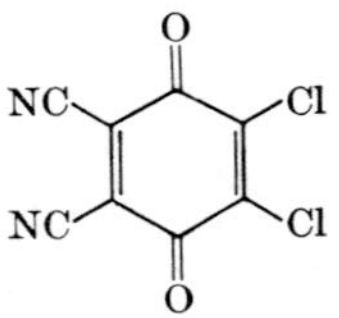	2,3-Dichloro-5,6-dicyano-*p*-benzoquinone (DDQ)	P. R. Hammond, *J. chem. Soc.* 3113 (1963)
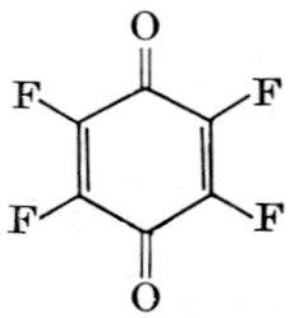	Fluoranil (tetrafluoro-*p*-benzo-quinone)	N. M. D. Brown, R. Foster and C. A. Fyfe, *J. chem. Soc.* (B), 406 (1967)

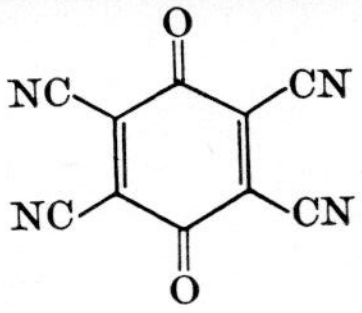

Compound	Structure	Reference
Tetracyano-*p*-benzoquinone		K. Wallenfels and G. Bachmann, *Angew. Chem.* **73**, 142 (1961); K. Wallenfels, G. Bachmann, D. Hofmann and R. Kern, *Tetrahedron* **21**, 2239 (1965)
Indanetrione		A. R. Lepley and J. P. Thelman, *Tetrahedron* **22**, 101 (1966)
2-Dicyanomethylene indane-1,3-dione		S. Chatterjee, *Science, N.Y.* **157**, 314 (1967)
7,7,8,8-Tetracyanoquinodimethane (TCNQ)		D. S. Acker, R. J. Harder, W. R. Hertler, W. Mahler, L. R. Melby, R. E. Benson and W. E. Mochel, *J. Am. chem. Soc.* **82**, 6408 (1960); D. S. Acker and W. R. Hertler, *J. Am. chem. Soc.* **84**, 3370 (1962)
11,11,12,12-Tetracyano-1,4-naphthaquinodimethane		S. Chatterjee, *J. chem. Soc.* (B), 1170 (1967)

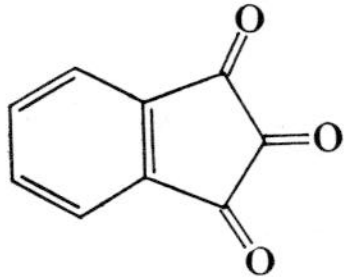

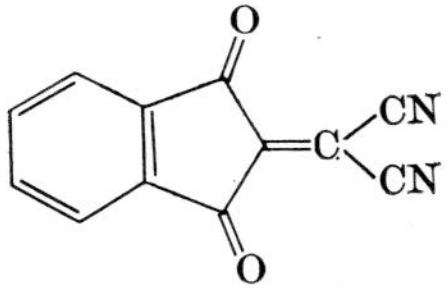

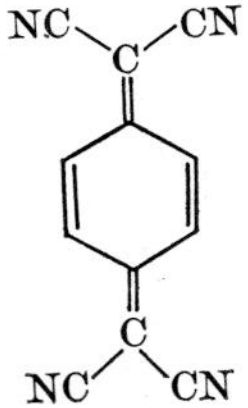

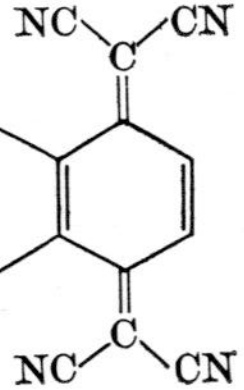

TABLE 1.1.—*continued*

Formula	Name*	Reference†
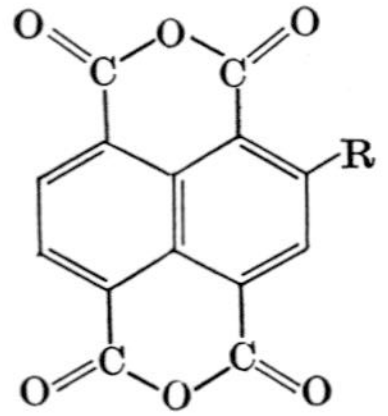	Naphthalene 1,4,5,8-tetra-carboxylic dianhydride (R = H) and the 2-bromo derivative (R = Br)	P. Jacquignon and N. P. Buu-Hoï, *Compt. Rend.* **249**, 717 (1959); P. Jacquignon, N. P. Buu-Hoï and M. Mangane, *Bull. Soc. chim. Fr.* 2517 (1964)
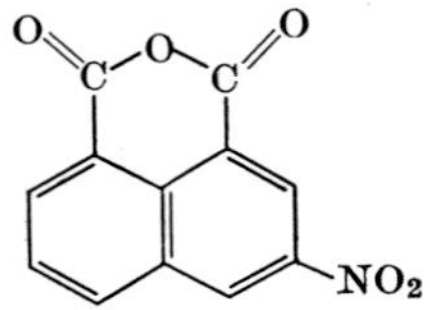	3-Nitro-1,8-naphthalic anhydride	I. Ilmet and S. A. Berger, *J. phys. Chem., Ithaca* **71**, 1534 (1967)
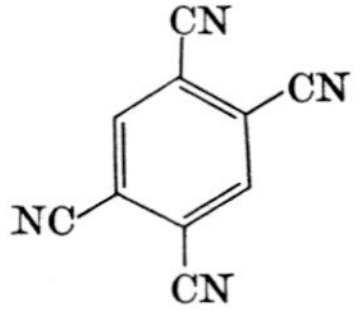	1,2,4,5-Tetracyanobenzene (pyromellitonitrile)	A. S. Bailey, B. R. Henn and J. M. Langdon, *Tetrahedron* **19**, 161 (1963); A. Zweig, J. E. Lehnsen, W. G. Hodgson and W. H. Dura, *J. Am. chem. Soc.* **85**, 3937 (1963); S. Iwata, J. Tanaka and S. Nagakura, *J. Am. chem. Soc.* **88**, 894 (1966)
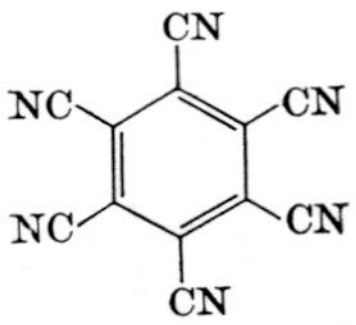	Hexacyanobenzene	K. Wallenfels and K. Friedrich, *Tetrahedron Lett.* 1223 (1963)

Structure	Compound	Reference
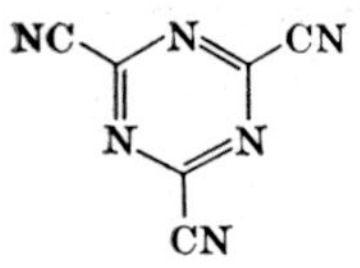 (structure: NC, N, N, N, CN, CN)	s-Tricyanotriazine	A. S. Bailey, B. R. Henn and J. M. Langdon, *Tetrahedron* **19**, 161 (1963)
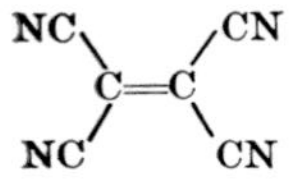 (structure: NC, NC, CN, CN)	Tetracyanoethylene (TCNE)	T. L. Cairns, R. A. Carboni, D. D. Coffman, V. A. Engelhardt, R. E. Heckert, E. L. Little, E. G. McGeer, B. C. McKusick, W. J. Middleton, R. M. Scribner, C. W. Theobald and H. E. Winberg, *J. Am. chem. Soc.* **80**, 2775 (1958); R. E. Merrifield and W. D. Phillips, *J. Am. chem. Soc.* **80**, 2778 (1958); K. Vasudevan and V. Ramakrishnan, *Rev. pure appl. Chem.* **17**, 95 (1967)
$CF_3C(CN){=}C(CN)CF_3$	*cis*- and *trans*-1,2-Dicyano-1,2-bis(trifluoromethyl)ethene	S. Proskow, H. E. Simmons and T. L. Cairns, *J. Am. chem. Soc.* **85**, 2341 (1963)
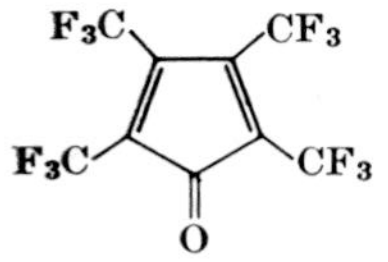 (structure: F_3C, CF_3, F_3C, CF_3, O)	Tetrakis(trifluoromethyl)cyclopentadienone	R. S. Dickson and G. Wilkinson, *J. chem. Soc.* 2699 (1964)
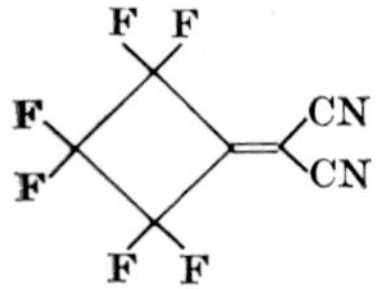 (structure: CN, CN, F F, F F)	Dicyanomethylenehexafluoro-cyclobutane	W. J. Middleton, *J. org. Chem.* **30**, 1402 (1965)

* Abbreviations which are widely used are also included.

† In general, the references given refer either to the original or to some particularly relevant, more recent paper describing the compound as a π-electron acceptor. Neither the list of compounds, nor the references, are intended to be exhaustive. Many references to these and other acceptors appear throughout the book, in particular see the Compound Index and Appendix Table 2, page 396.

acceptors.[12] Evidence for such behaviour has been obtained from interactions of pyrene, anthracene and perylene with the very strong electron
donor tetrakis(dimethylamino)ethylene (I).[13] Likewise 4,4'-dinitro-*p*-
terphenyl (II) has been shown to have the characteristic properties of an
electron donor and an electron acceptor as demonstrated by its various
charge-transfer complexes.[14, 15] 1,3-Dimethylalloxazine (III) has
also been shown to complex with electron donors and electron
acceptors.[16]

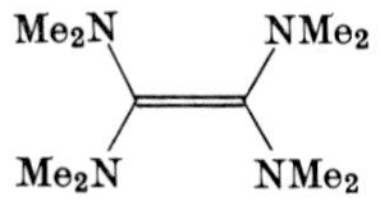

I

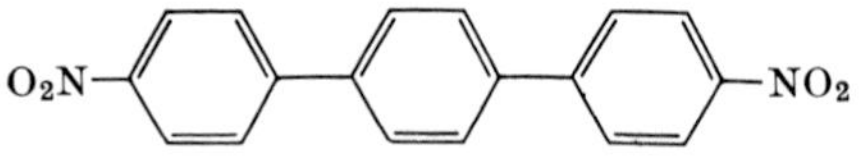

II

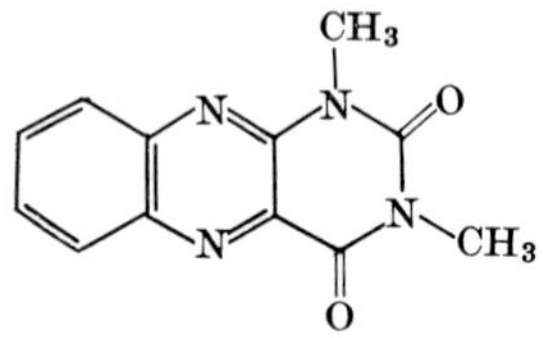

III

I.E. Limitations of the Present Study

The present discussion is primarily limited to charge-transfer complexes formed from organic species. By and large, complexes of metals
and of covalent inorganic components, such as the silver ion, aluminium
chloride, stannic chloride, boron trifluoride, are excluded. Exceptionally,
complexes of the halogens and pseudo-halogens are included. Complexes
of metal chelates with electron acceptors, in which the intermolecular
interaction is at least partially through the organic ligand, are discussed
(Chapter 8), as are the complexes of metallocenes with electron acceptors
in which the binding appears to be via the cyclopentadienyl moiety
rather than the metal atom (Chapter 10).

I.F. Historical Development[17]

For many years, chemists recognized that addition products may be obtained from stable compounds, the formation of which is contrary to the normal rules of valency. An early compilation of the properties of such products was made by Pfeiffer.[18] Many are highly coloured. Often the products persist in solution, in equilibrium with the component molecules formed by reversible dissociation. Many of these products we should now describe as charge-transfer complexes. Mulliken's now widely accepted description[1–3] of these complexes (Chapter 2) represents the evolutionary result of many, often apparently conflicting, earlier theories.

Sudborough[19] had depicted the products as having structures in which the component molecules were covalently bonded together. Pfeiffer[18] suggested that bonding occurred through the saturation of "residual valencies" in the component molecules. Bennett and Willis[20] considered that such an explanation could neither account for the integral, usually 1:1, stoichiometry shown by the components of the complex, nor for the colour. At a time when the nature of light absorption was not well-founded, they and others adhered to the thesis that colour formation implied compound formation, i.e. that the components were covalently bonded. For example, they denoted the nitro-compound–aromatic hydrocarbon complexes in such a way. They recognized that the two components were respectively a good electron acceptor and a good electron donor. However, the apparently instantaneous attainment of equilibrium in solution disfavoured the concept of covalent bonding.

A more specific expression of the residual valency theory was given by Briegleb[21, 22] who suggested that the hydrocarbon–nitro-compound adducts were the result of the electrostatic attraction between the localized dipoles of the nitro groups and the induced dipoles in the hydrocarbons. By assuming a separation of 3Å, he could justify the bonding energy of a few kcal/mol which had been observed. Briegleb and Schachowskoy[23] emphasized that their interpretation was in terms of localized dipoles so that a molecule such as 1,3,5-trinitrobenzene, while having an overall zero dipole moment, is nevertheless an effective polarizing component in the complex. However, the colouration is inexplicable in terms of this theory.

Pauling[24] suggested that the proximity of a hydrocarbon molecule to a molecule of picric acid, in a structure where the two molecules lay parallel and separated by a distance of ~3·5 Å, could account for the enhancement of optical absorption by the picric acid in the complex, in terms of the effect on the dielectric constant of the environment of the picric acid molecule. Pauling also suggested that the Ag^+–ethylene complex is stabilized by resonance between a structure involving no covalent bonding between the two moieties and a structure of the type

$$\begin{array}{c} >\!\!C^+\!\!-\!\!C\!\!< \\ | \\ Ag \end{array}$$

The distance between the component molecules in a complex was shown by crystallographic measurements to be only slightly less than the van der Waals distance.[25-28] This observation finally removed the possibility that any sort of normal covalent bonding could be responsible for these complexes.

Gibson and Loeffler[29] suggested that there might be sufficient transfer of charge in collisions of suitably orientated molecules of the two species. The intensification of colour of solutions of aromatic amines in nitro-aromatics when the hydrostatic pressure was increased, and the loss of colour when such solutions were frozen, were used as evidence in support of such collisional complexes.

Re-emphasis of the electron donor–electron acceptor nature of the two components was made by Weiss.[30] He took the extreme view that the complex was essentially ionic in character—a result of complete electron transfer from the donor to the acceptor molecule. He argued that the ionic bond could account for: (a) the generally observable greater strength of the binding than dipole interaction or dispersion forces would provide; (b) the very fast rate of formation of the complex; (c) the correlation of low ionization potential of the donor and high electron affinity of the acceptor with complexing ability; (d) the colour of the complex which could arise from electronic transitions of the two free radical ions which go to make up the complex; (e) the intermolecular separation of 3–3.5 Å, which is reasonable as an interionic distance; (f) the small but measurable dipole moments of 1,4-dinitrobenzene and 1,3,5-trinitrobenzene in benzene, naphthalene or dioxane, in which the difference between the observed dipole moment 0.7–0.9D and those calculated for ionic complexes (8–18D) is explained by very low equilibrium constants ($10^{-3} - 10^{-4}$ l/mol) at room temperature; (g) electrical conductivity. However, considerable enthalpy changes would be expected, whereas the experimental values are small (often ~4 kcal/mol or less). Determinations of the equilibrium constant for the interaction of 1,3,5-trinitrobenzene with benzene at room temperature indicate a value of an order 10^3 greater than that required to explain the observed values of the apparent dipole moment of 1,3,5-trinitrobenzene in benzene.

Weiss himself indicated that in general the electrical conductivities of solutions of molecular complexes in liquid sulphur dioxide were not high. The fact that molecular complexes are generally insoluble in, or are decomposed by, water would seem surprising if the complexes were ionic. Although the majority of complexes are now generally accepted to have only a small degree of ionic character in the ground state, there may be some systems which are predominantly ionic. They are, however, exceptional. Despite the obvious conflict of theory and fact, Weiss had probably brought theory to a point where, unrecognized by many, a hurdle had been cleared. Weiss's ideas were not without support— Woodward[31] proposed such an ionic structure as the first step in the interaction of diene and dienophile in the Diels-Alder reaction.

A different approach was made by Brackman.[32] He proposed that an adequate description of these complexes could be made by what he described as "complexresonance", [sic] whereby one of the canonical structures is the system of two non-bonded molecules, while in the second structure the two molecules are bonded together. He recognized that there was no simple connection between dipole moment of the polarizing component and the colour of the complex, and he was the first to propose that this absorption is a characteristic of the complex as a whole and is not a perturbed transition of either component. Although Brackman realized the general nature of the components as electron donors (which he termed nucleophilic substances) and electron acceptors (termed sextet substances), he did not suggest that the second of the two canonical structures had involved single electron transfer between the component molecules.

In 1949, Benesi and Hildebrand[33] reported that solutions containing an aromatic hydrocarbon and iodine had an electronic absorption band not present in either component alone. Mulliken[34] originally endeavoured to interpret this absorption in terms of spin-forbidden transition in the hydrocarbon. It was suggested that this transition became allowed in the complex, whose formation Benesi and Hildebrand postulated, because of the lowering of the symmetry of the donor on complex formation. However, in a "note added in proof" Mulliken[34] suggested that the characteristic absorption of such solutions could arise through an intermolecular charge-transfer transition. The consequences of his proposals led him to consider the implications in detail. The resulting valence bond treatment for complex formation between electron donors and electron acceptors appeared as a paper in 1952.[1] This, together with succeeding papers,[2, 3] provided an explanation for many previous observations and was the main stimulus to the extensive developments which have taken place in this field since 1952.

More recently, theoretical descriptions, other than the valence-bond charge-transfer model, have been given (Chapter 2).

Three books and a number of reviews have been written in the field of organic charge-transfer complexes. Some of these give a general coverage, others are primarily concerned with some specific aspect of these complexes. A list of these publications is given at the end of this chapter (Refs 35–61).

REFERENCES

1. R. S. Mulliken, *J. Am. chem. Soc.* **74**, 811 (1952).
2. R. S. Mulliken, *J. phys. Chem., Ithaca* **56**, 801 (1952).
3. R. S. Mulliken, *J. Chim. phys.* **61**, 20 (1964).
4. J. N. Murrell, S. F. A. Kettle and J. M. Tedder, "Valence Theory," Wiley, London, New York and Sydney (1965), p. 337.
5. M. J. S. Dewar and C. C. Thompson, Jr., *Tetrahedron* **Suppl. 7**, 97 (1966).
6. M. D. Bentley and M. J. S. Dewar, *Tetrahedron Lett.* 5043 (1967).
7. R. J. W. Le Fèvre, D. V. Radford, G. L. D. Ritchie and P. J. Stiles, *Chem. Commun.* 1221 (1967).
8. H. O. Hooper, *J. chem. Phys.* **41**, 599 (1964).
9. G. N. Lewis, "Valency and the Structure of Atoms and Molecules," Reinhold, New York (1923).
10. P. R. Hammond, *Nature, Lond.* **206**, 891 (1965).
11. K. E. Schuler, *J. chem. Phys.* **21**, 765 (1953).
12. B. Pullman and A. Pullman, *Nature, Lond.* **199**, 467 (1963).
13. P. R. Hammond and R. H. Knipe, *J. Am. chem. Soc.* **89**, 6063 (1967).
14. R. L. Hansen, *J. phys. Chem., Ithaca* **70**, 1646 (1966).
15. R. L. Hansen and J. J. Neumayer, *J. phys. Chem., Ithaca* **71**, 3047 (1967).
16. Y. Matsunaga, *Nature, Lond.* **211**, 182 (1966).
17. For a review of the earlier theories of such complexes see Ref. 38.
18. P. Pfeiffer, "Organische Molekülverbindungen," 2nd edition, Ferdinand Enke, Stuttgart (1927).
19. J. J. Sudborough, *J. chem. Soc.* **79**, 522 (1901).
20. J. W. Baker and G. M. Bennett, *Rep. Prog. Chem.* **28**, 138 (1931); G. M. Bennett and R. L. Wain, *J. chem. Soc.* 1108 (1936); G. M. Bennett and G. H. Willis, *J. chem. Soc.* 256 (1929).
21. G. Briegleb, *Z. phys. Chem.* **B16**, 249 (1932).
22. G. Briegleb, "Zwischenmolekulare Kräfte," G. Braun, Karlsruhe (1949), pp. 13, 55.
23. G. Briegleb and T. Schachowskoy, *Z. phys. Chem.* **B19**, 255 (1932).
24. L. Pauling, *Proc. natn. Acad. Sci. U.S.A.* **25**, 581 (1939).
25. J. S. Anderson, *Nature, Lond.* **140**, 583 (1937).
26. H. M. Powell and G. Huse, *Nature, Lond.* **144**, 77 (1939).
27. H. M. Powell and G. Huse, *J. chem. Soc.* 435 (1943).
28. H. M. Powell, G. Huse and P. W. Cooke, *J. chem. Soc.* 153 (1943).
29. R. E. Gibson and O. H. Loeffler, *J. Am. chem. Soc.* **62**, 1324 (1940).
30. J. J. Weiss, *Nature, Lond.* **147**, 512 (1941); *Trans. Faraday Soc.* **37**, 780 (1941); *J. chem. Soc.* 245 (1942); *Phil. Mag.* [8] **8**, 1169 (1963).
31. R. B. Woodward, *J. Am. chem. Soc.* **64**, 3058 (1942).
32. W. Brackman, *Recl Trav. chim. Pays-Bas Belg.* **68**, 147 (1949).

33. H. A. Benesi and J. H. Hildebrand, *J. Am. chem. Soc.* **71**, 2703 (1949).
34. R. S. Mulliken, *J. Am. chem. Soc.* **72**, 600 (1950).

BOOKS

35. G. Briegleb, "Elektronen-Donator-Acceptor-Komplexe," Springer-Verlag, Berlin (1961).
36. L. J. Andrews and R. M. Keefer, "Molecular Complexes in Organic Chemistry," Holden-Day, San Francisco (1964).
37. W. B. Person and R. S. Mulliken, "Molecular Complexes: A Lecture and Reprint Volume," Wiley, New York (1969).

REVIEWS

38. L. J. Andrews, *Chem. Rev.* **54**, 713 (1954).
39. A. N. Terenin, *Usp. Khim.* **24**, 121 (1955).
40. L. E. Orgel, *Quart. Rev. (London)* **8**, 422 (1954).
41. J. A. A. Ketelaar, *Chem. Weekbl.* **52**, 218 (1956).
42. S. P. McGlynn, *Chem. Rev.* **58**, 1113 (1958).
43. J. R. Platt, *A. Rev. phys. Chem.* **10**, 349 (1959).
44. D. Booth, *Sci. Prog., Lond.* **48**, 435 (1960).
45. G. Briegleb and J. Czekalla, *Angew. Chem.* **72**, 401 (1960).
46. S. P. McGlynn, *Radiat. Res. Suppl.* **2**, 300 (1960).
47. W. C. Price, *A. Rev. phys. Chem.* **11**, 133 (1960).
48. L. J. Andrews and R. M. Keefer, *Adv. inorg. Chem. Radiochem.* **3**, 91 (1961).
49. S. F. Mason, *Quart. Rev. (London)* **15**, 287 (1961).
50. J. N. Murrell, *Quart. Rev. (London)* **15**, 191 (1961).
51. D. A. Ramsay, *A. Rev. phys. Chem.* **12**, 255 (1961).
52. H. Tsubomura and A. Kuboyama, *Kagaku To Kôgyô (Tokyo)* **14**, 537 (1961).
53. G. Cauquis and J.-J. Basselier, *Annls Chim.* **7**, 745 (1962).
54. R. S. Mulliken and W. B. Person, *A. Rev. phys. Chem.* **13**, 107 (1962).
55. V. P. Parini, *Russ. chem. Revs* **31**, 408 (1962).
56. G. Briegleb, *Angew. Chem.* **76**, 326 (1964); *Angew. Chem.* (International Edition) **3**, 617 (1964).
57. E. M. Kosower, *in* "Progress in Physical Organic Chemistry," Vol. 3, eds. S. G. Cohen, A. Streitwieser, Jr. and R. W. Taft, Interscience, New York (1965).
58. A. Pullman and B. Pullman, *in* "Quantum Theory of Atoms, Molecules and the Solid State," ed. P.-O. Löwdin, Academic Press, New York and London (1966), p. 345.
59. R. Foster and C. A. Fyfe, *in* "Progress in Nuclear Magnetic Resonance Spectroscopy," Vol. 4, eds. J. W. Emsley, J. Feeney and L. H. Sutcliffe, Pergamon, Oxford (1969), Ch. 1.
60. C. K. Prout and J. D. Wright, *Angew. Chem.* **80**, 688 (1968).
61. H. A. Bent, *Chem. Rev.* **68**, 587 (1968).

Chapter 2

Theoretical Descriptions

2.A. Introduction

Mulliken's[1-9] valence-bond* (resonance) description of weak complexes formed between electron donors and electron acceptors is treated here in some detail. It provides a very adequate explanation of the characteristic electronic absorption in terms of an intermolecular charge-transfer transition.† At the same time it is felt by some workers that the importance of charge-transfer forces in stabilizing the ground state of such complexes has been exaggerated.[10-15] It has been suggested, for example, that an appreciable part of their stability may be the result of back-coordination.

Some alternative descriptions of weak electron donor–electron acceptor interactions are given in subsequent sections. For more detailed studies the reader is referred to a recent book by Person and Mulliken.[16]

2.B. Valence-Bond Treatment[4, 17-19]

2.B.1. General

Relatively weak interactions between an electron donor (D) and an electron acceptor (A) can be described in terms of a wave function of the form:

$$\psi_N(AD) = a\psi_0(A, D) + b\psi_1(A^- - D^+) \qquad (2.1)$$

* The term "valence-bond" is on occasion abbreviated to v.b., likewise "molecular orbital" is sometimes written as m.o.

† Because of the usually weak nature of the electronic interaction in the ground state, it is expected that the electronic transitions of the separated component species will also occur, though perhaps somewhat modified, in the spectrum of the complex.

The acceptor and the donor may in general be molecules, molecule-ions, atoms or atom-ions, with the restriction that they are both in their totally symmetric ground states. The wave function ψ_0 has been termed by Mulliken the "no bond" function. It corresponds to the structure in the complex in which binding results from such "physical" forces as dipole–dipole and dipole–induced-dipole interactions, London dispersion forces, and hydrogen bonding. The wave function ψ_1 has been termed the "dative" function. This corresponds to a structure of the complex where one electron has been completely transferred from the donor to the acceptor. In the case where A and D are neutral, this will involve covalent binding between the odd electrons in A^- and D^+. In general this binding will be relatively weak because of the large separation of A and D.

If ψ_N, ψ_0 and ψ_1 are all normalized so that $\int \psi_N^2 \, d\tau = 1, \int \psi_0^2 \, d\tau = 1$ and $\int \psi_1^2 \, d\tau = 1$, the coefficients a and b are related by the expression:

$$a^2 + 2abS_{01} + b^2 = 1 \tag{2.2}$$

where $S_{01} \equiv \int \psi_0 \psi_1 \, d\tau$.

For weak interactions, the ground-state energy (W_N) will be the energy of the two separated components (W_∞), modified by the "no bond" energy term (G_0) (which may be positive or negative) and by the resonance energy (X_0) of interaction between the states $\psi_0(A, D)$ and $\psi_1(A^- - D^+)$ thus:

$$W_N = W_\infty + G_0 - X_0 = W_0 - X_0 \tag{2.3}$$

For weak interactions W_N may be obtained approximately by second-order perturbation theory:

$$W_N \equiv \int \psi_N H \psi_N \, d\tau \approx W_0 - \frac{(H_{01} - S_{01} W_0)^2}{(W_1 - W_0)} \tag{2.4}$$

where $W_0 = \int \psi_0 H \psi_0 d\tau$; $W_1 = \int \psi_1 H \psi_1 d\tau$ is the energy of the dative structure $A^- D^+$; $H_{01} = \int \psi_0 H \psi_1 d\tau$; and H is the exact Hamiltonian for the entire set of nuclei and electrons.

The approximate relation of the coefficients a and b in equation (2.1) is given by second-order perturbation theory:

$$b/a \approx (H_{01} - S_{01} W_0)/(W_1 - W_0) \tag{2.5}$$

There will be an excited-state ψ_E, corresponding to the ground state ψ_N where:

$$\psi_E(AD) = a^* \psi_1(A^- - D^+) - b^* \psi_0(A, D) \tag{2.6}$$

If the function $\psi_E(AD)$ is normalized so that $\int \psi_E^2 d\tau = 1$, the coefficients a^* and b^* in equation (2.6) will be such that:

$$a^* - 2a^* b^* S_{01} + b^{*2} = 1 \tag{2.7}$$

For relatively weak interactions the energy of this excited state (W_E) will be greater than W_1 by the resonance energy (X_1):

$$W_E = W_1 + X_1 \tag{2.8}$$

By the second-order perturbation approximation:

$$W_E = W_1 + \frac{(H_{01} - S_{01} W_1)^2}{(W_1 - W_0)} \tag{2.9}$$

and the coefficients $a^{\pm}$ and $b^{\pm}$ in equation (2.6) are approximately related by:

$$b^{\pm}/a^{\pm} = -(H_{01} - S_{01} W_1)/(W_1 - W_0) \tag{2.10}$$

A diagrammatic representation of the relationship of these various energy terms is given in Fig. 2.1.

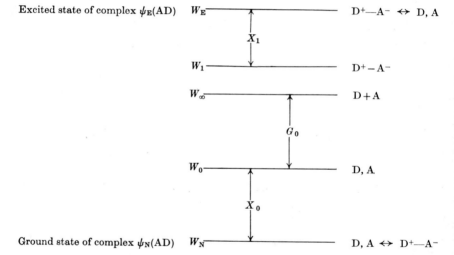

FIG. 2.1. A diagram showing the relationships between W_N, W_0, W_E, W_{∞}, G_0, X_0 and X_1.

For interaction between the structures ψ_0 and ψ_1 both S_{01} and H_{01} must be non-zero. This means that these two functions shall be of the same group-theoretic species. Except in cases where there is strong spin-orbit coupling, this means that the functions ψ_0 and ψ_1 are of the same spin-type and of the same orbital species.

For an interaction of electron donor–electron acceptor molecule pairs, the theoretical process $\psi_0 \rightarrow \psi_1$ involves the jump of one of a pair of electrons, initially occupying a m.o. in D, say ψ_D, to an unoccupied m.o.

in A, say ψ_A. This must occur in such a way that the excited electron still remains paired to its original electron partner. It may be shown that:

$$S_{01} = \sqrt{2}S_{AD}(1 + S_{AD}^2)^{1/2} \qquad (2.11)$$

where $S_{AD} = \int \psi_A \psi_D \, d\tau$. This is the overlap integral between the highest-filled molecular orbital of the donor and the lowest-unfilled molecular orbital of the acceptor (in the case of atom donors or acceptors, then the corresponding atomic orbitals are involved). For the usually small values of S_{AD}, S_{01} will vary linearly with S_{AD}. H_{01} will also vary linearly with S_{AD} for small values of S_{AD}. The electron donor and electron acceptor moieties will tend to orientate themselves relative to one another in such a way as to make S_{01}, or S_{AD}, a maximum (the overlap and orientation principle).

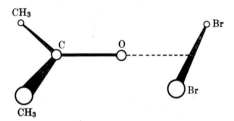

FIG. 2.2. The orientation of bromine in its complex with acetone which might be expected on the basis of the overlap and orientation principle. (After S. P. McGlynn, Ref. 20.)

This principle has been used to predict the geometry of certain complexes. An early success for the theory was the rationalization of the structure of benzene–silver perchlorate. The crystallographic determination showed that the silver ion is not on the six-fold axis, but rather over a C—C bond of the benzene ring. Since the electron will be donated from an e_{1g} m.o. orbital of the benzene (ψ_D), the silver ion must be off the six-fold axis of the benzene for S_{AD} to be non-zero.

However, there may be other conflicting requirements, for example: steric interference or strong localized interaction such as dipole-induced dipole, quadrupole, or hydrogen bonding so that maximum overlap may not be achieved. In some cases it may be that the modified highest-filled level of the donor, or the lowest-unfilled orbital of the acceptor, is not utilized. This has been proposed in the case of benzene–halogen complexes. If the axial symmetry observed in chains of alternating donor and acceptor molecules in the solid also occurs in the 1:1 complexes in

2

solution, then, since the overlap of the highest-filled π-orbital of benzene with the lowest-unfilled level in the halogen is zero when the complex has this geometry, the fact that the complex exists has been taken to imply that some lower orbital of benzene with a different symmetry is involved in complex formation.[9] In the acetone–bromine complex the atomic orbital of the oxygen from which the electron is donated is the $2p_y$: the acceptor orbital is a σ molecular orbital of the bromine molecule. Maximum overlap should therefore occur when the two components have the relative configuration shown in Fig. 2.2.[20] The experimentally determined crystal structure of the complex shows that the configuration is in fact as shown in Fig. 2.3. Other examples of the failure of the overlap and orientation principle are given in Chapter 8.

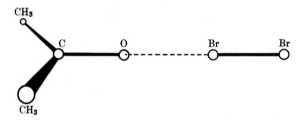

Fig. 2.3. The orientation of bromine in its complex with acetone as determined from X-ray crystallographic studies (O. Hassel and K. O. Strømme, *Nature, Lond.* **182**, 1155 (1958); *Acta Chem. Scand.* **13**, 275 (1959). After S. P. McGlynn, Ref. 20.)

2.B.2. Self-Complexes[4, 17]

Equation (2.1) is a reasonable representation of the ground state of a complex, the components of which are relatively well defined as an electron donor and electron acceptor respectively. However, one can envisage a levelling effect by suitably selecting the species D and A. There will come a time when contributions from structures where an electron has been donated *from* the acceptor *to* the donor become significant so that equation (2.1) will have to be replaced by the more general expression:

$$\psi_N = a\psi(A, D) + b\psi(A^- - D^+) + c\psi(A^+ - D^-) \tag{2.12}$$

In the limit where A and D are identical this expression becomes:

$$\psi_N = a\psi(A_1, A_{11}) + b[\psi(A_1^+ - A_{\bar{1}\bar{1}}) + \psi(A_{\bar{1}} - A_{11}^+)] \tag{2.13}$$

$$W_N \approx W_0 - \frac{2(H_{01} - S_{01} W_0)^2}{(W_1 - W_0)} \tag{2.14}$$

corresponding to equation (2.4), and

$$W_E \approx W_1 + \frac{2(H_{01} - S_{01} W_1)^2}{(W_1 - W_0)} \qquad (2.15)$$

corresponding to equation (2.9).

It is therefore expected that charge-transfer forces will exist between like components, although they might be expected to be relatively weak since generally a strong Lewis acid will show only weak Lewis base properties. There are some exceptions, however—for example, the larger polycyclic aromatic hydrocarbons should behave as strong acids and strong bases.

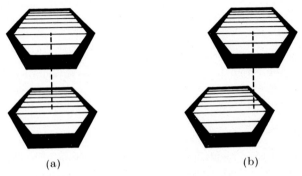

(a)   (b)

FIG. 2.4. Orientation of two benzene molecules for a non-zero overlap S_{AD}: (a) unacceptable; (b) acceptable. (After S. P. McGlynn, Ref. 20.)

For the particular example of the benzene–benzene-self-complex the overlap integral S_{AD} will be:

$$S_{AD} = \int \psi_{e_{1g}} \psi_{e_{2u}} \, d\tau \qquad (2.16)$$

Consequently the overlap and orientation principle would predict that any model in which the benzene rings are superimposed with their six-fold axes coincident is untenable, since for all such cases $S_{AD} = 0$. A possible structure is one in which the benzene rings are parallel but staggered with respect to one another (see Fig. 2.4).

2.B.3. Electronic Absorption Characteristic of the Complex

For simplicity the case is considered where there is only a single transition extra to the absorptions corresponding to the component moieties of the complex.* This absorption arises from the excitation from

---

* The components of the complex will generally be referred to as donor and acceptor *molecules*. Although this is generally the case, the components could include ions, molecular ions, etc.

the ground state of the complex ψ_N to the excited state ψ_E. For weak interactions this transition is effectively from the structure $\psi_0(A, D)$ to the structure $\psi_1(A^- - D^+)$, i.e. it is an *intermolecular charge-transfer transition* involving a one-electron jump from D to A. It is sometimes for this reason that complexes which give rise to this transition have been called charge-transfer complexes (see Chapter 1). It must be re-emphasized that the contribution of the charge-transfer structure in the ground state is usually very small, i.e. b/a in equation (2.1) is small.

A theoretical relationship between the energy of the charge-transfer transition E_{CT}, the ionization potential of the donor (I^D) and the electron affinity of the acceptor (E^A) may be derived in terms of the simple v.b. description. From equations (2.3) and (2.8), the energy of the transition (E_{CT}) is such that:

$$E_{CT} = h\nu_{CT} = (W_E - W_N) = (W_1 + X_1) - (W_0 - X_0) \qquad (2.17)$$

where ν_{CT} is the frequency characteristic of the intermolecular charge-transfer transition. Now equation (2.3) may be re-expressed as:

$$W_0 = W_\infty - G_0$$

where W_∞ is the energy of the separated components (usually in solution) and G_0 is the energy resulting from all the "no bond" interactions. The energy W_1 of the dative structure $\psi_1(A^- - D^+)$ relative to the energy of the separated components (W_∞) is the work involved in removing an electron from the donor (= ionization potential of the donor, I^D),*† followed by the acceptance of this electron by the electron acceptor (= electron affinity of the acceptor, E^A)* and finally the bringing together of the two so-formed ions A^-, D^+ to their equilibrium distance of separation; this will be mainly the coulombic attractive energy of the A^-, D^+ ions and is represented by G_1. Thus:

$$W_1 = W_\infty + I^D - E^A - G_1 \qquad (2.18)$$

---

* Mulliken has emphasized that the ionization potentials and the electron affinities used in these expressions should be "vertical" values rather than adiabatic values. By *vertical* ionization potential is meant the energy required to remove an electron from a molecule in its lowest vibrational state via a hypothetical Franck–Condon transition which results in a positive ion which still has the original ground-state atomic configuration. The so-called *adiabatic* ionization potential is the energy required to remove an electron by a process in which the resulting ion is in its lowest vibrational state. A corresponding difference obtains between the vertical and adiabatic electron affinity (see also Chapter 3).

† Strictly speaking, since the work involved is described in energy terms (usually in eV), the expression *ionization energy*, rather than *ionization potential*, should be used. However, since the latter term is so widely used in this sense, it will be retained throughout this book.

Since $W_0 = W_\infty - G_0$, we may write

$$W_1 - W_0 = (W_\infty + I^D - E^A - G_1) - (W_\infty - G_0)$$

or

$$W_1 - W_0 = I^D - (E^A + G_1 - G_0) \qquad (2.19)$$

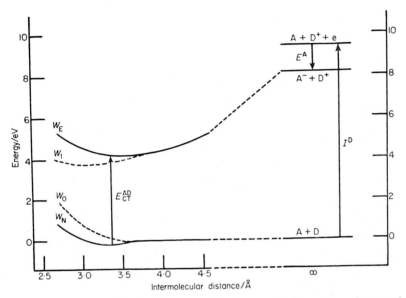

FIG. 2.5. A diagram showing the energetics for a weak charge-transfer complex between a donor (D) and an acceptor (A). W_N represents the energy corresponding to the ground-state wave function $\psi_N(AD)$ and W_E represents the energy corresponding to the excited-state wave function $\psi_E(AD)$. W_0 and W_1 represent these energies if there were no resonance stabilization. The energy of the optical intermolecular charge-transfer transition is marked E_{CT}^{AD}. (After R. S. Mulliken, Ref. 4.)

Equation (2.17) may be expressed:

$$E_{CT} = h\nu_{CT} = W_E - W_N = (W_1 - W_0) + (X_1 + X_0)$$
$$= I^D - E^A + (G_0 - G_1) + (X_1 + X_0) \qquad (2.20)$$

These energy relationships are shown diagrammatically in Fig. 2.5. For weak interactions the terms X_0 and X_1 are given by equations (2.4) and (2.5). If the matrix elements in these equations are abbreviated:

$$\beta_0 \equiv (H_{01} - S_{01} W_0) \qquad \text{and} \qquad \beta_1 = (H_{01} - S_{01} W_1)$$

equation (2.20) may now be written:

$$h\nu_{CT} = I^D - (E^A + G_1 - G_0) + \frac{\beta_0^2 + \beta_1^2}{I^D - (E^A + G_1 - G_0)} \qquad (2.21)$$

The relationship described in equation (2.21) is only applicable to interactions which are sufficiently weak to justify the application of simple perturbation theory. Yada *et al.*[21] have shown that for the iodine–aliphatic amine interactions, where the binding energy is fairly high, the variation method should be used to calculate the resonance

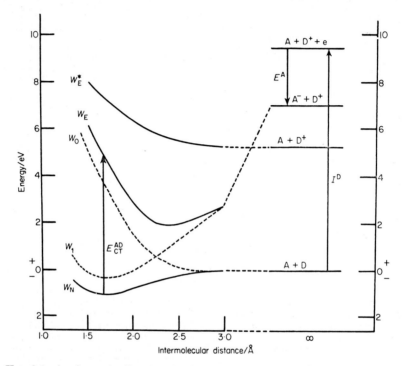

Fig. 2.6. A schematic diagram showing the energetics for a strong interaction between an electron donor and an electron acceptor, for example $BF_3 + NMe_3$. The symbols have the same significance as in Fig. 2.5. The diagram is largely qualitative. (After R. S. Mulliken, Ref. 4.)

energies. The resulting expression for the energy of the charge-transfer band is:

$$h\nu_{CT} = \frac{(W_1 - W_0)}{(1 - S_{01}^2)}\left[1 + \frac{4\beta_0\beta_1}{(W_1 - W_0)^2}\right]^{1/2} \qquad (2.22)$$

The schematic energy diagram for a very strongly interacting electron donor–electron acceptor pair such as trimethylamine–boron trifluoride (Fig. 2.6) may be compared with the diagram for a weakly accepting system (Fig. 2.5). Here the trend of decrease in energy of the transition

$\psi_N \to \psi_E$ with decrease in ionization potential of the donor is no longer valid. This absorption will now occur at relatively high energy.

2.C. Simple Molecular Orbital Treatment

An alternative approach to the energetics of the intermolecular charge-transfer transitions has been made, particularly by Dewar and his co-workers,[10, 22-26] using a simple molecular orbital description. For weak

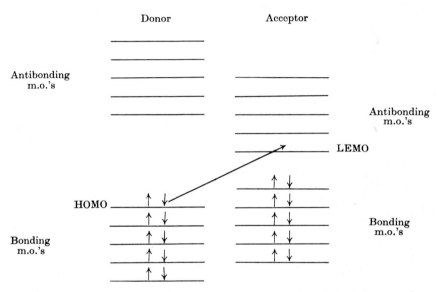

FIG. 2.7. Lowest charge-transfer transition represented as the transference of an electron from the highest-occupied molecular orbital (HOMO) of the donor to the lowest-empty molecular orbital (LEMO) of the acceptor. (After M. J. S. Dewar and A. R. Lepley, Ref. 10.)

interactions the problem has been treated using perturbation theory.[10] For such systems the interaction energies in the ground state are small compared with the transition energies to excited states of the complex, in particular the intermolecular charge-transfer excited states. Each such transition may be considered as arising from the transfer of an electron from a filled orbital in the donor to an empty orbital in the acceptor. For a single charge-transfer band, or for the lowest-energy charge-transfer band in cases where multiple transitions are observed in π–π complexes, this is generally assumed to involve the highest-occupied molecular orbital (HOMO) of the donor and the lowest-empty molecular orbital (LEMO) of the acceptor (Fig. 2.7).

In polycyclic aromatics the energy of the highest-occupied orbital in the ground state (B_i) may be expressed in a simple Hückel treatment as:

$$B_i = \alpha_0 + \chi_i \beta \qquad (2.23)$$

where α_0 is the coulomb integral, β is the resonance integral for two neighbouring carbon atoms and χ_i is the Hückel parameter for this orbital (the electron overlap integral and electron repulsion are both assumed to be zero). For complexes of these donors with a common acceptor species, we may write equation (2.24), which corresponds to equation (2.20), in the v.b. description:

$$\begin{aligned} h\nu_{CT} &= B_j - B_i + P \\ &= B_j - \alpha_0 - \chi_i \beta + P \end{aligned} \qquad (2.24)$$

where B_j represents the energy of the lowest-unfilled acceptor orbital and P is an energy term which corresponds to a perturbation of the appropriate energy levels in the donor and acceptor. The term P is analogous to the expression $(G_1 - G_0) + (X_1 + X_0)$ in the valence-bond description. Dewar[10] omits this term in his simple treatment. A more detailed treatment has been given by Murrell[27, 28] who has attempted to demonstrate a relationship between the stability of a charge-transfer complex and the intensity of the charge-transfer band.

A semi-empirical linear combination of molecular orbitals (LCMO) description has been used by Flurry[29] in order to predict the ground-state stabilization energies of charge-transfer complexes. The problem has been simplified by omitting any consideration of solvent or steric effects on the ground state stabilization. The degree of correlation of the calculated relative stabilities for sets of complexes, each with a common acceptor, is probably as large as might be expected in view of the approximations made. Simple Hückel molecular orbital calculations have also been made by Rourke.[30] Extended Hückel calculations on the tetracyanoethylene–benzene complex by Wold[31] have been unsuccessful in predicting the stability of the complex.

Other molecular orbital treatments include the delocalization method of Fukui et al.,[32] and the consideration by Iwata et al.,[33] of the configuration interaction among ground, charge transfer, and locally excited states.

2.D. Free-Electron Model

The free-electron model for conjugated molecules, which was developed by various workers including Kuhn[34, 35] and Bayliss,[36–39] treats the π-electrons in a conjugated molecule as a free-electron gas which moves in the potential field of the molecule. This concept has been applied to charge-transfer complexes by Shuler,[40] and, more recently, by Boeyens.[41] Both authors have used the simplified one-dimensional case, rather than the more complicated three-dimensional case.

For a single linear molecule, the electron energies may be represented by horizontal lines in a square-well potential, as in Fig. 2.8. The height of the sides of this well (V) corresponds to the energy of the K-electrons plus the zero-point energy. For two such molecules, say an electron donor (D)

Fig. 2.8. A one-dimensional potential model for a linear conjugated molecule. (After K. E. Shuler, Ref. 40.)

and an electron-acceptor (A), which interact, the one-dimensional model may be represented as a double-minimum potential[40] (Fig. 2.9). The potential barrier (of height V^{AD} relative to the molecule D) and width d, is an inverse measure of the delocalization of electrons between D and A. The lowering of the potential of D from V^D to V^{AD} becomes greater as the electron affinity (D^A) of A increases and the ionization potential (I^D) of D decreases. The width of the potential barrier (d) can be related to the intermolecular distance between A and D.

To ease mathematical computation, a somewhat simpler model has been used in which it is assumed that the delocalized electrons have the same energy in both the donor and the acceptor parts of the complex. This is shown in Fig. 2.10. Since it is assumed that only electrons from the donor molecule are delocalized, the potential box is drawn so that only the highest energy electrons of the donor are included. Also, for

2*

simplicity, it is assumed that $V^{AD} = I^D - E^A$. From the observed experimental relationship between the energy of the charge-transfer transition(s) and I^D (see Chapter 8), which closely resembles the behaviour of the energy of the lowest-frequency absorption bands of free aromatic

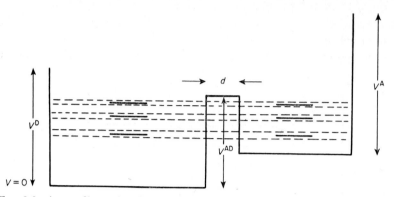

Fig. 2.9. A one-dimensional model for a charge-transfer complex of two linear conjugated molecules. The potential barrier corresponds to the non-conjugated linkage. The dotted lines show the resulting splitting of the potential energy levels. (After K. E. Shuler, Ref. 40.)

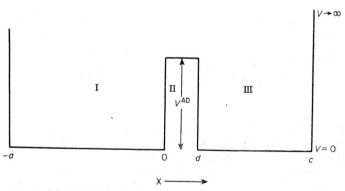

Fig. 2.10. A simplified one-dimensional potential for a charge-transfer complex. (After K. E. Shuler, Ref. 40.)

hydrocarbon donors, Boeyens[41] has suggested that the number of electrons in the energy levels characteristic of the complex shall be $(4n + 2)$. The energy levels, for $B < V^{AD}$, may be calculated as the eigenvalues for the wave equations:[40]

$$\frac{d^2 \psi_I}{dx^2} + \kappa B^{AD} \psi_I = 0 \tag{2.25}$$

$$\frac{d^2\psi_{II}}{dx^2} + \kappa(B^{AD} - V^{AD})\psi_{II} = 0 \qquad (2.26)$$

$$\frac{d^2\psi_{III}}{dx^2} + \kappa B^{AD}\psi_{III} = 0 \qquad (2.27)$$

Where $\kappa = 8\pi^2 m/h^2$, the wave functions (ψ) of equations (2.25)–(2.27) vanish at $x = -a$ and $x = c$. The potential (V) is zero from $x = -a$ to $x = 0$ (region I) and from $x = d$ to $x = c$ (region III), and $V = \infty$ outside the range $x = -a$ to c (Fig. 2.9). The solution to equations (2.25)–(2.27), subject to the condition that ψ and $d\psi/dx$ are continuous throughout the regions I, II and III, for the case where $B < V$, is:

$$\frac{\beta^2 - \alpha^2}{2\alpha\beta}\cos\alpha(c - d + a) - \frac{\beta^2 + \alpha^2}{2\alpha\beta}\cos\alpha(c - d - a) =$$
$$\sin\alpha(c - d + a)\coth\beta a \qquad (2.28)$$

with $\alpha = (\kappa B^{AD})^{1/2}$ and $B = [\kappa(V^{AD} - B^{AD})]^{1/2}$.

The corresponding expression when $B^{AD} > V$ is:

$$\frac{\beta^2 - \alpha^2}{2\alpha\beta}\cos\alpha(c - d - a) - \frac{\beta^2 + \alpha^2}{2\alpha\beta}\cos\alpha(c - d + a)$$
$$= \sin\alpha(c - d + a)\cot\beta d \qquad (2.29)$$

Solutions for various values of α have been calculated by Boeyens[41] for chloranil–aromatic hydrocarbon complexes. The degree of delocalization $(c - d)$ was adjusted to give the best fit for the experimentally observed charge-transfer excitation energy for one complex in the series. The parameter a was initially calculated according to Rüdenberg and Parr.[42] Numerical agreement is improved by a slightly different definition of a and d in which interaction between these quantities is allowed. Although the calculations are semi-empirical, they do provide a different approach to the treatment of this type of complex formation. The free-electron model may be a convenient description for correlating the experimentally observed effects of pressure on charge-transfer complexes, since the shortening of the intermolecular distance can be directly related to d and hence to the probability of tunnelling through this smaller potential barrier.

REFERENCES

1. R. S. Mulliken, *J. Am. chem. Soc.* **72**, 600 (1950).
2. R. S. Mulliken, *J. Am. chem. Soc.* **72**, 4493 (1950).
3. R. S. Mulliken, *J. chem. Phys.* **19**, 514 (1951).
4. R. S. Mulliken, *J. Am. chem. Soc.* **74**, 811 (1952).
5. R. S. Mulliken, *J. Phys. Chem., Ithaca* **56**, 801 (1952).

6. R. S. Mulliken, *J. Chim. phys.* **51**, 341 (1954).
7. R. S. Mulliken, *J. chem. Phys.* **23**, 397 (1955).
8. R. S. Mulliken, *Proc. Int. Conf. on Co-ordination Compounds*, Amsterdam (1955), p. 336; *Recl Trav. chim. Pays-Bas Belg.* **75**, 845 (1956).
9. R. S. Mulliken, *J. Chim. phys.* **61**, 20 (1964).
10. M. J. S. Dewar and A. R. Lepley, *J. Am. chem. Soc.* **83**, 4560 (1961).
11. H. O. Hooper, *J. chem. Phys.* **41**, 599 (1964).
12. M. J. S. Dewar and C. C. Thompson, Jr., *Tetrahedron*, **Suppl. 7**, 97 (1966).
13. M. D. Bentley and M. J. S. Dewar, *Tetrahedron Lett.* 5043 (1967).
14. R. J. W. Le Fèvre, D. V. Radford, G. L. D. Ritchie and P. J. Stiles, *Chem. Commun.* 1221 (1967).
15. M. W. Hanna, *J. Am. chem. Soc.* **90**, 283 (1968).
16. W. B. Person and R. S. Mulliken, "Molecular Complexes: A Lecture and Reprint Volume," Wiley, New York (1969).
17. S. P. McGlynn, *Chem. Rev.* **58**, 1113 (1958).
18. G. Briegleb, "Elektronen-Donator-Acceptor-Komplexe," Springer-Verlag, Berlin (1961), p. 7.
19. W. B. Person and R. S. Mulliken, *A. Rev. phys. Chem.* **13**, 107 (1962).
20. S. P. McGlynn, *Radia. Res.* **Supp. 2**, 300 (1960).
21. H. Yada, J. Tanaka and S. Nagakura, *Bull. chem. Soc. Japan* **33**, 1660 (1960).
22. M. J. S. Dewar and H. Rogers, *J. Am. chem. Soc.* **84**, 395 (1962).
23. A. R. Lepley, *J. Am. chem. Soc.* **84**, 3577 (1962).
24. A. R. Lepley, *J. Am. chem. Soc.* **86**, 2545 (1964).
25. A. R. Lepley, *Tetrahedron Lett.* 2823 (1964).
26. A. R. Lepley and C. C. Thompson, Jr., *J. Am. chem. Soc.* **89**, 5523 (1967).
27. J. N. Murrell, *J. Am. chem. Soc.* **81**, 5037 (1959).
28. J. N. Murrell, *Quart. Rev. (London)* **15**, 191 (1961).
29. R. L. Flurry, Jr., *J. phys. Chem., Ithaca* **69**, 1927 (1965).
30. T. A. Rourke, Thesis, University of St. Andrews (1966).
31. S. Wold, *Acta chem. scand.* **20**, 2377 (1966).
32. K. Fukui, A. Imamura, T. Yonezawa and C. Nagata, *Bull. chem. Soc. Japan* **34**, 1076 (1961); *Bull. chem. Soc. Japan* **35**, 33 (1962).
33. S. Iwata, J. Tanaka and S. Nagakura, *J. Am. chem. Soc.* **88**, 894 (1966).
34. H. Kuhn, *J. chem. Phys.* **16**, 840 (1948).
35. H. Kuhn, *J. chem. Phys.* **17**, 1198 (1949).
36. N. S. Bayliss, *J. chem. Phys.* **16**, 287 (1948).
37. N. S. Bayliss, *J. chem. Phys.* **17**, 1353 (1949).
38. N. S. Bayliss, *Aust. J. Sci.* **3A**, 109 (1950).
39. N. S. Bayliss, *Quart. Rev. (London)* **6**, 319 (1952).
40. K. E. Shuler, *J. chem. Phys.* **20**, 1865 (1952).
41. J. C. A. Boeyens, *J. phys. Chem., Ithaca* **71**, 2969 (1967).
42. K. Rüdenberg and R. G. Parr, *J. chem. Phys.* **19**, 1268 (1951).

Chapter 3

Electronic Spectra

3.A. Absorption Spectra

3.A.1. Introduction

In general, a complex formed between an electron donor and an electron acceptor still retains the absorptions of the components modified to a greater or lesser extent, together with one or more absorption bands characteristic of the complex *as a whole* (Fig. 3.1). The recognition of this fact by Brackman[1] was important historically because it led to the realization that the absorption is the result of an intermolecular charge-transfer transition and not a modified transition of one or other component (Chapter 1).

In practice, the absorption characteristic of the complex in solution may not be easily observed since the complex will be partially dissociated into its component species. It may be particularly difficult to measure those absorption bands due to "local excitations" in the donor and acceptor moieties of the complex corresponding to absorption bands in the separated components: such absorptions will usually be masked by the absorptions of the free components. This problem can sometimes be overcome experimentally by making the optical measurements at low temperatures which will more favour complex formation.[2] Generally, the extra absorption* characteristic of the complex as a whole is more easily observed than those resulting from "local excitations", when the interaction is between a strong donor and a strong acceptor. In such cases the transition usually appears as a separate band at considerably longer wavelengths than the absorptions of the component

---

* In the immediate discussion, this extra absorption will be referred to as a single band, although, as stated above, there may be two or more bands present (see also pp. 58 and 67).

molecules. The intensity of the absorption band of a complex is usually determined as the molar absorptivity (extinction coefficient) at the wavelength of maximum absorption. A direct determination of intensity cannot normally be made, because the degree of dissociation of the complex in solution is usually significant. Experimental methods for the evaluation of molar absorptivity of charge-transfer complexes are discussed in Chapter 6. Transmission [3-24] and reflection [25-27] spectra,

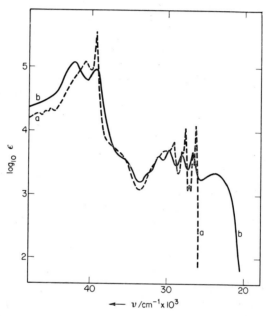

FIG. 3.1. Absorption spectra at −180°C in propyl ether/methylcyclohexane (4:1) of: (a) the estimated sum of the absorptions of tetrachlorophthalic anhydride and anthracene; (b) the estimated absorption of the charge-transfer complex tetrachlorophthalic anhydride–anthracene. (Adapted from J. Czekalla, Ref. 2.)

including specular reflection from a single crystal,[27] of solid charge-transfer complexes have been measured. In general, they show the same general optical absorption properties as the complexes in liquid solution (Fig. 3.2). Although the experimental techniques are more difficult in the solid phase, there is the advantage that usually all the molecules are complexed. Some anisotropic effects of orientated crystals are discussed in the next section. On occasion the electronic absorption spectrum of a solid has been used to distinguish between weak charge-transfer structures which show an absorption characteristic of the complex as a whole, and ionic structures in which there has been transfer of an electron or electrons from the donor to the acceptor in the ground state, in which case the electronic absorption spectrum is due to the two ionic species.[5-15, 20] In the chloranil–N,N,N′,N′-tetramethyl-p-phenylenediamine complex there is evidence of a "molionic lattice" consisting of a *ca.* 20% concentration of doubly charged A^{--} and D^{++} ions, together with a small concentration of A^- in a predominantly molecular lattice,

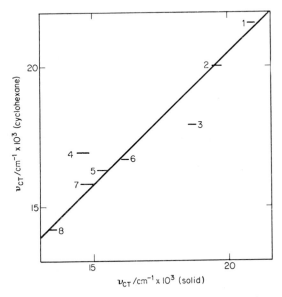

[R. Foster and T. J. Thomson, Ref. 5]

Fig. 3.2. Plot of the frequency of charge-transfer transitions of complexes of N,N,N′,N′-tetramethyl-p-phenylenediamine with the acceptors: (1) 2,4-dinitrotoluene; (2) 1,3-dinitrobenzene; (3) 2,4,6-trinitrotoluene; (4) 1,4-dinitrobenzene; (5) 1,3,5-trinitrobenzene; (6) methyl-p-benzoquinone; (7) p-benzoquinone; (8) chloro-p-benzoquinone in the solid phase, against the frequencies of these transitions for the same complexes in cyclohexane. The lengths of the lines represent the uncertainties of the position of ν_{CT} in the solid phase.

when these crystals are obtained in a very pure state.[16,17] In the presence of a very small excess of the donor over and above a 1:1 A:D stoichiometry, A^- and D^+ ions are observed [5, 7, 16,17] (see also Chapter 9).

The origin of the band which appears in addition to the absorption of the components in the spectrum of the complex has been discussed in terms of several theories in the previous chapter. As originally described by Mulliken, the band is the result of intermolecular charge-transfer transition from the ground state N to an excited state E.

Perturbations of locally excited transitions have been observed most frequently in iodine complexes. The visible absorption of iodine is shifted to higher energies in the complex. This may be a direct result of the partial transfer of an electron to an anti-bonding orbital in the acceptor, which would increase the effective size of the acceptor so that the excitation energy is increased.[28] In such cases it is expected that the greater the degree of charge transfer, the larger will be the hypsochromic shift. With other complexes, however, for example those of 7,7,8,8-tetracyanoquino-dimethane with methylbenzenes, the lowest-unfilled level of the acceptor is in some cases decreased, and in other cases increased, on complex formation.[29] For a more detailed general discussion the reader is referred to Briegleb's book.[30]

3.A.2. Polarization of the Charge-Transfer Absorption Band

Soon after the publication of Mulliken's charge-transfer theory, Nakamoto[3] provided an experimental observation which could be well

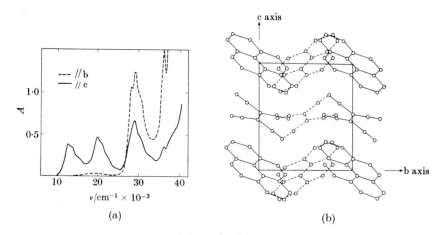

(a)   (b)

[H. Kuroda, I. Ikemoto and H. Akamatu, Ref. 24]

Fig. 3.3. The solid complex pyrene–tetracyanoethylene: (a) the polarized absorption spectra of a single crystal; (b) a projection of the crystal structure along the a-axis on the bc plane showing the orientation of the b- and c-axes.

explained in terms of this theory. Whereas the absorption of polarized light by orientated crystals of pure aromatics shows a stronger low-energy absorption when the electric vector is parallel than when it is perpendicular to the ring, the opposite obtains for the lowest energy intermolecular charge-transfer transition of a complex between two planar molecules; for example: quinhydrone; picryl chloride–hexamethylbenzene; chloranil–hexamethylbenzene. Similar observations have since been made on other systems such as: 1,2,4,5-tetracyanobenzene–naphthalene;[23] tetracyanoethylene–pyrene;[24] 1,3,5-trinitrobenzene–anthracene;[4, 20, 22] and chloranil with various aromatic amines[15] (Fig. 3.3). It should be pointed out that donor and acceptor molecules complexed in the solid state will not in general be arranged in discrete pairs. Normally the crystals consist of stacks of alternate donor and acceptor molecules so that each donor has two acceptor neighbours and vice versa (see Chapter 7). However, observations of polarized triplet emission from the donor moieties of hydrocarbon–tetrachlorophthalic anhydride complexes, randomly orientated in an alcohol glass at 77°K, also indicate that the charge-transfer *absorption* in donor–acceptor complex pairs is polarized along the intermolecular axis.[18] This is to be expected in Mulliken's theory, since the charge-transfer transition is effectively from the donor to the acceptor molecule.

The polarization of the corresponding charge-transfer fluorescence spectra of solid complexes is discussed in Section 3.B.

3.A.3. Energy of the Charge-Transfer Band ($E_{CT} = h\nu_{CT}$)

The property of charge-transfer complexes which is normally most readily, and certainly most frequently, measured is the energy of the (usually lowest) intermolecular charge-transfer transition of the complex in solution. A useful compilation of many such values obtained up to 1961 has been given by Briegleb.[31] References to other sources are given in Table 3.1. Some selected values are given in Table 3.2.

In Chapter 2 it was shown that, for weak interactions, the energy of the charge-transfer transition (E_{CT}) could be theoretically related to the ionization potential of the donor (I^D) and the electron affinity of the acceptor (E^A). By simple perturbation theory:

$$E_{CT} = I^D - (E^A + G_1 - G_0) + \frac{\beta_0^2 + \beta_1^2}{I^D - (E^A + G_1 - G_0)} \qquad \begin{array}{r}(3.1)\\(=2.21)\end{array}$$

where β_0 and β_1 are the matrix elements ($H_{01} - S_{01} W_0$) and ($H_{01} - S_{01} W_1$) respectively, G_0 is the energy of the "no bond" interaction and G_1 is the interaction of D^+ and A^- in the excited state. For stronger interactions, an expression for the E_{CT} may be obtained[32] by the variation method:

$$E_{CT} = \frac{(W_1 - W_0)}{(1 - S_{01}^2)} \left[1 + \frac{4\beta_0 \beta_1}{(W_1 - W_0)^2} \right]^{1/2} \qquad \begin{array}{r}(3.2)\\(=2.22)\end{array}$$

TABLE 3.1. Some references listing positions of intermolecular charge-transfer bands for complexes of a given acceptor with a series of donor molecules.

Acceptor	Reference
Tetracyanoethylene	A. R. Lepley, *J. Am. chem. Soc.* **86**, 2545 (1964)
	R. E. Merrifield and W. D. Phillips, *J. Am. chem. Soc.* **80**, 2778 (1958)
	H. Kuroda, M. Kobayashi, M. Kinoshita and S. Takemoto, *J. chem. Phys.* **36**, 457 (1962)
	E. M. Voigt and C. Reid, *J. Am. chem. Soc.* **86**, 3930 (1964)
	P. G. Farrell and J. Newton, *Tetrahedron Lett.* 5517 (1966)
	P. G. Farrell and J. Newton, *J. phys. Chem., Ithaca* **69**, 3506 (1965)
	A. Zweig, *Tetrahedron Lett.* 89 (1964)
	E. M. Voigt, *J. Am. chem. Soc.* **86**, 3611 (1964)
	M. Nepraš and R. Zahradník, *Colln Czech. chem. Commun. Engl. Edn* **29**, 1545, 1555 (1964)
	H. M. Rosenberg and D. Hale, *J. phys. Chem., Ithaca* **69**, 2490 (1965)
	W. M. Moreau and K. Weiss, *Nature, Lond.* **208**, 1203 (1965)
	M. J. S. Dewar and C. C. Thompson, Jr., *Tetrahedron* **Supp. 7**, 97 (1966)
	M. J. S. Dewar and H. Rogers, *J. Am. chem. Soc.* **84**, 395 (1962)
	A. R. Lepley and C. C. Thompson, Jr., *J. Am. chem. Soc.* **89**, 5523 (1967)
	A. Zweig, *J. phys. Chem., Ithaca* **67**, 506 (1963)
	T. Matsuo and H. Aiga, *Bull. chem. Soc. Japan* **41**, 271 (1968)
	G. Briegleb, J. Czekalla and G. Reuss, *Z. phys. Chem. Frankf. Ausg.* **30**, 316 (1961)
Chloranil	G. Briegleb and J. Czekalla, *Z. Elektrochem.* **63**, 6 (1959)
	G. Briegleb and J. Czekalla, *Z. phys. Chem. Frankf. Ausg.* **24**, 37 (1960)
	G. Briegleb, J. Czekalla and G. Reuss, *Z. phys. Chem. Frankf. Ausg.* **30**, 316 (1961)
	R. Beukers and Z. Szent-Györgyi, *Recl Trav. chim. Pays-Bas Belg.* **81**, 255 (1962)
	M. Nepraš and R. Zahradník, *Colln Czech. chem. Commun. Engl. Edn* **29**, 1545, 1555 (1964)

TABLE 3.1.—*continued*

Acceptor	Reference
Bromanil	M. Nepraš and R. Zahradník, *Colln Czech. chem. Commun.* **29**, 1545 (1964) M. Kinoshita, *Bull. chem. Soc. Japan* **35**, 1609 (1964)
Other quinones ＂ ＂	M. Nepraš and R. Zahradník, *Colln Czech. chem. Commun. Engl. Edn* **29**, 1545 (1964) T. K. Mukherjee, *J. phys. Chem.*, *Ithaca* **71**, 2277 (1967) P. R. Hammond, *J. chem. Soc.* 3113 (1963); 471 (1964) R. D. Srivastava and G. Prasad, *Spectrochim. Acta* **22**, 1869 (1966) R. Beukers and A. Szent-Györgyi, *Recl Trav. chim. Pays-Bas Belg.* **81**, 255 (1962)
7,7,8,8-Tetracyano-quonodimethane	M. Nepraš and R. Zahradník, *Colln Czech. chem. Commun. Engl. Edn* **29**, 1545 (1964) W. Damerau, *Z. Naturf.* **21b**, 937 (1966) R. Beukers and A. Szent-Györgyi, *Recl Trav. chim. Pays-Bas Belg.* **81**, 255 (1962)
1,3,5-Trinitrobenzene	A. Bier, *Recl Trav. chim. Pays-Bas Belg.* **75**, 866 (1956) H. Kuroda, K. Yoshihara and H. Akamatu, *Bull. chem. Soc. Japan* **35**, 1604 (1962) M. J. S. Dewar and A. R. Lepley, *J. Am. chem. Soc.* **83**, 4560 (1961) M. Nepraš and R. Zahradník, *Colln Czech. chem. Commun. Engl. Edn* **29**, 1545 (1964) R. Beukers and A. Szent-Györgyi, *Recl Trav. chim. Pays-Bas Belg.* **81**, 255 (1962) G. Briegleb and J. Czekalla, *Z. phys. Chem. Frankf. Ausg.* **24**, 37 (1960)
2,4,7-Trinitrofluorenone	A. R. Lepley, *J. Am. chem. Soc.* **84**, 3577 (1962) M. Nepraš and R. Zahradník, *Colln Czech. chem. Commun. Engl. Edn* **29**, 1545 (1964).
1,2,4,5-Tetracyano-benzene	A. Zweig, J. E. Lehnsen, W. G. Hodgson and W. H. Jura, *J. Am. chem. Soc.* **85**, 3937 (1963) S. Iwata, J. Tanaka and S. Nagakura, *J. Am. chem. Soc.* **88**, 894 (1966)

TABLE 3.2. Energies of the lowest intermolecular charge-transfer transitions

Acceptor	1,3,5-Trinitrobenzene			Chloranil			Iodine		
Donor	ν_{CT}	Solvent	Ref.	ν_{CT}	Solvent	Ref.	ν_{CT}	Solvent	Ref.
Benzene	35·2	CHCl$_3$	142	29·9	CCl$_4$	a	34·2	CCl$_4$	b
Toluene	33·0	CCl$_4$	a	27·0	CCl$_4$	59	33·1	CCl$_4$	b
o-Xylene	31·4	CH$_2$Cl$_2$	c	24·7	CH$_2$Cl$_2$	c	31·6	CCl$_4$	b
m-Xylene	31·6	CCl$_4$	a	24·6	CCl$_4$	c	31·4	CCl$_4$	b
p-Xylene	32·0	CHCl$_3$	142	23·4	CCl$_4$	a	32·9	CCl$_4$	b
Mesitylene	29·1	CH$_2$Cl$_2$	c	23·3	CH$_2$Cl$_2$	c	30·1	CCl$_4$	b
Durene	29·2	CCl$_4$	a	20·9	CCl$_4$	a	30·1	CCl$_4$	b
Pentamethylbenzene	27·2	CH$_2$Cl$_2$	c	20·2	CH$_2$Cl$_2$	c	28·0	CCl$_4$	b
Hexamethylbenzene	25·4	CCl$_4$	a	19·4	CCl$_4$	a	26·7	CCl$_4$	b
Naphthalene	27·1	CCl$_4$	a	21·0	CCl$_4$	a	27·8	CCl$_4$	b
Anthracene	21·7	CHCl$_3$	142	16·1	CCl$_4$	a	23·3	CCl$_4$	a
Chrysene	24·0	CHCl$_3$	62	18·5	CCl$_4$	150	25·4	CCl$_4$	i
Phenanthrene	27·0	CHCl$_3$	142	21·6	CCl$_4$	150	27·5	CCl$_4$	j
Perylene	20·8	CHCl$_3$	62	14·1	CCl$_4$	a			
Pyrene	22·5	CHCl$_3$	62	16·6	CCl$_4$	150	23·8	CCl$_4$	i
Triphenylene	26·0	CHCl$_3$	k	20·7	CCl$_4$	150	25·4	CCl$_4$	i
Fluorene				20·0	CCl$_4$	150			
Benzo[a]pyrene	21·0	CHCl$_3$	l	14·7	CHCl$_3$	l			
Benz[a]anthracene	23·0	CHCl$_3$	l	16·9	CCl$_4$	a			
trans-Stilbene	25·7	CCl$_4$	m	19·4	CCl$_4$	150	26·8	CCl$_4$	b
Biphenylene	23·4	CH$_2$Cl$_2$	n	16·9	CH$_2$Cl$_2$	n			
Azulene	21·8	CH$_2$Cl$_2$	h	15·5	CH$_2$Cl$_2$	h	24·5	CH$_2$Cl$_2$	h
Aniline	25·0	CHCl$_3$	142	18·9	CHCl$_3$	62	28·9	CHCl$_3$	o
N,N-dimethylaniline	20·5	CHCl$_3$	142	15·4	CCl$_4$	a	27·9	CHCl$_3$	o
N,N,N',N'-tetramethyl-p-phenylene-diamine	18·0	CCl$_4$	59						
Anisole	29·1	CH$_2$Cl$_2$	c	22·3	CH$_2$Cl$_2$	c			

* DDQ = 2,3-dichloro-5,6-dicyano-p-benzoquinone.   † TCNE = tetracyanoethylene.
‡ DCNQ = 2,3-dicyano-p-benzoquinone.   § TNF = 2,4,7-trinitrofluorenone.

Refs.: [a] G. Briegleb and J. Czekalla, Z. phys. Chem. Frankf. Ausg. **24**, 37 (1960); [b] R. M. Keefer and L. J. Andrews, J. Am. chem. Soc. **74**, 4500 (1952); [c] P. H. Emslie, unpublished work; [d] P. R. Hammond, J. chem. Soc. 3113 (1963); [e] P. G. Farrell and J. Newton, J. phys. Chem., Ithaca **69**, 3506 (1965); [f] B. Dodson, unpublished work; [g] I. Horman, unpublished

where W_0 is the energy of the "no bond" state, W_1 is the energy of the dative state and S_{01} is the overlap integral for these two states (see Chapter 2).

An alternative relationship derived from simple molecular orbital considerations, in terms of the Hückel coefficient (χ_i) of the resonance

$(\nu_{CT}/cm^{-1} \times 10^3)$ for various electron donor–electron acceptor pairs.

*DDQ *			TCNE†			DCNQ‡			TNF§		
ν_{CT}	Solvent	Ref.	ν_{CT}	Solvent	Ref.	ν_{CT}	Solvent	Ref.	ν_{CT}	Solvent	Ref.
24·6	CH_2Cl_2	c	25·8	CH_2Cl_2	153	25·7	CH_2Cl_2	c			
22·7	CH_2Cl_2	c	24·2	$CHCl_3$	e	23·8	CH_2Cl_2	d			
21·1	CH_2Cl_2	c	22·8	$CHCl_3$	e	22·2	CH_2Cl_2	c			
21·1	CH_2Cl_2	c	22·8	$CHCl_3$	e	22·2	CH_2Cl_2	c			
19·6	CH_2Cl_2	c	21·4	$CHCl_3$	e	20·7	CH_2Cl_2	c			
19·6	CH_2Cl_2	c	21·4	CH_2Cl_2	153	21·0	CH_2Cl_2	c			
17·2	CH_2Cl_2	c	20·8	CH_2Cl_2	146	18·5	CH_2Cl_2	c			
16·8	CH_2Cl_2	c	19·2	CH_2Cl_2	148	18·2	CH_2Cl_2	c			
16·0	CH_2Cl_2	c	18·4	CH_2Cl_2	148	17·4	CH_2Cl_2	c			
16·2	CH_2ClCH_2Cl	f	18·2	CH_2Cl_2	146				23·2	$CHCl_3$	34
12·1	CH_2ClCH_2Cl	g	13·5	CCl_4	150	14·1	CH_2Cl_2	h	18·5	$CHCl_3$	34
			15·9	CCl_4	150				20·7	$CHCl_3$	34
17·1	CH_2Cl_2	g	18·5	CH_2Cl_2	148				23·0	$CHCl_3$	34
			11·2	CCl_4	150	11·9	CH_2Cl_2	h	16·1	$CHCl_3$	34
12·5	CH_2Cl_2	g	13·7	CH_2Cl_2	148	14·6	CH_2Cl_2	h	19·2	$CHCl_3$	34
16·1	CH_2Cl_2	g	18·0	CCl_4	150				23·5	$CHCl_3$	34
16·0	CH_2Cl_2	g	17·7	CCl_4	150						
			12·2	$CHCl_3$	33				16·9	$CHCl_3$	34
12·0	CH_2ClCH_2Cl	g	13·9	$CHCl_3$	33				19·2	$CHCl_3$	34
14·6	CH_2ClCH_2Cl	g	16·3	$CHCl_3$	c				21·2	$CHCl_3$	34
13·6	CH_2Cl_2	n	14·6	CH_2Cl_2	n	15·1	CH_2Cl_2	c			
			13·5	$CHCl_3$	33	15·6	CH_2Cl_2	h	18·7	$CHCl_3$	34
16·0	CH_2Cl_2	c	17·0	$CHCl_3$	86	17·2	CH_2Cl_2	c			
			14·9	$CHCl_3$	e						
			10·4	Et_2O	p	12·0	CH_2Cl_2	59			
18·1	CH_2Cl_2	c	19·7	CH_2Cl_2	153						

work; [h] A. C. M. Finch, *J. chem. Soc.* 2272 (1964); [i] R. Bhattacharya and S. Basu, *Trans. Faraday Soc.* **54**, 1286 (1958); [j] J. Peters and W. B. Person, *J. Am. chem. Soc.* **86**, 10 (1964); [k] R. Foster, unpublished work; [l] A. Szent-Györgyi, "Introduction to a Submolecular Biology," Academic Press, New York (1960); [m] G. Briegleb and J. Czekalla, *Z. Elektrochem.* **59**, 184 (1955); [n] P. H. Emslie, R. Foster and R. Pickles, *Can. J. Chem.* **44**, 9 (1966); [o] A. K. Chandra and D. C. Mukherjee, *Trans. Faraday Soc.* **60**, 62 (1964); [p] W. Liptay, G. Briegleb and K. Schindler, *Z. Elektrochem.* **66**, 331 (1962).

integral β of the donor component, was also described in Chapter 2, namely:

$$E_{CT} = B_j - \alpha_0 - \chi_i \beta + P \qquad (3.3)$$

$$(=2.24)$$

where B_j represents the energy of the lowest-unfilled acceptor orbital and P is an energy term corresponding to the perturbation of the energy levels in A and D.[33–37]

DEPENDENCE OF THE ENERGY OF THE CHARGE-TRANSFER BAND ON THE IONIZATION POTENTIAL OF THE DONOR

The relationship between the energy ($h\nu_{CT}$) of the lowest-energy intermolecular charge-transfer band and the ionization potential of the donor (I^D) for a series of complexes with a common acceptor species has been the source of much discussion. The ionization potentials of many of the electron donors employed have been determined experimentally. These include values obtained from direct photo-ionization experiments, particularly by Watanabe and his co-workers,[38, 39] and more recently by Price et al.[40, 41] One disadvantage of the method is that the positive ion formed is not identified, so that a process other than the intended one may be measured. This can be circumvented by combining the experiment with mass-spectrophotometric indentification of the product ion.[42] By and large, there is remarkably good agreement between photo-ionization values and those obtained from ultraviolet Rydberg series measurements. These latter are a measure of the appropriate electronic transition involving the 0, 0 vibrational levels and represent the so-called adiabatic ionization potentials [I(ad)].* By contrast, electron-impact experiments tend to yield higher values. They may exceed the values obtained from Rydberg series or photo-ionization experiments by as little as 0·1 eV or less; on the other hand, the difference may be more than 1 eV. The claim has been made[43, 44] that the electron-impact experiments yield vertical ionization potentials I(vert). However, it has been pointed out[45] that, although low-resolution electron-impact experiments should yield vertical ionization potentials, high-resolution experiments should give adiabatic values. Values of ionization potentials from electron impact data obtained by different workers are unfortunately often not in agreement with one another. The recent technique of photoelectron spectroscopy, pioneered by Turner,[46–51] has provided a further series of reliable values of ionization potentials. These include higher as well as first ionization potentials. Although in principle the method yields both adiabatic and vertical ionization potentials, in practice only adiabatic values are readily obtained. A compilation of ionization potentials obtained by various methods has recently been published.[52]

The correlations of ionization potential with the energy of the charge-transfer band have involved mainly adiabatic values, particularly those derived from photo-ionization data.

* See footnote on p. 24.

If it is supposed that the terms $(G_0 - G_1)$ and $(\beta_0^2 + \beta_1^2)$ in equation (3.1) are constant for a series of complexes of different donors with the same acceptor species, then this equation may be written:[53, 54]

$$h\nu_{CT} = I^D - C_1 + \frac{C_2}{I^D - C_1} \tag{3.4}$$

where C_1 and C_2 are constants for a given acceptor. Typical parameters for sets of weak complexes are given in Table 3.3. They are the values

TABLE 3.3. Coefficients of equation (3.4) ($h\nu_{CT}$ and I^D in eV)

Acceptor	Solvent	C_1/eV	C_2/eV^2	Ref.*
Iodine (I$_2$)	n-heptane†	6	3·4	53
	CCl$_4$	5·2	1·5	54
	—	5·45 ± 0·26	1·63 ± 0·43	a
1,3,5-Trinitrobenzene	CCl$_4$	5·00	0·70	54
	—	5·29 ± 0·11	1·12 ± 0·20	a
Chloranil	CCl$_4$	5·70	0·44	54
	—	5·85 ± 0·09	0·58 ± 0·11	a
2,4,7-Trinitrofluorenone	—	5·61 ± 0·11	0·74 ± 0·17	a
Tetracyanoethylene	CCl$_4$	6·10	0·54	b
	—	6·06 ± 0·33	0·32 ± 0·20	a
Maleic anhydride	CCl$_4$	4·4		c
	—	4·59 ± 0·65	0·37 ± 2·1	a
Tetrachlorophthalic anhydride	CCl$_4$	4·9		c
	—	4·72 ± 0·58	0·05 ± 1·8	a
p-Benzoquinone	—	5·15 ± 0·30	0·8 ± 0·7	a
7,7,8,8-Tetracyanoquino-dimethane	—	6·10 ± 0·09	0·25 ± 0·09	a

* Refs.: [a] R. S. Becker and E. Chen, *J. chem. Phys.* **45**, 2403 (1966); [b] G. Briegleb, J. Czekalla and G. Reuss, *Z. phys. Chem. Frankf. Ausg.* **30**, 333 (1961); [c] G. Briegleb, "Elektronen-Donator-Acceptor-Komplexe," Springer-Verlag, Berlin (1961), p. 77.
† Also 2,2,4-trimethylpentane.

which best correlated the then-available experimental data. As the range of donors is increased, some change in these values can be expected. Slightly prior to the application of equation (3.4), McConnell *et al.*[55] showed that there is an approximately linear relationship [equation (3.5)] between I^D and ν_{CT} for complexes of iodine with a wide range of relatively weak donors:

$$h\nu_{CT} = I^D - E^A - W \tag{3.5}$$

where E^A is the electron affinity of the acceptor and W is the dissociation energy of the charge-transfer excited state. Since the publication of that paper, similar linear relationships have been described for complexes of many other acceptors. In general:

$$h\nu_{CT} = aI^D + b \qquad (3.6)$$

The coefficients a and b for various acceptors are listed in Table 3.4. These experimentally observed linear relationships are not so conflicting

TABLE 3.4. Coefficients of equation (3.6)*

Acceptor	Solvent	a	b	Ref.†
Tetracyanoethylene	CH_2Cl_2	0·486	−1·31	146
Tetracyanoethylene	CH_2Cl_2	{ 0·87‡ 0·92§	{ −4·86‡ −5·12§	148
Tetracyanoethylene	CH_2Cl_2	0·83	−4·42	a
Tetracyanoethylene	$CHCl_3$	0·82	−4·28	b
Chloranil	CCl_4	0·89	−5·13	57
Bromanil	CCl_4	{ 0·928‡ 0·973‖	{ −4·95‡ −5·27‖	c
1,3,5-Trinitrobenzene	CCl_4	0·89	−4·25	57
Iodine		0·67	−1·9	55
Iodine		0·87	−3·6	56
Pyromellitic dianhydride	CCl_4	0·87	−3·91	151

* Caution must be exercised in using relationships of this type. See text. The particular parameters are at least somewhat dependent on the chemical nature of the donor component.

† Ref.: ^{a} E. M. Voigt and C. Reid, *J. Am. chem. Soc.* **86**, 3930 (1964); ^{b} P. G. Farrell and J. Newton, *J. phys. chem., Ithaca* **69**, 3506 (1965); cf. *Tetrahedron Lett.* 5517 (1966); ^{c} M. Kinoshita, *Bull. chem. Soc. Japan* **35**, 1609 (1962).

‡ For polycyclic hydrocarbons.

§ Data from Ref. 146 incorporated.

‖ For substituted aromatic compounds, particularly amines. The relationship $h\nu = 0.85\,I^D - 4.32$ eV also gives a good fit.

with equation (3.4) as would appear. This is partly because of experimental difficulties in evaluating ν_{CT} owing to the broadness of the band and, on occasion, because of multiple overlapping bands (see p. 67). It is also partly due to the fact that the range of measured values of I^D is relatively small and does not in general extend much below 7 eV; consequently the segment of the parabola described by equation (3.4), over which the results have been applied, approximates to a straight line represented by equation (3.6).[56] The value of a in equation (3.6) has no direct significance; the line itself is only an approximation of equation (3.1) or (3.4).

Typical plots of $h\nu_{CT}$ against I^D from photo-ionization data, or from Rydberg analysis, are shown in Fig. 3.4. Corresponding plots which use I^D values from selected electron impact data [assumed to approximate to $I^D(\text{vert})$] are very similar. As far as the experimental correlation of

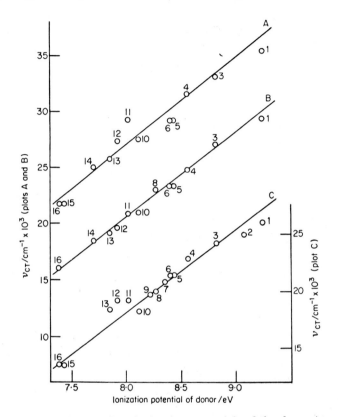

FIG. 3.4. Plots of ν_{CT} against ionization potentials of the donor (measured by the photo-ionization technique or from Rydberg analysis) for the series of donors: (1) benzene; (2) chlorobenzene; (3) toluene; (4) o- or m-xylene; (5) p-xylene; (6) mesitylene; (7) styrene; (8) biphenyl; (9) anisole; (10) naphthalene; (11) durene; (12) pentamethylbenzene; (13) hexamethylbenzene; (14) aniline; (15) azulene; (16) anthracene, with the acceptors: (A) 1,3,5-trinitrobenzene; (B) chloranil; (C) tetracyanoethylene.

I^D with $h\nu_{CT}$ is concerned, any possible advantage of using electron impact data is outweighed by the uncertainty of the data. The plots shown in Fig. 3.4 are characteristic of the many weak interactions which appear in the literature. Although some of the points which deviate from the least-squares line may be accounted for in terms of experimental inaccuracies in evaluating ν_{CT} and $I^D(ad)$, there seems little doubt that many of the discrepancies are real. It has often been assumed, erroneously, that there *must* be a linear relationship between ν_{CT} and I^D for a series of complexes with a common acceptor species. There is no theoretical reason to expect that this should be the case.[56] The approximate linearity arises because of the relative magnitudes of the various terms in equation (3.1). Even the assumption that $(G_0 - G_1)$ and $(\beta_1 + \beta_2)$ are each constant for such a set of complexes can only be at most approximately correct. Complexes which show deviations from these correlations are therefore not necessarily exceptional. Incidentally, the stronger the resonance interaction, the smaller will be the coefficient a in equation (3.6).

Despite the errors which may occur, both the linear relationship[57] [equation (3.6)] and the more refined parabolic relationship [equation (3.4)] have been used to evaluate unknown ionization potentials.[54, 58] It should now be obvious that all values of I^D obtained by interpolation from such plots must be treated with circumspection. Values of I^D obtained by extrapolation to low values are particularly questionable; yet the correspondence of such values with theoretically obtained values is often surprisingly good (Table 13.4). Although values of I^D estimated from empirical linear relationships of I^D and ν_{CT} are often virtually independent of the particular acceptor used, this does not necessarily give more weight to an estimate of I^D. The variation in ν_{CT} throughout a set of complexes of different donors with a given acceptor may be paralleled in a corresponding set with a second acceptor without ν_{CT} necessarily being proportional to I^D in either set of complexes.[59, 60] Indeed, this might be expected since there will be other factors besides I^D which will affect the electron donor potentialities of a component of a complex and some may be relatively independent of the choice of acceptor. This is shown by a plot of ν_{CT} for a set of complexes of a series of donors with one acceptor species against the corresponding values of ν_{CT} for the set of complexes of the same donors with a second acceptor (Fig. 3.5). The correlation is considerably better than that of either plot of ν_{CT} against I^D (cf. Fig. 3.4). Plots of the type illustrated in Fig. 3.5 provide useful evidence for the presence of charge-transfer complexes when the appropriate ionization potential values of the donors are not available.[61-63] In many cases where both acceptors are π-acceptors, the lines have unit gradient (Fig. 3.5). These correspond to equal values of the parameter a in equation (3.6), (cf. Table 3.4). However, in some cases the gradient is significantly different from unity. Such an example is the plot of ν_{CT} for iodine complexes against ν_{CT} for the corresponding complexes with a π-acceptor.[59] Theoretically this case can be readily understood in terms of the difference in overlap for these two types of acceptor.

If the overlap is very strong, as for example in the series of complexes of iodine with aliphatic amines where donation occurs from the localized lone-pair electrons on the nitrogen, then for a plot of I^D against ν_{CT} the points fall wide of the curve which correlated the ν_{CT} values for complexes

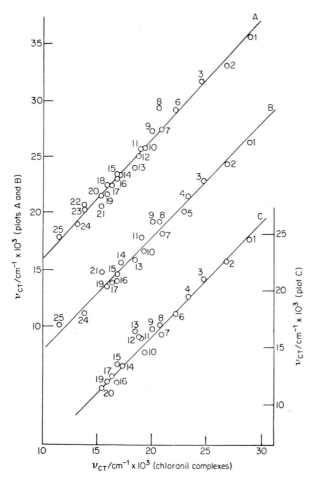

FIG. 3.5. Plots of ν_{CT} for a series of complexes of (A) 1,3,5-trinitrobenzene; (B) tetracyanoethylene; (C) 2,3-dichloro-5,6-dicyano-p-benzoquinone against ν_{CT} for the corresponding complexes of chloranil. The donors are: (1) benzene; (2) toluene; (3) m-xylene; (4) mesitylene; (5) biphenyl; (6) anisole; (7) naphthalene; (8) durene; (9) pentamethylbenzene; (10) *trans*-stilbene; (11) hexamethylbenzene; (12) aniline; (13) chrysene; (14) acenaphthene; (15) biphenylene; (16) benz[a]anthracene; (17) pyridine; (18) 2-naphthylamine; (19) anthracene; (20) 1-naphthylamine; (21) N,N-dimethylaniline; (22) perylene; (23) N,N,N',N'-tetramethyl-p-diaminodiphenylmethane (tetrabase); (24) tetracene; (25) N,N,N',N'-tetramethyl-p-phenylenediamine. (Mainly carbon tetrachloride or dichloromethane solvent.)

of iodine with weaker donors (Fig. 3.6).[56] Indeed in this case the inter-action is sufficiently high to make the simple perturbation theory (p. 24) inapplicable. For these stronger interactions the variation principle must be used (see p. 26). In such cases equation (3.2) replaces equation (3.1). A reasonable fit for the aliphatic amine–molecular iodine complexes as a group is obtained when the parameters $C_1 = 6 \cdot 9$ eV, $S_{01} = 0 \cdot 3$ and

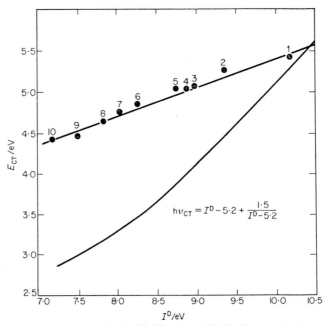

[R. S. Mulliken and W. B. Person, Ref. 56]

FIG. 3.6. Dependence of the charge-transfer transition energy ($h\nu_{CT}$) upon the ionization potential of the donor (I^D) for complexes of I_2. Values of I^D are from Ref. 39. The lower curve is that found by Briegleb to fit the data for most I_2–π-donor complexes (Ref. 30, p. 78). The data for the aliphatic amines from Ref. 32 are shown in the upper curve: (1) ammonia; (2) pyridine; (3) methylamine; (4) ethylamine; (5) n-butylamine; (6) dimethylamine; (7) diethylamine; (8) trimethyl-amine; (9) triethylamine; (10) tri(n-propyl)amine.

$\beta_0 = -2 \cdot 5$ eV are used in conjunction with equation (3.2) (Fig. 3.6). By comparison, $C_1 = 5 \cdot 2$ eV, $S_{01} = 0 \cdot 1$ and $\beta_0 = 0 \cdot 6$ eV for the correlation using equation (3.4) for the weaker iodine complexes in Fig. 3.6.[64] The smaller values of the parameters for these latter complexes are in harmony with their expected much lower interaction energy. For a series of iodine *atom* complexes, Person[64] has suggested the parameters $C_1 = 8$ eV, $S_{01} = 0 \cdot 2$ and $\beta_0 = -1$ eV in equation (3.2). These complexes are therefore intermediate in strength between the "weak" complexes, the energetics of which have been discussed in terms of the second order perturbation theory, and the "strong" complexes where the variation principle has been applied. It re-emphasizes the dangers of assuming that a single linear relationship between ν_{CT} and I^D may necessarily be applied to a set of complexes of a given acceptor.

Attempts have been made to determine the mode of interaction of certain π-acceptors with aza-aromatics which, potentially, can act either as n-donors or as π-donors. In some of the systems, only weak, poorly resolved bands are observed even after the compensation of absorption by the free components. It is difficult to make deductions from such data. In more favourable cases, some conclusions have been drawn. Kearns et al.[65] have made measurements on complexes with fairly strong π-acceptors at 77°K in which the charge-transfer bands are well separated from other absorptions. From predictions of the position of the charge-transfer bands using the first ionization potentials in conjunction with the correlation expressed by equation (3.4), combined with the expected substituent effects on the energy of the charge-transfer bands, it has been concluded that quinoline, 6-bromo-, 8-bromo-, 1,2,3,4-tetrahydro- and 2-methylquinoline and isoquinoline act as n-donors, whilst acridine and 8-hydroxyquinoline act as π-donors. Their arguments presume that the estimated ionization potentials of the donors relate to an electron in the highest occupied π-orbital. This is probably correct. The conclusions made should only be applied in the first place to the complexes of the π-acceptors actually studied, namely chloranil, p-benzoquinone and 1,3,5-trinitrobenzene. However, it is likely that the behaviour of these donors towards other π-acceptors is similar. By contrast, there is considerable evidence that aza-aromatics usually act as n-donors rather than π-donors towards σ-acceptors such as molecular iodine.[66-70] This conclusion is supported by theoretical considerations of maximum overlap of the appropriate orbitals of the donor and acceptor moieties in the complex.

Polarographic oxidations may provide, by their experimentally determined half-wave oxidation potentials $\mathscr{E}_{1/2}^{OX}$, a measure of the ionization potential of a compound. It is necessary that there is electro-chemical equilibrium at the electrode and that the wave is "reversible". If the effect of the difference in diffusion of the compound and its positive ion is small and entropy changes are small, we may write[71, 72]

$$\mathscr{E}_{1/2}^{OX} = I^D + \Delta F_{solv} + \text{constant} \tag{3.7}$$

ΔF_{solv} is the difference in solvation energy between the compound and its positive ion. If it is assumed that, for a series of compounds, variations in I^D are much greater than ΔF_{sol}, the $\mathscr{E}_{1/2}^{OX}$ may be used as a measure of the electron-donating ability of a donor. Because of this restriction it is obviously advisable to compare compounds which are chemically related. Values of $\mathscr{E}_{1/2}^{OX}$ for donor molecules, obtained from polarographic measurements of acetonitrile solutions, and plotted against ν_{CT} values for complexes of a common acceptor, yield good straight lines (Fig. 3.7). The results are of interest in that the polarographic data extend to

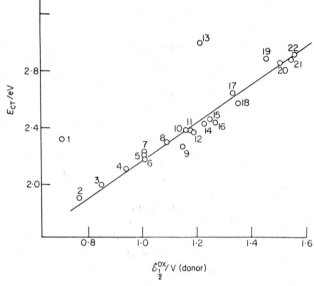

[E. S. Pysh and N. C. Yang, Ref. 72]

FIG. 3.7. Plot of transition energies of 2,4,7-trinitrofluorenone complexes of: (1) azulene; (2) tetracene; (3) perylene; (4) benzo[a]pyrene; (5) benzo[b]chrysene; (6) benzo[ghi]perylene; (7) dibenzo[a,e]pyrene; (8) anthracene; (9) dibenzo[b,e]-pyrene; (10) pyrene; (11) benz[a]anthracene; (12) dibenz[a,h]anthracene; (13) acenaphthylene; (14) coronene; (15) dibenz[a,c]anthracene; (16) benzo[e]pyrene; (17) picene; (18) chrysene; (19) fluoranthene; (20) phenanthrene; (21) naphthalene; (22) triphenylene, against $\mathscr{E}_{1/2}^{OX}$ for the respective donors.

compounds for which no direct ionization-potential data are as yet available. The deviations shown by azulene and acenaphthylene in Fig. 3.7 are probably due to large changes in the solvation term in equation (3.7).[72]

DEPENDENCE OF THE ENERGY OF THE CHARGE-TRANSFER BAND ON THE ELECTRON AFFINITY OF THE ACCEPTOR

The complementary variation of ν_{CT} with the electron affinity (E) of the acceptor has been much less widely studied.* This has been due

* When the electron affinity (E) refers specifically to an electron acceptor species, it will be written as E^A.

mainly to the paucity of electron affinity data. Few direct determinations of E have been made on species which we may consider to be electron acceptors. Mulliken[73] has estimated the electron affinity of the iodine molecule to be about $1\cdot7 \pm 1$ eV. More recently, from a theoretical calculation of the potential energy curves for I_2 and I_2^- using the Morse function, and utilizing the interatomic equilibrium distance, dissociation energy and fundamental vibrational frequency, Person[64] has suggested a value of $1\cdot7 \pm 0\cdot5$ eV for I_2. In a similar fashion, he has estimated the electron affinities of Br_2, Cl_2 and ICl as $1\cdot2 \pm 0\cdot5$ eV, $1\cdot3 \pm 0\cdot4$ eV and $1\cdot7 \pm 0\cdot6$ eV respectively.

Electron affinities have been obtained from electron absorption co-efficients.[74] These have been restricted to hydrocarbons which are normally considered as electron donors in charge-transfer complexes. The high electron affinity values of some hydrocarbons suggest that they could also act as electron acceptors.

By analogy with one-electron oxidation potentials (p. 49) determin-ations of polarographic one-electron reduction potentials $\mathscr{E}_1^{red}$, whilst not directly measuring E, nevertheless provide one of the few inde-pendent experimental estimates of E which are available for a range of acceptor species.[75-82] Maccoll[83] has shown that there should be a linear relationship between $\mathscr{E}_1^{red}$ and E for angularly condensed aromatic hydrocarbons. The half-wave reduction potentials of a wide range of electron acceptors in aprotic solvents, particularly acetonitrile, have been studied by Peover and his co-workers.[75-82] The single-electron process is effectively diffusion-controlled and reversible. Within experimental error, the reduction potential ($\mathscr{E}_1^{red}$) is equal to the half-wave potential ($\mathscr{E}_{1/2}^{red}$). For measurements against a standard calomel electrode:

$$\mathscr{E}_{1/2}^{red} = \mathscr{E}_1^{red} = E - \Delta F_{solv} - \phi_{Hg} - \mathscr{E}_{Hg:Hg^{++}}^{\ominus} \qquad (3.8)$$

where ΔF_{solv} is the difference in solvation energy between the compound and its negative ion, and represents mainly the solvation of the anion. ϕ_{Hg} is the work function $e^- $ (in Hg) $\rightarrow Hg_{liq} + e^-$ (equal to $4\cdot54$ eV) and $\mathscr{E}_{Hg:Hg^{++}}^{\ominus}$ is the absolute value of the calomel electrode (equal to $0\cdot53$ V). For two different acceptors A_1 and A_2, if ΔF_{solv} is assumed to be constant, then:

$$E^A(A_1) - E^A(A_2) = \mathscr{E}_1(A_1) - \mathscr{E}_1(A_2) \qquad (3.9)$$

Thus, if the electron affinity of one acceptor is known, then others may be derived from a knowledge of their single-electron reduction potentials. Unfortunately there seems at present to be little certainty about the E^A values of any species which might be taken as a standard. Briegleb[58] has suggested that an average value of $\Delta F_{\text{solv}} = -3 \cdot 66$ eV might be taken, so that equation (3.7) becomes

$$E^{\text{A}} = -\mathcal{E}_{1/2}^{\text{red}} + 1 \cdot 41 \qquad (3.10)$$

An alternative course is simply to use values of $\mathcal{E}_1$ (assumed to be equal to $\mathcal{E}_{1/2}$) as a measure of electron affinity. Experimental values of $\mathcal{E}_{1/2}$ for various acceptor molecules are given in Table 3.5.

TABLE 3.5. Half-wave reduction potentials $\mathcal{E}_{1/2}^{\text{red}}$ for some electron acceptors in acetonitrile at 25°C

Acceptor	$\mathcal{E}_{1/2}^{\text{red}}/\text{V}$	Ref.*
1. p-Benzoquinone	$-0 \cdot 51$	78
2. 2,3-Dichloro-5,6-dicyano-p-benzoquinone	$+0 \cdot 51$	78
3. Bromanil	$0 \cdot 00$	78
4. Chloranil	$+0 \cdot 01$	78
5. Fluoranil	$-0 \cdot 04$	78
6. Trichloro-p-benzoquinone	$-0 \cdot 08$	78
7. 2,5-Dichloro-p-benzoquinone	$-0 \cdot 18$	78
8. 2,6-Dichloro-p-benzoquinone	$-0 \cdot 18$	78
9. Chloro-p-benzoquinone	$-0 \cdot 34$	78
10. Fluoro-p-benzoquinone	$-0 \cdot 37_4$	80
11. Bromo-p-benzoquinone	$-0 \cdot 32_5$	80
12. Iodo-p-benzoquinone	$-0 \cdot 33$	80
13. Nitro-p-benzoquinone	$-0 \cdot 40_5$	80
14. Trifluoromethyl-p-benzoquinone	$-0 \cdot 22_8$	80
15. Acetyl-p-benzoquinone	$-0 \cdot 30_6$	80
16. Methoxy-p-benzoquinone	$-0 \cdot 61_5$	80
17. Dimethylamino-p-benzoquinone	$-0 \cdot 74_5$	80
18. Tetramethyl-p-benzoquinone	$-0 \cdot 84$	78
19. Trimethyl-p-benzoquinone	$-0 \cdot 75$	78
20. 2,5-Dimethyl-p-benzoquinone	$-0 \cdot 67$	78
21. Methyl-p-benzoquinone	$-0 \cdot 58$	78
22. Cyano-p-benzoquinone	$-0 \cdot 12_0$	80
23. Phenyl-p-benzoquinone	$-0 \cdot 50$	78
24. o-Benzoquinone	$-0 \cdot 31$	78
25. 1,4-Naphthoquinone	$-0 \cdot 71$	78
26. 2-Methyl-1,4-naphthoquinone	$-0 \cdot 77$	78
27. 2-Amino-1,4-naphthoquinone	$-0 \cdot 92$	78
28. 5-Hydroxy-1,4-naphthoquinone	$-0 \cdot 52$	78
29. 1,2-Naphthoquinone	$-0 \cdot 56$	78
30. 1,4-Anthraquinone	$-0 \cdot 75$	78
31. 9,10-Anthraquinone	$-0 \cdot 94$	78
32. 9,10-Phenanthraquinone	$-0 \cdot 66$	78
33. Diphenoquinone	$-0 \cdot 24$	78
34. Pyromellitic dianhydride	$-0 \cdot 55$	77
35. Dibromopyromellitic dianhydride	$-0 \cdot 32$	77
36. Phthalic anhydride	$-1 \cdot 31$	77
37. Tetrachlorophthalic anhydride	$-0 \cdot 86$	77
38. Maleic anhydride	-0.84	77
39. Tetracyanoethylene	$+0 \cdot 24$	77

TABLE 3.5.—*continued*

Acceptor	$\mathscr{E}_{1/2}^{\text{red}}/V$	Ref.*
40. 1,2,4,5-Tetracyanobenzene	−0·71	77
41. 7,7,8,8-Tetracyanoquinodimethane	+0·19	77
42. Nitrobenzene	−1·15	a
43. 1,2-Dinitrobenzene	−0·83	82
44. 1,3-Dinitrobenzene	−0·91	82
45. 1,4-Dinitrobenzene	−0·70	82
46. 4-Chloronitrobenzene	−1·08	82
47. 1,3,5-Trinitrobenzene	−0·60	82
48. 1-Fluoro-2,4-dinitrobenzene	−0·88	82
49. 1,3-Difluoro-2,4-dinitrobenzene	−0·86	82

* Ref.: *a*A. H. Maki and D. H. Geske, *J. Am. chem. Soc.* **83**, 1852 (1961).

From the relationship of the energy of the charge-transfer transition for a weak interaction described in equation (3.1) namely:

$$h\nu_{CT} = I^D - E^A - G_1 + G_0 + \frac{\beta_0^2 + \beta_1^2}{I^D - E^A - G_1 + G_0} \qquad (3.1)$$

for a set of complexes of a given donor with a series of acceptors, if the terms $(\beta_0^2 + \beta_1^2)$ and $(G_1 - G_0)$ are constant, a parabolic relationship between ν_{CT} and E^A analogous to equation (3.4) would be expected, thus:

$$h\nu_{CT} = - E^A - g_1 + \frac{g_2}{-E^A + C_1} \qquad (3.11)$$

where g_1 and g_2 are constants for a given donor moiety. Evidence that such a relationship may sometimes hold is shown by the plot of ν_{CT} for complexes of hexamethylbenzene against $\mathscr{E}_{1/2}^{\text{red}}$ (as a measure of E^A) for various acceptors, although the data are just as well correlated by a linear function (Fig. 3.8). There is a large scatter of the points. This may in part be due to the approximation: $\Delta F_{\text{solv}} = \text{constant}$, in equation (3.8).

Plots of ν_{CT} for a set of complexes of a series of acceptors with a given donor, against ν_{CT} for another set of complexes of the same series of acceptors with a second donor, are often straight lines of unit gradient[60] (Fig. 3.9). Such plots are the counterpart of those illustrated in Fig. 3.5. However, the deviations tend to be greater, and in many cases the gradient is not equal to unity. This reflects a generally greater sensitivity of the term $(G_0 - G_1 + X_1 + X_0)$ [see equations (2.19), (2.20)] to changes in acceptor than to changes in donor. Where regular behaviour exists,

3

then obviously from equation (3.1), relative values of "apparent" electron affinity may be obtained which will be independent of the particular donor. For example, if for complexes of a given donor with the acceptors A_1 and A_2, $(G_0 - G_1) + (X_1 + X_0)$ is constant, then:

$$E^A(A_1) = E^A(A_2) - h\nu_{CT}(A_1) - h\nu_{CT}(A_2) \tag{3.12}$$

This relationship has been used [84] to calculate E^A for the iodine molecule by comparison with E^A for the iodine atom. In view of the difference

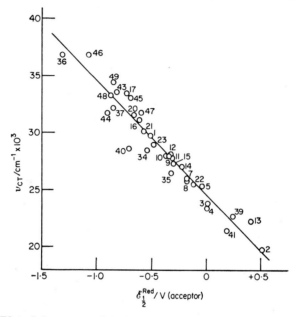

FIG. 3.8. Plot of the energy of the lowest charge-transfer transition for a series of hexamethylbenzene–electron acceptor complexes against $\mathscr{E}^{red}_{1/2}$ values for the acceptors. The numbering corresponds to the acceptors listed in Table 3.5.

between these two acceptors, it seems unlikely that $(G_0 - G_1) + (X_1 + X_0)$ is in fact independent of the acceptor, although for a structurally related set of donors, the term may to a first approximation be independent of the donor. Therefore, although the values of E^A for the iodine molecule so obtained are all nearly equal to one another, irrespective of the particular donor, this consistency could be due merely to a constant difference in the term $(G_0 - G_1) + (X_1 + X_0)$ between iodine atom and iodine molecule complexes, so that the estimated value of E^A for I_2 is in error by this difference. If this semi-empirical method of estimating E^A is used, then it is prudent to make the comparisons between complexes of the same bond type and the same configuration. In principle, E^A may be obtained from the parabolic plot of $h\nu_{CT}$ against I^D for a series of complexes of a given acceptor, using equation (3.4).[58] The constant C_1 which is evaluated from the plot is, by definition, equal to $E^A - (G_0 - G_1)$. Now the term $(G_0 - G_1)$ is predominantly the Coulombic attraction between D^+ and

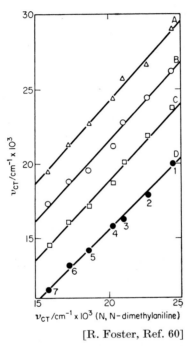

[R. Foster, Ref. 60]

FIG. 3.9. Plots of ν_{CT} of a series of N,N-dimethylaniline–acceptor complexes against ν_{CT} for the corresponding complexes of: (A) hexamethylbenzene; (B) N-methylaniline; (C) N,N,N′,N′-tetramethyl-p-diaminodiphenylmethane (tetrabase); (D) N,N,N′,N′-tetramethyl-p-phenylenediamine. The acceptors are: (1) 1,3-dinitrobenzene; (2) 2,4,6-trinitrotoluene; (3) 1,3,5-trinitrobenzene; (4) p-benzoquinone; (5) chloro-p-benzoquinone; (6) 2,6-dichloro-p-benzoquinone; (7) chloranil. (Cyclohexane solvent.)

A^-. For iodine complexes, the term $(G_0 - G_1)$ has been estimated as $-4\cdot3$ eV, based on a geometry in which the iodine molecule lies parallel to the plane of the ring of the aromatic donor; and as $-3\cdot5$ eV for a complex in which the iodine molecule is perpendicular to the plane of the ring of the aromatic donor. Obviously, the final value assigned to E^A is strongly dependent on the estimate of the term $(G_0 - G_1)$. A similar but more composite energy term needs to be estimated if the linear approximation of equation (3.4) is used, namely, equation (3.5), for a given acceptor.

The plots of $\nu_{CT}(A_1)$ against $\nu_{CT}(A_2)$ for complexes of two acceptors, which have the same a value in equation (3.5) (as is often the case, see Table 3.5) are equivalent to equation (3.12) and should give correct values of the *differences* in E^A provided the term $(G_0 - G_1) + (X_1 - X_0)$ remains constant. The present problem with any of these empirical methods of estimating E^A is the paucity of, and doubt attached to, values of E^A obtained by independent methods, which might otherwise provide the standards for comparison.

For a comprehensive literature survey (up to 1963) of measurements relating to the electron affinities of organic molecules, the reader is referred to a review by Briegleb.[58]

DEPENDENCE OF THE ENERGY OF THE CHARGE-TRANSFER BAND ON MOLECULAR ORBITAL COEFFICIENTS

Correlations analogous to equation (3.5) have been obtained by the application of equation (3.3) and other related expressions. For a series of complexes with a common acceptor species, if the term P is effectively constant, then a plot of $h\nu_{CT}$ against the Hückel coefficients χ_i for the donors should be linear.[33-37] Typical results are shown in Fig. 3.10. Deviations may in part be due to the rudimentary nature of the calculation of the Hückel parameter. The relationship given by equation (3.3) with $P = 0$, has been used to estimate Hückel parameters for various molecules. It is exactly analogous to the estimates of I^D which have been made using empirical relationships such as equation (3.5). The resonance integral β is evaluated from the slope of the plot of $h\nu_{CT}$ against χ_i. For various complexes Nepraš and Zahradník[85, 86] have shown that the value of β is somewhat dependent on both the class of donor and on the particular acceptor. Instead of using the simple relationship, equation (3.3), allowance may be made for the decreased electron repulsion in the positive-ion state, relative to the uncharged state, as was suggested by Wheland and Mann[87] and developed by Streitwieser and Nair[88, 89] and by Ehrenson.[90] Ionization potentials thus calculated show a reasonably good linear correlation with ν_{CT} (Fig. 3.11).

In principle, values of the energy of the lowest-unfilled orbital of the acceptor (B_j) may be obtained from equation (3.3) by extrapolation to $\chi_i = 0$. However, this requires not only the evaluation of α_0 but also an estimation of the term P in equation (3.3). Calculations have been made in some cases on the assumption that $P = 0$. Alternatively, where estimates of these energy terms are imprecise, it may prove more satisfactory to use equation (3.3) to determine *differences* in B_j for various acceptors. These acceptors should be of the same type $(\pi, \sigma \ldots)$ and should be complexed with donors of a homogeneous class (prefer-

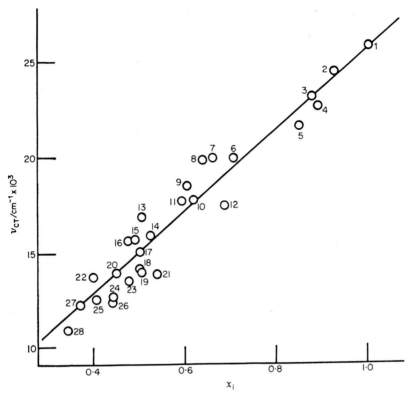

FIG. 3.10. Plot of charge-transfer transition energies for a series of tetracyano-ethylene–donor complexes against the Hückel coefficients (χ_i) for the donors. (Spectral data are mainly those of Dewar and Rogers, Ref. **33**, and of Merrifield and Phillips, Ref. **146**.) The donors represented are: (1) benzene; (2) toluene; (3) o-xylene; (4) m-xylene; (5) p-xylene; (6) biphenyl; (7) m-terphenyl; (8) acenaphthylene; (9) phenanthrene; (10) naphthalene: (11) p-terphenyl; (12) triphenylene; (13) picene; (14) chrysene; (15) dibenz[a,j]anthracene; (16) dibenz[a,h]anthracene; (17) dibenz[a,c]anthracene; (18) benzo[e]pyrene; (19) benz[a]anthracene; (20) pyrene; (21) coronene; (22) dibenzo[b,e]pyrene; (23) azulene; (24) dibenzo[a,e]-pyrene; (25) benzo[b]chrysene; (26) benzo[ghi]perylene; (27) benzo[a]pyrene; (28) perylene.

ably π, so that relatively weak complexes will be formed). If for such complexes the term P in equation (3.3) is constant, then for complexes of two acceptors A_1 and A_2 with a common donor species:

$$h\nu_{CT}(A_1) - h\nu_{CT}(A_2) = B_j(A_1) - B_j(A_2) \qquad (3.13)$$

This corresponds to the relationship equation (3.12) obtained from valence-bond considerations.

Lepley and Thompson[37] have shown that a correlation exists for tetracyanoethylene–aromatic hydrocarbon complexes, not only between χ_i for the highest occupied orbitals of the donor components and ν_{CT} for the first charge-transfer transition, but also between χ_i for the penultimate occupied orbital and ν_{CT} for the second charge-transfer

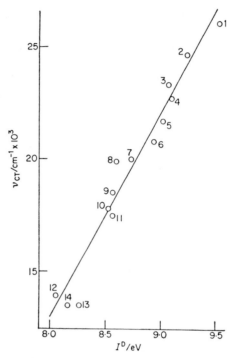

FIG. 3.11. Plot of charge-transfer transition energies for the tetracyanoethylene complexes of: (1) benzene; (2) toluene; (3) o-xylene; (4) m-xylene; (5) p-xylene; (6) styrene; (7) biphenyl; (8) acenaphthylene; (9) phenanthrene; (10) naphthalene; (11) fluorene; (12) pyrene; (13) azulene; (14) anthracene, against the ionization potentials of these donors calculated using the ω-technique. (Values mainly from Refs. 88, 89 and 90.)

band. The poorer correlation in the second set may be the result of lack of consideration of non-bonding states.

CORRELATION OF THE ENERGY OF THE CHARGE-TRANSFER BAND WITH TRANSITIONS IN THE DONOR

The lowest energy intramolecular transition in the donor component may be written as:

$$h\nu_D = m_j(D) - m_i(D) \qquad (3.14)$$

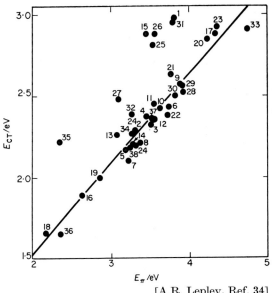

[A.R. Lepley, Ref. 34]

FIG. 3.12. Plot of the transition energies (E_{CT}) for a series of 2,4,7-trinitro-fluorenone complexes of the hydrocarbons: (1) acenaphthylene; (2) anthracene; (3) azulene; (4) benz[a]anthracene; (5) benzo[ghi]perylene; (6) benzo[e]pyrene; (7) benzo[a]pyrene; (8) benzo[b]chrysene; (9) chrysene; (10) coronene; (11) dibenz[a,c]anthracene; (12) dibenz[a,h]anthracene; (13) dibenzo[a,l]pyrene; (14) dibenzo[a,e]pyrene; (15) fluoranthene; (16) naphthacene; (17) naphthalene; (18) pentacene; (19) perylene; (20) phenanthrene; (21) picene; (22) pyrene; (23) triphenylene; (24) benzocoronene; (25) benzo[b]fluoranthene; (26) benzo[ghi]-fluoranthene; (27) benzo[k]fluoranthene; (28) benzo[a]fluorene; (29) benzo[b]-fluorene; (30) benzo[c]fluorene; (31) 9-benzylidenefluorene; (32) decacylene; (33) fluorene; (34) o-phenylenepyrene; (35) rubicene; (36) rubrene; (37) dibenzo[g,p]-chrysene; (38) tribenzo[a,e,i]pyrene; against the first π-π^* singlet transitions (E_π) of the donor molecules.

where $m_j(D)$ is the energy of the lowest-unfilled orbital of the donor to which a transition is permitted, and $m_i(D)$ is the energy of the highest-filled orbital of the donor. If the donor is a polycyclic aromatic hydrocarbon, then:

$$m_j(D) = \alpha_0 - \chi_i \beta \qquad (3.15)$$

where α is the Coulomb integral and β the resonance integral for the appropriate atom. Whence from equations (3.3), (3.14) and (3.15) the energy of the transition is:

$$h\nu_D = -2\chi_i \beta \qquad (3.16)$$

If this is compared with the energy of the intermolecular charge-transfer transition obtained in equation (3.3), it is seen that:

$$h\nu_{CT} = m_j(A) - \alpha_0 + P + \tfrac{1}{2}h\nu_D \qquad (3.17)$$

where $m_j(A)$ is the energy of the lowest-unfilled orbital of the acceptor, and P is a perturbation term (see p. 28). For a series of complexes with an acceptor for which P might be expected to be constant (see p. 28), a plot of ν_{CT} against ν_D should be linear with gradient equal to one-half. Figure 3.12 shows an example where these assumptions are reasonably justified. This argument is effectively the same as that used to relate I^D with $h\nu_D$, for which Matsen[91] proposed the relationship:

$$I^D = 4{\cdot}39 + 0{\cdot}857h\nu_D \text{ (eV)} \qquad (3.18)$$

and more recently Kuroda[92] has suggested:

$$I^D = 5{\cdot}10 + 0{\cdot}700h\nu_D \text{ (eV)} \qquad (3.19)$$

since it yields I^D values which are consistent with values obtained by other methods, particularly from charge-transfer complexes.

CORRELATION OF THE ENERGY OF THE CHARGE-TRANSFER BAND WITH HAMMETT SUBSTITUTION CONSTANTS

Hammett substitution constants (σ) provide an invaluable empirical measure of the electron-attracting ability of a substituent.[93] Hammond[94]

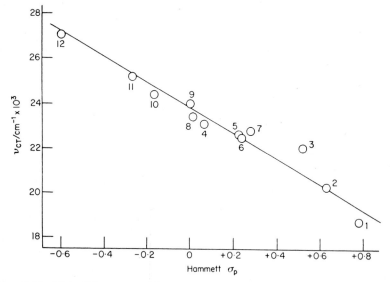

FIG. 3.13. Plot of charge-transfer transition energies for a series of complexes of mono-X-substituted p-benzoquinones with hexamethylbenzene in carbon tetrachloride at 25°C, where —X is: (1) —NO$_2$; (2) —CN; (3) —COCH$_3$; (4) —F; (5) —Cl; (6) —Br; (7) —I; (8) —C$_6$H$_5$; (9) —H; (10) —CH$_3$; (11) —OMe; (12) —NMe$_2$. (Adapted from P. R. Hammond, Ref. 94.)

has shown that for hexamethylbenzene complexes of a series of mono-substituted p-benzoquinones there is a near-linear relationship between ν_{CT} and the Hammett para-constant (σ_p) for the substituent in the acceptor (Fig. 3.13). The high degree of correlation may itself be fortuitous.[94] A corresponding plot is obtained for a series of complexes of

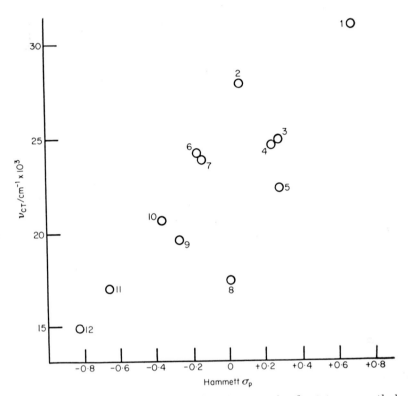

FIG. 3.14. Plot of charge-transfer transition energies for tetracyanoethylene complexes with mono-substituted benzenes, C_6H_5X, where X is: (1) —CN; (2) —F; (3) —Cl; (4) —Br; (5) —I; (6) —CH_3; (7) —$CH(CH_3)_2$; (8) —SH_3; (9) —OCH_3; (10) —OH; (11) —NH_2; (12) —$N(CH_3)_2$, against the corresponding Hammett σ_p substituent constants.

substituted donors with a common acceptor species, tetracyanoethylene (Fig. 3.14), although in this case the correlation is of a lower order. Kosower et $al.$[95] have shown that ν_{CT} values, for a series of complexes of iodide ion with 1-X-benzyl-4-carbomethoxypyridinium ions, are similarly related. All these observations are compatible with the Hammett parameter being an index of electron availability in the molecule concerned, ν_{CT} decreasing with increasing values of σ when the substituent is

3*

in the acceptor and ν_{CT} increasing with σ when the substituent is in the donor.

EFFECT OF SOLVENT ON THE ENERGY OF THE CHARGE-TRANSFER BAND

Kosower[96-99] has shown that ν_{CT} for the complex between the iodide ion and a pyridinium ion is extremely solvent-sensitive. For complexes

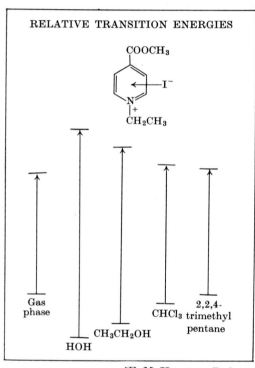

[E. M. Kosower, Ref. 96]

FIG. 3.15. Diagram showing the relative effect of different solvents on the intermolecular charge-transfer transition in a pyridinium–iodide complex.

in which the components are oppositely charged species we may express the ground state as predominantly an ion pair with a small admixture of a structure involving a pyridinium radical and an iodine atom with the two odd electrons coupled, i.e. $a^2 \ll b^2$ in equation (2.1), in contrast to the ground state of a weak complex formed from two neutral species where $a^2 \gg b^2$. The excited state of the pyridinium complex will be given by equation (3.20):

$$\psi_{\text{E}} = a^{\neq}\,\psi_0(\text{A, D}) - b^{\neq}\,\psi_1(\text{A}^- - \text{D}^+) \qquad (3.20)$$

where $a^{\neq 2} \gg b^{\neq 2}$.

In the ground state of the complex there is a large dipole perpendicular to the plane of the pyridinium ring. In the excited state the main dipole lies in the plane of the ring. Solvating molecules will orientate themselves about the complex (the so-called cybotactic region) so that the energy is minimized in the ground state. By the Franck–Condon principle there will be no solvent reorientation at the instant of excitation. Thus the cybotactic region, orientated to minimize the energy in the ground state, will tend to increase the energy of the excited state. The greater the solvating power of the solvent, the greater will be the decrease in the energy of the ground state and the greater the increase in the excited state compared with the ideal system of a single isolated complexed pair *in vacuo*, and hence the greater the hypsochromic shift (Fig. 3.15). The changes in the charge-transfer absorption spectra are so large and their measurement so readily made experimentally, that Kosower[96-99] has suggested that the position of the charge-transfer band may be used as an estimate of the ionizing power of a solvent. In these ways it is a preferable measure to the Y-value[100-102] based on the relative rates of hydrolysis of *tert*-butyl chloride in different solvents. Kosower has defined the Z-value of a solvent as the energy in kcal/mol of the peak maximum of the charge-transfer band of 4-carbomethoxy-1-ethylpyridinium iodide in that solvent. This particular pyridinium iodide was chosen because of its greater solubility in the less-solvating solvents. Some Z-values, along with E_T values, are given in Table 13.3. These are the energies (in kcal mol^{-1}) of the peak maximum of the intramolecular charge-transfer absorption band of the pyridinium-N-phenolbetaine (I) which is a

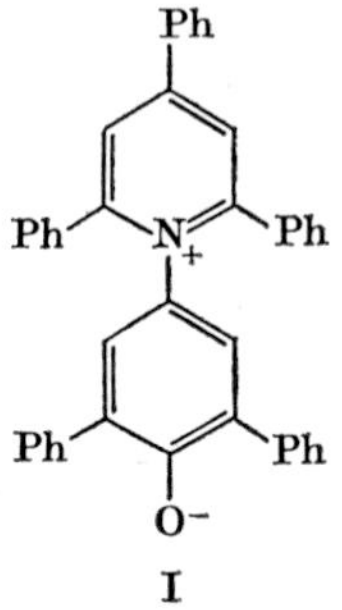

Ph

Ph — N⁺ — Ph

Ph — Ph

O⁻

I

solvent-sensitive transition and has been proposed as an alternative measure of solvent ionizing power.[103, 104]

Complexes of iodide ion with other planar aromatic cations such as tropylium show a solvent-dependence of ν_{CT} similar to that of pyridinium iodide[105] (see Chapter 10). The energies of the charge-transfer maxima for complexes of bromide and iodide ion with the neutral acceptor chloranil in different solvents show a linear dependence on the Z-value of the solvent.[106] By contrast, complexes of the acceptor tropylium with neutral donors such as pyrene or phenothiazine have ν_{CT} values which are considerably less sensitive to solvent change.[107]

The relative independence of the energies of charge-transfer fluorescence of some alkylpyridinium–halide ion complexes from the nature of the solvent lends support to Kosower's explanation of the solvent-dependencies of charge-transfer absorption bands[108] (see Section 3.B).

The majority of charge-transfer complexes involve uncharged component species. In these complexes the dependence of $h\nu_{CT}$ on solvent is relatively small. De Maine[109] has suggested that this is a result of the formation of a *liquid lattice penetration complex* in which the complex is surrounded by a cage of donor molecules, which acts as a buffer to any influence of the bulk solvent. If a wide range of solvents is considered, there is no immediately obvious correlation between $h\nu_{CT}$ and the dielectric constant of the solvent for many complexes formed from neutral components.[110–114] This suggests that the Onsager model of a dipole in a dielectric continuum is not applicable to these systems.[115] For strongly solvating solvents such as the alcohols, there is a hypsochromic shift as the Z-value of the solvent increases (Fig. 3.16). At first sight this is unexpected since for complexes formed from neutral species the excited state is more ionic. However, the complexes will be solvated in the ground state by such solvents. Apart from solvent–solute hydrogen bonding, many component species, particularly acceptors, have large localized dipoles. By the Franck–Condon principle there will be no reorientation of this solvent cage to minimize the energy of the excited state at the instant of excitation. The fact that the more ionizing solvents cause a hypsochromic shift suggests that the solvation in the ground state results mainly from the localized interactions with the constituent molecules, rather than from the dipole due to the contribution from the dative structure, since a cybotactic region caused by such an electric field is just the environment which would reduce the energy of the excited state, where a similar but much larger dipole will exist. This would have resulted in a bathochromic shift. In support of the idea of solvation of the components, it has been pointed out that the solvent shifts for the complexes are comparable in sign and magnitude with those of their components.[111]

The gross interaction shown by the solvents capable of specific solvation masks general solvent effects which are seen in solutions of the more inert solvents. In particular, the small shifts in the position of maximum intensity of the charge-transfer band have been related to dielectric properties of the solvent. Through the polarization of solvent molecules by the solute dipole(s), the increase in solute dipole in the excited state (for neutral ground-state components) will result in the upper state being more stabilized than the ground state by this non-specific solvent interaction.[116, 117] This is in agreement with the observed bathochromic shift as the polarizability (and Z-value) of this group of solvents increases. It is also in harmony with the observation that the

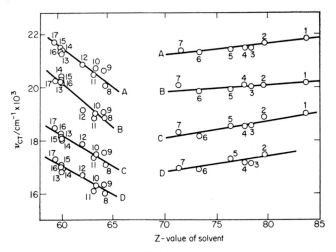

[P. H. Emslie and R. Foster, Ref. 110]

Fig. 3.16. Plots of charge-transfer transition energies for the complexes (A) 1,3,5-trinitrobenzene–N,N-dimethylaniline; (B) 1,2,4,5-tetracyanobenzene–N,N-dimethylaniline; (C) chloranil–acenaphthene; (D) chloranil–pyrene, in various solvents, against the Z-value of the solvent. Solvents: (1) methanol; (2) ethanol; (3) propan-1-ol; (4) butan-1-ol; (5) propan-2-ol; (6) decan-1-ol; (7) 2-methyl-propan-2-ol; (8) 1,1,2,2-tetrachloroethane; (9) dichloromethane; (10) 1,2-dichloroethane; (11) chloroform containing 0·13м ethanol; (12) 1,1-dichloroethane; (13) cyclohexane; (14) 2,2,4-trimethylpentane; (15) hexane; (16) heptane; (17) ethyl acetate.

energy of the charge-transfer band of some complexes decreases as the refractive index of the solvent increases.[118–120] Voigt[121, 122] has shown that, for a given tetracyanoethylene complex, a plot of ν_{CT} against the refractive index function [123] $(n^2 - 1)/(2n^2 + 1)$, where $n =$ refractive index of the solvent, gives a straight line for saturated hydrocarbons. However, the points for perfluorocarbon solvents lie along a different straight line. The two lines extrapolate to the same intercept when $(n^2 - 1)(2n^2 + 1) = 0$, for $n = 1$, which may be equated to the value of ν_{CT} in the gas phase for that complex. There is good agreement in the case of the p-xylene–tetracyanoethylene complex with the directly determined value of ν_{CT} for the gaseous phase.[124]

Several studies of gas-phase charge-transfer complex systems have been made recently[124–130] (Table 7.3). In all cases there is a blue shift (often ~ 2500 cm^{-1}) in the position of the intermolecular charge-transfer band relative to its position in the liquid phase. Prochorow and Tramer[129] have tried to explain this difference as being due to the combination of a dielectric shift and a "cage" effect in the liquid phase. The latter is the

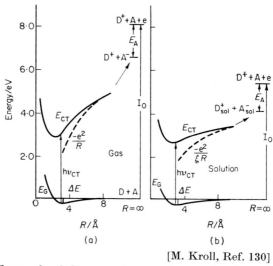

[M. Kroll, Ref. 130]

Fig. 3.17. Energy level diagrams for gas-phase and solution charge-transfer complexes. The vertical energy scale is somewhat arbitrary but approximates to the case of tetracyanoethylene complexes. R = intermolecular distance, ξ = dielectric constant of the solution.

description for the intermolecular forces which will tend to compress the complex. The force-constant for vibration of a complex along the intermolecular axis should be small; consequently a compression of 0·1–0·2 Å is not unreasonable. Trotter[131] has discussed the effect of such mechanical pressures on the properties of charge-transfer complexes in some detail. For the complex tetracyanoethylene–p-xylene he has used as the intermolecular distance in the vapour phase the value of 3·5 Å, which is somewhat larger than the interplanar spacing in the solid phase obtained crystallographically.[132] Estimates of the degree of electron transfer were made from values of I^D and E^A, using the empirical relationship[55] described by equation (3.5):

$$h\nu_{CT} = I^D - E^A - W \tag{3.5}$$

where $W \approx ke^2/r \approx$ Coulomb energy of the charge-transfer transition (r_{AD} is the intermolecular distance in the complex, e is the fractional electronic charge-transferred, and $k = 1.44 \times 10^{-7} \, eV.cm.e^2$). Using this value of $e \, (= 0.90)$, in combination with the difference in energy of the charge-transfer transition in the vapour state and in solution in the liquid phase, Trotter has calculated that the intermolecular distance in the complex decreases by about 0.3 Å in going from the vapour phase to dichloromethane solution.

The effect of a dielectric medium has been considered in some detail by Kroll,[130] who has shown that the observed shifts could be accounted for in this way, although much depends on the intermolecular distances chosen (Fig. 3.17). For more detail concerning charge-transfer complexes in the gaseous phase, the reader is referred to Kroll's paper.[130]

EFFECT OF PRESSURE ON THE CHARGE-TRANSFER TRANSITION IN THE LIQUID PHASE[133-140]

The energies of the charge-transfer transitions of the interactions between nitroaromatics and aromatic amines are decreased as the hydrostatic pressure is increased. Similar effects have been observed for complexes of iodine and of tetracyanoethylene.[140] However, although the charge-transfer band of the hexamethylbenzene–tetracyanoethylene complex shifts to longer wavelengths at intermediate pressures, at high pressures (ca. 7×10^3 atm.) the shifts are to the short wavelength side of the absorption at 1 atm.[140] It has been suggested that this reversal may be due to the steric hindrance between the nitrogen atoms of the tetracyanoethylene molecule and the methyl groups of the hexamethylbenzene molecule. In a close-packed arrangement of the donor and acceptor molecules lying parallel, there may be a significant increase in repulsion at very high pressures. Isomeric complexes could also account for frequency shifts,[140] although this cannot by itself be taken as strong evidence for such isomerization.

MULTIPLE CHARGE-TRANSFER BANDS

Not infrequently, more than one intermolecular charge-transfer transition is obtained.[6, 24, 36, 61, 63, 140-153] On occasion, the absorption, instead of being resolved into two separate bands, may be observed as a single asymmetric band with an exceptionally large half band-width. The multiplicity may arise from electron donation from more than one energy level in the donor, from acceptance at more than one energy level in the acceptor, from differences in the interaction energy, or from combinations of these possibilities. Frequently it is due to electron transitions from more than one level in the donor. This was first suggested by Orgel[154] in the case of the methylbenzene complexes of chloranil and of 1,3,5-trinitrobenzene. Double charge-transfer bands are observed in complexes of those methylbenzenes in which the degeneracy of the $^2E_{1g}$ level of the parent benzene is removed by substitution. By a first-order perturbation calculation, Orgel obtained the following relationship for the ionization potentials $I^D(\mathrm{I})$ and $I^D(\mathrm{II})$ corresponding to the two energy levels in the donor:

$$I^D(\mathrm{I}) = I^B - n\gamma_s - p_I\,\epsilon_s \tag{3.21}$$

$$I^D(\mathrm{II}) = I^B - n\gamma_s - p_{II}\,\epsilon_s \tag{3.22}$$

where I^B is the ionization potential of benzene, n the number of substituents (S) in the ring, ϵ_s and γ_s are parameters and depend on the nature of the substituent S, ϵ_s is a measure of the total interaction of S with the π-electron system of the ring, and γ_s is a function of the interaction of S with the σ-system. p_I and p_{II} are constants which depend on n and on the relative positions of the substituents. The degree of splitting should increase as the electron-donating property of the substituent(s) increases and should be largest for p-di- and sym-tetra-substituted donor molecules. Observations by Voigt[143] show that these predictions are largely fulfilled.

Where the multiplicity arises from the participation of more than one energy level in the donor, we might expect the energy difference between the two charge-transfer transitions in the complex to be equal to the energy difference between the two appropriate levels in the donor. This assumes that the donor–acceptor interaction is weak so that the difference in the interaction energy is small. For example, the energy difference between the two observed transitions in crystalline 1,3,5-trinitro-benzene–anthracene complex is 8000 cm^{-1}.[4] This agrees well with the calculated energy difference of the two highest-filled orbitals of anthra-cene, namely 8210 cm^{-1}.[155] The difference in frequency between the charge-transfer bands of complexes of a given donor should be independent of the particular acceptor. This has been observed in several systems (Table 3.6) where on occasion more than two maxima are observed. By contrast, in a series of complexes of 1,2,4,5-tetracyano-benzene, double maxima are observed in which the energy difference between the two maxima is *independent* of the donor (Table 3.6). This suggests that transitions are occurring from the highest-filled level of the donor to two vacant levels in the acceptor.[152] Hexamethylbenzene is amongst the donors which gives a double maximum; the multiplicity could not arise through transitions from two donor levels since the highest-filled level of this donor is doubly degenerate. Furthermore, the

TABLE 3.6. Multiple charge-transfer band maxima

Donor	Acceptor*	Solvent	ν_1/cm^{-1}	ν_2/cm^{-1}	ν_3/cm^{-1}	$\Delta\nu_{12}/$ cm^{-1}	Ref.†
p-Xylene	TCNE	CH_2Cl_2	21,800	24,100		2300	146
Biphenyl	TCNE	CCl_4	20,000	25,700		5700	150
Phenanthrene	TCNE	CCl_4	19,000		28,200		150
Naphthalene	TCNE	CCl_4	18,200	23,400		5200	150
Triphenylene	TCNE	CCl_4	18,000	22,100		4100	150
Fluorene	TCNE	CCl_4	17,700	23,800	24,700	6100	150
cis-Stilbene	TCNE	CH_2Cl_2	18,900	25,400		6500	61
$trans$-Stilbene	TCNE	CH_2Cl_2	16,800	25,800		9000	61
Chrysene	TCNE	CCl_4	15,900	18,800	22,900	2900	150
Dibenz[a,h]-anthracene	TCNE	CCl_4	14,500	17,100	21,900	2600	150
Benz[a]anthracene	TCNE	CCl_4	14,300	18,500	25,100	4200	150
Pyrene	TCNE	CCl_4	14,000	20,500	25,800	6500	150
		‡	12,700	20,000		7300	24
Anthracene	TCNE	CCl_4	13,500	21,500		8000	150
Perylene	TCNE	CH_2Cl_2	11,110	11,500		390	148
Anthanthrene	TCNE	CH_2Cl_2	11,370	11,630		260	148
Acenaphthene	TCNE	CH_2Cl_2	15,300	22,750		7450	148
Anisole	TCNE	CH_2Cl_2	19,600	26,100		6500	143
m-Dimethoxy-benzene	TCNE	CH_2Cl_2	17,900	22,700		4800	143
p-Dimethoxy-benzene	TCNE	CH_2Cl_2	15,700	26,300		10,600	143
N,N,N′,N′-Tetramethyl-p-phenylenediamine	TCNE	Et_2O	10,400	23,500		13,100	a
1,2,4,5-Tetra-methoxybenzene	TCNE	CH_2Cl_2	12,500	22,700		10,200	143
Biphenyl	Chloranil	CCl_4	23,000	28,700		5700	150
Naphthalene	Chloranil	CCl_4	20,900	26,000		5100	150
Chrysene	Chloranil	CCl_4	18,500	21,500		3000	150
Dibenz[a,h]-anthracene	Chloranil	CCl_4	17,200	19,800		2600	150
Benz[a]anthracene	Chloranil	CCl_4	16,900	21,150		4250	150
Pyrene	Chloranil	CCl_4	16,600	23,100		6500	150
Benz[a]anthracene	DDQ	CH_2Cl_2	12,000	16,400		4400	b
Pyrene	DDQ	CH_2Cl_2	12,500	18,200		5700	b
Anthracene	DDQ	CH_2ClCH_2Cl	12,100	20,200		8100	b
Acenaphthene	DDQ	CH_2ClCH_2Cl	13,500	19,200		5700	b
Mesitylene	TCNB	‡	28,300	33,700		5400	152
Hexamethylbenzene	TCNB	$CHCl_2CHCl_2$	23,500	29,900		6400	152
Naphthalene	TCNB	$CHCl_2CHCl_2$	25,000	>31,300		>6300	152
		‡	24,600	31,500		6900	23
N,N-Dimethylaniline	TCNB	$CHCl_2CHCl_2$	18,300	25,000		6700	152
N,N,N′N′-Tetra-methyl-p-phenylenediamine	TCNB	$CHCl_2CHCl_2$	13,900	20,300		6400	152
Anisole	I_2	CCl_4	29,000	33,900		4900	145
o-Phenylenediamine	TNB	$CHCl_3$	21,900	24,000	28,400	2100	142

TABLE 3.6—continued

Donor	Acceptor*	Solvent	ν_1/cm^{-1}	ν_2/cm^{-1}	ν_3/cm^{-1}	$\Delta\nu_{12}$/ cm^{-1}	Ref.†
p-Phenylenediamine	TNB	CHCl$_3$	20,000	26,000		6000	142
Toluene	PMDA	CCl$_4$	30,300§	34,000§		3700	151
p-Xylene	PMDA	CCl$_4$	28,100§	32,500§		4400	151
Mesitylene	PMDA	CCl$_4$	27,300§	32,900§		5600	151
Durene	PMDA	CCl$_4$	25,600§	28,600§	32,200§	3000	151
Pentamethyl-benzene	PMDA	CCl$_4$	24,400§	29,000§	32,100§	4600	151
Hexamethylbenzene	PMDA	CCl$_4$	23,000§	28,100§	31,700§	5100	151

* TCNE = tetracyanoethylene; DDQ = 2,3-dichloro-5,6-dicyano-p-benzoquinone; TCNB = 1,2,4,5-tetracyanobenzene; TNB = 1,3,5-trinitrobenzene; PMDA = pyromellitic dianhydride.

† Refs.: [a] W. Liptay, G. Briegleb and K. Schindler, *Z. Elektrochem.* **66**, 331 (1962). [b] I. Horman, unpublished work.

‡ Measured on pure solid complexes.

§ Fluorescence-excitation maxima (effectively equivalent to absorption maxima).

calculated energy difference between the two lowest-unfilled levels in 1,2,4,5-tetracyanobenzene agrees well with the observed energy of separation of the two charge-transfer bands.[152]

The possibility that it is the relative orientation of the donor and acceptor in an interaction which gives rise to two charge-transfer bands,[6, 24, 36, 140, 143, 149, 153] will now be considered. The two highest-filled orbitals in the donor differ in the positions of their nodal planes (Fig. 3.18). If Mulliken's maximum-overlap principle is applied (Chapter 2), two isomeric complexes with orientations shown in Fig. 3.19 might be expected. For example, although the intensities of the two charge-transfer bands of the tetracyanoethylene complexes with naphthalene

(a)   (b)

FIG. 3.18. The two highest-filled π-molecular orbitals in p-xylene. The lines between areas of positive and negative sign are the vertical nodal planes.

FIG. 3.19. The two orientations of tetracyanoethylene with respect to p-xylene which permit allowed transitions from the two highest-filled levels in the donor indicated in Fig. 3.18.

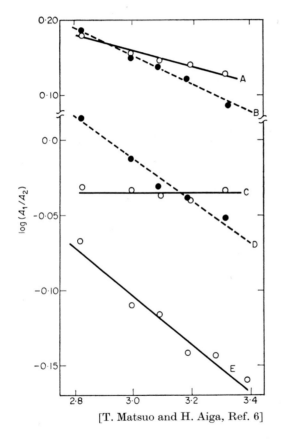

[T. Matsuo and H. Aiga, Ref. 6]

FIG. 3.20. The temperature dependences of the intensity ratios between the first and second charge-transfer bands for the tetracyanoethylene complexes of: (A) p-terphenyl; (B) *trans*-α-methylstilbene; (C) naphthalene; (D) *trans*-stilbene; (E) diphenylamine. The absorbances at the absorption maxima of the first and second charge-transfer bands are given by A_1 and A_2 respectively.

and substituted naphthalenes are almost equal, they show little variation in relative intensity with temperature.[36] This *may* be the result of two isomers with nearly equal oscillator strengths and equal heats of formation, but it does not provide direct evidence for the existence of two isomeric complexes. Examples of a variation of relative intensity of two charge-transfer bands of a single complex with temperature have been observed, however[6] (Fig. 3.20). A corresponding effect, caused by an increase in hydrostatic pressure, has also been reported for the naphthalene–tetracyanoethylene interaction[140] (see p. 77). By contrast, the enthalpies of formation of some complexes are markedly temperature dependent. This could be accounted for by the presence of two isomeric complexes which have different heats of formation. Estimates of the two hypothetical association constants, based on *direct* measurements of the intensities of each of the two bands, is not possible. Even if one band were due entirely to one complex, and the second band to the second complex, evaluations based on either band would result in an estimate of the sum of the two association constants (see Chapter 6). However, it may be that the interaction is a compromise: even if isomers such as those illustrated in Fig. 3.19 exist, they could be taken as representing extreme conformer structures in which the relative rotation of one molecule with respect to its partner readily occurs. Such molecular motion has been suggested from the basis of crystallographic evidence[156] and n.m.r. measurements.[157, 158] The barrier to rotation appears in some cases to be very low (1–2 kcal mol^{-1}) and of the same order as the enthalpy of formation of such complexes. Thus, to describe such systems as two isomeric complexes or a single complex is partly a matter of terminology.

3.A.4. Intensity of Intermolecular Charge-Transfer Transitions

The integrated intensity of an absorption band, termed the oscillator strength (**f**), is given by the theoretical expression:[159]

$$\mathbf{f} = 4 \cdot 704 \times 10^{-7} \nu_{\max} \mu_{\mathrm{EN}}^2 \qquad (3.23)$$

where $\nu_{\max}$ is the frequency in cm^{-1} of the maximum intensity of the band and μ_{EN} is the transition dipole:

$$\mu_{\mathrm{EN}} = -e \int \psi_{\mathrm{E}} \sum_{i} \mathrm{r}_i \psi_{\mathrm{N}} \, d\tau \qquad (3.24)$$

where r_i is the vector distance of the ith electron from some convenient origin. It may be approximated to:

$$\mu_{\mathrm{EN}} = a^{\mp} b(\mu_1 - \mu_0) + (aa^{\mp} - bb^{\mp})(\mu_{01} - \mu_0 S_{01}) \qquad (3.25)$$

where a, b, $a^{\mp}$ and $b^{\mp}$ are the coefficients in equations (2.1) and (2.6); S_{01} is the overlap integral ($\equiv \int \psi_0 \psi_1 d\tau$); μ_0, μ_1 are the dipole moments of the "no bond" and dative structures; and μ_{01} is the transition moment between the "no bond" and dative structures $(-e \int \psi_0 \sum_i r_i \psi_1 d\tau)$. Since S_{01} is usually small, equation (3.25) may be further approximated to:

$$\mu_{EN} = a^{\ddagger} b(\mu_1 - \mu_0) \qquad (3.26)$$

In principle, oscillator strengths may be estimated experimentally. Thus:

$$\mathbf{f} = 4 \cdot 32 \times 10^{-9} \int \epsilon \, d\nu \qquad (3.27)$$

$$\mathbf{f} = 1 \cdot 35 \times 10^{-8} \epsilon_{max}(\nu_{max} - \nu_{1/2}) \qquad (3.28)$$

where the frequencies are measured in cm^{-1}, $\nu_{1/2}$ is the half-width of the band, ϵ is the molar absorptivity (extinction coefficient) and ϵ_{max} is the maximum molar absorptivity of the band. In practice, there are experimental difficulties with charge-transfer complexes. Only rarely is the band not overlapped by other bands; in fact the band is often so poorly resolved that not even equation (3.28) may be applied. Partly as a consequence, and partly through expediency, the molar absorptivity, usually the maximum value, is generally taken as the measure of intensity of the absorption band, although experimental determinations of oscillator strength have been made.[130, 136, 140, 160-162] Because the complexes are dissociated to some extent in solution, a direct experimental measure of ϵ^{AD} for a complex cannot be obtained from a single solution. In the majority of methods employed, optical measurements are made on a series of solutions. From such observations it is now generally agreed that, whilst the product $K^{AD} \cdot \epsilon^{AD}$ may be evaluated satisfactorily, the separation of the two terms is not so readily achieved. This problem was not fully appreciated when the intensities of intermolecular charge-transfer transitions were first discussed. Some aspects of the experimental determination of molar absorptivity are discussed in Chapter 6.

From equation (3.26) it is seen that, as the resonance interaction increases (an increase in b), so μ_{EN}, and consequently $\mathbf{f}$ and ϵ_{max}^{AD}, should increase. Thus, for a set of complexes of a series of chemically related donors with a common acceptor, we should expect ϵ_{max}^{AD} to increase as the interaction increases (decreased I^D, whence, in the absence of large-steric effects, an increase in the association constant). Early experimental work suggested that, in some cases, the opposite was the case (Table 3.7). More recent determinations of ϵ_{max}^{AD} using n.m.r. data in conjunction with optical measurements indicate that, in one series of complexes at least, ϵ_{max}^{AD} may in fact increase with increasing donor strength,[163] contrary to earlier reports based entirely on optical measurements[164] (Table 3.8).

The effect of specific solvation of the donor, acceptor and complex on the value of ϵ_{max}^{AD}, as usually estimated experimentally, has also been considered latterly.[165] These problems are discussed in detail in Chapter 6. The possibility that the experimental values of ϵ_{max}^{AD} might be an

TABLE 3.7. Molar absorptivities and association constants for a series of iodine complexes in carbon tetrachloride, 25°C*

Donor	λ_{max}/nm	ϵ_{max}^{AD}/l mol^{-1} cm	K_c^{AD}/l mol^{-1}
Benzene	292	16,400	0·15
Toluene	302	16,700	0·16
o-Xylene	316	12,500	0·27
m-Xylene	318	12,500	0·31
p-Xylene	304	10,100	0·31
Mesitylene	332	8850	0·82
Durene	332	9000	0·63
Pentamethylbenzene	357	9260	0·88
Hexamethylbenzene	375	8200	1·35

* From R. M. Keefer and L. J. Andrews, *J. Am. chem. Soc.* **74**, 4500 (1952).

TABLE 3.8. Molar absorptivities of a series of 1,3,5-trinitrobenzene complexes calculated from optical measurements alone and calculated by dividing the product term ($K_c^{AD} . \epsilon^{AD}$) obtained optically by K_c^{AD} obtained from n.m.r. measurements.*

Donor†	$(K_c^{AD} . \epsilon^{AD} \times 10^{-3})$‡/ l^2 mol^{-2} cm^{-1}	K_c^{AD}§/ l mol^{-1}	ϵ^{AD}∥/ l mol^{-1} cm^{-1}	$\epsilon^{AD'}$¶/ l mol^{-1} cm^{-1}
Mesitylene	1·74	0·79$_7$	3000	2180
Durene	3·33	1·32$_0$	2390	2520
Pentamethyl-benzene	5·69	1·93$_5$	2440	2950
Hexamethylbenzene	9·92	3·30$_8$	2500	3100

* In carbon tetrachloride solution at 33·5°C under the condition $[D]_0 \geqslant [A]_0$, Ref. 163.
† Optical maxima for the benzene, toluene and xylene complexes could not be resolved.
‡ From optical measurements.
§ From n.m.r. measurements.
∥ From optical measurements, Ref. 164.
¶ $\epsilon^{AD'} = (K_c^{AD} . \epsilon^{AD})$ opt./(K_c^{AD}) n.m.r.

artifact of the method employed was not considered when these problems were first discussed. Furthermore, some weak donor–acceptor pairs, for example, aliphatic hydrocarbon–iodine in perfluoroheptane, whilst showing strong charge-transfer bands in solution, appear to show no molecular association.[166, 167] Evans[166, 167] suggested that this might be the result of charge-transfer transitions in collisional complexes.

This idea was developed by Mulliken and Orgel [168, 169] who proposed that the charge-transfer transitions could occur not only in the 1:1 complex, but also during the random collision of a donor and an acceptor molecule, providing that, at the normal van der Waals distance, the overlap integral S_{AD} is non-zero. For such a donor–acceptor system in solution, let the association constant for the 1:1 complex be:

$$K_c^{AD} = \frac{[AD]}{([A]_0 - [AD])([D]_0 - [AD])} \tag{3.29}$$

where $[A]_0$, $[D]_0$ are the total (free and complexed) concentrations of A and D respectively and $[AD]$ is the concentration of the 1:1 complex. Let us assume that the molecules which are involved in such complexes are not involved in "contact" collision with other "free" D or A molecules. Let us further suppose that the concentration of pairs of D and A molecules, in "contact" [pair], as opposed to those which are complexed together, is measured by the product of the concentrations of "free" D and "free" A molecules, thus:

$$[\text{pair}] = \alpha([A]_0 - [AD])([D]_0 - [AD]) \tag{3.30}$$

where α is a measure of the relative number of contacts one species can make simultaneously with one molecule of the second component. (Experimentally the condition $[D]_0 \gg [A]_0$ is usually used; in this case α is the average number of contact sites for a D molecule around a single A molecule.) Combining equation (3.29) and equation (3.30), we obtain:

$$[\text{pair}] = \frac{\alpha}{K_c} - [AD] \tag{3.31}$$

Now the absorbance (A) for a 1-cm path length of solution at a given wavelength is:

$$A = \log_{10} I/I_0 = \sum_i C_i \epsilon_i \tag{3.32}$$

where I_0 and I are the intensities of the incident beam and the beam after passing through the solvent, C_i is the molar concentration of the ith component, and ϵ_i is the molar absorptivity of this component. If for simplicity we assume that the solvent and the free components have no optical absorption at the wavelength of measurement at which the molar absorptivity of the 1:1 charge-transfer complex is ϵ^{AD}, and the "molar" absorptivity of the contact pair is ϵ^{pair}, then:

$$A = [AD]\epsilon^{AD} + [\text{pair}]\epsilon^{\text{pair}} = [AD]\epsilon^{\text{app}} \tag{3.33}$$

where ϵ^{app} is the apparent molar absorptivity for the system when normal procedures are used; for example, Benesi and Hildebrand's method [170] (see Chapter 6) to evaluate K_c^{AD} and ϵ^{AD}, where it is assumed that all the charge-transfer absorption is due to transitions in the discrete 1:1 complexes. For convenience let $p = \alpha(\epsilon^{pair}/\epsilon^{AD})$. Then from equations (3.31) and (3.33)

$$\epsilon^{app} = \epsilon^{AD}\left(1 + \frac{p}{K_c^{AD}}\right) \tag{3.34}$$

Thus, if contact charge-transfer occurs, the apparent molar absorptivity of the complex will be larger than if there were no contact charge-transfer. The difference will be accentuated by small values of K_c^{AD}. Mulliken has suggested that for some systems the contribution from contact charge transfer may be 75% of the total charge-transfer absorption of the system. Recent measurements of the lower intensity of the charge-transfer band in the gas phase compared with the intensity in the liquid phase in a diluting solvent (Table 7.3) have been used as evidence in support of contact charge-transfer. It has been assumed that "contact" charge-transfer is negligible in the gas phase.* The observation that the oscillator strength of the charge-transfer band of the tetracyano-ethylene–hexamethylbenzene complex remains constant over a temperature range provides some evidence against significant contributions by isomeric 1:1 complexes or contact charge-transfer in this particular system. [161]

An alternative explanation of the apparent decrease in the intensities of the charge-transfer bands of some complexes with a common acceptor with donors of increasing strength (Table 3.7) was suggested by Murrell. [171] Suppose that the donor has an excited state D^* which can interact with the excited state ψ_E of the complex, thus modifying it to $\psi_{E'}$; whence:

$$\psi_{E'} = a^* \psi_1(A^+, D^-) - b^* \psi_D(A, D) + c^* \psi_2(A, D^*) \tag{3.35}$$

The charge-transfer band due to the transition $\psi_N \rightarrow \psi_{E'}$ will now involve some intensity-borrowing from the transition $\psi_D \rightarrow \psi_D^*$. The amount of borrowing will depend on the overlap S_{AD^*} between ψ_D^* and the vacant acceptor orbital. Since ψ_D^* is an excited-state orbital, it should be larger than the ground-state orbital ψ_D. Consequently S_{AD^*} should be larger than S_{AD}. It is this latter overlap which determines the stability of the complex. It could be that S_{AD} is sufficiently small or zero for a particular

---

* In terms of dispersion theory ϵ^{app} is expected to be somewhat larger in the liquid phase than in the gas phase. However, some of the increases seem remarkably large. The increased intensity might be accounted for in terms of the compression effect [131] in the liquid phase, referred to on p. 66. This will increase the ground-state–excited-state overlap and therefore increase the transition probability. The decrease in the transition dipole due to the diminished intermolecular distance in the liquid phase is relatively small and its effect on the intensity of the transition will be small.

acceptor pair to allow effectively no binding between D and A, whilst S_{AD^+} is large enough for a charge-transfer transition to occur. Murrell[171] has pointed out that this is an alternative explanation of contact charge-transfer.

For interaction between $\psi_D^{\pm}$ and $\psi_{E'}$, these orbitals must be of appropriate symmetry. For example, in a quinone–aromatic hydrocarbon complex in which the planes of the two molecules are parallel, there can be no borrowing from transitions involving those excited states of either component, which are polarized in the plane of the benzene ring, since these will be perpendicular to the intermolecular charge-transfer transition between the rings. The absence of such borrowing has been demonstrated experimentally in the anthracene–1,3,5-trinitrobenzene crystal.[4] Observations on the charge-transfer excitation of the phosphorescence of polyacene–tetrachlorophthalic anhydride complexes in glasses at 77°K also suggest the absence of intensity-borrowing.[18] By contrast, it has been argued that there is a significant degree of energy-borrowing in the naphthalene complexes of tetrachlorophthalic anhydride, of tetrabromophthalic anhydride, and of tetraiodophthalic anhydride.[162]

The charge-transfer absorption at 22,000 cm^{-1} in the case of the anthracene–1,3,5-trinitrobenzene complex is about twice as strong in the crystal state as in solution.[4] This may be accounted for by the fact that, in the solid, every donor molecule has two acceptor molecules as neighbours, and vice versa, whereas it is presumed that only donor–acceptor pairs are formed in solution. Consequently there should be twice the transition probability in the former case compared with the latter.[4]

Estimates of the change in intensity of the charge-transfer absorption with pressure have been made.[140] However, there are obvious experimental difficulties, and, since the evaluation of ϵ under much more nearly ideal conditions is still under discussion, the values which have been quoted may be open to some doubt (Chapter 6). For example ϵ_{CT}, for the benzene–tetracyanoethylene complex in dichloromethane, appears to double as the pressure increases from 1 atm. to 4×10^3 atm. This may be explained in terms of the Mulliken theory since the increased pressure should increase the overlap between the appropriate filled orbital of the donor and vacant orbital of the acceptor.

As opposed to the uncertainty of the absolute values of intensity, there can be no doubt concerning the relative intensities where a particular donor–acceptor system shows more than one absorption band. For the biphenyl–tetracyanoethylene system, the two charge-transfer bands do not alter in relative intensity as the pressure is altered. By contrast, the intensities of the two bands of the naphthalene–tetracyanoethylene complex do alter with respect to one another with pressure. This latter observation lends some support to the theory that each band is associated with an isomeric complex.[140] An alternative explanation could be found in terms of two transitions of a single complex, the transitions of one being favoured at the expenses of the other as the increased pressure alters the configuration of the complex.

Finally, it should be re-emphasized that there is uncertainty as to the correctness of published experimental values of molar absorptivity (see Chapter 6), which confounds much of the discussion concerning absorption intensities of these complexes.

3.A.5. Intramolecular Charge-Transfer Interactions across Space

Certain molecules contain an effective electron-donor site, whilst in another part of the same molecule a potential electron-acceptor site exists. In the absence of a conjugated pathway between these two sites, interaction between them may still occur if the molecule can assume a conformation so that there is a finite overlap between the highest-filled orbital of the donor moiety and the lowest-unfilled orbital of the acceptor moiety. This type of intramolecular charge-transfer transition *across space* may be contrasted with intramolecular charge-transfer transitions occurring within a conjugated system, for example, in *p*-nitroaniline.

Various reports have been given of optical absorptions which have been assigned to such intramolecular, across-space, charge-transfer transitions. The fact that these absorptions obey Beer's law is good evidence that they are the result of intramolecular interaction rather than a head-to-tail intermolecular interaction between the donor site of one molecule and the acceptor site of another molecule. White[172] has reported intramolecular interaction in a series of compounds with the general formula II.* The charge-transfer band moves slightly to higher

$$H_2N-\langle\!\!\!\!\bigcirc\!\!\!\!\rangle-X-\langle\!\!\!\!\bigcirc\!\!\!\!\rangle-NO_2$$

II

energies as n increases (Table 3.9). The detection of such a band in 4-amino-4′-nitrodiphenylmethane (II, $n = 1$) is remarkable in that the planes of the two aromatic rings cannot be less than ~109° 28′ to one another. The much more favoured conformation, which may be achieved when two methylene groups separate the two aromatic systems, is reflected in the observations made on the intramolecular charge-transfer interaction in substituted phenylalkylpyridinium chlorides of the type III.[173] The pyridinium moiety acts as the electron acceptor. The

---

* But under other conditions there is evidence for intermolecular complexing (p. 298).

TABLE 3.9. The wavelength of maximum absorption, molar absorptivity and oscillator strength of aqueous methanolic solutions of various compounds with the general formula II, relative to the absorptions of the corresponding separated acceptor and donor halves of the molecules.*

Central group in II	λ'_{max}†/nm	ϵ'_{max}†/ 1 mol^{-1} cm^{-1}	$f^{\ddagger}_{\ddagger}$ (rel.)
IIa CH$_2$	324	1620	115
IIb (CH$_2$)$_2$	313	1330	114
IIc (CH$_2$)$_3$	310	1480	100
IId *cis*(1,2-cyclopentano)	312	2420	155
IIe *trans*(1,2-cyclopentano)	308	2470	168

* Ref. 172.

† From difference in absorption between the compound and the sum of the absorptions of 4-nitrotoluene and *p*-toluidine.

‡ Relative integrated intensity over the range 280–650 nm.

TABLE 3.10. The wavelengths of maximum absorption and molar absorptivities of aqueous solutions of compounds with the general structure III, relative to the absorptions of the corresponding separated acceptor and donor halves of the molecules.*

Structural group†	n†	$\lambda'_{max\ddagger}$/nm	$\epsilon'_{max\ddagger}$/l mol^{-1} cm^{-1}	$\lambda_{max\ddagger}$/nm	$\epsilon_{max\ddagger}$/l mol^{-1} cm^{-1}
a	1	294	945		
	2	296	1575		
	3	296	570		
	4	293	200		
b	1	295	478	326	475
	2	297	823	328	755
	3	298	738		
	4	293	280		
c	1	281	1083		
	2	282	1691		
	3	281	633		
	4	279	405		
d	1	281§	80§	291	725
	2	282§	830§	290	1275
	3	282§	350§	291	696
	4	280	245	288	180

* Ref. 173.

† See structure III.

‡ From the difference in absorption between the compound and the sum of the appropriate molecule [toluene or γ-(*p*-methoxyphenyl) propyl chloride] and acceptor molecule [3- or 4-cyano-N-methylpyridinium chloride].

§ Shoulder.

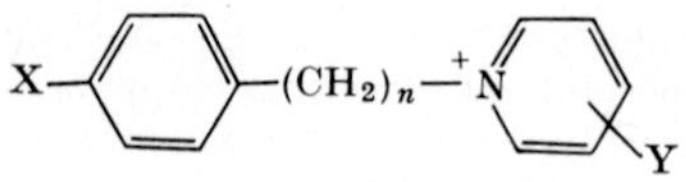

	X	Y
a	H	4-CN
b	OMe	4-CN
c	H	3-CN
d	OMe	3-CN

intensities of the charge-transfer bands for given substituents X and Y are a maximum when $n = 2$ in structure III (Table 3.10), cf. Table 3.9. The introduction of a p-methoxy group into the phenyl ring gives rise to a double charge-transfer absorption similar to equivalent inter-molecular charge-transfer systems.

Another group which shows similar optical properties are the compounds IV.[174] Shifrin has studied the same type of behaviour in N-(β-indolylethyl)-3-carbamoylpyridinium chloride (V)[175] and various N-(β-p-X-phenylethyl)-3-carbamoylpyridinium chlorides[176] (VI) where

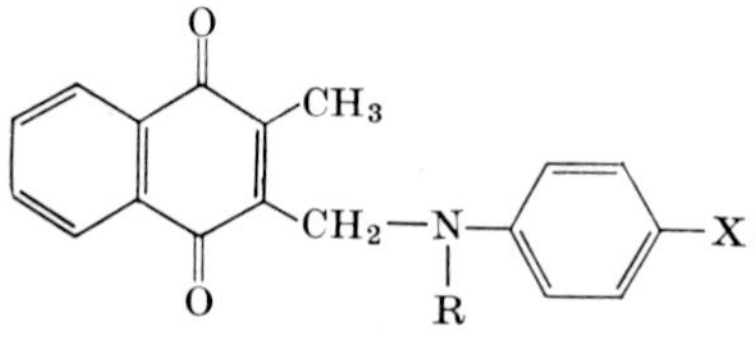

R	X
Me	H
Me	Cl
H	H
H	Me
H	COOMe

IV

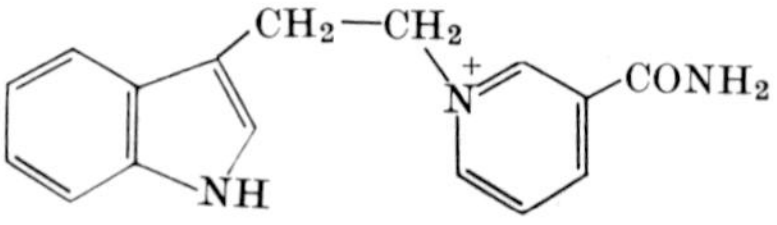

V

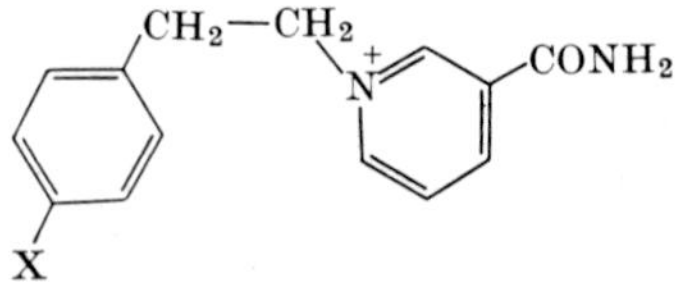

VI

$X = —H, —NH_2, —OCH_3, —OH, —CH_3, —Cl$. In the latter series of compounds, the energies corresponding to the observed transitions (Table 3.11) were found to correlate well with the ionization potentials of

TABLE 3.11. Spectral properties of the intramolecular transition of N-(β-p-X-phenylethyl)-3-carbamoylpyridinium halides (structure VI).*

—X	ν_{max}/cm^{-1}	$\epsilon_{max}/l\ mol^{-1}\ cm^{-1}$	f
—NH$_2$	$31{,}750 \pm 150$	910	0·030
—OH	$33{,}600 \pm 150$	1040	0·022
—OCH$_3$	$33{,}850 \pm 150$	1160	0·023
—CH$_3$	$34{,}500 \pm 50$	1010	0·013
—Cl	$35{,}000 \pm 80$	950	0·020
—H	$35{,}400 \pm 70$	1080	0·026

* Ref. 176.

the corresponding substituted benzene C_6H_5X, thus providing support for the hypothesis that these absorptions are the result of charge-transfer transitions.

Of particular interest is [2,2]paracyclophanequinone (VII), which has an ultraviolet absorption including a band at 29,000 cm^{-1} ($\epsilon = 597$),

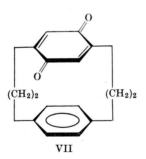

VII

which the authors suggest is due to an intramolecular transannular charge-transfer transition.[177] The band is not observed in the corresponding [4.4]paracyclophane, in which compound the benzene and quinone rings are more widely separated. The ultraviolet spectrum of 9,10-dihydro-9,10-ethano-11,12-dioxoanthracene has an absorption band at 39,600 cm^{-1} ($\epsilon = 46{,}000$), which Cookson and Lewin[178] suggest is due to a similar charge-transfer transition (see also Chapter 10).

The copolymers of 2,4,6-trinitrostyrene with 2-vinylpyridine, with 4-vinylpyridine and with p-dimethylaminostyrene all exhibit optical absorption bands at relatively long wavelengths.[179] These bands, which obey Beer's law, have been attributed to intramolecular charge-transfer transitions involving the donor and acceptor moieties in the polymer (see also Chapter 10, p. 297).

3.B. Emission Spectra

A fluorescence band, or series of bands, due to the transitions $\psi_E \rightarrow \psi_N$, which is the reverse of the intermolecular charge-transfer absorption band, may be observed in the emission spectrum of a charge-transfer complex. It was unfortunate that in the first systems to be studied, namely 1,3,5-trinitrobenzene complexes of polycyclic aromatic hydrocarbons in glasses at $-180°C$, a phosphorescence emission band of the donor virtually coincided with the charge-transfer fluorescence band.[180–182] The effect of the superposition in these systems was to produce emission spectra with obvious, though not highly resolved, fine structure which closely resembled the triplet emission spectra of the hydrocarbon in the absence of 1,3,5-trinitrobenzene. The very reasonable postulate was made that the *total* emission of the complex was the phosphorescence (triplet $\rightarrow$ singlet) transition of the donor. It would appear that the coincidence of the two emission spectra over the series of donors used is accidental. It could be accounted for if the energy difference between the lowest triplet level and the first ionization potential is approximately constant in the series of donors used. Bier and Ketelaar[183] later showed that, for the 1,3,5-trinitrobenzene complexes of anthracene and phenanthrene, there is a mirror-image relationship between the emission and the charge-transfer absorption spectra. This relationship was then observed[23, 26, 162, 184–194] in many electron donor–electron acceptor systems both at room temperature and at $-190°C$. Czekalla and his co-workers showed that, for a set of complexes of a given donor with a series of acceptors of increasing strength, the energy of the emission band decreased parallel to the decrease in the energy of the charge-transfer absorption band (Table 3.12).[26, 184–188] This observation has since been confirmed by several workers,[151, 162] (Fig. 3.21). The result is to be expected if the emission is the fluorescence process $\psi_E \rightarrow \psi_N$. Had the emission been the phosphorescence of the donor, then the energy of the transition would have been independent of the acceptor. Nevertheless, there is no doubt that, in many systems, phosphorescence emission of the donor occurs along with the fluorescence from the excited charge-transfer state.[192] McGlynn and his co-workers[162, 189–191] accounted for the phosphorescence by proposing that the energy of the excited charge-transfer state (ψ_E) could be lost by

TABLE 3.12. Fluorescence and absorption maxima for a series of complexes in an n-propyl ether–isopentane glass at −190°C.

Acceptor	Donor	Fluorescence ν_{max}/cm^{-1}	Absorption ν_{max}/cm^{-1}	Ref.
Chloranil	Durene	15,000	20,600	187
2,5-Dichloro-p-benzoquinone	Durene	16,300	22,700	187
1,3,5-Trinitrobenzene	Durene	19,500	28,000	187
Tetrachlorophthalic anhydride	Durene	21,700	29,400	187
Chloranil	Hexamethyl-benzene	13,500	19,900	186
2,5-Dichloro-p-benzoquinone	Hexamethyl-benzene	14,750	21,300	186
2,4,6-Trinitrobenzoic acid	Hexamethyl-benzene	17,450	25,700	186
Picryl chloride	Hexamethyl-benzene	17,600	26,900	186
1,3,5-Trinitrobenzene	Hexamethyl-benzene	17,000	25,900	186
2,4,6-Trinitroanisole	Hexamethyl-benzene	18,200	26,600	186
2,4,6-Trinitrotoluene	Hexamethyl-benzene	18,000	27,000	186
Tetrachlorophthalic anhydride	Hexamethyl-benzene	20,000	26,500	186
Chloranil	Naphthalene	15,600	20,700	187
2,5,-Dichloro-p-benzoquinone	Naphthalene	17,300	22,400	187
1,3,5-Trinitrobenzene	Naphthalene	19,500	26,500	187
Tetrachlorophthalic anhydride	Naphthalene	22,500	27,800	187
Trimesyl trichloride	Naphthalene	23,000	28,700	187
Chloranil	Phenanthrene	14,800	21,400	187
2,5-Dichloro-p-benzoquinone	Phenanthrene	15,800	23,500	187
1,3,5-Trinitrobenzene	Phenanthrene	19,200	26,800	187
Tetrachlorophthalic anhydride	Phenanthrene	21,500	28,500	187
Trimesyl trichloride	Phenanthrene	22,500	29,300	187
1,3,5-Trinitrobenzene	Anthracene	16,400	21,600	187
Tetrachlorophthalic anhydride	Anthracene	19,000	23,500	187
Trimesyl trichloride	Anthracene	19,200	24,000	187
Chloranil	Benz[a]-anthracene	13,500	17,000	187
1,3,5-Trinitrobenzene	Benz[a]-anthracene	17,000	22,400	187
Tetrachorophthalic anhydride	Benz[a]-anthracene	19,600	23,700	187
Trimesyl trichloride	Benz[a]-anthracene	20,000	25,000	187

transfer to a lower energy triplet state of the complex (T). Since this triplet level is dissociative at the energy of the intersystem crossing, McGlynn suggested that probably the complex would largely dissociate to yield the acceptor in the ground state and the donor in its lowest-excited triplet state. The donor would then return to its ground state (ψ_N) with phosphorescence emission. In solution there is some loss of vibrational structure, and small but measurable red shifts are observed

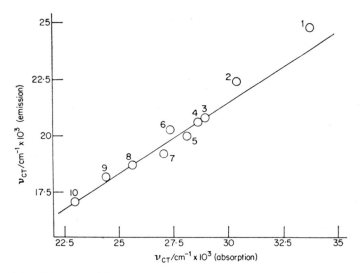

FIG. 3.21. Plot of the energies of the fluorescence emission for a series of pyro-mellitic dianhydride–donor complexes in carbon tetrachloride at room temperature against the energies of the charge-transfer absorption energies (from fluorescence-*excitation* maxima); where the donors are: (1) benzene; (2) toluene; (3) *o*-xylene; (4) *m*-xylene; (5) *p*-xylene; (6) 1,2,4-trimethylbenzene; (7) mesitylene; (8) durene; (9) pentamethylbenzene; (10) hexamethylbenzene. (From the data of H. M. Rosenberg and E. C. Eimutis, Ref. 151.)

in the phosphorescence of donor–acceptor mixtures, compared with the phosphorescence of the pure donor. The relaxation time is often decreased.[188] In the crystalline solid complexes, these effects are accentuated. It suggests that the triplet emission may occur to a considerable extent before dissociation, or after reassociation in the excited triplet state (T).[162]

In a similar fashion, the fluorescence of the complex 1,2,4,5-tetra-cyanobenzene–naphthalene has been accounted for as emission from the lowest-excited singlet state of the complex, whilst the observed phosphorescence is assigned to the transition from "locally excited" donor molecules in the triplet state.[23]

The marked quenching effect, by the addition of electron acceptors, on the fluorescence spectra of donor molecules is readily accountable by the competition of the above processes in the energy-degrading mechanisms. The quantum yield ratio of phosphorescence to fluorescence of the electron donor in various charge-transfer complexes has been studied.[162, 188, 192] The results confirm that the intersystem crossing probability makes the process an important fluorescence-quenching mechanism.[162] It is suggested [195] that the large decrease in the observed phosphorescence lifetime of the donor (complexed or free) in complexes with acceptors containing heavy atoms is due to an increase in the radiative-transition probability and not to an increase in non-radiative processes.

Whereas the energies of charge-transfer absorption of ion-pair complexes, such as pyridinium iodide, show a strong solvent-dependence, the energies of the corresponding fluorescence spectra appear to be very insensitive to solvent change.[108] This is expected on the basis of Kosower's[96] explanation of the solvent behaviour of the energies of the absorption transition, if there is time before fluorescence emission for solvent–solute reorientation to an equilibrium configuration in the charge-transfer excited state. In this state the complex components are neutral species and in consequence will not be strongly solvated. For a Franck–Condon transition, there will be insufficient time for solvent-reorientation during emission; consequently the fluorescence energy of such a system should show little dependence on the nature of the solvent, despite the ionic character of the ground state. By the same argument, the fluorescence emission from an essentially ionic excited charge-transfer state originating from a charge-transfer complex consisting of neutral components in the ground state should be strongly solvent-dependent. The few relevant values which have been published involve too small a range of solvents for any definite conclusions to be drawn.[193]

The fluorescence emission of single crystals of anthracene–1,3,5-trinitrobenzene, polarized parallel and perpendicular to the c-axis of the crystal, has been measured and compared with the corresponding polarized absorption spectra.[21] The decrease in energy of 1700 cm^{-1} from absorption to emission of the bands assigned to the lowest intermolecular charge-transfer transition is observed. From this, it is concluded that the intermolecular distance of the complex is shortened by ~0.2 Å on excitation. The decrease in the dichroic ratio for this transition between absorption and emission is explained by a larger contribution for the in-plane transition of anthracene in the excited state of the complex. The polarization of charge-transfer fluorescence, excited by irradiation in the charge-transfer bands and in the donor absorption-

4

bands, has been measured for various tetracyanoethylene–hydrocarbon and other complexes in organic glasses at 77°K.[194] Whereas the fluorescence excited in the charge-transfer band is positively polarized, the fluorescence excited in the donor-absorption band has a negative or zero polarization. This has been interpreted in terms of the accepted structure for π-π-complexes involving a strongly overlapping parallel arrangement of donor and acceptor pairs, in which there is some disorientation.

TABLE 3.13. Energies of fluorescence and lowest energy absorption maxima for a series of complexes in mobile liquid solution (carbon tetrachloride solvent).

Donor	Acceptor	Fluorescence ν_{max}/cm^{-1}	Absorption ν_{max}/cm^{-1}	Ref.
1. Benzene	Pyromellitic dianhydride	24,800	33,600*	151
2. Toluene		22,400	30,300*	151
3. o-Xylene		20,800	28,900*	151
4. m-Xylene		20,600	28,600*	151
5. p-Xylene		20,000	28,100*	151
6. 1,2,4-Trimethylbenzene		19,200	27,000*	151
7. Mesitylene		20,300	27,300*	151
8. Durene		18,700	25,600*	151
9. Pentamethylbenzene		18,200	24,400*	151
10. Hexamethylbenzene		17,100	23,000*	151
		16,900	22,900	193
11. Naphthalene		18,900	24,900	193
12. Hexamethylbenzene	Tetrachlorophthalic anhydride	18,600	25,600	184
		19,000	25,000	193
13. N-Phenylcarbazole		17,400	~23,800	193

* Fluorescence—excitation maxima (effectively equivalent to absorption maxima).

Fluorescence of charge-transfer complexes in mobile liquid solutions at room temperature, as opposed to complexes in solid glass-like matrices, generally at low temperatures, has been observed in a few cases;[151, 184, 193] for example, tetrachlorophthalic anhydride complexes of hexamethylbenzene and of N-phenylcarbazole, and the complexes of pyromellitic dianhydride with methylbenzenes and with naphthalene (Table 3.13). Good correlations of the energy of the fluorescence band with the ionization potential of the donor and with the energy of the absorption band, for a series of complexes with a given acceptor,[151] confirms the assignment of the fluorescence to the $\psi_E \rightarrow \psi_N$ transition.

The effect of pressure on the fluorescence of various tetrachlorophthalic anhydride complexes has been studied.[138] The emission band shows only a 500 cm^{-1} red shift at 30 kbar, compared with a red shift of

~1500 cm^{-1} in the absorption spectrum at 25 kbar, relative to the positions of these transitions at atmospheric pressure. This is readily accounted for by the fact that the slope of the energy–D–A distance

TABLE 3.14. Energies of fluorescence and phosphorescence maxima for a series of complexes in a diethyl ether–isopentane glass at −196°C.*

Donor	Acceptor	Fluorescence ν_{max}/cm	Phos-phorescence ν_{max}/cm	I^D/eV†
Hexamethylbenzene		20,000‡	18,300	7·85
Durene		21,500‡	19,500	8·03
Mesitylene		22,500	19,800	8·39
Toluene	1,2,4,5-Tetra-	24,700	20,500	8·82
Benzene	cyanobenzene	26,300	20,700	9·245
Triphenylene		19,600‡	19,300	7·9
Phenanthrene		20,200	19,200	8·02
—			21,200§	
Hexamethylbenzene		19,800	19,800	7·85
Durene	Pyromellitic	21,500‡	20,300	8·03
Mesitylene	dianhydride	22,500	20,000	8·39
—			22,300	
Hexamethylbenzene			20,200	7·85
Durene	Tetrachloro-	21,000‡	20,000	8·03
Mesitylene	phthalic	21,500‡	20,200	8·39
—	anhydride		22,200‖	
Hexamethylbenzene		22,300	20,900	7·85
Durene	Phthalic		22,400	8·03
Mesitylene	anhydride		22,700	8·39
—			23,600¶	

* From Ref. 196.
† From Ref. 52.
‡ Since the phosphorescence is relatively strong compared with the fluorescence and the separation of the fluorescence and phosphorescence maxima is small, it is difficult to determine accurately the fluorescence maxima.
§ 0–0 band at 22,650 cm^{-1}.
‖ 0–0 band at 23,500 cm^{-1}.
¶ 0–0 band at 25,800 cm^{-1}.

curve is greater for the excited state than for the ground state energy of the complex at the positions corresponding to Franck–Condon transitions, accepting that the equilibrium excited-state separation distance D–A is shorter than the corresponding distance in the ground state (see Tanaka and Yoshihara,[21] p. 85 above).

By contrast with the phosphorescence previously observed and assigned to a "locally excited" triplet state of the donor moiety within the charge-transfer complex, phosphorescence emission, attributable to a triplet state of the charge-transfer complex as a whole, has recently been reported. Iwata et al.[196] have shown that, for 1,2,4,5-tetracyano-benzene complexes of mesitylene, durene and hexamethylbenzene, the energy of the phosphorescence band decreased parallel to the fluorescence emission band, the latter showing a mirror-image relationship with the charge-transfer absorption spectrum (Table 3.14). In these complexes the zero-order triplet level of the charge-transfer state must be of lower

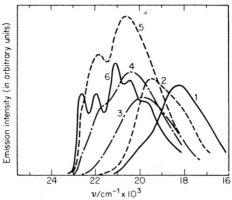

[S. Iwata, J. Tanaka and S. Nagakura, Ref. 196]

FIG. 3.22. The phosphorescence spectra of 1,2,4,5-tetracyanobenzene complexes in an ethyl ether-isopentane glass at 77°K with: (1) hexamethylbenzene; (2) durene; (3) mesitylene; (4) toluene; (5) benzene, compared with (6) the phosphorescence spectrum of 1,2,4,5-tetracyanobenzene in the absence of a donor.

energy than the triplet level of either the localized excited triplet state of the donor or of the acceptor in the complex. However, in the case of the benzene and toluene complexes of 1,2,4,5-tetracyanobenzene, it appears that the acceptor triplet is the lowest-excited triplet level which gives rise to the observed phosphorescence. In these cases the phosphorescence corresponds closely with that of the free acceptor (Fig. 3.22). The phosphorescence emissions of the tetrachlorophthalic anhydride, phthalic anhydride and pyromellitic dianhydride complexes of mesitylene, durene and hexamethylbenzene (Table 3.14) have also been assigned to transitions from the charge-transfer triplet level, rather than to transitions from either "locally excited" component.[196] The lifetime of the charge-transfer phosphorescences decreases with increasing donor ability (Table 3.15).*

* *Note in proof.* It has now been suggested that these emission spectra are the result of delayed fluorescence (Ref. 197).

TABLE 3.15. Lifetime (in secs.) of phosphorescence of some complexes in a degassed ethyl ether–isopentane glass at −196°C.*

Donors \ Acceptors†	TCNB	TCPA	PA
Mesitylene	3·3		
Durene	1·9	0·25 (0·25)‡	0·78
Hexamethylbenzene	0·41	0·004 (0·008)‡	0·3
(Acceptor alone)	3·2	0·21	1·4

* From Ref. 196.
† TCNB = 1,2,4,5-tetracyanobenzene; TCPA = tetrachlorophthalic anhydride; PA = phthalic anhydride.
‡ Values for these complexes in an isopentane–n-propylether glass (Ref. 188).

REFERENCES

1. W. Brackman, *Recl Trav. chim. Pays-Bas Belg.* **68**, 147 (1949).
2. J. Czekalla, *Z. Elektrochem.* **68**, 1157 (1959).
3. K. Nakamoto, *J. Am. chem. Soc.* **74**, 390, 392, 1739 (1952).
4. R. M. Hochstrasser, S. K. Lower and C. Reid, *J. molec. Spectrosc.* **15**, 257 (1965).
5. R. Foster and T. J. Thomson, *Trans. Faraday Soc.* **59**, 296 (1963).
6. T. Matsuo and H. Aiga, *Bull. chem. Soc. Japan* **41**, 271 (1968).
7. H. Kainer and A. Überle, *Chem. Ber.* **88**, 1147 (1955).
8. P. L. Kronick, H. Scott and M. M. Labes, *J. chem. Phys.* **40**, 890 (1964).
9. Y. Matsunaga, *Nature, Lond.* **211**, 183 (1966).
10. Y. Matsunaga, *Nature, Lond.* **205**, 72 (1965).
11. M. M. Labes, R. Sehr and M. Bose, *J. chem. Phys.* **32**, 1570 (1960).
12. M. M. Labes, R. Sehr and M. Bose, *J. chem. Phys.* **33**, 868 (1960).
13. Y. Matsunaga, *J. chem. Phys.* **40**, 3453 (1964).
14. Y. Matsunaga, *J. chem. Phys.* **42**, 1982 (1965).
15. T. Amano, H. Kuroda and H. Akamatu, *Bull. chem. Soc. Japan*, **41**, 83 (1968).
16. G. T. Pott, "Molionic Lattices," Thesis, University of Groningen (1966).
17. G. T. Pott and J. Kommandeur, *Molec. Phys.* **13**, 373 (1967).
18. M. Chowdhury and L. Goodman, *J. Am. chem. Soc.* **86**, 2777 (1964).
19. K. Nakamoto, *Bull. chem. Soc. Japan*, **26**, 70 (1953).
20. S. K. Lower, R. M. Hochstrasser and C. Reid, *Molec. Phys.* **4**, 161 (1961).
21. J. Tanaka and K. Yoshihara, *Bull. chem. Soc. Japan* **38**, 739 (1965).
22. R. M. Hochstrasser, S. K. Lower and C. Reid, *J. chem. Phys.* **41**, 1073 (1964).
23. S. Iwata, J. Tanaka and S. Nagakura, *J. Am. chem. Soc.* **89**, 2813 (1967).
24. H. Kuroda, I. Ikemoto and H. Akamatu, *Bull. chem. Soc. Japan* **39**, 1842 (1966).
25. G. Briegleb and H. Delle, *Z. phys. Chem.* **24**, 359 (1960).
26. J. Czekalla, A. Schmillen and K. J. Mager, *Z. Elektrochem.* **63**, 623 (1959).
27. B. G. Anex and E. B. Hill, Jr., *J. Am. chem. Soc.* **88**, 3648 (1966).
28. R. S. Mulliken, *Recl Trav. chim. Pays-Bas Belg.* **75**, 845 (1956).

29. R. D. Holm, W. R. Carper and J. A. Blancher, *J. phys. Chem., Ithaca* **71**, 3960 (1967).
30. G. Briegleb, "Elektronen-Donator-Acceptor-Komplexe," Springer-Verlag, Berlin (1961), p. 54.
31. G. Briegleb, "Elektronen-Donator-Acceptor-Komplexe," Springer-Verlag, Berlin (1961), p. 30.
32. H. Yada, J. Tanaka and S. Nagakura, *Bull. chem. Soc. Japan* **33**, 1660 (1960).
33. M. J. S. Dewar and H. Rogers, *J. Am. chem. Soc.* **84**, 395 (1962).
34. A. R. Lepley, *J. Am. chem. Soc.* **84**, 3577 (1962).
35. A. R. Lepley, *J. Am. chem. Soc.* **86**, 2545 (1964).
36. A. R. Lepley, *Tetrahedron Lett.* 2823 (1964).
37. A. R. Lepley and C. C. Thompson, Jr., *J. Am. chem. Soc.* **89**, 5523 (1967).
38. K. Watanabe, *J. chem. Phys.* **26**, 542 (1957); K. Watanabe, F. F. Marmo and E. C. Y. Inn, *Phys. Rev.* **91**, 1155 (1956); K. Watanabe and T. Nakayama, *J. chem. Phys.* **29**, 48 (1958); K. Watanabe and J. R. Mottl, *J. chem. Phys.* **26**, 1773 (1957).
39. K. Watanabe, T. Nakayama and J. R. Mottl, "Final Report on Ionization Potentials of Molecules by a Photoionisation Method," Dept. Army 5B99-01-004 Ordnance R and D-TB2-0001 OOR-1624 University of Hawaii (1959).
40. W. C. Price, R. Bralsford, P. V. Harris and R. G. Ridley, *Spectrochim. Acta* **14**, 45 (1959).
41. R. Bralsford, P. V. Harris and W. C. Price, *Proc. R. Soc.* **A258**, 459 (1960).
42. H. Hurzeler, M. G. Inghram and J. D. Morrison, *J. chem. Phys.* **28**, 76 (1958).
43. J. Momigny, *Nature, Lond.* **199**, 1179 (1963).
44. P. G. Wilkinson, *Astrophys. J.* **138**, 778 (1963).
45. A. J. C. Nicholson, *J. chem. Phys.* **43**, 1171 (1965).
46. M. I. Al-Joboury and D. W. Turner, *J. chem. Soc.* 5141 (1963).
47. M. I. Al-Joboury, D. P. May and D. W. Turner, *J. chem. Soc.* 616 (1965).
48. T. N. Radwan and D. W. Turner, *J. chem. Soc.* (A), 85 (1966).
49. M. I. Al-Joboury and D. W. Turner, *J. chem. Soc.* (B), 373 (1967).
50. A. D. Baker, D. P. May and D. W. Turner, *J. chem. Soc.* (B), 22 (1968).
51. D. W. Turner, *in* "Advances in Physical Organic Chemistry," Vol. 4, ed. V. Gold, Academic Press, London and New York (1966), p. 31.
52. V. I. Vedeneyev, L. V. Gurvich, V. N. Kondrat'yev, V. A. Medvedev and Y. L. Frankevich, "Bond Energies, Ionization Potentials and Electron Affinities," Arnold, London (1966).
53. S. H. Hastings, J. L. Franklin, J. C. Schiller and F. A. Matsen, *J. Am. chem. Soc.* **75**, 2900 (1953).
54. G. Briegleb and J. Czekalla, *Z. Elektrochem.* **63**, 6 (1959).
55. H. M. McConnell, J. S. Ham and J. R. Platt, *J. chem. Phys.* **21**, 66 (1953).
56. R. S. Mulliken and W. B. Person, *A. Rev. phys. Chem.* **13**, 107 (1962).
57. R. Foster, *Nature, Lond.* **183**, 1253 (1959).
58. G. Briegleb, *Angew. Chem.* **76**, 326 (1964); *Angew. Chem.* (International Edition) **3**, 617 (1964).
59. R. Foster, *Nature, Lond.* **181**, 337 (1958).
60. R. Foster, *Tetrahedron* **10**, 96 (1960).
61. W. H. Laarhoven and R. J. F. Nivard, *Recl Trav. chim. Pays-Bas Belg.* **84**, 1478 (1965).
62. R. Beukers and A. Szent-Györgyi, *Recl Trav. chim. Pays-Bas Belg.* **81**, 255 (1962).
63. R. Foster and T. J. Thomson, *Trans. Faraday Soc.* **59**, 2287 (1963).

64. W. B. Person, *J. chem. Phys.* **38**, 109 (1963).
65. D. R. Kearns, P. Gardner and J. Carmody, *J. phys. Chem.*, *Ithaca* **71**, 931 (1967).
66. E. K. Plyler and R. S. Mulliken, *J. Am. chem. Soc.* **81**, 823 (1959).
67. V. G. Krishna and M. Chowdhury, *J. phys. Chem.*, *Ithaca* **67**, 1067 (1963).
68. C. Reid and R. S. Mulliken, *J. Am. chem. Soc.* **76**, 3869 (1954).
69. J. N. Chaudhuri and S. Basu, *Trans. Faraday Soc.* **55**, 898 (1959).
70. M. Chowdhury and S. Basu, *Trans. Faraday Soc.* **56**, 335 (1960).
71. G. J. Hoijtink, *Recl Trav. chim. Pays-Bas Belg.* **77**, 555 (1958).
72. E. S. Pysh and N. C. Yang, *J. Am. chem. Soc.* **85**, 2124 (1963).
73. R. S. Mulliken, *J. Am. chem. Soc.* **72**, 600 (1950).
74. W. E. Wentworth and R. S. Becker, *J. Am. chem. Soc.* **84**, 4263 (1962).
75. M. E. Peover, *Nature, Lond.* **191**, 702 (1961).
76. M. E. Peover, *Nature, Lond.* **193**, 475 (1962).
77. M. E. Peover, *Trans. Faraday Soc.* **58**, 2370 (1962).
78. M. E. Peover, *J. chem. Soc.* 4540 (1962).
79. M. E. Peover, *Trans. Faraday Soc.* **58**, 1656 (1962).
80. K. M. C. Davis, P. R. Hammond and M. E. Peover, *Trans. Faraday Soc.* **61**, 1516 (1965).
81. M. E. Peover, *Trans. Faraday Soc.* **60**, 417 (1964).
82. M. E. Peover, *Trans. Faraday Soc.* **60**, 479 (1964).
83. A. Maccoll, *Nature, Lond.* **163**, 178 (1949).
84. J. Jortner and U. Sokolov, *Nature, Lond.* **190**, 1003 (1961).
85. M. Nepraš and R. Zahradník, *Tetrahedron Lett.* 57 (1963).
86. M. Nepraš and R. Zahradník, *Colln Czech. chem. Commun. Engl. Edn* **29**, 1545, 1555 (1964).
87. G. W. Wheland and D. E. Mann, *J. chem. Phys.* **17**, 264 (1949).
88. A. Streitwieser and P. M. Nair, *Tetrahedron* **5**, 149 (1959).
89. A. Streitwieser, *J. Am. chem. Soc.* **82**, 4123 (1960).
90. S. Ehrenson, *J. phys. Chem.*, *Ithaca* **66**, 706 (1962).
91. F. A. Matsen, *J. chem. Phys.* **24**, 602 (1956).
92. H. Kuroda, *Nature, Lond.* **201**, 1214 (1964).
93. L. P. Hammett, "Physical Organic Chemistry," McGraw-Hill, New York (1940), Ch. 7; H. H. Jaffé, *Chem. Rev.* **53**, 191 (1953).
94. P. R. Hammond, *J. chem. Soc.* 471 (1964).
95. E. M. Kosower, D. Hofmann and K. Wallenfels, *J. Am. chem. Soc.* **84**, 2755 (1962).
96. E. M. Kosower, *J. Am. chem. Soc.* **80**, 3253, 3261 (1958).
97. E. M. Kosower and G.-S. Wu, *J. Am. chem. Soc.* **83**, 3142 (1961).
98. E. M. Kosower, G.-S. Wu and T. S. Sorensen, *J. Am. chem. Soc.* **83**, 3147 (1961).
99. E. M. Kosower, W. D. Closson, H. L. Goering and J. C. Gross, *J. Am. chem. Soc.* **83**, 2013 (1961).
100. E. Grunwald and S. Winstein, *J. Am. chem. Soc.* **70**, 846 (1948).
101. A. H. Fainberg and S. Winstein, *J. Am. chem. Soc.* **78**, 2770 (1956).
102. S. Winstein, E. Grunwald and H. W. Jones, *J. Am. chem. Soc.* **73**, 2700 (1951).
103. K. Dimroth, C. Reichardt, T. Siepmann and F. Bohlmann, *Justus Liebigs Annln Chem.* **661**, 1 (1963).
104. C. Reichardt, *Angew. Chem.* **77**, 30 (1965); *Angew. Chem.* (International Edition) **4**, 29 (1965).
105. E. M. Kosower, *J. org. Chem.* **29**, 956 (1964).

106. K. M. C. Davis, *J. chem. Soc.* (B), 1128 (1967).
107. M. Feldman and B. G. Graves, *J. phys. Chem., Ithaca* **70**, 955 (1966).
108. J. S. Brinen, J. G. Koren, H. D. Olmstead and R. C. Hirt, *J. phys. Chem., Ithaca* **69**, 3791 (1965).
109. P. A. D. de Maine, *J. chem. Phys.* **26**, 1199 (1957).
110. P. H. Emslie and R. Foster, *Recl Trav. chim. Pays-Bas Belg.* **84**, 255 (1965).
111. K. M. C. Davis and M. C. R. Symons, *J. chem. Soc.* 2079 (1965).
112. H. W. Offen and M. S. F. A. Abidi, *J. chem. Phys.* **44**, 4642 (1966).
113. R. Foster and D. Ll. Hammick, *J. chem. Soc.* 2685 (1954).
114. N. S. Isaacs, *J. chem. Soc.* (B), 1351 (1967).
115. L. Onsager, *J. Am. chem. Soc.* **58**, 1486 (1936).
116. N. S. Bayliss, *J. chem. Phys.* **18**, 292 (1950).
117. N. S. Bayliss and L. Hulme, *Aust. J. Chem.* **6**, 257 (1953).
118. H. M. Rosenberg and D. Hale, *J. phys. Chem., Ithaca* **69**, 2490 (1965).
119. H. M. Rosenberg, *Chem. Commun.* 312 (1965).
120. H. M. Rosenberg, E. Eimutis and D. Hale, *Can. J. Chem.* **44**, 2405 (1966).
121. E. M. Voigt, *J. phys. Chem., Ithaca* **70**, 598 (1966).
122. E. M. Voigt, unpublished work.
123. E. G. McRae, *J. phys. Chem., Ithaca* **61**, 562 (1957).
124. M. Kroll and M. L. Ginter, *J. phys. Chem., Ithaca* **69**, 3671 (1965).
125. F. T. Lang and R. L. Strong, *J. Am. chem. Soc.* **87**, 2345 (1965).
126. J. M. Goodenow and M. Tamres, *J. chem. Phys.* **43**, 3393 (1965).
127. M. Tamres and J. M. Goodenow, *J. phys. Chem., Ithaca* **71**, 1982 (1967).
128. J. Prochorow, *J. chem. Phys.* **43**, 3394 (1965).
129. J. Prochorow and A. Tramer, *J. chem. Phys.* **44**, 4545 (1966).
130. M. Kroll, *J. Am. chem. Soc.* **90**, 1097 (1968).
131. P. J. Trotter, *J. Am. chem. Soc.* **88**, 5721 (1966).
132. J. C. A. Boeyens and F. H. Herbstein, *J. phys. Chem., Ithaca* **69**, 2160 (1965).
133. D. R. Stephens and H. G. Drickamer, *J. chem. Phys.* **30**, 1518 (1959).
134. R. B. Aust, G. A. Samara and H. G. Drickamer, *J. chem. Phys.* **41**, 2003 (1964).
135. H. W. Offen, *J. chem. Phys.* **42**, 430 (1965).
136. W. H. Bentley and H. G. Drickamer, *J. chem. Phys.* **42**, 1573 (1965).
137. H. W. Offen and A. H. Kadhim, *J. chem. Phys.* **45**, 269 (1966).
138. H. W. Offen and J. F. Studebaker, *J. chem. Phys.* **47**, 253 (1967).
139. J. S. Ham, *J. Am. chem. Soc.* **76**, 3881 (1954).
140. J. R. Gott and W. G. Maisch, *J. chem. Phys.* **39**, 2229 (1963).
141. N. Smith, Ph.D. Dissertation, University of Chicago (1954).
142. A. Bier, *Recl Trav. chim. Pays-Bas Belg.* **75**, 866 (1956).
143. E. M. Voigt, *J. Am. chem. Soc.* **86**, 3611 (1964).
144. G. Briegleb, *Z. Elektrochem.* **61**, 537 (1957).
145. P. A. D. de Maine, *J. chem. Phys.* **26**, 1189 (1957).
146. R. E. Merrifield and W. D. Phillips, *J. Am. chem. Soc.* **80**, 2778 (1958).
147. A. Kuboyama, *J. chem. Soc. Japan* **81**, 558 (1960).
148. H. Kuroda, M. Kobayashi, M. Kinoshita and S. Takemoto, *J. chem. Phys.* **36**, 457 (1962).
149. A. Zweig, *J. phys. Chem., Ithaca* **67**, 506 (1963); *Tetrahedron Lett.* 89 (1964).
150. G. Briegleb, J. Czekalla and G. Reuss, *Z. phys. Chem. Frankf. Ausg.* **30**, 316 (1961).
151. H. M. Rosenberg and E. C. Eimutis, *J. phys. Chem., Ithaca* **70**, 3494 (1966).
152. S. Iwata, J. Tanaka and S. Nagakura, *J. Am. chem. Soc.* **88**, 894 (1966).

153. E. M. Voigt and C. Reid, *J. Am. chem. Soc.* **86**, 3930 (1964).
154. L. E. Orgel, *J. chem. Phys.* **23**, 1352 (1955).
155. J. R. Hoyland and L. Goodman, *J. chem. Phys.* **36**, 12 (1962).
156. R. J. Prosen and K. N. Trueblood, *Acta crystallogr.* **9**, 741 (1956).
157. D. F. R. Gilson and C. A. McDowell, *J. chem. Phys.* **39**, 1825 (1963).
158. D. F. R. Gilson and C. A. McDowell, *Can. J. Chem.* **44**, 945 (1966).
159. R. S. Mulliken, *J. Am. chem. Soc.* **74**, 811 (1952).
160. G. Briegleb, "Elektronen-Donator-Acceptor-Komplexe," Springer-Verlag, Berlin (1961), p. 62.
161. R. Foster and I. B. C. Matheson, *Spectrochim. Acta* **23A**, 2037 (1967).
162. N. Christodouleas and S. P. McGlynn, *J. chem. Phys.* **40**, 166 (1964).
163. P. H. Emslie, R. Foster, C. A. Fyfe and I. Horman, *Tetrahedron* **21**, 2843 (1965).
164. C. C. Thompson, Jr. and P. A. D. de Maine, *J. phys. Chem., Ithaca* **69**, 2766 (1965).
165. S. Carter, J. N. Murrell and E. J. Rosch, *J. chem. Soc.* 2048 (1965).
166. D. F. Evans, *J. chem. Phys.* **23**, 1424, 1426 (1955).
167. D. F. Evans, *J. chem. Soc.* 4229 (1957).
168. R. S. Mulliken, *Recl Trav. chim. Pays-Bas Belg.* **75**, 845 (1956).
169. L. E. Orgel and R. S. Mulliken, *J. Am. chem. Soc.* **79**, 4839 (1957).
170. H. A. Benesi and J. H. Hildebrand, *J. Am. chem. Soc.* **71**, 2703 (1949).
171. J. N. Murrell, *J. Am. chem. Soc.* **81**, 5037 (1959).
172. W. N. White, *J. Am. chem. Soc.* **81**, 2912 (1959).
173. J. W. Verhoeven, I. P. Dirkx and T. J. de Boer, *Tetrahedron Lett.* 4399 (1966).
174. R. Carruthers, F. M. Dean, L. E. Houghton and A. Ledwith, *Chem. Commun.* 1206 (1967).
175. S. Shifrin, *Biochim. biophys. Acta* **81**, 205 (1964).
176. S. Shifrin, *Biochim. biophys. Acta* **96**, 173 (1965).
177. D. J. Cram and A. C. Day, *J. org. Chem.* **31**, 1227 (1966).
178. R. C. Cookson and N. Lewin, *Chemy Ind.* 984 (1956).
179. N. C. Yang and Y. Gaoni, *J. Am. chem. Soc.* **86**, 5022 (1964).
180. C. Reid, *J. chem. Phys.* **20**, 1212 (1952).
181. M. M. Moodie and C. Reid, *J. chem. Phys.* **20**, 1510 (1952).
182. M. M. Moodie and C. Reid, *J. chem. Phys.* **22**, 252 (1954).
183. A. Bier and J. A. A. Ketelaar, *Recl Trav. chim. Pays-Bas Belg.* **73**, 264 (1954).
184. J. Czekalla and K.-O. Meyer, *Z. phys. chem. Frankf. Ausg.* **27**, 185 (1961).
185. J. Czekalla, A. Schmillen and K. J. Mager, *Z. Elektrochem.* **61**, 1053 (1957).
186. J. Czekalla, G. Briegleb, W. Herre and R. Glier, *Z. Elektrochem.* **61**, 537 (1957).
187. J. Czekalla, G. Briegleb and W. Herre, *Z. Elektrochem.* **63**, 712 (1959).
188. J. Czekalla and K. J. Mager, *Z. Elektrochem.* **66**, 65 (1962).
189. S. P. McGlynn and J. D. Boggus, *J. Am. chem. Soc.* **80**, 5096 (1958).
190. S. P. McGlynn, J. D. Boggus and E. Elder, *J. chem. Phys.* **32**, 357 (1960).
191. S. P. McGlynn, *Chem. Rev.* **58**, 1113 (1958).
192. J. Czekalla, G. Briegleb and H. J. Vahlensieck, *Z. Elektrochem.* **63**, 715 (1959).
193. G. D. Short and C. A. Parker, *Spectrochim. Acta* **23A**, 2487 (1967).
194. J. Prochorow and A. Tramer, *J. chem. Phys.* **47**, 775 (1967).
195. K. B. Eisenthal and M. A. El Sayed, *J. chem. Phys.* **42**, 794 (1965).
196. S. Iwata, J. Tanaka and S. Nagakura, *J. chem. Phys.* **47**, 2203 (1967).
197. G. D. Short, *Chem. Commun.* 1500 (1968).

4*

Infrared Spectra

4.A. Introduction

When there is only weak interaction between the components of a charge-transfer complex, the infrared spectrum of the complex often shows only small differences, compared with the sum of the spectra of the two components.[1-6] Some of these differences which are observed in the solid phase may be the result of crystal-packing effects.[7] The general lack of change in the infrared spectrum on complex formation has been used to distinguish between weak charge-transfer complexes and the products of electron-transfer [3, 6]* or proton-transfer reactions [8, 9] (see also Chapter 9). Nevertheless, changes in intensity and energy of the vibrational bands of either or both donor and acceptor may be observed in weak interactions. These are generally larger in complexes involving σ-donors and σ-acceptors, such as the amine–iodine complexes, than in π-π-interactions. In some of the former cases, a low energy band associated with an intermolecular vibrational mode has been observed.

4.B. Complexes with π-Donors

For homopolar diatomic acceptors such as I_2, Br_2 or Cl_2, the infrared forbidden stretching vibrations of the free molecules become infrared active and decrease in frequency on interaction with an electron donor. A similar decrease in frequency is shown by asymmetric acceptors. This effect is also observed in complexes with n-donors (see Section 4.C). A selection of the many observations made, including some involving Raman spectroscopy, are listed in Table 4.1. The fact that such an infrared absorption band could be observed for the Cl–Cl stretching mode of molecular chlorine when complexed with benzene led to the

---

* In the case of complete electron transfer from the donor to the acceptor in the ground state, the infrared corresponds closely to the sum of the spectra of the two ions.[3, 6]

TABLE 4.1. The effect of complexing on the fundamental stretching frequency of some diatomic electron-acceptor molecules.

System	State*	ν/cm^{-1}	Type of spectrum†	Ref.‡
Cl_2§	Liquid	541	R	a
	CCl_4 soln.	540·5	R	a
Cl_2§–benzene		531	ir	18
		526	ir	10
		525	R	a
Br_2	Gas	317	R	46
	$CHCl_3$ soln.	312	R	a
Br_2–benzene		305	ir	18
		307	ir	24
	Solid	296	ir	24
		302	R	46
Br_2–toluene		300	R	46
Br_2–chlorobenzene		306	ir	18
Br_2–benzophenone	CCl_4 soln.	307·5	ir	18
Br_2–biphenyl	CCl_4 soln.	307·5	ir	18
Br_2–pyridine		281	R	46
Br_2–acetonitrile		308	R	46
I_2	Gas	213	bs	46
	$CHCl_3$ soln.	207	R	a
I_2–benzene		201	R	a
		205	R	46
I_2–toluene		204	R	46
I_2–o-xylene		202	R	46
I_2–pyridine	Cyclohexane soln.	183	ir	45
	n-Hexane soln.	180	ir	43
	n-Heptane soln.	184	ir	35
	Benzene soln.	174	ir	35
		167	R	46, a
		167	ir	43
I_2–2-methylpyridine	Cyclohexane soln.	182	ir	45
I_2–4-methylpyridine	Cyclohexane soln.	181	ir	45
I_2–trimethylamine	Solid (nujol mull)	185	ir	42
I_2–dimethylsulphoxide		189	R	46
I_2–acetonitrile		207	R	46
ICl‖	Gas	381·5	ir	b
	CCl_4 soln.	375	ir	18
ICl–benzene		355	ir	16
		355	R	46
		352	R	a
ICl–toluene		353	R	46
ICl–biphenyl	CCl_4 soln.	358	ir	18
ICl–acetonitrile		347	R	46
ICl–pyridine		277	ir	43
	Benzene soln.	270	R	a
	Solid	274, 266	ir	36
IBr	CCl_4 soln.	261	ir	36
IBr–benzene		249	ir	36

TABLE 4.1. (*continued*)

System	State*	ν/cm^{-1}	Type of spectrum†	Ref.‡
IBr–toluene		247	ir	36
IBr–*o*-xylene		247	ir	36
IBr–acetonitrile		244·5	ir	36
IBr–pyridine	⎰	205·0	ir	36
	⎰ Solid (nujol mull)	200	ir	44
	⎱ Solid (nujol mull)	199, 190	ir	36
	⎱ Benzene soln.	204	ir	43

* Liquid solution in excess donor unless otherwise stated.
† R = Raman; bs = from band spectra; ir = infrared.
‡ Refs.: *ᵃ* H. Stammreich, R. Forneris and Y. Taveres, *Spectrochim. Acta* **17**, 1173 (1961); *ᵇ* W. V. F. Brooks and B. Crawford, *J. chem. Phys.* **23**, 363 (1955).
§ Cl^{35}Cl^{37}.
‖ ICl^{35}.

suggestion [10, 11] that the halogen molecule must be orientated with respect to the benzene molecule in such a way that the two halogen atoms are in different electronic environments. Such spectra were used to argue in favour of the axial model for the benzene–iodine complex in which the iodine molecule lies along the benzene C_6 axis, and against the "resting" model in which the iodine molecule lies parallel to the benzene ring, with the centre of its molecule on the C_6 axis, [12] (Fig. 8.19, p. 231). In fact, any unsymmetrical complex would have been an acceptable model. The observed increased intensities of the benzene A_{1g} fundamental at 992 cm^{-1} and the E_{1g} fundamental at 850 cm^{-1} were originally taken as added evidence in favour of the axial model. [13–15]

There is a linear correlation between the added effective charge resulting from complex formation, and the relative change in the force constant of the bond of diatomic halogen and halogen-like acceptors on complex formation. [16–18] This has been interpreted as support for the charge-transfer resonance model for a large group of complexes. [16–18] However, the consequent calculations of intensity increases do not account for the observed intensity changes.

Ferguson and Matsen [19–21] have proposed that the electron affinity of the acceptor changes during the stretching vibration of the acceptor (and correspondingly for the donor). This means that the interaction energy changes; consequently, there will be an oscillating dipole moment with a frequency equal to the stretching frequency of the acceptor. This changing dipole will potentiate the absorption of the acceptor. A similar explanation has been applied to O–H stretching vibration in H-bonded complexes.[22] Likewise, the intensity of the forbidden A_{1g} benzene band at 992 cm^{-1} is accounted for in the complex by the change in vertical ionization energy of benzene during the vibration. These conclusions invalidate the earlier arguments concerning the geometry of the benzene halogen complexes.[13–15]

This theory of vibronic interaction has been developed further by Friedrich and Person.[23] By making the electronic states dependent on the vibrational coordinates, they have derived an expression for the intensities of vibrations in charge-transfer complexes. Factors which have to be included are: the relative contributions of the "no bond" and dative structures to the ground state wave functions; the electronic transition moment of the charge-transfer band; and the derivative of the vertical electron affinity of the acceptor molecule with respect to the internuclear distance for the particular stretching vibration (or the corresponding derivative of the vertical ionization potential of the donor for donor vibrations). Calculations of the intensity enhancements of X–Y stretching vibrations in X–Y halogen acceptor molecules when complexed with donors have been made. All the results may be accounted for in terms of the postulate of electronic re-orientation during the vibration. A relationship has also been obtained for the frequency shift of the vibration in a molecule on complex formation from the potential energy functions for the "no bond" and dative states of the complex. These may be only approximate; for example, in the case of the stretching frequency of a halogen acceptor X–Y, these two structures are reasonably represented by the free halogen XY and the free ion (XY)$^-$. In terms of force constant, Friedrich and Person[23] have shown that:

$$\frac{(k^0 - k_N)}{k_0} = \frac{\Delta k}{k} = 1 - \frac{k_1}{k_0}(b^2 + abS_{01}) \qquad (4.1)$$

where k_0 and k_1 are the X–Y stretching force constants in free XY and XY$^-$ molecules, k_N is the force constant for the X–Y bond in the complex, a and b are the coefficients of the "no bond" and dative functions in the ground state wave function of the complex [equation (2.1), p. 18], and S_{01} is the overlap integral between these two states. This expression has been used to estimate b.

Friedrich and Person's[23] explanation of the intensity enhancements and frequency shifts in the components of a complex in terms of vibronic interaction reaffirm Ferguson and Matsen's[19–21] more recent conclusions regarding the geometry of charge-transfer complexes; namely, that infrared measurements would appear to provide no information as to the relative orientation of the donor and acceptor in the complex solution. It has been pointed out that the benzene–halogen complexes are only weakly bound. The benzene–iodine stretching frequency has been estimated as 75–100 cm^{-1}. As a consequence, the various other benzene–iodine vibrational frequencies will be very low and the excited states of these vibrations will be well populated at normal temperatures, so that normal symmetry conditions will not apply. This re-emphasizes the assertion that infrared measurements cannot normally distinguish between the various possible geometries of a complex. Arguments concerning the relative orientation of donor and acceptor in complexes in solution, based on crystallographic studies of solid crystalline adducts, are of doubtful value, since there may be interaction of one donor molecule with two or more acceptor molecules (and vice versa) in infinite arrays in the solid phase, compared with the discrete 1:1 complexes in solution. Nevertheless, observations of the infrared spectrum of the solid bromine–benzene complex[24] have enabled some modification to be made to the crystal structure description determined by X-ray diffraction. The infrared spectrum, which includes a Br–Br stretching vibration, is inconsistent with the crystal structure[25] having an axial configuration with chains of alternating benzene and halogen molecules, in which the halogens are equidistant from two neighbouring benzene molecules in the chain and are at centres of symmetry (see Chapter 8). Person et al.[24] suggest that these conflicting observations can be reconciled if, in the crystals, the halogen molecules are not, in fact, exactly equidistant from their immediate benzene neighbours in the chain, but are somewhat closer to one than the other. If there is a random choice of neighbour, the X-ray measurements will reflect an average structure. Alternatively, the low energy of the intermolecular benzene–halogen vibration will average out the intermolecular distances as estimated by X-ray diffraction.

The dichroic behaviour of orientated crystals of the tetracyanoethylene complexes of hexamethylbenzene and pentamethylbenzene has been reasonably explained[26] in terms of the recent theories of Ferguson and Matsen.[19-21] Comparison of the C≡N and C=C vibration modes in the complexes was made with the values for the acceptor anion, studied as the potassium salt. An estimate of 5–10% electron transfer in the ground state has been made on the basis of the relative shifts of the C≡N and C=C vibration modes compared with the shifts observed in the anion.

A second solid complex of tetracyanoethylene and hexamethylbenzene with a molecular ratio of 1:2 has also been prepared.[27] The out-of-plane and in-plane infrared spectra of the 1:1 and 1:2 complexes are compared in Fig. 4.1. The spectra of the 1:2 complex may be interpreted in terms of a structure which consists of stacks which contain sequences of

$\cdots$D$\cdot$A$\cdot$D$\cdots$D$\cdot$A$\cdot$D$\cdots$D$\cdot$A$\cdot$D$\cdots$ molecules. In the 1:1 complex the out-of-plane infrared spectrum shows no absorption at 1295 cm^{-1} (an absorption assigned to the totally symmetric C—CH$_3$ stretching mode of hexamethylbenzene, which is forbidden in the free molecule). This is to be expected in terms of Ferguson and Matsen's[19-21] and Person and

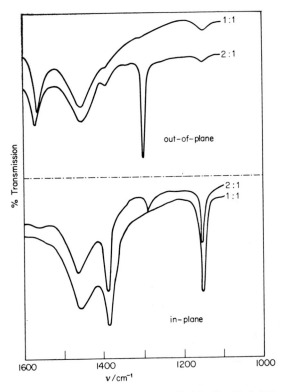

[B. Hall and J. P. Devlin, Ref. 27]

FIG. 4.1. Infrared spectra in the 1100–1600 cm^{-1} range for the 1:1 and 2:1 crystalline complexes of hexamethylbenzene and tetracyanoethylene.

Friedrich's[23] theory since the charge oscillation between D and A, characteristic of a vibronic interaction, is effectively zero in the symmetrical $\cdot$D$\cdot$A$\cdot$D$\cdot$A$\cdot$ sequence of the 1:1 complex. However, in the 2:1 complex, the lower symmetry of the environment of the donor molecules permits the charge oscillation, and a strong absorption is observed at 1295 cm^{-1} (Fig. 4.1). By contrast, the symmetry of the environment of the tetracyanoethylene molecule should be comparable in the 1:1 and

1:2 complexes. The similar absorption of the totally symmetric double-bond mode of tetracyanoethylene at 1560 cm^{-1} in the two complexes provides support for this conclusion.

4.C. Complexes of σ-Acceptors with *n*-Donors

The majority of infrared measurements on such complexes have been concerned with the shifts (usually to lower energies) and the changes in intensity (generally increases) of bands in one or other component of the

TABLE 4.2. Fundamental stretching vibration of the intermolecular N$\cdots$I bond in various complexes.

Complex	State	ν/cm^{-1}	Ref.
Iodine–trimethylamine	Solid (nujol mull)	145	42
Iodine–pyridine	Cyclohexane soln.	94	45
Iodine–2-methylpyridine	Cyclohexane soln.	87·5	45
Iodine–3-methylpyridine	Cyclohexane soln.	88·5	45
Iodine–4-methylpyridine	Cyclohexane soln.	88	45
Iodine–2,4-dimethylpyridine	Cyclohexane soln.	84·5	45
Iodine–2,6-dimethylpyridine	Cyclohexane soln.	74·5	45
Iodine–3-chloropyridine	Cyclohexane soln.	73	45
Iodine–3-bromopyridine	Cyclohexane soln.	65	45
Iodine monochloride–pyridine	Pyridine soln.	160	43
	Pyridine soln.	163	46
	Solid (nujol mull)	170	44
	Benzene soln.	147	43
Iodine monobromide–pyridine	Solid (nujol mull)	160	44
	Pyridine soln.	144	43
	Benzene soln.	134	43
Iodotrifluoromethane–trimethyl-amine	Gas	77	47
	Solid (−185°C)	126	47

complex, rather than with features relating to an intermolecular mode. Components studied include, for example, carbonyl compounds,[28, 29] ethers,[28] sulphoxides,[30, 31] nitriles,[32, 33] and amines.[16, 28, 34–37] Superficial observation of spectral changes on complex formation may be misleading. Relatively small effects may be the result of large changes in the force constant or effective charge of some vibrations.[38] The theories of intensity enhancements and frequency shifts discussed in the previous section,[19–22] are equally applicable to complexes between σ-acceptors and *n*-donors. The frequencies of some of these complexes are included in Table 4.1.

Decreases in the vibration frequency of a particular bond have been used as evidence for a particular site of a charge-transfer interaction. For example, it is argued that the interaction of iodine with N,N-dimethyl-acetamide[39] or N,N-dimethylbenzamide[40] is through the oxygen atom since the C=O frequency of the free amide decreases on complex formation in both cases. Similarly, the large shift in the C—N stretching frequency in N,N-dimethylaniline on forming a complex with sulphur dioxide has been used to argue that the complexing is via a localized orbital on the nitrogen atom, rather than a de-localized π-orbital.[41]

In a few systems in the far infrared, a band has been observed which could not be assigned to either the donor or the acceptor moiety,[36, 42–47] (Table 4.2). This appears to be due to an *intermolecular* stretching vibration. Thus, two bands have been observed[42] in the spectrum of the solid complex of trimethylamine–iodine at 184 and 145 cm^{-1}. Whereas the higher energy band has been assigned to the intramolecular I—I stretching vibration, it is suggested that the band at 145 cm^{-1} is due to the intermolecular stretching vibration between one iodine atom of the acceptor molecule and the nitrogen atom of the donor molecule. Calculations of the energies of these two transitions, based on a linear complex model (I), are in good agreement with the experimental values for these

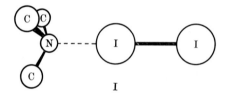

I

two vibrations. Corresponding absorptions have been measured in the far infrared spectra of the solid complexes of pyridine with iodine monochloride[44] and with iodine monobromide.[44] Bands similarly assigned to the intermolecular stretching mode of N···I have been observed in *solutions* of iodine,[43, 45] (Fig. 4.2), iodine monochloride,[43] and iodine monobromide[36, 43] complexes of pyridine and substituted pyridines (Table 4.2). A Raman band at 163 cm^{-1} for iodine monochloride solutions in pyridine has been likewise assigned to an intermolecular vibration mode of N···I.[46] The experimental method, employed in this determination by Klaeboe, uses a laser Raman spectrophotometer. The Raman frequencies for the halogens in halogen–donor complexes were observed to be considerably lower than in the free halogens. Some of these results are recorded in Table 4.1. There seems little doubt that this technique will be further exploited as this type of instrumentation becomes more available.

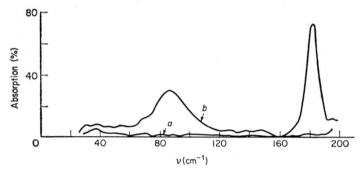

[R. F. Lake and H. W. Thompson, Ref. 45]

FIG. 4.2. Far infrared spectra in cyclohexane: (a) 2-methylpyridine 0·00954 mol/l; (b) 2-methylpyridine, 0·00954 mol/l, plus iodine, 0·01 mol/l (path-length 9·35 mm).

The recent comparison of the fundamental stretching vibration of the intermolecular bond of the iodotrifluoromethane–trimethylamine complex in the gas phase with the same vibration in the solid phase[47] (Table 4.2) emphasizes the care needed in interpreting the data obtained from solution spectra. A very similar problem exists with the electronic spectra (see Chapter 3).

REFERENCES

1. R. E. Gibson and O. H. Loeffler, *J. Am. chem. Soc.* **62**, 1324 (1940).
2. W. R. Burton and R. E. Richards, *J. chem. Soc.* 1316 (1950).
3. H. Kainer and W. Otting, *Chem. Ber.* **88**, 1921 (1955).
4. G. Briegleb, "Elektronen-Donator-Acceptor-Komplexe," Springer-Verlag, Berlin (1961), pp. 94, 103.
5. L. J. Andrews and R. M. Keefer, "Molecular Complexes in Organic Chemistry," Holden-Day, San Francisco (1964), p. 32.
6. Y. Matsunaga, *J. chem. Phys.* **41**, 1609 (1964); *Nature, Lond.* **205**, 72 (1965).
7. R. A. Friedel, *J. phys. Chem., Ithaca* **62**, 1341 (1958).
8. R. D. Kross and V. A. Fassel, *J. Am. chem. Soc.* **79**, 38 (1957).
9. G. Briegleb and H. Delle, *Z. Elektrochem.* **64**, 347 (1960).
10. J. Collin and L. D'Or, *J. chem. Phys.* **23**, 397 (1955).
11. L. D'Or, R. Alewaeters and J. Collin, *Recl Trav. chim. Pays-Bas Belg.* **75**, 862 (1956).
12. R. S. Mulliken, *J. chem. Phys.* **23**, 397 (1955).
13. E. E. Ferguson, *J. chem. Phys.* **25**, 577 (1956).
14. E. E. Ferguson, *J. chem. Phys.* **26**, 1265, 1357 (1957).
15. E. E. Ferguson, *Spectrochim. Acta* **10**, 123 (1957).
16. W. B. Person, R. E. Humphrey, W. A. Deskin and A. I. Popov, *J. Am. chem. Soc.* **80**, 2049 (1958).

17. W. B. Person, R. E. Humphrey and A. I. Popov, *J. Am. chem. Soc.* **81**, 273 (1959).

18. W. B. Person, R. E. Erickson and R. E. Buckles, *J. Am. chem. Soc.* **82**, 29 (1960).

19. E. E. Ferguson and F. A. Matsen, *J. chem. Phys.* **29**, 105 (1958).

20. E. E. Ferguson and F. A. Matsen, *J. Am. chem. Soc.* **82**, 3268 (1960).

21. E. E. Ferguson, *J. Chim. phys.* **61**, 257 (1964).

22. P. C. McKinney and G. M. Barrow, *J. chem. Phys.* **31**, 294 (1959).

23. H. B. Friedrich and W. B. Person, *J. chem. Phys.* **44**, 2161 (1966).

24. W. B. Person, C. F. Cook and H. B. Friedrich, *J. chem. Phys.* **46**, 2521 (1967).

25. O. Hassel and K. O. Strømme, *Acta chem. scand.* **12**, 1146 (1958).

26. J. Stanley, D. Smith, B. Latimer and J. P. Devlin, *J. phys. Chem., Ithaca* **70**, 2011 (1966).

27. B. Hall and J. P. Devlin, *J. phys. Chem., Ithaca* **71**, 465 (1967); B. Mosyńska and A. Tramer, *J. chem. Phys.* **46**, 820 (1967).

28. D. L. Glusker, H. W. Thompson and R. S. Mulliken, *J. chem. Phys.* **21**, 1407 (1953).

29. E. Augdahl and P. Klaeboe, *Acta chem. scand.* **16**, 1637, 1647, 1655 (1962).

30. R. Dahl, T. Gramstad and P. Klaeboe, *Acta chem. scand.* **19**, 2248 (1965).

31. E. Augdahl and P. Klaeboe, *Acta chem. scand.* **18**, 18 (1964).

32. K. F. Purcell and R. S. Drago, *J. Am. chem. Soc.* **88**, 919 (1966).

33. E. Augdahl and P. Klaeboe, *Spectrochim. Acta* **19**, 1665 (1963).

34. R. A. Zingaro and W. B. Witmer, *J. phys. Chem., Ithaca* **64**, 1705 (1960).

35. E. K. Plyler and R. S. Mulliken, *J. Am. chem. Soc.* **81**, 823 (1959).

36. Y. Yagi, A. I. Popov and W. B. Person, *J. phys. Chem., Ithaca* **71**, 2439 (1967).

37. I. Haque and J. L. Wood, *Spectrochim. Acta* **23A**, 959, 2523 (1967).

38. R. S. Mulliken and W. B. Person, *A. Rev. phys. Chem.* **13**, 107 (1962).

39. C. D. Schmulbach and R. S. Drago, *J. Am. chem. Soc.* **82**, 4484 (1960).

40. R. L. Carlson and R. S. Drago, *J. Am. chem. Soc.* **84**, 2320 (1962).

41. W. E. Byrd, *Inorg. Chem.* **1**, 762 (1962).

42. H. Yada, J. Tanaka and S. Nagakura, *J. molec. Spectrosc.* **9**, 461 (1962).

43. S. G. W. Ginn and J. L. Wood, *Trans. Faraday Soc.* **62**, 777 (1966).

44. F. Watari, *Spectrochim. Acta* **23A**, 1917 (1967).

45. R. F. Lake and H. W. Thompson, *Proc. R. Soc.* **A297**, 440 (1967).

46. P. Klaeboe, *J. Am. chem. Soc.* **89**, 3667 (1967).

47. N. F. Cheetham and A. D. E. Pullin, *Chem. Commun.* 233 (1967).

Chapter 5

Nuclear Magnetic Resonance Spectra

5.A. Liquid Phase

5.A.1. General

The observed chemical shift (δ) for a given nucleus in a molecule in dilute solution is not the same as the shift in the gaseous phase (δ_{gas}).[2, 3] The difference, sometimes called the solvent screening constant (δ_{solv}), is due in part to contributions from the bulk diamagnetic-susceptibility difference of the solution and reference samples (δ_b), the solvent magnetic anisotropy (δ_a) and weak dispersion (van der Waals) interaction between solute and solvent molecules (δ_W). For polar solutes a polar effect (δ_E) is also experienced; this is generally interpreted in terms of a reaction field; that is, the result of a secondary electric field in the solvent owing to its polarization by the permanent dipole of the solute.[3-7] It has been suggested by Kuntz and Johnston[8] that at least certain observed contributions to solvent shifts could be better explained in terms of solvent–solute complex rather than by the reaction field theory. However, any specific complex formation in solution will also make its contribution (δ_s) to the observed chemical shift. In general, we may therefore write:

$$\delta_{solv} = \delta_{gas} - \delta = \delta_b + \delta_W + \delta_a + \delta_E + \delta_s \qquad (5.1)$$

Measurements using an internal reference eliminate the bulk-suscepti-bility term, δ_b. Nevertheless, although one of the criteria in the choice of an internal reference is its lack of reactivity with other species present, there may be some degree of interaction and this could be reflected in chemical shifts measured against such a reference. However, the problems of providing sufficiently accurate susceptibility corrections[9] when an external reference is used, together with the tedium of the technique, are great enough to dissuade most workers from using the method. If there are doubts concerning the "ideal" behaviour of a reference substance, the most useful advice is to seek another substance. In favourable cases, the effects of weak, non-bonding interactions and reaction fields are small compared with effects due to specific complex formation. Elsewhere, when the total shift may itself be small, great caution must be exercised before assigning the effect to any particular cause.

The appearance of the magnetic resonance spectrum of a nucleus which can exist in more than one chemical environment is dependent on its lifetimes in these different environments.[10-18] Let us take the simple case of a nucleus which can exist in two environments, I and II, in either of which alone it appears as a singlet. If the lifetimes in the two states are long compared with $\sqrt{2}(\pi|\delta_I - \delta_{II}|H)^{-1}$ (where $|\delta_I - \delta_{II}|H$ is the frequency difference in Hz between the two absorptions at δ_I and δ_{II} if each appeared singly) then two lines are observed. If the lifetimes are short compared with $\sqrt{2}(\pi|\delta_I - \delta_{II}|H)^{-1}$, then only a single, time-averaged line is observed[12, 17] (Fig. 5.1). In the latter case, the position of the line (δ) is dependent on the relative populations, P_I and P_{II} in the two environments:

$$\delta = P_I \delta_I + P_{II} \delta_{II} \qquad (5.2)$$

For a nucleus rapidly exchanging between n different environments, a corresponding relationship holds. In the case of weak complex formation, under conditions for which equation (5.2) is valid, the environment I could represent an unassociated molecule and environment II the same molecule in a complexed state. In general, any change in the system, which will cause an alteration in the relative populations of P_I, P_{II}, will affect the observed time-averaged chemical shift (δ). Such change will include concentration changes, temperature variations, and alterations in the nature of the second component of the complex. Although qualitative and quantitative evidence for complex-formation may be obtained from such chemical shift changes, the results will not in themselves distinguish those complexes which we have described as charge-transfer complexes (see Chapter 1) from those complexes in which there is no contribution from charge-transfer interaction. Indeed, the results which have been obtained emphasize the very arbitrary limits which are set on the term "charge-transfer complex". It is impossible to make a clear demarcation between such complexes and some other types of complex; for example, those involving dipole–induced-dipole or hydrogen-bonding forces. For nearly all the interactions under present discussion, the rates of exchange between different environments at normal experimental temperatures fall within the category of very fast reactions. However, it must be anticipated that for systems in which the binding approaches that of a normal covalent bond, the rate of chemical exchange may be only of the same order as, or less than, the rate of exchange between the magnetic states of the nucleus, for which line-broadening, or separate lines, will be observed (Fig. 5.1.).

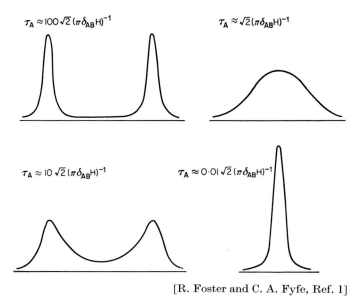

$$\tau_A \approx 100 \sqrt{2} \, (\pi \delta_{AB} H)^{-1}$$

$$\tau_A \approx \sqrt{2} (\pi \delta_{AB} H)^{-1}$$

$$\tau_A \approx 10 \sqrt{2} \, (\pi \delta_{AB} H)^{-1}$$

$$\tau_A \approx 0{\cdot}01 \sqrt{2} \, (\pi \delta_{AB} H)^{-1}$$

[R. Foster and C. A. Fyfe, Ref. 1]

Fig. 5.1. Changes in the shape of the resonance for an increasing exchange rate between two positions, A and B, with equal populations. (After H. S. Gutowsky and C. H. Holm, Ref. 12.)

5.A.2. Interactions of Carbonyl Compounds and of Ethers with Aromatic Compounds

Various groups of workers have studied the effect of the presence of aromatic species on the nuclear magnetic resonance spectra of particular carbonyl compounds and ethers. Hatton and Richards[19, 20] observed that the dilution of dimethylformamide with benzene caused a large upfield shift of the methyl resonances. The signal due to the methyl group *trans* to the carbonyl group shifts more than the signal due to the *cis*-methyl group. These results were interpreted in terms of a specific solvent–solute interaction involving the structure I in which the association is between the positive charge on the nitrogen in the dipolar–amide

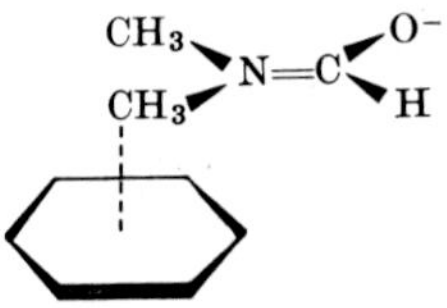

I

structure and the π-electrons of the benzene. The oxygen atom containing the negative end of the amide dipole is in a position as far away as possible from the benzene ring with its high π-electron density. Further evidence for specific solvent–solute interaction, rather than a general solvent effect, includes the large temperature coefficients of the observed chemical shifts,[21-23] and the fact that the temperature at which intra-molecular free rotation of the amide occurs, together with the activation energy for rotation, are both solvent-[22] and concentration-dependent.[23] Similar conclusions have been made concerning the interaction of N-methyl-N-cyclohexylamide,[24] of methylformamide[25] and of N-methyl-lactams[26] with benzene.

The behaviour of other carbonyl-containing compounds with aromatic molecules appears to be similar in principle to the amide interactions. However, whereas with the amides the binding appears to be through the nitrogen atom, in the case of esters and ketones it is generally suggested that the interaction* is through the carbon atom of the carbonyl group.[3, 27-50]

Most workers have taken the difference in chemical shifts of various proton resonances in the carbonyl compound (or other solute) when dissolved in an aromatic solvent and when dissolved in an isotropic solvent, generally carbon tetrachloride, as an indication of such solute–aromatic solvent interactions. The shift differences involved have been used to a large extent to make structural assignments of various absorptions in the n.m.r. spectra of the solutes or, alternatively, to assign structural features in the molecule from the particular solvent-shift behaviour. Such shifts are sometimes termed "aromatic solvent-induced shifts" (A.S.I.S.). It was in 1960 that Slomp and MacKellar[51] observed the differential chemical shifts in the methyl resonances of various steroids when the solvent is changed from deuterochloroform to pyridine. The potentialities of the method were developed mainly by Bhacca and Williams,[30] using to a large extent benzene as the aromatic solvent. The subject has been reviewed recently by Laszlo;[3] other references may be found in a paper by Ronayne and Williams.[40] For a given solvent pair,

---

* But see below as to the nature of the interaction.

the shift difference in the non-aromatic and in the aromatic solvent ($\delta_{\text{non-arom}} - \delta_{\text{arom}}$) is approximately constant for protons in a particular position in the molecule, relative to the carbonyl group. Reference planes for predicting solvent shifts have been proposed for various carbonyl compounds, for example II and III. Similar arguments have been used

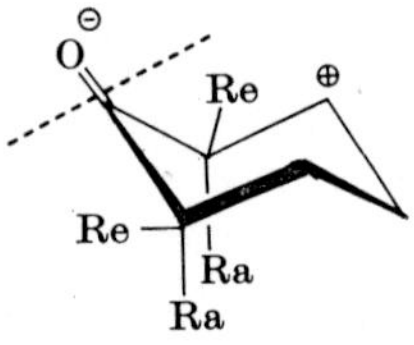

II 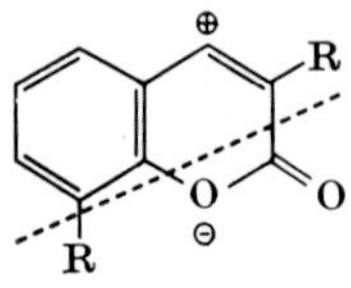   III

for the solvent shifts in other types of molecule, for example: ethers,[52-54] amines,[55-57] lactones,[58,59] acid anhydrides,[60,61] coumarins,[62] and cyanine dyes.[63] Where comparisons between the solvent shifts of protons of a solute in two different aromatic solvents have been made, for example, benzene and pyridine, the results sometimes indicate that different references planes (IV, V) have to be used for the two systems.[64]

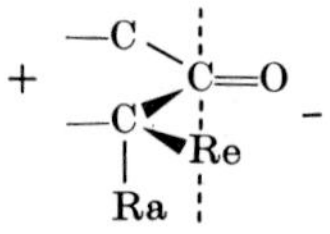

plane for solvent pair
$CHCl_3$—C_6H_6

IV

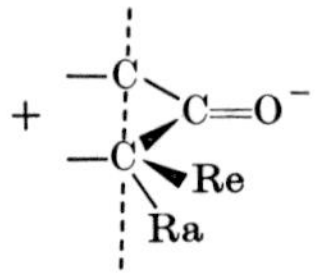

plane for solvent pair
CCl_4—C_5H_5N

V

Many other systems have been, and continue to be, studied. Recent reviews provide a number of references.[1,3,30]

To some extent such correlations must be empirical since, if these solvent-induced shifts are the result of specific complex formation, then they are really composite terms dependent, not only on the chemical shift of the pure complex in solution, measured for example by δ_0, but also on the fraction of solute present in the complexed state. For an infinitely dilute solution, the measured shift, δ_∞, will not be equal to δ_0, unless K, the association constant of the solvent-solute complex, is infinite. For finite values of K, $\delta_\infty = \delta_0 K (1 + K)^{-1}$.[65] Because associations of this type are relatively weak, the measured shift will not approach the value of δ_0. Consequently, unless K and δ_0 vary in a parallel fashion, comparisons of a single function of both K and δ_0 may not be meaningful although in series of closely related compounds some correlation appears to obtain. It must be emphasized that such criticisms do not apply to the relative shifts of atoms in the same molecule since these will all be functions of the same association constant, and consequently the measured shift values for dilute solutions will be proportional to the corresponding δ_0 values. These facts give at least a partial reinterpretation of Hatton and Richards'[19, 20] results for the interaction of dimethylformamide with various methylbenzenes. They extrapolated their measured shift values to values which would correspond to infinitely dilute solutions (δ_∞). It was found that δ_∞ decreased as the degree of methylation of the benzene increased. This result was originally explained in terms of steric interference to complex formation by the aromatic methyl groups. However, if the two factors, K and δ_0, upon which δ_∞ is dependent, are separated and evaluated by studying the behaviour of the system in a diluting solvent, as has been done by Sandoval and Hanna,[26] it appears that the degree of association does in fact increase as the degree of methylation of the aromatic component is increased. It is the greater rate of decrease in the values of δ_0 throughout this series which accounts for the original observations of Hatton and Richards.[19, 20]

Laszlo and Williams[34] have measured the shifts for steroidal ketones in toluene solution as a function of temperature. The value of the chemical shift obtained by extrapolation to absolute zero was considered to be equal to δ_0. The enthalpy change on complex formation, assuming $1:1$ complex formation, has been estimated as $\Delta H^{\ominus m}* = -0 \cdot 65 \pm 0 \cdot 15$ kcal mol^{-1}. Although this is only claimed to represent the order of magnitude of $\Delta H^{\ominus},*$ it does nevertheless indicate the weakness of the binding in this type of interaction. A similar value has been obtained for the association of toluene-d$_8$ with 4-nitrobenzaldehyde.[66] Attempts are being made at present to determine the degree of association for several model systems in the presence of a third "inert" diluting solvent.[67, 68] Preliminary results indicate that, if the shifts are interpreted in terms of $1:1$ associations, then the formation constants are very low.

It was stated at the outset that in the field of charge-transfer complexes it is impossible to define sharply the range of interactions which is reasonably described by this term (Chapter 1). The examples cited in this section may be a case in point, although the general opinion at present is that there is little evidence of a significant charge-transfer component in these complexes. Diehl[69] showed that the solvent-induced shift is less in certain solutes where there is steric hindrance to the

---

* The significance of the subscript and superscripts of the term for the enthalpy change are explained in Chapter 6, p. 173.

approach of the aromatic solvent molecule, than in an analogous solute lacking the steric congestion. These results have been used to argue the presence of specific complex formation: for example, the larger shift for the *meta*-protons in nitrobenzene than for the corresponding protons in 2,4,6-trimethylnitrobenzene (nitromesitylene), is explained[69] in terms of the restricted approach of the solvent benzene in the latter case compared with the former. This and similar examples could be explained alternatively in terms of charge-transfer complexing in which the acceptor (nitrobenzene) is weakened by the introduction of electron-donating methyl groups into the molecule. Either argument by itself is tenuous. Other examples may be cited against the steric argument. Thus, the benzene shifts for simple methylcyclohexanones[31, 50] are nearly identical to those for various sterically hindered steroidal ketones.[27–29, 32] Indeed, much of the value of "aromatic solvent-induced shift values" for ketones is due to their independence of the steric environment of the atom concerned, and dependence only on its position in space with respect to the carbonyl group. The major contribution to the stabilization of the complexes may well be through local-dipole–induced-dipole inter-actions.[3, 40] Although reference has been made to "specific" complexes, and structures representing a particular orientation of the components in the complex, such diagrams should be interpreted as approximating to the most probable, or the average, of many possible configurations. The degree of solvent–solute association for some ketones dissolved in benzene has been estimated using a dilution technique.[70] The method allows the stoichiometry of the complex to be determined unambiguously.

Certain complexes, namely those of substituted p-benzoquinones with benzene, which have been treated[33] as dipolar complexes with a possible structure VI, have solvent shifts which are large compared with corre-sponding systems involving saturated ketones.[34] It may be that these interactions are better looked upon as 1:1 π-π-charge-transfer complexes in which the structure approximates to VII rather than VI. Quantitative studies of chemical shifts of related systems, in an "inert" diluting

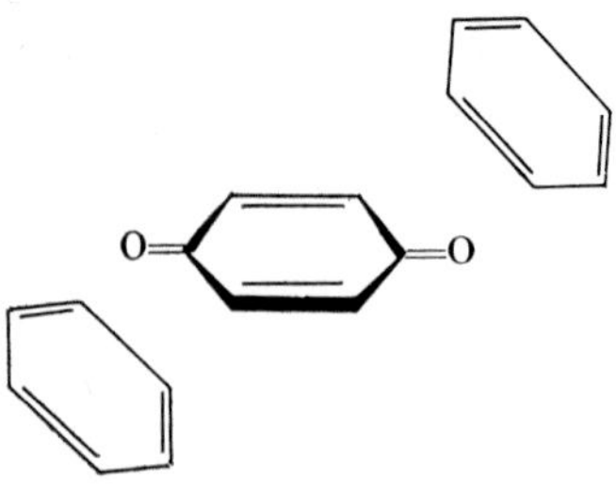

VI    VII

solvent (Chapter 7) suggest [71, 72] that the degree of association for such interactions would be significantly greater than for structures of the type VI.* This would explain the relatively large shifts. Baker and Davis [70] have suggested that the 2-isopropyl-5-methyl-p-benzoquinone–benzene complex, for which they have obtained evidence of 2:2 stoichiometry, has possibly the structure VIII.

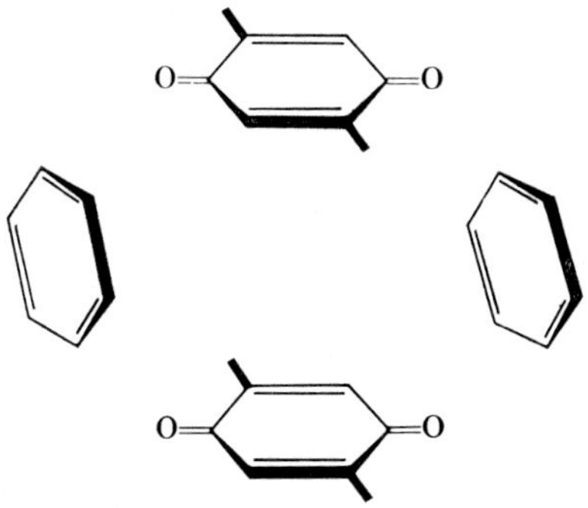

VIII

5.A.3. Other Systems

Apart from the systems described in Section 5.A.2, some of which can only marginally qualify for the description "charge-transfer complex", there is a whole range of interactions for which this term is generally acceptable, in particular, those complexes which have observable optical absorptions assigned as intermolecular charge-transfer transitions (Chapter 3). A number of these systems have now been studied by n.m.r. spectroscopy. Not all were described as charge-transfer interactions by their original investigators.

It is apparent from equation (5.2) that, unless a significant fraction of the species which contains the nucleus being measured is in the complexed state, then, at most, only a small change in the observed shift (δ) will occur between solutions containing no complex and those containing equilibrium mixtures of components and complex. This fact may be the reason why some workers observed little or no change in δ for various donor solvents; for example, the addition of iodine to benzene or other aromatic solvents. [73–75] However, when the electron donor–electron acceptor system is in a suitable solvent under such concentration conditions that a significant proportion of the donor or acceptor species being measured is in the form of the complex, then a concentration-dependence of the measured chemical shift often, though not always, is observed. Some of the earlier examples include mixtures of 1,3,5-

* Note in proof. Very recent experiments (Ref. 88) are more in support of structure VI.

trinitrobenzene and hexamethylbenzene in deuterated chloroform;[76] the changes in the chemical shift of p-benzoquinone in carbon tetrachloride on the addition of relatively large amounts of methylbenzenes or naphthalene;[77] and the addition of iodine to various amines including pyridine.[73]

As already pointed out, the measured position of the chemical shift (δ) of a nucleus in a molecular species which is partially complexed depends not only on the amount of complex present, and therefore on K^{AD}, but also on the hypothetical chemical shift for the measured nucleus in the pure complex alone in solution. This shift is often expressed as the difference in shifts between the nucleus in the component molecule when fully complexed and in the free component molecule, and is here given the symbol Δ_0 (see Chapter 6). Obviously, if Δ_0 is small, the observed shift differences of a nucleus in the uncomplexed molecule in solution and in the equilibrium mixture will be small, even though the fraction of the molecules which are complexed may be large. This situation is the n.m.r. analogue of an optical absorption in which the difference in molar absorptivity between the component and complex is small: it results in only small differences in the measured absorbance of equilibrium mixtures compared with that of the component, although the degree of complexing (K^{AD}) may be high.

Determinations of K^{AD} and Δ_0 on a recognized charge-transfer system were first made by Hanna and Ashbaugh.[78] They applied a method which had previously been used to study hydrogen-bonded systems[79-81] in order to measure the position of equilibrium for the interactions of a series of methylbenzenes with 7,7,8,8-tetracyanoquinodimethane in dioxane solution. They noted that their results were dependent on the particular solvent scale used. More recently they have accounted for at least part of the observed shift in terms of solvent interaction[82] (see also Chapter 6). The same method has been applied to a study of the 1,3,5-trinitrobenzene–1,5-pentamethylenetetrazole (IX)

IX

interaction in carbon tetrachloride solution.[83] Solvent–solute interactions in this solvent should be very much less than for dioxane solutions, and the results obtained do show a reasonable agreement with the association constant obtained from optical data.[83] An alternative algebraic method[84] has been used to determine association constants for a wide range of donors with various aromatic nitrocompounds,[71, 84-91] aromatic nitriles,[71] and fluorine-containing electron acceptors [92-94] from chemical-shift measurements of nuclei in the electron-acceptor species in the presence of large and varying excess amounts of the electron-donor species (Chapter 6). Such is the case with some complexes of iodine with alkyl sulphides in which proton resonances in the donor have been measured.[95] The degree of interaction of iodine with p-methoxythioanisole in carbon tetrachloride has been determined by Larsen and Allred [96] from measurements of the position of absorption of the methyl protons in the donor in the presence of varying amounts of iodine. The value of Δ_0 obtained in the evaluation of K^{AD} was found to be temperature-independent. The interaction of iodine with 2,4,6-trimethyl-pyridine in mixtures of nitrobenzene and carbon tetrachloride has also been studied by the same authors.[97] Larsen and Allred [97] rationalized their results in terms of the formation of an ionic product and a charge-transfer complex. Two sets of methyl signals were observed, namely those arising from the pyridinium cation and those representing the time-averaged value for the absorptions in the free and charge-transfer-complexed pyridine. The association constant for the ionic product was estimated from the band areas. From the value of this constant and the particular concentration-dependence of the position of the time-averaged signal, the association constant for the charge-transfer complex was evaluated.

A discussion of equilibrium constants, evaluated largely from n.m.r. data, is given in Chapter 7.

5.A.4. Chemical Shift Values in Complexed Species

The difference (Δ_0) between the chemical shift of a nucleus in the free component and in the complexed component can be obtained from measurements of chemical shift for equilibrium mixtures at two or more concentrations in dilute solution (see previous section and Chapter 6). It is reasonable to seek a relationship between values of Δ_0 and the con-stituents of complexes. However, the various attempts to interpret the precise chemical shifts of simple molecules have not been wholly success-ful, neither has there been complete agreement between different workers.[98-104] It is therefore perhaps understandable that a detailed interpretation of the values of Δ_0 for nuclei in various complexes has yet to be given. Features of charge-transfer complexes which might con-tribute to the magnitude of Δ_0 include the effect of the ring current in one (aromatic) component on a nucleus in the second component of the complex; the direct effect of transfer of charge in the complex; and the differences in anisotropy of polar groups, resulting from the transfer of

TABLE 5.1. A selection of chemical shift values for protons in pure complexes in solution, relative to the shift of the same nucleus in the free component in solution (Δ_0^{AD}) at 33·5°C, 60·004 MHz.

Donor (D)	Acceptor (A)	Solvent	Proton measured	$\Delta_0^{AD}/$ Hz*	Ref.
Benzene	1,3,5-trinitrobenzene	CCl_4	$H_{2,4,6}$ in A	75	84
Toluene	1,3,5-trinitrobenzene	CCl_4	$H_{2,4,6}$ in A	72	84
p-Xylene	1,3,5-trinitrobenzene	CCl_4	$H_{2,4,6}$ in A	68	84
Mesitylene	1,3,5-trinitrobenzene	CCl_4	$H_{2,4,6}$ in A	61	84
Mesitylene	1,3,5-trinitrobenzene	CH_2ClCH_2Cl	$H_{2,4,6}$ in A	58†	90
Mesitylene	1,3,5-trinitrobenzene	CH_2ClCH_2Cl	$H_{2,4,6}$ in D	36‡	90
Durene	1,3,5-trinitrobenzene	CCl_4	$H_{2,4,6}$ in A	62	84
Durene	1,3,5-trinitrobenzene	CH_2ClCH_2Cl	$H_{2,4,6}$ in A	71†	90
Durene	1,3,5-trinitrobenzene	CH_2ClCH_2Cl	$H_{3,6}$ in D	35‡	90
Hexamethyl-benzene	1,3,5-trinitrobenzene	CCl_4	$H_{2,4,6}$ in A	65	84
Hexamethyl-benzene	1,2,3,5-tetranitro-benzene	CH_2ClCH_2Cl	$H_{4,6}$ in A	99	71
Hexamethyl-benzene	3,5-dinitrobenzo-nitrile	CCl_4	H_4 in A	75	91
Hexamethyl-benzene	3,5-dinitrobenzo-nitrile	CCl_4	$H_{2,6}$ in A	69	91
Hexamethyl-benzene	1,4-dinitrobenzene	CCl_4	$H_{2,3,5,6}$ in A	91	84
Hexamethyl-benzene	4,-nitrobenz-aldehyde	CCl_4	$H_{3,5}$ in A	92	88
Hexamethyl-benzene	4-nitrobenz-aldehyde	CCl_4	$H_{2,6}$ in A	94	88
Hexamethyl-benzene	4-nitrobenz-aldehyde	CCl_4	H_{CHO} in A	74	88
Hexamethyl-benzene	7,7,8,8-tetracyano-quinodimethane	CH_2ClCH_2Cl	$H_{2,3,5,6}$ in A	64†	71
Hexamethyl-benzene	2,3-dicyano-p-benzoquinone	CH_2ClCH_2Cl	$H_{5,6}$ in A	72†	71
N,N-dimethyl-aniline	1,3,5-trinitrobenzene	CCl_4	$H_{2,4,6}$ in A	61	84
Indole	1,3,5-trinitrobenzene	CH_2ClCH_2Cl	$H_{2,4,6}$ in A	97†	85
Indole	1,3,5-trinitrobenzene	CH_2ClCH_2Cl	H_3 in D	43†	85
Indole	1,3,5-trinitrobenzene	CH_2ClCH_2Cl	H_{arom} in D‖	19‡	85

* Probable error ±5% except where stated.
† Probable error ±10%.
‡ Probable error ±20%.
‖ Line of maximum intensity in aromatic multiplet.

charge. The variation in the magnitude of Δ_0 with the structure of the complex shows marked differences for ^{19}F resonances compared with ^1H resonances. For this reason, the Δ_0 values corresponding to these two nuclei are discussed separately.

Δ_0 *values for* ^1H *resonances.* For interactions between benzenoid donors and benzenoid acceptors, the Δ_0 values are generally within the range 60–80 Hz at 60·004 MHz.* For a series of complexes having a common acceptor species, the values of Δ_0 appear to be effectively constant (Table 5.1).† This lack of dependence of Δ_0 on the nature of the donor component suggests that transfer of charge is not a significant factor in determining the magnitude of Δ_0. Further support for this conclusion is obtained from the observation that ^1H chemical shifts are probably changed by only about 10 p.p.m. for a one-electron charge on an adjacent carbon atom in a simple molecule,[105, 106] and it is known that the degree of transfer of charge in this type of charge-transfer complex is probably small.

The smallness of the range of Δ_0 values suggests that ring-current effects may be important. There are several observations which support this suggestion. For some benzenoid donor–benzenoid acceptor systems, it has been possible to determine Δ_0 values for protons in both the donor and the acceptor. The values are not only numerically similar (Table 5.1) but are also of the *same sign*, namely, upfield from the absorptions of the parent molecules. If transfer of charge had made a significant contribution to Δ_0, then a downfield shift of proton absorptions in the donor on complex formation would have been expected. A second observation in favour of ring-current effects is that the Δ_0 values of 1,3,5-trinitrobenzene when complexed with excess naphthalene in dichloromethane are much larger (138 Hz) than when complexed with alkylbenzenes ($\sim$80 Hz).‡ In the naphthalene complex the acceptor protons are closer to the centre of the ring current.[85] Further, when in this system the acceptor is in excess, the structure of the naphthalene multiplet changes (Fig. 5.2). It has been shown that, although the coupling constants remain unaltered, the chemical shift difference between the H_α and H_β protons in naphthalene decreases.[76,85] This means that the H_α resonances are shifted more than the H_β resonances, in agreement with the ring-current hypothesis if the geometry of the complex indicated in structure X is assumed, in which the H_α protons are closer than the H_β protons to the acceptor ring current. Furthermore, the Δ_0 values for both H_α and H_β are smaller than the Δ_0 values for 1,3,5-trinitrobenzene in the same complex, or for the donor protons in the corresponding benzene–1,3,5-trinitrobenzene complex. These observations provide further support for the importance of ring-current effects. Results similar to

---

* Unless otherwise stated, all chemical shifts for ^1H resonances expressed in Hz are for resonances measured at 60·004 MHz; corresponding fluorine resonances are based on a resonance frequency of 56·462 MHz.

† See also Tables 7.6, 7.8, 7.9 and 7.10.

‡ At 60·004 MHz.

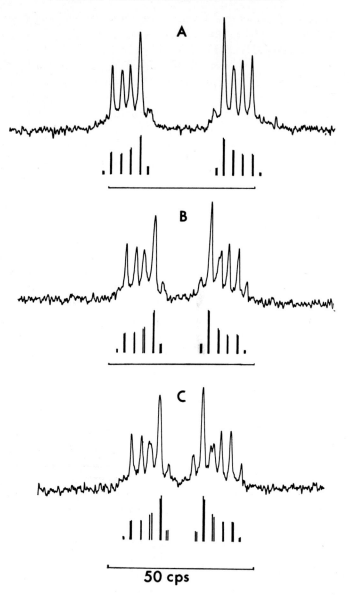

[C. A. Fyfe, Ref. 107]

FIG. 5.2. Proton magnetic resonance spectra at 100 MHz of naphthalene in 1,2-dichloroethane and the theoretical spectra; (A) 0·05M naphthalene ($J_A = 0·0$ Hz; $J_B = 6·0$ Hz; $J' = 1·4$ Hz; $J'' = 8·6$ Hz; $V_0\delta = 37·3$ Hz); (B) 0·05M naphthalene plus 0·95M 1,3,5-trinitrobenzene ($V_0\delta = 27·5$ Hz), couplings as in (A); (C) 0·05M naphthalene plus 2·37M 1,3,5-trinitrobenzene ($V_0\delta = 22·5$ Hz), couplings as in (A).

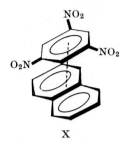

X

those described for the naphthalene–1,3,5-trinitrobenzene complex have been observed for the corresponding anthracene complex[107] (see Fig. 5.3). These may be explained in terms of the symmetrical structure XI.

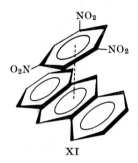

XI

Although the ring-current hypothesis provides an explanation of many observations of aromatic donor–aromatic acceptor systems, it does not appear to be wholly satisfactory. In particular, the quantitative theoretical calculations made by Johnson and Bovey[108] indicate that, at the intermolecular distances encountered in charge-transfer complexes, the maximum shifts (Δ_0) would be of the order of 20 Hz at 60 MHz. It should also be pointed out that, although no trend in Δ_0 for ^{1}H resonance with ionization potential of the donor has been observed by Foster *et al.*,[84–90] such a correlation has been reported by Hanna and his co-workers.[78, 82] Here Δ_0 is reported to decrease as the degree of complexing of 7,7,8,8-tetracyanoquinodimethane with a series of alkylbenzenes increases. This variation was attributed to changes in the paramagnetic contribution to the proton shifts in the acceptor component. However, the authors have demonstrated that the values of Δ_0 which they obtained are concentration-scale dependent,[82] probably because the association constants in these cases are relatively small (see Chapter 6). Although there may be some doubt as to the absolute values of Δ_0 in this series of complexes, the trend in Δ_0 to lower values as the donor strength is increased cannot be denied.

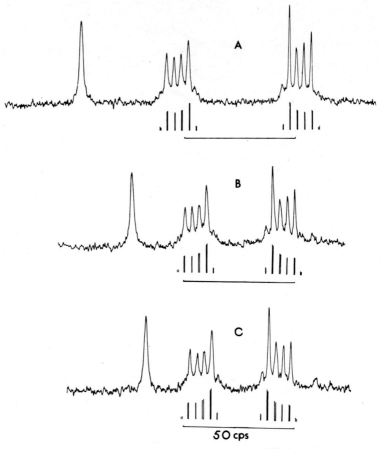

[C. A. Fyfe, Ref. 107]

Fig. 5.3. Proton magnetic resonance spectra at 100 MHz of anthracene in 1,2-dichloroethane and the theoretical spectra; (A) 0·05M anthracene ($J_A = 0·2$ Hz. $J_B = 6·5$ Hz, $J' = 1·4$ Hz, $J'' = 8·6$ Hz, $V_0\delta = 54·5$ Hz); (B) 0·05M anthracene plus 0·94M 1,3,5-trinitrobenzene ($V_0\delta = 39$ Hz), couplings as in (A); (C) 0·05M anthracene plus 1·96M 1,3,5-trinitrobenzene ($V_0\delta = 34·5$ Hz), couplings as in (A).

By contrast, in the considerably stronger interactions of iodine and of iodine monochloride with various methylpyridines, *downfield* shifts of the methyl protons in the donor are observed when the acceptor is added.[109] Such shifts are explicable in terms of a transfer-of-charge lowering the methyl-proton shielding constants. A similar downfield shift of donor protons has also been observed in iodine–alkyl sulphide complexes.[95]

Δ_0 *values for* ^{19}F *resonances.* ^{19}F chemical shift measurements rather than ^1H measurements can be used to advantage in charge-transfer systems. Not only are the association constants of fluorine-containing acceptors with electron donors large compared with the corresponding complexes of the non-fluorinated acceptors, but also the Δ_0 values for ^{19}F are often at least double the values for ^1H.[92, 93]

In contrast to proton shift of π-π interactions, the Δ_0 values for ^{19}F in acceptors complexed with a series of donors, increase significantly as the

TABLE 5.2. A selection of chemical shift values for fluorine nuclei in the acceptor moiety in pure complexes in solution in relation to the shift of the same nucleus in the free acceptor in solution (Δ_0^{AD}) at 33·5°C, at 56·462 MHz.*

Donor	Acceptor	Solvent	Δ_0^{AD}*	Ref.
Benzene	Fluoranil	CCl_4	99	93
Toluene	Fluoranil	CCl_4	127	93
o-Xylene	Fluoranil	CCl_4	149	88
m-Xylene	Fluoranil	CCl_4	161	88
p-Xylene	Fluoranil	CCl_4	161	93
Mesitylene	Fluoranil	CCl_4	172	93
Durene	Fluoranil	CCl_4	208	93
Pentamethylbenzene	Fluoranil	CCl_4	229	93
Hexamethylbenzene	Fluoranil	CCl_4	261	93
Hexamethylbenzene	Fluoranil	$CHCl_3$	293	93
Hexamethylbenzene	Fluoranil	CH_2ClCH_2Cl	192	93
Hexamethylbenzene	Fluoranil	CH_2Cl_2	205	93
Toluene	p-$C_6F_4(CN)_2$†	CCl_4	82	93
p-Xylene	p-$C_6F_4(CN)_2$†	CCl_4	100	93
Mesitylene	p-$C_6F_4(CN)_2$†	CCl_4	111	93
Durene	p-$C_6F_6(CN)_2$†	CCl_4	137	93
Pentamethylbenzene	p-$C_6F_4(CN)_2$†	CCl_4	154	93
Hexamethylbenzene	p-$C_6F_4(CN)_2$†	CCl_4	169	93

* The values from Ref. 93 which were measured at 37·644 MHz have been recalculated to correspond to the higher energy.
† p-$C_6F_4(CN)_2$ = 1,4-dicyano-2,3,5,6-tetrafluorobenzene.

ionization potential of the donor decreases (Table 5.2). It is know that ^{19}F chemical shifts are more dependent on electron density than are ^1H chemical shifts,[110] and it would seem plausible to assign at least a part of the observed change in the chemical shift in the acceptor to a transfer of charge in the complex. Confirmation of this is provided by qualitative measurements of the ^{19}F chemical shift in the donor 4-fluoroaniline in the presence of excess 1,3,5-trinitrobenzene in 1,2-dichloroethane solution: a greatly reduced upfield shift is observed.[111] Further convincing examples are provided by the complexes of 4-fluorobenzonitrile acting as a donor with various boron halide acceptors.[112]* Although with the very strong acceptors B_2Cl_4, BCl_3, BBr_3 the complexes were so stable that the exchange rate was too low to observe a single population-averaged signal, such lines were observed in complexes involving the weaker acceptors $B(CH_3)_3$, BF_3 and $p\text{-}FC_6H_4BCl_2$. In these latter cases, the ^{19}F resonance in the donor molecule always moved downfield as the degree of complex formation increased. In the interaction represented by the structure XII, the ^{19}F resonance of the donor moved downfield

Acceptor   Donor

XII

$(\Delta_0 = -10\cdot9$ p.p.m.$)$ and the ^{19}F resonance of the acceptor moved upfield $(\Delta_0 = +12 \pm 1$ p.p.m.$)$ on forming the complex.

It would be of interest to study the effect of charge-transfer complex formation on ^{13}C magnetic resonances in both donor and acceptor moieties.

5.B. Nuclear Magnetic Resonance and Quadrupole Measurements of Solid Charge-Transfer Complexes

Relatively few n.m.r. measurements have been made on solid charge-transfer systems. However, one major application has been an investigation to determine whether the component molecules in a solid charge-transfer complex are rigidly held in the crystal lattice or whether molecular rotation is possible. The proton magnetic resonance of the solid benzene–silver perchlorate complex has been measured at 298°K and at 77°K by Gilson and McDowell.[113] From the spectra, they obtained values for the second moments of $0\cdot9$ G^2 at 298°K and $6\cdot10$ G^2 at 77°K. From the structure of the complex obtained by standard X-ray diffraction techniques,[114] Gilson and McDowell calculated a value of $5\cdot9$ G^2 for the second moment, in good agreement with their experimentally determined low-temperature value. They concluded that, at 77°K, there is no rotation of the molecules of benzene in the crystal lattice. If allowance is made for the rotation of the benzene molecules about their six-fold axes, then calculation of the second moment for the complex leads to a value of $1\cdot02$ G^2. This is very close to the value observed experimentally

* Although the types of complex generally discussed in this book exclude complexes involving boron compounds (see Chapter 1), this example of the effect of charge-transfer on the sign of the shift is considered to be sufficiently important to warrant its inclusion.

at room temperature. Gilson and McDowell[113, 115] therefore concluded that the benzene molecule is more or less freely rotating at room temperature. They point out, however, that the n.m.r. measurements will be consistent with free rotation if the rate of reorientation is greater than 10^4–10^5 times per second, whilst X-ray diffraction measurements will lead to a picture of a stationary system if the molecule spends most of its time at the bottom of a potential-energy well for rotational motion. Nevertheless Smith[116] argues that his crystallographic data do not substantiate a model for this complex in which the benzene ring is freely rotating. Smith and Rundle[114] had suggested that in fact the benzene ring was distorted.

Using the same technique, Gilson and McDowell[117] have studied the hexafluorobenzene complexes of benzene, mesitylene and durene, using both ^1H and ^{19}F absorptions. The complexes of tetrabromomethane with benzene and with durene were also investigated. Again, the results can be explained in terms of molecular rotation at room temperature. There is similar evidence for rotation of hexamethylbenzene in the solid hexamethylbenzene–chloranil complex at room temperature.[118] By contrast, the fact that the ^{35}Cl nuclear quadrupole resonance in the same complex shows two absorptions[119] suggests that there is little or no orientation of the chloranil molecules in the complex. It is also concluded that, for this complex the degree of transfer of charge in the ground state is not more than 5–10%. This estimate is based on the assumption that the nuclear quadrupole moment of chlorine depends on the electron density on the carbon atoms, and is in general agreement with earlier estimates of ionic character based on dipole moment measurements.[120] Nuclear quadrupole resonance measurements on various chloroform–electron donor systems suggest a similar low degree of charge transfer.[121]

No ^{35}Cl nuclear quadrupole resonance absorption was observed[119] for either of the complexes picryl chloride–hexamethylbenzene or p-chloroaniline–1,3,5-trinitrobenzene. Douglass[119] suggested that this might be the result of disordering in the crystal lattices.

REFERENCES

1. R. Foster and C. A. Fyfe, in "Progress in Nuclear Magnetic Resonance Spectroscopy," Vol. 4, eds. J. W. Emsley, J. Feeney and L. H. Sutcliffe, Pergamon, Oxford and New York (1969), Ch. 1.
2. J. A. Pople, W. G. Schneider and H. J. Bernstein, "High Resolution Nuclear Magnetic Resonance," McGraw-Hill, New York, Toronto, London (1959), Ch. 16.
3. P. Laszlo, in "Progress in Nuclear Magnetic Resonance Spectroscopy," Vol. 3, eds. J. W. Emsley, J. Feeney and L. H. Sutcliffe, Pergamon, Oxford and New York (1969), Ch. 6.

4. A. D. Buckingham, T. Schaefer and W. G. Schneider, *J. chem. Phys.* **32**, 1227 (1960).
5. A. D. Buckingham, *Can. J. Chem.* **38**, 300 (1960).
6. P. Diehl and R. Freeman, *Molec. Phys.* **4**, 39 (1961).
7. W. T. Raynes, A. D. Buckingham and H. J. Bernstein, *J. chem. Phys.* **36**, 3481 (1962).
8. I. D. Kuntz, Jr. and M. D. Johnston, Jr., *J. Am. chem. Soc.* **89**, 6008 (1967).
9. C. Lussan, *J. Chim. phys.* **61**, 462 (1964).
10. H. S. Gutowsky, D. W. McCall and C. P. Slichter, *J. chem. Phys.* **21**, 279 (1953).
11. H. S. Gutowsky and A. Saika, *J. chem. Phys.* **21**, 1688 (1963).
12. H. S. Gutowsky and C. H. Holm, *J. chem. Phys.* **25**, 1228 (1956).
13. E. Grunwald, A. Loewenstein and S. Meiboom, *J. chem. Phys.* **27**, 630, 641 (1957).
14. A. Loewenstein and S. Meiboom, *J. chem. Phys.* **27**, 1067 (1957).
15. M. Takeda and E. O. Stejskal, *J. Am. chem. Soc.* **82**, 25 (1960).
16. H. M. McConnell, *J. chem. Phys.* **28**, 430 (1958).
17. J. A. Pople, W. G. Schneider and H. J. Bernstein, "High Resolution Nuclear Magnetic Resonance," McGraw-Hill, New York, Toronto, London (1959), p. 222.
18. A. Allerhand, H. S. Gutowsky, J. Jonas and R. A. Meinzer, *J. Am. chem. Soc.* **88**, 3185 (1966).
19. J. V. Hatton and R. E. Richards, *Molec. Phys.* **3**, 253 (1960).
20. J. V. Hatton and R. E. Richards, *Molec. Phys.* **5**, 153 (1962).
21. J. V. Hatton and W. G. Schneider, *Can. J. Chem.* **40**, 1285 (1962).
22. A. G. Whittaker and S. Siegel, *J. chem. Phys.* **42**, 3320 (1965).
23. A. G. Whittaker and S. Siegel, *J. chem. Phys.* **43**, 1575 (1965).
24. R. M. Moriarty, *J. org. Chem.* **28**, 1296 (1963).
25. L. A. LaPlanche and M. T. Rogers, *J. Am. chem. Soc.* **86**, 337 (1964).
26. A. A. Sandoval and M. W. Hanna, *J. phys. Chem., Ithaca* **70**, 1203 (1966).
27. R. M. Moriarty and J. M. Kliegman, *J. org. Chem.* **31**, 3007 (1966).
28. N. S. Bhacca and D. H. Williams, *Tetrahedron Lett.* 3127 (1964).
29. D. H. Williams and N. S. Bhacca, *Tetrahedron* **21**, 1641, 2021 (1965).
30. N. S. Bhacca and D. H. Williams, "Applications of N.M.R. Spectroscopy in Organic Chemistry," Holden-Day, San Francisco (1964), Ch. 7.
31. S. Bory, M. Fétizon, P. Laszlo and D. H. Williams, *Bull. Soc. chim. Fr.* 2541 (1965).
32. D. H. Williams and D. A. Wilson, *J. chem. Soc.* (B), 144 (1966).
33. J. Ronayne and D. H. Williams, *Chem. Commun.* 712 (1966).
34. P. Laszlo and D. H. Williams, *J. Am. chem. Soc.* **88**, 2799 (1966).
35. J. H. Bowie, D. W. Cameron, P. E. Schütz and D. H. Williams, *Tetrahedron* **22**, 1771 (1966).
36. J. D. Connolly and R. McCrindle, *Chemy. Ind.* 379 (1965).
37. J. D. Connolly and R. McCrindle, *Chemy. Ind.* 2066 (1965).
38. C. J. Timmons, *Chem. Commun.* 576 (1965).
39. F. H. Cottee and C. J. Timmons, *J. chem. Soc.* (B) 326 (1968).
40. J. Ronayne and D. H. Williams, *J. chem. Soc.* (B) 540 (1967), and refs. therein.
41. C. R. Narayanan and N. K. Venkatasubramanian, *Tetrahedron Lett.* 3639 (1965).
42. J. Seyden-Penne, T. Strzalko and M. Plat, *Tetrahedron Lett.* 4597 (1965).
43. P. S. Wharton and T. I. Bair, *J. org. Chem.* **30**, 1681 (1965).

44. J. Seyden-Penne, T. Strzalko and M. Plat, *Tetrahedron Lett.* 3611 (1966).
45. D. W. Boykin, Jr., A. B. Turner and R. E. Lutz, *Tetrahedron Lett.* 817 (1967).
46. R. Freeman and N. S. Bhacca, *J. chem. Phys.* **45**, 3795 (1966).
47. J. Ronayne, M. V. Sargent and D. H. Williams, *J. Am. chem. Soc.* **88**, 5288 (1966).
48. F. Johnson, N. A. Starkovsky and W. D. Gurowitz, *J. Am. chem. Soc.* **87**, 3492 (1965).
49. B. Hampel and J. M. Kraemer, *Tetrahedron* **22**, 1601 (1966).
50. M. Fétizon, J. Goré, P. Laszlo and B. Waegell, *J. org. Chem.* **31**, 4047 (1966).
51. G. Slomp and S. L. MacKellar, Jr., *J. Am. chem. Soc.* **82**, 999 (1960).
52. J. E. Anderson, *Tetrahedron Lett.* 4713 (1965).
53. J. H. Bowie, J. Ronayne and D. H. Williams, *J. chem. Soc.* (B), 785 (1966).
54. J. H. Bowie, J. Ronayne and D. H. Williams, *J. chem. Soc.* (B), 535 (1967).
55. J. N. Murrell and V. M. S. Gil, *Trans. Faraday Soc.* **61**, 402 (1965).
56. J. Ronayne and D. H. Williams, *J. chem. Soc.* (B), 805 (1967).
57. D. J. Barraclough, P. W. Hickmott and O. Meth-Cohn, *Tetrahedron Lett.* 4289 (1967).
58. Y. Ichikawa and T. Matsuo, *Bull. chem. Soc. Japan* **40**, 2030 (1967).
59. G. Di Maio, P. A. Tardella and C. Iavarone, *Tetrahedron Lett.* 2825 (1966).
60. T. Matsuo, *Can. J. Chem.* **45**, 1829 (1967).
61. C. Ganter, L. G. Newman and J. D. Roberts, *Tetrahedron* **Supp. 8**, 507 (1966).
62. R. Grigg, J. A. Knight and P. Roffey, *Tetrahedron* **22**, 3301 (1966).
63. R. Radeglia and S. Dahne, *Ber. Bunsenges physik. Chem.* **70**, 745 (1966).
64. D. H. Williams, *Tetrahedron Lett.* 2305 (1965).
65. S. K. Alley, Jr. and R. L. Scott, *J. phys. Chem., Ithaca* **67**, 1182 (1963).
66. R. E. Klink and J. B. Stothers, *Can. J. Chem.* **44**, 37 (1966).
67. R. Foster, unpublished work.
68. D. R. Twiselton, unpublished work.
69. P. Diehl, *J. Chim. phys.* **61**, 199 (1964).
70. K. M. Baker and B. R. Davis, *J. chem. Soc.* (B), 261 (1968).
71. R. Foster and C. A. Fyfe, *Trans. Faraday Soc.* **62**, 1400 (1966).
72. Y. Nakayama and T. Matsuo, *Kogyo Kagaku Zasshi* **69**, 1925 (1966).
73. A. Fratiello, *J. chem. Phys.* **41**, 2204 (1963).
74. S. Matsuoka, A. Mori and S. Hattori, *Sci. Rep. Kanazawa Univ.* **8**, 45 (1962).
75. S. Matsuoka and S. Hattori, *J. phys. Soc. Japan* **17**, 1073 (1962).
76. W. Koch and Hch. Zollinger, *Helv. chim. Acta* **48**, 554 (1965).
77. H. H. Perkampus and U. Krüger, *Z. phys. Chem. Frankf. Ausg.* **48**, 379 (1966).
78. M. W. Hanna and A. L. Ashbaugh, *J. phys. Chem., Ithaca* **68**, 811 (1964).
79. P. J. Berkeley, Jr. and M. W. Hanna, *J. phys. Chem., Ithaca* **67**, 846 (1963).
80. C. M. Huggins, G. C. Pimentel and J. N. Shoolery, *J. chem. Phys.* **23**, 1244 (1955).
81. E. D. Becker, U. Liddell and J. N. Shoolery, *J. molec. Spectrosc.* **2**, 1 (1958).
82. P. J. Trotter and M. W. Hanna, *J. Am. chem. Soc.* **88**, 3724 (1966).
83. T. C. Nehman and A. I. Popov, *J. phys. Chem., Ithaca* **70**, 3688 (1966).
84. R. Foster and C. A. Fyfe, *Trans. Faraday Soc.* **61**, 1626 (1965).
85. R. Foster and C. A. Fyfe, *J. chem. Soc.* (B), 926 (1966).
86. R. Foster and C. A. Fyfe, *Biochim. biophys. Acta* **112**, 490 (1966).
87. J. W. Morris, unpublished work.
88. D. R. Twiselton, unpublished work.
89. M. I. Foreman, unpublished work.
90. R. Foster and C. A. Fyfe, unpublished work.

91. R. Foster and M. I. Foreman, unpublished work.
92. R. Foster and C. A. Fyfe, *Chem. Commun.* 642 (1965).
93. N. M. D. Brown, R. Foster and C. A. Fyfe, *J. chem. Soc.* (B), 406 (1967).
94. J. W. Morris, unpublished work.
95. E. T. Strom, W. L. Orr, B. S. Snowden, Jr. and D. E. Woessner, *J. phys. Chem., Ithaca* **71**, 4017 (1967).
96. D. W. Larsen and A. L. Allred, *J. Am. chem. Soc.* **87**, 1216 (1965).
97. D. W. Larsen and A. L. Allred, *J. Am. chem. Soc.* **87**, 1219 (1965).
98. J. A. Elridge and L. M. Jackman, *J. chem. Soc.* 859 (1961).
99. B. P. Dailey, *J. chem. Phys.* **41**, 2304 (1964).
100. R. J. Abraham, R. C. Sheppard, W. A. Thomas and S. Turner, *Chem. Commun.* 43 (1965).
101. J. A. Elvidge, *Chem. Commun.* 160 (1965).
102. J. I. Musher, *J. chem. Phys.* **43**, 4081 (1965).
103. J. M. Gaidis and R. West, *J. chem. Phys.* **46**, 1218 (1967).
104. J. I. Musher, *J. chem. Phys.* **46**, 1219 (1967).
105. J. R. Leto, F. A. Cotton and J. S. Waugh, *Nature, Lond.* **180**, 978 (1957).
106. G. Fraenkel, R. E. Carter, A. McLachlan and J. H. Richards, *J. Am. chem. Soc.* **82**, 5846 (1960).
107. C. A. Fyfe, unpublished work.
108. C. E. Johnson and F. A. Bovey, *J. chem. Phys.* **29**, 1012 (1958).
109. J. Yarwood, *Chem. Commun.* 809 (1967).
110. A. Saika and C. P. Slichter, *J. chem. Phys.* **22**, 26 (1954).
111. C. A. Fyfe, unpublished work.
112. R. W. Taft and J. W. Carten, *J. Am. chem. Soc.* **86**, 4199 (1964).
113. D. F. R. Gilson and C. A. McDowell, *J. chem. Phys.* **39**, 1825 (1963).
114. H. G. Smith and R. E. Rundle, *J. Am. chem. Soc.* **80**, 5075 (1958).
115. D. F. R. Gilson and C. A. McDowell, *J. chem. Phys.* **40**, 2413 (1964).
116. H. G. Smith, *J. chem. Phys.* **40**, 2412 (1964).
117. D. F. R. Gilson and C. A. McDowell, *Can. J. Chem.* **44**, 945 (1966).
118. J. E. Anderson, *J. phys. Chem., Ithaca* **70**, 927 (1966).
119. D. C. Douglass, *J. chem. Phys.* **32**, 1882 (1960).
120. G. Briegleb and J. Czekalla, *Z. Elektrochem.* **59**, 184 (1955).
121. R. A. Bennett and H. O. Hooper, *J. chem. Phys.* **47**, 4855 (1967).

Chapter 6

Determination of the Position of Equilibrium—Methods

6.A. Introduction

Many studies have been made of the equilibria between charge-transfer complexes and their free component molecules in the liquid phase. The great majority of these systems involve a diluting solvent. The concentration of the various solute species in the solvent is generally assumed to be sufficiently low for the systems to be treated as dilute solutions. A variety of methods has been used in attempts to determine the position of equilibrium. The results should provide

5*

measures of the free energies of formation of the complexes, and, from their temperature variation, the corresponding enthalpy and entropy changes.

In most optical methods, the evaluation of the equilibrium constant is tantamount to a determination of the extinction coefficient of the complex species. Various methods used to evaluate equilibrium constants will be described. It has become apparent that a too simple interpretation of the experimental data can yield serious anomalies. Some of the assumptions which have been made, either explicitly or tacitly, are questioned in an attempt to resolve some of these problems.

Hitherto, the evaluation of association constants has mainly been made by graphical methods.[1, 2] In particular, linear relationships of functions are often used. Parameters for these lines are evaluated from experimental data. In many cases the best fit is obtained by a simple least-squares procedure. Recently, however, non-graphical trial-and-error computer methods involving some iteration process have become more common.[3-8] These latter techniques have so far been applied mainly to optical data (see p. 137). Obviously, the principle is applicable to the treatment of other physical data, for example, to infrared[9] and to n.m.r. data.[10]

Most of the examples of methods for evaluating association constants (Section 6.B) have been chosen from studies directly involving organic charge-transfer complexes. However, methods which have been used for other categories of complex but which could be applied to organic charge-transfer complexes are also described. For a more general text on the determination of the position of equilibria the reader is referred to the monograph by Rossotti and Rossotti.[11]

6.B. Methods of Evaluating Equilibrium Constants

6.B.1. General

Although the interaction of an electron donor with an electron acceptor may give rise to more than one species of complex in solution, most methods of evaluating the degree of association have been based, at least initially, on the assumption that a single complex species with a definite stoichiometry is formed. Unless otherwise stated, this assumption underlies the methods described below. For want of information, it is generally assumed that the activity coefficients for the species D, A and AD are all unity.* Thus for an equilibrium:

$$D + A \ \rightleftharpoons\ AD \tag{6.1}$$

---

* But see Section 6.C.2.

a thermodynamic equilibrium constant K_c^{AD} is equated to the quotient Q_c:

$$K_c^{AD} = Q_c = [AD]/[A][D] \qquad (6.2)$$

where [AD], [A] and [D] represent the equilibrium concentrations of AD, A and D respectively. As an alternative, the mole-fraction equilibrium constant K_X^{AD} may be used, where:

$$K_X^{AD} = X_{AD}/X_A X_D = [AD]/[A] X_D \qquad (6.3)$$

where X_{AD}, X_A and X_D are the mole-fractions of AD, A and D respectively at equilibrium. For solutions where the donor is in large excess over the acceptor, K_c^{AD} and K_X^{AD} are related by the expression:

$$K_X^{AD} = K_c^{AD} \left(\frac{1000 d_s}{M_s} - \frac{[D]_0 M_D d_s}{M_s d_D} + [D]_0 \right) \qquad (6.4)$$

where d is the density, M the molecular weight of the solvent (subscript s) or donor (subscript D). For dilute solutions in which K_c^{AD} is large compared with $(1 - M_D d_s/M_s d_D)$, equation (6.4) approximates to $K_X^{AD} \approx K_c^{AD} . 1000 d_s/M_s$. Equilibrium constants may also be expressed on a molal scale:

$$K_m^{AD} = M_{AD}/M_A M_D = [AD]/M_D[A] \qquad (6.5)$$

where M_{AD}, M_A and M_D are the molal concentrations of AD, A and D respectively. In systems where the donor is in large excess over the acceptor, K_m^{AD} and K_c^{AD} are related by the expression:

$$K_m^{AD} = K_c^{AD} \left(d_s - \frac{[D]_0 M_D d_s}{1000 d_D} \right) \qquad (6.6)$$

For dilute solutions in systems where K_m^{AD} is large compared with $M_D d_s/1000 d_D$, equation (6.6) approximates to $K_m^{AD} \approx K_c^{AD} . d_s$.* Some consequences of the effects of different concentration scales when the degree of association is small are discussed in Section 6.C.1.

There will be competition by the solvent for one or other or both components in the complex by acting as an electron donor or acceptor (or possibly both). In general, if the solvent (S) interacts with both A and D, the observed association constant K_c^{AD}(obs) will be less than the correct association constant K_c^{AD}(corr) = [AD]/[A][D] for the system, had there been no such participation by the solvent. If it is assumed that only free A molecules may complex with free D molecules, it can be shown that:

$$K_c^{AD}(corr) = K_c^{AD}(obs)(1 + K_c^{AS}[S])(1 + K_c^{DS}[S]) \qquad (6.7)$$

---

* Some published association constants are expressed in terms of kg solution per mol. These are represented as K_r^{AD}. For dilute solutions K_r^{AD} approximates to K_m^{AD}.

where $K_c^{AS} = [AS]/[A][S]$ is the association constant for the equilibrium :

$$A + S \rightleftharpoons AS \qquad (6.8)$$

and K_c^{DS} is the corresponding association constant for the species DS. For, say, a series of complexes of various donors with a common acceptor species, this has the effect of reducing the equilibrium constants of the complexes by a constant fraction. Such competition will be present to some extent in all experimental systems. Very few of the values of association constants which have appeared in the literature take into account such solvent competition.[12, 13] In the following sections the values of equilibrium constants, ostensibly for the solute equilibria, will likewise include any effect of solvent competition. Algebraical relationships will be expressed for K_c^{AD}, rather than for K_m^{AD} or K_x^{AD}.

6.B.2. Ultraviolet and Visible Spectrophotometry

Generally, solutions containing interacting D and A species show not only the absorptions of D and A, sometimes noticeably modified, but also a new band or bands which are assigned to intermolecular charge-transfer transition(s) of the complex as a whole (see Chapter 3). The intensity of this band is often used as a measure of the concentration of complex in a given solution : exceptionally, other absorptions may be used. For example, in iodine complexes, the molecular transition which occurs at about 520 nm in free iodine and is shifted to higher energies when complexed with a donor, has been employed, Table 6.1.[14, 15] Alternatively, the decrease in the absorption of one component as the concen-

TABLE 6.1. A comparison of association constants (K_X^{AD}) for some iodine-donor complexes obtained from measurements of the modified iodine transition in the complex in the visible region and from measurements of the charge-transfer transition in the ultraviolet (all at 25°C, n-heptane solutions).*

Donor	Visible		Ultraviolet	
	Wavelength/nm	K_X^{AD}	Wavelength/nm	K_X^{AD}
Trimethylene oxide	452	26·1	248	25·2
2-Methyltetrahydrofuran	454	20·2	252	23·5
Tetrahydrofuran	455	17·2	249	20·0
Tetrahydropyran	456	17·0	253	16·9
Propylene oxide	460	6·4	232	6·4†
Diethyl ether	462	5·9	252	6·6

* Calculated or quoted from the data given in Refs. 14 and 15.
† Measured at 22·2°C.

tration of the second component is increased may be used to estimate the association constant of the complex formed.

Measurements are usually made in regions where the component species absorb only weakly. Let us consider the case of an ideal system in which a 1:1 complex AD is formed:

$$K_c^{AD} = \frac{[AD]}{[D][A]} = \frac{[AD]}{([D]_0 - [AD])([A]_0 - [AD])} \tag{6.9}$$

where $[D]_0$ and $[A]_0$ indicate the total, i.e. free and complexed, concentrations of D and A respectively. These will correspond with the "weighed out" amounts of D and A if the solutions are made up from the component D and A species. Equation (6.9) may be rearranged:

$$\frac{[A]_0}{[AD]} = \frac{1}{K_c^{AD}} \cdot \frac{1}{[D]_0} + \frac{[A]_0}{[D]_0} + 1 - \frac{[AD]}{[D]_0} \tag{6.10}$$

For simplicity let us at first assume that the optical absorption in the region of measurement is due entirely to the complex AD. Then for a 1-cm path-length of solution,* if Beer's law is obeyed, the absorbance (optical density) A is:

$$A = \epsilon_\lambda^{AD} \cdot [AD] \tag{6.11}$$

where ϵ_λ^{AD} is the molar absorptivity (extinction coefficient) of AD at the wavelength of measurement, λ. From equations (6.10) and (6.11):[16]

$$\frac{[A]_0}{A} = \frac{1}{K_c^{AD} \epsilon_\lambda^{AD}} \cdot \frac{1}{[D]_0} + \frac{[A]_0}{[D]_0 \epsilon_\lambda^{AD}} + \frac{1}{\epsilon_\lambda^{AD}} - \frac{[AD]}{\epsilon_\lambda^{AD}[D]_0} \tag{6.12}$$

If $[D]_0 \gg [A]_0$ and $[D]_0$ and $[A]_0$ are sufficiently small, the second and fourth terms on the right-hand side in equation (6.12) may be neglected so that:

$$\frac{[A]_0}{A} = \frac{1}{K_c^{AD} \epsilon_\lambda^{AD}} \cdot \frac{1}{[D]_0} + \frac{1}{\epsilon_\lambda^{AD}} \tag{6.13}$$

Equation (6.13) is generally called the Benesi–Hildebrand equation.[17] Typical plots using this equation are shown in Fig. 6.1. For a series of solutions in which $[D]_0 \gg [A]_0$ a plot of $[A]_0/A$ against $[D]_0^{-1}$ should be linear. The intercept of the line with the ordinate is $(\epsilon_\lambda^{AD})^{-1}$ and the gradient is equal to $(\epsilon_\lambda^{AD} K_c^{AD})^{-1}$ so that K^{AD} may be evaluated. The linearity of this plot for values derived from experimental data is not, however, a good test for the absence of any termolecular complexes[18] (see Section 6.C.6).

Variants of equation (6.13) have been used by other workers. Scott[19]

* Unless otherwise stated, a path-length of 1 cm will be assumed in the following discussion. For other path-lengths (l cm), the term A should be replaced by A/l wherever it occurs.

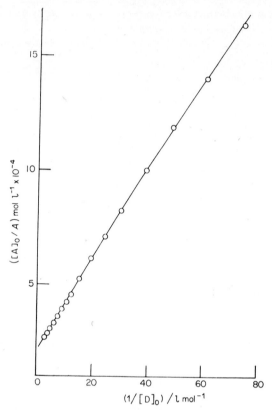

FIG. 6.1. Benesi–Hildebrand equation: plot of $[A]_0/A$ against $1/[D]_0$ for a series of 1,2-dichloroethane solutions containing hexamethylbenzene (D) and tetra-cyanoethylene (A); $[A]_0 = 1\cdot185 \times 10^{-4}$M, $[D]_0 = 0\cdot3135$ to $0\cdot01411$M, measured at 540 nm, 33·2°C. (Experimental data from Ref. 58.)

has argued that the Benesi–Hildebrand method requires an extra-polation to concentrated solutions. As an alternative he has suggested a rearrangement of equation (6.13):

$$\frac{[D]_0[A]_0}{A} = \frac{[D]_0}{\epsilon_\lambda^{AD}} + \frac{1}{K_c^{AD}\epsilon_\lambda^{AD}} \tag{6.14}$$

A plot of the left-hand term against $[D]_0$ should be linear. In this case the evaluation of ϵ_λ^{AD} and K_c^{AD} involves an extrapolation to dilute solution (Fig. 6.2).

Further rearrangement of equation (6.13) gives:

$$\frac{A}{[D]_0} = -K_c^{AD} A + K_c^{AD}[A]_0 \epsilon_\lambda^{AD} \tag{6.15}$$

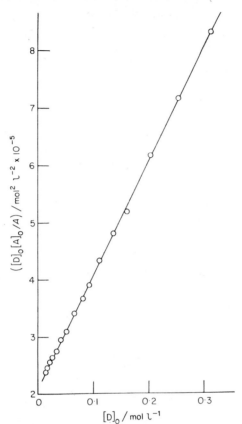

Fig. 6.2. Scott equation: plot of $[D]_0[A]_0/A$ against $[D]_0$, the same experimental data as in Fig. 6.1.

whence, for a series of solutions containing different concentrations of D but with constant concentration of A, still with the condition $[D]_0 \gg [A]_0$, a plot of $A/[D]_0$ against A should be linear.[20] From the gradient of the line, K_c^{AD} may be evaluated directly, without recourse to an extrapolated intercept. Extrapolation, but to an infinitely dilute solution, is required to obtain ϵ_λ^{AD} (Fig. 6.3).

Equation (6.15) may be modified by dividing throughout by $[A]_0$. Evaluation of K_c^{AD} is now obtained as the negative gradient of the plot of $A/[D]_0[A]_0$ against $A/[A]_0$. This may be applied to a set of solutions in which $[D]_0 \gg [A]_0$ without the restrictive condition that $[A]_0$ has to be constant for the solutions.

A dilution method has been described for evaluating K_c^{AD} from solutions where one component is in large excess.[21] It has the advantage that relatively small amounts of material are required. If m_{AD}, m_A and m_D are the mole masses of the solutions corresponding to the concentrations [AD], [A]$_0$ and [D]$_0$ respectively and if v is the volume of the

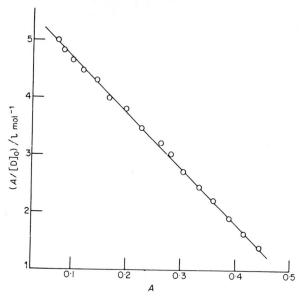

FIG. 6.3. Foster–Hammick–Wardley equation: plot of $A/[D]_0$ against A, the same experimental data as in Fig. 6.1.

solution in litres, then for solutions where $[D]_0 \gg [A]_0$, from equation (6.13):

$$v = \frac{K_c^{AD} \epsilon_\lambda^{AD} m_A m_D}{vA} - K_c^{AD} m_D \qquad (6.16)$$

A plot of v against $v^{-1}A^{-1}$ should be linear. From the intercept of the line with the ordinate, and from the gradient, K_c^{AD} and ϵ_λ^{AD} may be obtained.

The methods so far described are only applicable to systems where the optical absorbance A is due only to the complex. For systems where $[D]_0 \gg [A]_0$, any decrease in [D] through complex formation may usually be ignored, so that a sufficient compensation for any small absorption by D may be made either by an arithmetical correction or by having a solution of D of concentration $[D]_0$ in the reference beam of the spectrophotometer. If A has a significant absorption, then for such solutions a relationship such as that devised by Ketelaar and his co-workers may be used.[22] Let the absorbance of the solution be already compensated for any absorption due to D which is in large excess; if ϵ_λ^A is the molar absorptivity of A at the wavelength of measurement, then for a 1-cm path-length of solution the absorbance A' is:

$$A' = \epsilon_\lambda^{AD}[AD] + \epsilon_\lambda^A[A] = \epsilon_\lambda^{AD}[AD] + \epsilon_\lambda^A([A]_0 - [AD]) \qquad (6.17)$$

An apparent molar absorptivity of A at a wavelength λ may be defined as

$$\epsilon_\lambda^a = A'/[A]_0 \qquad (6.18)$$

If equations (6.17), (6.18) and (6.10) are combined, and the approximation is made that in equation (6.10) the terms $[A]_0/[D]_0$ and $[AD]/[D]_0$ are zero for solutions where $[D]_0 \gg [A]_0$, then:

$$\frac{1}{\epsilon_\lambda^a - \epsilon_\lambda^A} = \frac{1}{K_c^{AD}(\epsilon_\lambda^{AD} - \epsilon_\lambda^A)} \cdot \frac{1}{[D]_0} + \frac{1}{\epsilon_\lambda^{AD} - \epsilon_\lambda^A} \qquad (6.19)$$

This equation described by Ketelaar and his co-workers[22] has the same form as the Benesi–Hildebrand equation (6.13) and reduces to it when $\epsilon_\lambda^A = 0$. It may also be used in forms corresponding to equations (6.14) and (6.15), namely:

$$\frac{[D]_0}{(\epsilon_\lambda^a - \epsilon_\lambda^A)} = \frac{[D]_0}{(\epsilon_\lambda^{AD} - \epsilon_\lambda^A)} + \frac{1}{K_c^{AD}(\epsilon_\lambda^{AD} - \epsilon_\lambda^A)} \qquad (6.20)$$

$$\frac{(\epsilon_\lambda^a - \epsilon_\lambda^A)}{[D]_0} = -K_c^{AD}(\epsilon_\lambda^a - \epsilon_\lambda^A) + K_c^{AD}(\epsilon_\lambda^{AD} - \epsilon_\lambda^A) \qquad (6.21)$$

A method of evaluating K_c^{AD} which utilizes virtually the same basic relationships as the Ketelaar method has been described by Nash.[1]

An alternative approach has been made in the determination of K_c^{AD} for systems where the components absorb strongly in the region of the charge-transfer band.[23, 24] For example, if two donor species D and D' are mixed with the acceptor A such that $[D]_0 = [D']_0 \gg [A]_0$, then, with an obvious notation:

$$K_c^{AD} = \frac{[AD]}{[D]_0[[A]_0 - [AD] - [AD']]} \qquad (6.22)$$

$$K_c^{AD'} = \frac{[AD']}{[D]_0([A]_0 - [AD] - [AD'])} \qquad (6.23)$$

whence

$$[AD'] = \frac{K_c^{AD'}[D]_0[A]_0}{1 + K_c^{AD}[D]_0 + K_c^{AD'}[D]_0} \qquad (6.24)$$

In the simple case where A, D, D' and AD have negligible absorption at a wavelength where the molar absorptivity of AD' is $\epsilon_\lambda^{AD'}$, the measured absorbance for a 1-cm path-length will be:

$$A = \epsilon_\lambda^{AD'}[AD'] \qquad (6.25)$$

If the absorbance of a solution containing A and D' of concentrations $[A]_0$ and $[D]_0$, but containing none of the species D is A_1 at this wavelength, and if the absorbance of a solution containing A, D and D' at concentrations $[A]_0$, $[D]_0$ and $[D]_0$ respectively is A_2 at the same wavelength, then by a combination of equations (6.13), (6.24) and (6.25):

$$K_c^{AD'} = \frac{A_1 - A_2}{A_2}\left(K_c^{AD} + \frac{1}{[D]_0}\right) \qquad (6.26)$$

whence, from a knowledge of $K_c^{AD'}$, K_c^{AD} may be evaluated.[23, 24]

de Maine[25] has suggested an algebraic method for the analysis of spectrophotometric data of systems in which the components A and D as well as AD absorb at the wavelength of measurement. The separate determination of the molar absorptivities of the two components each measured singly in an inert solvent is not required. The method avoids an assumption which is normally made though often not justified,[26] namely that the molar absorptivity of each component is unchanged when the second component is added. The procedure is rather complicated and appears to have been little used and is now superseded by trial-and-error computer methods (see below).

All the methods described thus far have depended on the experimental condition that one of the two component species is present in large excess, so that its concentration is virtually unaltered on complex formation. Often this is experimentally convenient, since many acceptors are only slightly soluble in the solvents generally used. Furthermore, under these conditions a larger fraction of one component is in the complexed state than if near-equimolar quantities of A and D are taken. Person[27] has shown that this can be a very important factor in the experimental evaluation of K^{AD}. However, if one component is in large excess, it normally means that the total solute concentration is high. Therefore there may be occasions when it is desirable to use solutions which are equimolar or near-equimolar with respect to A and D. If the experimental condition $[A]_0 = [D]_0$ is applied, then in the simple case in which only the complex absorbs, from equation (6.12):

$$\frac{[A]_0}{A} = \frac{1}{K_c^{AD}\epsilon_\lambda^{AD}} \cdot \frac{1}{[A]_0} + \frac{2}{\epsilon_\lambda^{AD}} - \frac{[AD]}{\epsilon_\lambda^{AD}[D]_0} \qquad (6.27)$$

It may be possible to neglect the final term in equation (6.27). In such cases the relationship assumes the form of the Benesi–Hildebrand equation [28, 29] [which may be expressed in an alternative form equivalent to equation (6.14)]. In cases where this approximation is not justified, an estimate of the final term may be obtained by an iterative procedure.

A linear relationship which may be used for any constant ratio of component concentrations including $[A]_0 = [D]_0$ may be obtained by expressing the total concentration of one component in terms of the second, say $[D]_0 = n[A]_0$. For a system where at the wavelength of measurement (λ) only the complex absorbs, from equation (6.12) it may be shown that: [30]

$$\frac{[A]_0}{A} = \frac{(n+1)}{n} \cdot \frac{1}{\epsilon_\lambda^{AD}} + \frac{1}{nK_c^{AD}\,\epsilon_\lambda^{AD}} \cdot \frac{1}{[A]_0} \qquad (6.28)$$

if $[AD] \ll [A]_0$, whence, a plot of $[A]_0/A$ against $1/[A]_0$ should be a straight line, from the intercept and gradient of which K_c^{AD} and ϵ_λ^{AD} may be evaluated.

In the more general case where D and A both have significant aborptions at the wavelength of measurement and for which no special conditions are made regarding the values of $[D]_0$ and $[A]_0$, the absorbance of a given solution at a wavelength λ will be:

$$A = \epsilon_\lambda^A[A] + \epsilon_\lambda^D[D] + \epsilon_\lambda^{AD}[AD] \qquad (6.29)$$

where ϵ_λ^A, ϵ_λ^D are the molar absorptivities of A and D at a wavelength λ. If we write: $A_A = \epsilon_\lambda^A[A]_0$, and $A_D = \epsilon_\lambda^D[D]_0$ then:

$$A = [AD](\epsilon_\lambda^{AD} - \epsilon_\lambda^A - \epsilon_\lambda^D) + A_A + A_D \qquad (6.30)$$

whence from equations (6.10) and (6.30):

$$\frac{1}{K_c^{AD}} = \frac{(A - A_A - A_D)}{(\epsilon_\lambda^{AD} - \epsilon_\lambda^A - \epsilon_\lambda^D)} - [A]_0 - [D]_0 + \frac{[A]_0[D]_0(\epsilon_\lambda^{AD} - \epsilon_\lambda^A - \epsilon_\lambda^D)}{(A - A_A - A_D)} \qquad (3.31)$$

This is the Rose-Drago equation. [31] Under conditions where:

$$\frac{A - A_A - A_D}{\epsilon_\lambda^{AD} - \epsilon_\lambda^A - \epsilon_\lambda^D} - [A]_0 \gg \frac{[A]_0[D]_0(\epsilon_\lambda^{AD} - \epsilon_\lambda^A - \epsilon_\lambda^D)}{A - A_A - A_D} - [D]_0$$

equation (6.31) reduces to the Ketelaar equation (6.19), and to the Benesi–Hildebrand equation (6.13) when $\epsilon_\lambda^A = \epsilon_\lambda^D = 0$. Rose and Drago [31] suggest that, to obtain ϵ_λ^{AD} and K_c^{AD} from an application of experimental data to equation (6.31), values of ϵ_λ^{AD} should be selected at random, and, for one set of experimental data, the corresponding values of $(K_c^{AD})^{-1}$

calculated. These are then plotted using Cartesian coordinates, and a curve is constructed. Further curves are likewise drawn from other sets of experimental data. Ideally, all these curves should intersect at a single point which defines the correct values of K_c^{AD} and ϵ_λ^{AD}. Lines from experimental data in fact give intersections over a finite area. The size of the area is an inverse measure of the accuracy of the determination (Fig. 6.4).

A simple pairwise evaluation of K_c^{AD} and ϵ_{AD}^λ from similar data has been devised by Nagakura.[32] If A_0, A' and A" are the measured absorbances of solutions whose total concentration of one component, say A, is

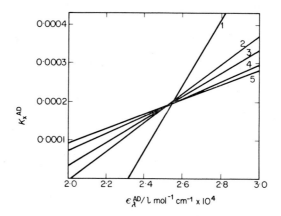

[R. S. Drago and N. J. Rose, Ref. 31]

FIG. 6.4. Rose–Drago equation: plot of hypothetical values of ϵ_λ^{AD} against the corresponding value of K_X^{AD} calculated using equation (6.31) for a series of iodine-triethylamine solutions in n-heptane. (Experimental data from S. Nagakura, *J. Am. chem. Soc.* **80**, 520 (1958).)

constant and equal to $[A]_0$, and whose total concentrations of donor are zero, $[D]_0'$ and $[D]_0''$ respectively, and for these two latter solutions the concentrations of the complex at equilibrium are $[AD]'$ and $[AD]''$, then:

$$K_c^{AD} = \frac{[AD]'}{([D]_0' - [AD]')([A]_0 - [AD]')} = \frac{[AD]''}{([D]_0'' - [AD]'')([A]_0 - [AD]'')}$$

(6.32)

By the elimination of $[A]_0$ from equation (6.32) and by the assumption that any terms which contain [AD] to any power greater than unity may be ignored, it follows that:

$$K_c^{AD} = \frac{[AD]'[D]_0'' - [AD]''[D]_0'}{([AD]'' - [AD]')[D]_0'[D]_0''}$$

(6.33)

In the simple case where the optical absorption measured is due to the species AD and A only, then:

$$A' = [AD]'(\epsilon_\lambda^{AD} - \epsilon_\lambda^A) + A_0 \tag{6.34}$$

$$A'' = [AD]''(\epsilon_\lambda^{AD} - \epsilon_\lambda^A) + A_0 \tag{6.35}$$

by combining equations (6.33), (6.34) and (6.35):

$$K_c^{AD} = \frac{[D]_0'(A_0 - A'') + [D]_0''(A' - A_0)}{[D]_0[D]_0''(A'' - A')} \tag{6.36}$$

Although optical measurements are often made only at the wavelength at which the intensity of the intermolecular charge-transfer band is a maximum, it is obviously desirable that measurements are made over a range of wavelengths. Liptay[33] has suggested a general method for treating such results. If A_{ik} is the absorbance, due to complex formation, of a given solution k at a frequency i, then all values of A_{ik} may be arranged in the form of a matrix:

$$A_{ik} = \begin{vmatrix} A_{11} & A_{12} & \cdots & A_{1q} \\ A_{21} & A_{22} & \cdots & A_{2q} \\ A_{p1} & A_{p2} & \cdots & A_{pq} \end{vmatrix} \tag{6.37}$$

The matrix is normalized by making $A_{mk} = 1$ for all solutions at some frequency m, usually the frequency of maximum absorption. A new matrix is obtained by dividing all the absorbances for a given solution by the absorbance of that solution at the particular frequency m.

$$\zeta_{ik} = \frac{A_{ik}}{A_{mk}} \tag{6.38}$$

whence:

$$\zeta_{ik} = \begin{vmatrix} \zeta_{11} & \zeta_{12} & \zeta_{13} & \cdots & \zeta_{1q} \\ \zeta_{21} & \zeta_{22} & \zeta_{23} & \cdots & \zeta_{2q} \\ \zeta_{p1} & \zeta_{p2} & & \cdots & \zeta_{pq} \end{vmatrix} \tag{6.39}$$

If only one complex is present and if the shape of the absorption band is the same for all solutions, then $\zeta_{11} = \zeta_{12} = \zeta_{13} = \ldots \zeta_{iq}$ at all frequencies.[33] The relationship would also hold if there were two or more isomeric complexes, or two complexes having different compositions were present, provided that the values of ζ_j for both complexes were proportional to one another or $\zeta_j = 0$ for one or other complex.[34]

As opposed to the various graphical methods, trial-and-error procedures[3–9] using iterative techniques have recently been proposed to enable the evaluation of K^{AD} and ϵ^{AD}. Such methods have various advantages. The approximations which generally have to be made in graphical procedures in order to obtain a linear relationship are no longer necessary. Secondly, allowance can more readily be made for extra complicating factors such as the presence of termolecular complexes, or absorption by component species. Thirdly, extreme concentration conditions, such as $[D]_0 \gg [A]_0$, do not have to be chosen. Fourthly, estimates of errors can be included in certain of the computations.

de Maine [3] has considered the case of a 1 : 1 reaction, to which equation (6.12) may be applied. If A' is the total absorbance due to the complex, A, and D (which species have molar absorptivities ϵ_λ^{AD}, ϵ_λ^A and ϵ_λ^D respectively at the wavelength of measurement) then the absorbance (A) due to AD is such that:

$$A = \epsilon_\lambda^{AD}[AD] = A' - \epsilon_\lambda^A([A]_0 - [AD]) - \epsilon_\lambda^D([D]_0 - [AD]) \quad (6.40)$$

Equation 6.9 is solved for [AD] using an arbitrary value K_i^{AD}. A is then calculated from equation 6.40 and the value obtained is introduced into a rearranged form of equation (6.12) in order to obtain a new value for K_c^{AD}. The process is repeated to obtain a convergent value for K_c^{AD}.

Conrow et al., [4] in their procedure, calculate [AD] from an initial guess of K_c^{AD}. This is then used to find the best fit for ϵ_λ^{AD} in equation (6.40). The fit for K_c^{AD} and ϵ_λ^{AD} is measured by calculating the absorbance at each point and accumulating the sum of the squares of the deviations of the calculated from the experimental values. An arbitrary change in the value of K_c^{AD} is then made and the whole process repeated. If the fit is improved, further changes are made in the same direction. When poorer results are obtained then a change in the value of K_c^{AD} in the opposite direction is made. Thereafter, the size of the change is decreased and the process is repeated. The calculation is terminated when the change becomes arbitrarily small. On the basis of calculations from synthetic data, it is concluded that the final values of K_c^{AD} and ϵ_λ^{AD} are more sensitive to errors in concentrations than to errors in optical absorbance data. As a measure of the reliability of the final results, values of K_c^{AD} 10% on either side of the best value of K_c^{AD} were calculated. The ratio of the percentage change in fit to $10\times$ (the % change in K_c^{AD}), termed the *sharpness*, is a measure of the reliability of K_c^{AD}. The larger the sharpness the more reliable the value of K_c^{AD}. [4, 9]

More recently, non-linear methods have been employed in order to evaluate K_c^{AD} and ϵ_λ^{AD}. [5, 6] The general least-squares adjustment described by Deming, [35] has been used. For details the reader is referred to Deming's book and a paper by Wentworth et al. [5] In outline (after

Rossiensky and Kellawi[6]), values of K_c^{AD} and ϵ_λ^{AD} are required which fulfil the conditions:*

$$\frac{\partial}{\partial K} \sum (A_c - A_m)^2 = 0 \qquad (6.41)$$

$$\frac{\partial}{\partial \epsilon} \sum (A_c - A_m)^2 = 0 \qquad (6.42)$$

where A_c are the calculated values of absorbance and A_m are the measured values of absorbance for the individual experimental determinations involving particular values of $[A]_0$ and $[D]_0$.

Let the initial guesses K_1 and ϵ_1 yield a value of absorbance A_{c1}. The value of A_c to be substituted in equations (6.41) and (6.42) is expressed in terms of a truncated Taylor series:

$$A_c = A_{c1} + \Delta K_1 (\partial A / \partial K)_1 + \Delta \epsilon_1 (\partial A / \partial \epsilon)_1 \qquad (6.43)$$

where ΔK_1 and $\Delta \epsilon_1$ are the corrections to K_1 and ϵ_1 respectively. These correction terms may be calculated from the "normal" equations:

$$\sum \{A_{c1} + \Delta K_1 (\partial A / \partial K)_1 + \Delta \epsilon_1 (\partial A / \partial \epsilon)_1 - A_m\}(\partial A / \partial K)_1 = 0 \quad (6.44)$$

$$\sum \{A_{c1} + \Delta K_1 (\partial A / \partial K)_1 + \Delta \epsilon_1 (\partial A / \partial \epsilon)_1 - A_m\}(\partial A / \partial \epsilon)_1 = 0 \quad (6.45)$$

where the differential coefficients with subscript 1 mean the values calculated using the values K_1 and ϵ_1. Iteration of the calculation is continued until $\Delta \epsilon$ and ΔK reach some arbitrary limit of smallness.

6.B.3. Infrared Determinations

In principle, some of the analytical procedures for electronic absorptions described in the previous section may be applied to infrared absorptions of solutions containing electron donors and electron acceptors in equilibrium with the complex species.[9, 36-42] Since the intermolecular stretching frequency is not normally accessible (see Chapter 4), measurements are made on a band which occurs in the spectrum of one or other component and in the spectrum of the complex. There must be a sufficient difference in the intensity of absorption by these two species at some wavelength. Many of the electron donor–acceptor systems studied have also involved localized intermolecular hydrogen bonding and are therefore atypical. Several determinations have been made from measurements of the I–I stretching frequency in iodine complexes.[41, 42] If the extinction coefficient for a transition in, say,

---

* The normal sub- and superscripts to K and ϵ are omitted for clarity. In the immediate discussion they refer to the species AD and are represented elsewhere by K_c^{AD} (for molar concentration) and ϵ_λ^{AD}.

the donor is ϵ_λ^D and the extinction coefficient at the same wavelength (λ) in the complex is ϵ_λ^{AD}, then for a path-length l cm of solution:

$$\frac{A}{l} = \epsilon_\lambda^D([D]_0 - [AD]) + \epsilon_\lambda^{AD}[AD] \tag{6.46}$$

where A is the absorbance for a l-cm path-length of solution. As an alternative to equation (6.46) it may be necessary to use an expression which involves the integrated band intensities, rather than peak intensities. The equilibrium constant K_c^{AD} for such a system may be evaluated from measurements on A for a series of solutions in which $[A] \gg [D]$ by employing a relationship of the type described by Ketelaar et al.[22] [see equation (6.19), p. 133], viz.:[36]

$$\frac{[A]_0[D]_0}{(A - A_0)} = \frac{1}{l(\epsilon_\lambda^{AD} - \epsilon_\lambda^D) K_c^{AD}} + \frac{[A]_0}{l(\epsilon_\lambda^{AD} - \epsilon_\lambda^D)} \tag{6.47}$$

where A_0 is the absorbance of a solution of path-length l containing D at a concentration $[D]_0$ in the absence of A.

An alternative method for evaluating K_c^{AD} is by trial. Values of $[AD]$ for various values of K_c^{AD} are calculated from the expression:[40]

$$[AD] = \frac{[D]_0 + [A]_0 + 1/K_c^{AD}}{2} - ([D]_0 + [A]_0 + 1/K_c^{AD})^2 - [D]_0[A]_0 \ldots \tag{6.48}$$

From equation (6.46) a plot of A/l against the estimated concentration $[AD]$ will be linear only when the correct value of K_c^{AD} has been used in equation (6.48).

Estimates of association constant may also be obtained from intensities of Raman lines. Hitherto, the method appears to have been applied only to complexes of inorganic Lewis acids.[43, 44]

6.B.4. Nuclear Magnetic Resonance [45]

When the molecular environment of a nucleus undergoes rapid reversible change, the position of the magnetic resonance absorption of the nucleus will represent a time-averaged resultant of its behaviour in the different environments (see Chapter 5). The measured chemical shift of such a nucleus is:

$$\delta = \sum P_i \delta_i \tag{6.49}$$

where P_i is the population of the nucleus in the ith environment relative to the total population of the nucleus (i.e. $\sum P_i = 1$), and δ_i is the chemical shift the nucleus would have if it were purely in the ith environment.

In the simple case of an equilibrium between a 1 : 1 complex AD and its components D and A, in which the forward and back reactions are very fast, the measured chemical shift of a nucleus in, say, the acceptor moiety is, from equation (6.49):

$$\delta = P_A \delta_A + P_{AD} \delta_{AD} \qquad (6.50)$$

where P_A is the fraction of acceptor molecules uncomplexed, and P_{AD} is the fraction of complexed acceptor molecules. Since $(P_A + P_{AD}) = 1$, equation (6.50) may be written:

$$\delta = P_{AD}(\delta_{AD} - \delta_A) + \delta_A \qquad (6.51)$$

Association constants of various types of complex have been evaluated using equation (6.51). A plot of P_{AD} against δ should be linear. This requires the insertion of K_c^{AD} as a variable parameter which is adjusted until the values of P_{AD} obtained from it vary linearly with the corresponding values of δ.[46] The method is suitable for solution by computer methods, programs for which are available.[10, 47]

Alternatively, by combination of equations (6.9) and (6.51), it may be shown that for an observed signal in component A, a plot of $[D]_0/(\delta_A - \delta)$ against $([A]_0 + [D]_0 + [AD])$ should be linear. If in fact the plot is made against $([A]_0 + [D]_0)$, the gradient of the plot $\sim 1/(\delta_{AD} - \delta_A)$ can be used to obtain an approximate value of $[AD]$, and hence of K_c^{AD}. The method has been applied to hydrogen-bonded complexes.[48]

In the case of systems with large association constants, the value of $(\delta_{AD} - \delta_A)$ may, to a good approximation, be obtained by adding successive quantities of the second component to the system until further additions produce no increase in the observed shift of the measured nucleus in the first component.[49] Such a procedure has been used in the case of iodine–alkyl sulphide complexes.[49] The method, when applicable, has the advantage that a large excess of one or other component is not required. In practice many workers prefer to use computationally simpler methods.

Under certain conditions, K_c^{AD} may be evaluated by a simple graphical method. Let $(\delta_A - \delta_{AD})$, the shift for the pure complex relative to the shift for the pure acceptor in solution, be represented by Δ_0^{AD}. Also let $(\delta_A - \delta)$, the observed shift for the system in equilibrium relative to the shift for the pure acceptor in solution, be represented by Δ. These terms are depicted in Fig. 6.5. From equation (6.51):

$$P_{AD} = \frac{\delta_A - \delta}{\delta_A - \delta_{AD}} = \frac{[AD]}{[A]_0} = \frac{\Delta}{\Delta_0^{AD}} \qquad (6.52)$$

For solution showing ideal behaviour and where $[D]_0 \gg [A]_0$:

$$K_c^{AD} = \frac{[AD]}{[D]_0([A]_0 - [AD])} \qquad (6.53)$$

From equations (6.52) and (6.53) the following relationships may be obtained:[50-54]

$$\frac{1}{\Delta} = \frac{1}{K_c^{AD}} \cdot \frac{1}{\Delta_0^{AD}} \cdot \frac{1}{[D]_0} + \frac{1}{\Delta_0^{AD}} \qquad (6.54)$$

This is the analogue of the Benesi–Hildebrand equation for optical absorption. For a series of solutions in which $[D]_0 \gg [A]_0$, a plot of $1/\Delta$ against $1/[D]_0$ should be linear. The intercept with the ordinate yields the product $K_c^{AD} \Delta_0^{AD}$ and from the gradient Δ_0^{AD} may be obtained. This type of evaluation has some disadvantages in common with the

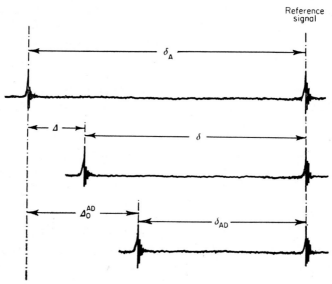

FIG. 6.5. The relationship between Δ, Δ_0^{AD}, δ, δ_A and δ_{AD}, cf. equation (6.52).

analogous optical relationship, namely, that to evaluate K_c^{AD}, extrapolation to concentrated solutions is required, an intercept has to be evaluated, and the value of the intercept is relatively small. The method, originally used to study hydrogen-bonded complexes,[50-53] has been applied to charge-transfer complex equilibria by Hanna et al.[54,55]

As an alternative to equation (6.54), the equation may be rearranged in the form:[56]

$$\Delta/[D]_0 = -\Delta K_c^{AD} + \Delta_0^{AD} K_c^{AD} \qquad (6.55)$$

For a series of solutions in which $[D]_0 \gg [A]_0$, a plot of $\Delta/[D]_0$ against Δ should be linear, with the gradient of the line equal to $-K_c^{AD}$. The value of

Δ_0^{AD} may be obtained from the intercept with the ordinate. This is the n.m.r. analogue of the optical relationship expressed by equation (6.15), and shares with it the advantage that K_c^{AD} may be obtained directly from the gradient of the line without recourse to an extrapolation to an infinitely concentrated solution as is required with the Benesi–Hildebrand equation (6.13) and with equation (6.54).

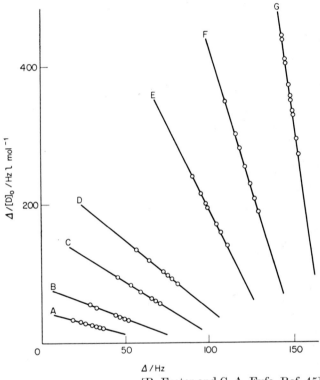

[R. Foster and C. A. Fyfe, Ref. 45]

FIG. 6.6. Plots of $\Delta/[D]_0$ against Δ for the complexes of fluoranil with (A) benzene; (B) toluene; (C) p-xylene; (D) mesitylene; (E) durene; (F) pentamethylbenzene; and (G) hexamethylbenzene in carbon tetrachloride at 33·5°C. (From data in Ref. 63.)

In order to apply either equation (6.54) or equation (6.55), the concentration of one component of the complex must be small compared with the second component. The lower limit for the first component is a concentration sufficient for the absorption signal to be detectable above the noise. It is therefore desirable, though not essential, that the observed signal shall be a singlet. In this respect 1,3,5-trinitrobenzene and 1,4-dinitrobenzene are convenient acceptor molecules. For solubility reasons it is generally more suitable to use the condition $[D]_0 \gg [A]_0$ rather than the converse. Plots of $\Delta/[D]_0$ against Δ by the application of equation (6.55) show remarkably small deviations from linearity. Some examples are given in Fig. 6.6.

For a given system, identical values of K_c^{AD} have been obtained using either the condition $[D]_0 \gg [A]_0$, and measuring the shift of a nucleus in the acceptor moiety; or $[D]_0 \ll [A]_0$ and measuring the shift of a nucleus in the donor moiety.[57, 58] In choosing suitable systems it is desirable to select donors and acceptors both of which have symmetrical molecules, in order to take advantage of the extra sensitivity resulting from singlet

TABLE 6.2. Values of association constant (K_r^{AD}) from p.m.r. chemical shift measurements determined under the conditions $[D]_0 \gg [A]_0$ and $[D]_0 \ll [A]_0$ (all at 33·5°C).

System	Rel. conc.	Proton measured	$K_r^{AD}/$ kg mol^{-1}	Ref.*
1,3,5-Trinitrobenzene–mesitylene in 1,2-dichloroethane	$[A]_0 \gg [D]_0$ $[A]_0 \ll [D]_0$	$H_{2,4,6}$ in D $H_{2,4,6}$ in A	$0·2_6$ $0·2_8$	a
1,3,5-Trinitrobenzene–durene in 1,2-dichloroethane	$[A]_0 \gg [D]_0$ $[A]_0 \ll [D]_0$	$H_{3,6}$ in D $H_{2,4,6}$ in A	$0·4_0$ $0·3_3$	a
1,3,5-Trinitrobenzene–indole in 1,2-dichloroethane	$[A]_0 \gg [D]_0$ $[A]_0 \gg [D]_0$ $[A]_0 \ll [D]_0$	H_3 in D H_{arom} in D† $H_{2,4,6}$ in A	$0·9_3$ $1·0_0$ $0·8_5$	57

* Ref.: ^{a} R. Foster and C. A. Fyfe, unpublished work.
† Band of maximum intensity in the aromatic multiplet.

absorptions. Suitable systems are 1,3,5-trinitrobenzene–mesitylene in 1,2-dichloroethane and 1,3,5-trinitrobenzene–durene in 1,2-dichloroethane. Because of the lower solubility of 1,3,5-trinitrobenzene, it is not possible to obtain such a wide concentration range when the acceptor is in excess compared with determinations where the donor is in excess. Values of K_c^{AD} obtained from such plots are compared in Table 6.2 with the corresponding values of K_c^{AD} obtained under the condition $[D]_0 \gg [A]_0$. The agreement provides a reasonable vindication of the method. No correlation is to be expected between the two Δ_0^{AD} values for a given complex, of course, since these are measures of shifts of different nuclei in the complex.

Further support for the method has been obtained from measurements on the interaction of 2,4-difluoro-1,3,5-trinitrobenzene with hexamethylbenzene ($[D]_0 \gg [A]$).[59] Independent ^1H and ^{19}F resonance measurements yield the same value of K_c^{AD} for the complex. Similar results have been obtained for other systems (Table 6.3). However, recently some anomalous observations have been made in which different values of K^{AD} and $\Delta H^\ominus$ are obtained from measurements of non-equivalent nuclei[60–62] (Table 6.3) (see Sections 6.C.2 and 6.C.3).

TABLE 6.3. Values of association constant (K_r^{AD}) from n.m.r. chemical shift measurements of non-equivalent nuclei in the donor or acceptor component (all at $33.5°C$).

System	Component containing measured nuclei	Nucleus*	$K_r^{AD}/$ kg mol^{-1}	Ref.
1,3-Difluoro-2,4,6-trinitrobenzene–hexamethylbenzene in CHCl$_3$	A	^1H ^{19}F	0.4_7 0.4_5	59
1,3,5-Trinitrobenzene–indole in CH$_2$ClCH$_2$Cl	D	$^1H_3$ $^1H_{arom}$†	0.9_3 1.0_0	57
3,5-Dinitrobenzonitrile	A	$^1H_{1,6}$ $^1H_4$	3.5_6 3.3_1	62
3,5-Dinitrobenzotrifluoride–hexamethylbenzene in CCl$_4$	A	$^1H_{1,6}$ $^1H_4$ ^{19}F	1.73 1.76 1.19	62
4-Nitrobenzaldehyde–hexamethyl-benzene in CCl$_4$	A	$^1H_{3,5}$ $^1H_{2,6}$ $^1H_{CHO}$	0.5_8 0.7_3 0.9_3	61

* Subscript refers to position of nucleus in the molecule.
† Band of maximum intensity in the aromatic multiplet.

The use of ^{19}F resonance in acceptors such as fluoranil, as opposed to ^1H resonance in the more usual proton-containing acceptors, has certain advantages.[59, 63] There is no interference from proton absorption by the donor or the solvent. Secondly, the ^{19}F chemical shift differences tend to be large, so that the accuracy of the method is increased.

It is possible to extend the application of equation (6.55) to systems containing several acceptors A′, A″, A‴ ... simultaneously in the presence of a single donor and so to evaluate the equilibrium constants corresponding to the various complexes A′D, A″D, A‴D ... from the data from a single set of solutions.[64] Apart from the justified assumption that the reactions shall be very fast, the only condition which has to be fulfilled is that $\sum [A_i]_0 \ll [D]_0$. The practical problem of overlapping bands may be minimized by a suitable choice of acceptors combined

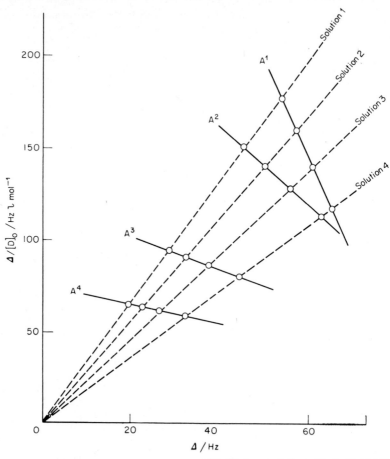

[R. Foster and C. A. Fyfe, Ref. 45]

FIG. 6.7. Plots of $\Delta/[D]_0$ against Δ for the complexes of: (A^1) 1,3,5-trinitrobenzene; (A^2) 2,5-dichloro-p-benzoquinone; (A^3) 1,4-dinitrobenzene; (A^4) p-benzoquinone with hexamethylbenzene in carbon tetrachloride at 33·5°C. The measurements were made on solutions containing all the acceptors together. (From data in Ref. 64.)

with the narrowness of the n.m.r. absorption bands. The method provides a near-ideal comparison of the complexing ability of different acceptors since the bulk donor–solvent environment is identical for all the acceptors used. Plots of $\Delta/[D]_0$ against Δ, utilizing equation (6.55), for a series of hexamethylbenzene solutions containing four different acceptors,[64] are shown in Fig. 6.7. The values of association constant obtained from these plots agree well with the values obtained from determinations involving each acceptor separately, or in pairs (Table 6.4).

TABLE 6.4. Values of association constant (K_r^{AD}) and the chemical shift* for the pure complex relative to the acceptor (Δ_0^{AD}) for complexes of hexamethylbenzene with: (A1) 1,3,5-trinitrobenzene; (A2) 2,5-dichloro-p-benzoquinone; (A3) 1,4-dinitrobenzene; (A4) p-benzoquinone in carbon tetrachloride at 33·5°C, measured simultaneously, in pairs and separately.†

Acceptor	A1		A2		A3		A4	
Combination	K_r^{AD}/kg mol^{-1}	Δ_0^{AD}/Hz	K_r^{AD}/kg mol^{-1}	Δ_0^{AD}/Hz	K_r^{AD}/kg mol^{-1}	Δ_0^{AD}/Hz	K_r^{AD}/kg mol^{-1}	Δ_0^{AD}/Hz
Pairs A1, A3; A2, A4 A1A2, A3, A4	5·2	65	2·1	87	0·9	96	0·7	81
together	4·8	66	2·1	86	0·9	97	0·6	90
Each singly	5·1	65	1·9	91	1·0	91	0·7	89

* Measured at 60·004 MHz.
† Ref. 64.

6.B.5. General Distribution Methods

Estimates of the stabilities of complexes have been made by measuring the change in the partition of one component of the complex between two immiscible liquids when the second component, soluble in only one of the two phases, is added to the system.[65] There is, however, usually a very significant salting-out effect.[66, 67] Thus, the addition of a potentially interacting hydrocarbon to picric acid, distributed between chloroform and water, has a two-fold effect. The concentration of the picric acid in the chloroform phase is increased through the formation of a complex with the hydrocarbon, but decreased because of the depression of the solubility through the presence of more solute. Estimates of the salting-out process have been made by a study of the effect of the addition of substances which are not expected to complex with the picric acid. Unfortunately, the consequent corrections which have to be applied are very large, and the final results may not be very meaningful. Details of the method have been described by Hammick and his co-workers.[66, 67]

An interesting variant of this method is the measurement of association constant for complex formation between a volatile and non-volatile reactant in a volatile solvent by measuring the partitioning between the liquid and the vapour phase.[68] The experiment involves the determination of the ratio of the two volatile components in the vapour over the ternary mixture, which is compared with that over the binary mixture in which the non-volatile component is omitted. The method appears to have been applied only to hydrogen-bonded systems.

Stabilities of complexes have been determined by measurements of solubility.* For example, complex formation between Ag^+ ion acceptor with various hydrocarbon donors has been studied by measuring the solubility of the donor in aqueous silver nitrate solutions of constant ionic strength.[69-71] The apparent association constant for such a complex is equal to the quotient $[AD]/([D]_0 - [AD])([A]_0 - [AD])$ [equation (6.9)] if it is assumed that the solutions are ideal and that only a 1:1 complex occurs. The total concentration (free and complexed) of the donor is determined, usually by extraction. This enables $[D]_0$ in equation (6.49) to be estimated. The concentration of uncomplexed donor in solution $([D]_0 - [AD])$ is obtained by assuming that it is equal to the concentration of donor in the solvent in the absence of the complexing agent. In order to minimize activity effects, a solution of potassium nitrate of constant ionic strength is used; for example, in determinations involving silver nitrate. The values of K_c^{AD}(app.) for many Ag^+ ion complexes are dependent on the concentrations of silver nitrate used. It has been suggested that this is due to the formation of significant quantities of the termolecular complex $(Ag^+)_2 D$.

When solid or liquid solutes are used as the saturating substances, it is not possible to vary their concentration for a given concentration of the second component at a given temperature. It is therefore not possible to determine the number of molecules of the first component in the complex species. However, estimates of association constant have been made on systems where the saturating material is a gas.[72, 73] For example, Henry's law constant has been measured for hydrogen chloride and hydrogen bromide in various "inert" solvents such as heptane and in solutions of various donors. The measurements were made at $-78 \cdot 5°C$ to avoid significant partial pressure corrections due to the donor and solvent. In such systems the total concentration of both donor and acceptor may be varied.

A recent development of this principle has been to use gas-liquid (or

---

* The relative stabilities of some *solid* complexes have been determined by measuring the concentration of acceptor in an aqueous solution saturated with respect to the complex and to the donor. [R. Behrend, *Z. phys. Chem.* **15**, 183 (1894); J. H. Brönsted, *Z. phys. Chem.* **78**, 184 (1911); O. Dimroth and C. Bamberger, *Justus Liebigs Annln. Chem.* **438**, 67 (1924); D. Ll. Hammick and H. P. Hutchison, *J. chem. Soc.* 89 (1955)]. The results for naphthalene picrate agree closely with those obtained from e.m.f. measurements [R. P. Bell and J. A. Fendley, *Trans. Faraday Soc.* **45**, 121 (1949)].

gas-solid) chromatographic techniques.[74-82] For example, the equilibrium constants of Ag^+ ions with olefins have been determined using ethylene glycol solutions of silver nitrate supported on various surfaces. At low concentrations where any salting-out effect is negligible, the following relationship holds:

$$H = H_0(1 + K_c^{AD}[A]_0) \qquad (6.56)$$

where H is the partition coefficient:

$$H = \frac{\text{conc. of sample in liquid phase}}{\text{conc. of sample in carrier gas}}$$

and H_0 is the value of H when $[A]_0 = 0$. The partition coefficients have been determined using the relationship

$$H = \frac{3}{2}\left[\frac{(P_i/P_0)^2 - 1}{(P_i/P_0)^3 - 1}\right] Ft_r/V_L \qquad (6.57)$$

where P_i and P_0 are the absolute inlet and outlet pressures respectively of the carrier gas; F is the carrier-gas flow-rate corrected for temperature; t_r is the difference in retention-time between the donor gas and air or methane used as a standard; V_L is the volume of the liquid phase at the column temperature.

6.B.6. Polarography

The reversible one-electron reduction of organic acceptor molecules in aprotic solvents is modified when an electron donor is added to the system, the first wave of the acceptor being shifted to more negative potentials.[83-85] This shift ($\Delta E_{1/2}$) is related to the association constants K_c^{AD}, $K_c^{AD_2}$ ... $K_c^{AD_n}$ of the equilibria:

$$D + A \rightleftharpoons AD \qquad K_c^{AD} = [AD]/[D][A]$$

$$AD + D \rightleftharpoons AD_2 \qquad K_c^{AD_2} = [AD_2]/[AD][D]$$

$$AD_{n-1} + D \rightleftharpoons AD_n \qquad K_c^{AD_n} = [AD_n]/[AD_{n-1}][D]$$

For systems in which $[D]_0 \gg [A]_0$:

$$\frac{I_s}{I_c}\,\text{antilog}_{10}\frac{F\,\Delta E_{1/2}}{2\cdot303\,RT} = 1 + K_c^{AD}[D]_0 + K_c^{AD_2}[D]_0^2 \ldots + K_c^{AD_n}[D]_0^n \ldots \qquad (6.58)$$

providing all activity coefficients are unity, where I_s/I_c is the ratio of the diffusion currents of uncomplexed and complexed acceptor at the donor concentration $[D]_0$. High concentrations of supporting electrolyte have been used to eliminate migration currents. The values of K_c^{AD} so far

6

published do not agree closely with optically determined values.[83, 84, 86, 87] The latter are themselves sensitive to the presence of 0·5M tetra-n-butylammonium perchlorate (used in the polarographic determinations as the supporting electrolyte).[83, 84, 86] To interpret some of the polarographic results in terms of equation (6.58), termolecular complexes with stabilities which are high compared with K_c^{AD} have to be postulated.[83, 84] In other similar systems, there is apparently only 1:1 complexing.

Holm et al.[87] have obtained association constants for a series of polymethylbenzene–7,7,8,8-tetracyanoquinodimethane complexes in chloroform solution. However, they noted that the position of the optical absorption corresponding to the lowest electronic transition in the acceptor is altered on complex formation. For the mesitylene and durene complexes the energy level is increased, whilst for the penta-methylbenzene and hexamethylbenzene complexes it is decreased. Such perturbations will obviously affect estimates of association constants determined polarographically, although, as in this case, the necessary corrections can be applied if the effect is recognized.

Despite the present uncertainty about the results, the polarographic method is valuable as a new approach to the determination of equilibrium constants of organic charge-transfer complexes.

6.B.7. Direct Calorimetry

A calorimetric method which allows the simultaneous measurement of both the equilibrium constant and the enthalpy of complex formation has been described.[88-90] If $\Delta H^{\ominus c}$ is the heat of formation of 1 mole of the complex AD,* then:

$$[AD] = \frac{\left(\dfrac{1000\, \Delta H}{v}\right)}{\Delta H^{\ominus c}} \qquad (6.59)$$

where v is the volume in millilitres of the solution and ΔH is the measured heat of formation for an unknown amount of complex in the volume v of solution. Since:

$$K_c^{AD} = \frac{[AD]}{([A]_0 - [AD])([D]_0 - [AD])} \qquad (6.9)$$

from equations (6.9) and (6.59) we may write:

$$(K_c^{AD})^{-1} = \frac{1000\, \Delta H}{v\, \Delta H^{\ominus c}} + \frac{[A]_0[D]_0\, v\, \Delta H^{\ominus c}}{1000\, \Delta H} - ([A]_0 + [D]_0) \qquad (6.60)$$

Equation (6.60) is analogous to equation (6.31) (p. 135) devised by Rose and Drago from the evaluation of K_c^{AD} from optical data. Values of K_c^{AD}

* For symbols used, see p. 173.

and $\Delta H^{\ominus}$ from a set of experimental data may be obtained by use of arbitrary values of $\Delta H^{\ominus}$ in the same manner in which arbitrary values of molar absorptivity are used in applications of equation (6.60).

A similar method has been used by Nelander[91] to determine the association constants and thermodynamic properties of a series of disulphide–iodine complexes.

6.B.8. Dielectric Constant Measurements

Estimates of association constants for hydrogen-bonded complexes have been obtained from measurements of the dielectric constants of solutions of one solute (X) in a solvent (S) in various constant mixtures of the second solute (Y) and solvent.[92, 93] The method has been developed by Few and Smith[92] from a suggestion made by Hammick et al.[94]

If the weight fractions of X, Y and S in solution are w_X, w_Y and w_S respectively and the amounts Δw_X and Δw_Y of X and Y respectively react to form a 1:1 complex XY, then the total specific polarization of the solution is given by

$$P = P_X(w_X - \Delta w_X) + P_Y(w_Y - \Delta w_Y) + P_{XY}(\Delta w_X + \Delta w_Y) + P_S w_S$$

where P_X, P_Y, P_{XY} and P_S are the specific polarizations of X, Y, XY and S respectively. The apparent polarization of X (P_X^*), calculated on the assumption that no complex is formed, is given by

$$w_X P_X^* = P - w_Y P_Y - w_S P_S$$

and hence $w_X(P_X^* - P_X) = P_{XY}(\Delta w_X + w_Y) - \Delta w_Y P_Y - \Delta w_X P_X$. Since $\Delta w_X / \Delta w_Y = M_X / M_Y$, where M_X and M_Y are the molecular weights of X and Y respectively, we obtain the expression:

$$P_X^* - P_X = \Delta P(\Delta w_X / w_X) \tag{6.61}$$

where $\Delta P = P_{XY} - P_X - P_Y$, the increase in the molecular polarization on the formation of 1 g mol of the complex XY. For such 1:1 complex formation:

$$K_c^{XY} = [XY]/[X][Y] = \Delta w_X/[Y](w_X - \Delta w_X) \tag{6.62}$$

From equations (6.61) and (6.62), since $[Y] = w_Y d/M_Y$, where d is the density of the solvent mixture:

$$\frac{\Delta P}{(P_X^* - P_X)} = 1 + \frac{M_X}{K_c^{XY} w_Y d} \tag{6.63}$$

Whence, from a plot of $(P_X^* - P_X)^{-1}$ against $w_Y^{-1} d^{-1}$, a straight line of gradient $M_X/K_c^{XY} \Delta P$ and intercept with the ordinate ΔP^{-1} should be obtained. The method is obviously applicable to charge-transfer systems.

6.B.9. Osmometry

The thermoelectric differential vapour-pressure method[95-97] which is normally used to determine the molecular weight of a single solute species in dilute solution can be applied to a solution in which the solute is an equilibrium mixture of complex and the constituent components.[98-100] From the known, weighed-out total concentrations of A and D, and the apparent molecular weight of the solution, the association constant may readily be calculated. A particularly simple case is where, for a 1:1 complex, $[A]_0 = [D]_0$. These solutions may be obtained by dissolution of the solid 1:1 complex for these systems where such complexes are isolable.

6.B.10. Kinetic Measurements of Further Reactions of the Species Involved in the Initial Charge-Transfer Equilibrium

Further irreversible chemical reaction may ensue from a charge-transfer complex equilibrium mixture. This might possibly occur by the reaction of the donor and acceptor moieties with each other, or by the reaction of one or more of the species present with some other added substance. On occasion, the kinetics of these relatively slow reactions are such that it is possible to calculate the position of the initial charge-transfer equilibrium from the observed rate-constants, as opposed to parameters of the system (such as absorbance) extrapolated back to the time of mixing.[101-109]

Consider the reaction of, say, an acceptor (A) with a reagent (R) to form a product by an irreversible process. Suppose, for simplicity, that when R is in large excess, the reaction shows pseudo first-order kinetics with a rate constant k_u. Let us also suppose that the acceptor forms a charge-transfer complex (AD) with some donor D, and that this reacts with R under pseudo first-order conditions with a rate constant k_c, and that the products A′ and A′D are possibly in equilibrium as indicated in the following scheme:[103-106]

$$
\begin{array}{ccc}
A+D & \overset{K_c{}^{AD}}{\rightleftharpoons} & AD \\[2mm]
R \downarrow k_u & & R \downarrow k_c \\[2mm]
A'+D & \overset{K_c{}^{A'D}}{\rightleftharpoons} & A'D \\
+X & & +X
\end{array}
\qquad (6.64)
$$

The observed rate constant (k_{obsd}) based on the total concentration of A, namely $[A]_0$ will relate to k_u and k_c by equation (6.65):

$$
k_{obsd} = -\frac{d[A]_0}{dt} \cdot \frac{1}{[A]_0} = k_u \frac{[A]}{[A]_0} + k_c \frac{[AD]}{[A]_0}
\qquad (6.65)
$$

where [A], as elsewhere, denotes the concentration of free, uncomplexed A. Equation (6.65) may be re-expressed:

$$k_{\text{obsd}} = k_{\text{u}} + (k_{\text{c}} - k_{\text{u}}) \frac{[\text{AD}]}{[\text{A}]_0} \tag{6.66}$$

The observed kinetics will be pseudo first-order if $[\text{AD}]/[\text{A}]_0$ remains constant. This will be the case under the condition $[\text{D}]_0 \gg [\text{A}]_0$. (It will also obtain provided $K_{\text{c}}^{\text{AD}} \simeq K_{\text{c}}^{\text{A'D}}$.) For solutions in which $[\text{D}]_0 \gg [\text{A}]_0$:

$$K_{\text{c}}^{\text{AD}} = \frac{[\text{AD}]}{[\text{D}]_0([\text{A}]_0 - [\text{AD}])} \tag{6.53}$$

Combination of equations (6.53) and (6.66) gives:

$$\frac{1}{(k_{\text{obsd}} - k_{\text{u}})} = \frac{1}{(k_{\text{c}} - k_{\text{u}})} + \frac{1}{K_{\text{c}}^{\text{AD}}[\text{D}]_0(k_{\text{c}} - k_{\text{u}})} \tag{6.67}$$

The values of k_{u} may be obtained from a separate experiment in which $[\text{D}]_0 = 0$. For a series of concentrations of D, a plot of $1/(k_{\text{obsd}} - k_{\text{u}})$ against $[\text{D}]_0^{-1}$ should be linear. From the gradient and the intercept of this line, K_{c}^{AD} may be calculated.

The method has been applied, for example, to the complexes of various poly-nitro-9-fluorenyl-p-toluenesulphonates with phenanthrene in glacial acetic acid.[103, 104, 106] In this system, both free and complexed acceptor undergo acetolysis. Other systems have been studied in a similar manner.[105] The values of K_{c}^{AD} obtained by this type of kinetic treatment are compared with values obtained by direct spectrophotometry in Table 6.5.

In other systems studied, the slow irreversible reaction occurs between the electron donor and acceptor species, either free or complexed (see Chapter 11). The exact nature of the reactants and of the reaction(s) (there may be two concurrent reactions involving complexed and uncomplexed molecules) is in some instances revealed by the kinetics of the reaction. The interactions of alkylanilines with tetracyanoethylene studied by Rappoport and Horowitz[110] showed initially very fast charge-transfer complex equilibria, followed by slow irreversible processes which involve reaction between the donor and acceptor species. A similar behaviour has been observed in the reaction of tetracyanoethylene with indole in solvents such as dichloromethane.[111] In these cases the kinetics do not permit a simple evaluation of the positions of the initial equilibria. By contrast, the kinetic behaviour of the reaction of 2,3-dichloro-5,6-dicyano-p-benzoquinone with hexamethylbenzene does appear to allow such a calculation to be made.[108] Under conditions

TABLE 6.5. A comparison of association constants (K_c^{AD}) calculated from kinetic data with values obtained by "normal" equilibrium methods.

System	Solvent	Temp./°C	Kinetic K_c^{AD}/l mol^{-1}	Equilib. K_c^{AD}/l mol^{-1}	Ref.
2,4,7-Trinitro-9-fluorenyl-p-toluenesulphonate–phenanthrene	Glacial CH$_3$COOH	55·85	2·8 ± 0·3	3·0 ± 1·0	104
N-(Indole-3-acryloyl)imidazole–3,5-dinitro-benzoate ion	Aqueous*	25·0	21·5	23·6	107
2,4-Dinitrochlorobenzene–aniline†	Ethanol:ethyl-acetate 1:1 v/v	24·4 ± 0·1	0·170 ± 0·004	0·12 ± 0·05‡	101
2,4-Dinitrochlorobenzene–aniline†	Ethanol	24·4 ± 0·1	0·49 ± 0·05	0·29 ± 0·13‡§ 0·34 ± 0·15‡||	101
Bromine–mesitylene	Carbon tetrachloride	25	0·308	0·381	102

* Borax buffer, pH = 10·40, μ = 0·50.
† May in fact refer to σ-complex formation. However the analysis is independent of the type of complex formed provided the formation from the components and the back reaction are fast compared with the rate of formation in the final product.
‡ Measured at 23·8 ± 0·1°C.
§ Measured at 444 nm.
|| Measured at 450 nm.

where one component is in large excess, the decomposition of the charge-transfer complex shows pseudo first-order kinetics. Reasonable mechanisms for this reaction include the monomolecular decomposition of the charge-transfer complex, the bimolecular reaction of uncomplexed A with uncomplexed D, or a combination of these two processes, occurring concurrently but with different rate constants (see Chapter 11). The relationship between the observed pseudo first-order rate constant (k_1), the association constant for the initial equilibrium (K_c^{AD}), and the concentration of the component in large excess $[X]_0$, is of the same form for any of these three mechanisms, namely:

$$\frac{k_1}{[X]_0} = \frac{F}{1 + K_c^{AD}[X]_0} \tag{6.68}$$

where F is a constant. From the variation of k_1 with $[X]_0$, the association constant K_c^{AD} may be evaluated. Experimentally, this particular system is difficult to study because of the high reactivity of 2,3-dichloro-5,6-dicyano-p-benzoquinone. Some of the values of association constant originally published for this system are in need of revision.[109]

Similar studies have been made of the reactions of aniline with 2,4-dinitrochlorobenzene under second-order rate conditions.[101] The value of the association constant obtained from the kinetic data has been compared with the value calculated by the normal technique of using absorbances which have been extrapolated to the time of mixing (Table 6.5.).

6.B.11. Relaxation Methods

The position of equilibrium can be calculated if the rates of the forward and back reactions are known. It may be possible to measure these very high rates by means of relaxation methods such as temperature-jump with microwave heating, electric-impulse, or dielectric loss,[112] although no values of association constants for charge-transfer complexes have as yet been obtained by such techniques.

6.C. Interpretation of Experimental Data used in the Determination of Association Constants

6.C.1. General

Criticisms of the evaluations of association constant, which have mainly involved optical methods, have been of two kinds: (a) the effect of experimental errors, (b) the effect of using too simple a description of complex formation, which fails to reflect sufficiently the actual system.

Hammond[113] has shown that the effect of experimental errors in the measurement of $[A]_0$, $[D]_0$ and A has a rapidly increasing effect as the value of K_c^{AD} decreases. This is based on an analysis of errors from two experimental points using the Benesi–Hildebrand equation [equation (6.13)] or one of its alternative forms. Person[27] has argued that the most accurate values of K_c^{AD} are obtained when [AD] is the same order of magnitude as the concentration of the more dilute component; for K_c^{AD} determined by a Benesi–Hildebrand or similar procedure, this requires that the component in excess must have a concentration greater than $0 \cdot 1 \times (1/K_c^{AD})/\text{mol } l^{-1}$. If it has a smaller value, the Benesi–Hildebrand plot will have an intercept with the ordinate which will tend towards zero. This may account, at last in part, for Benesi–Hildebrand plots which appear to pass through the origin (see p. 74). Other workers have shown, using synthetic data, that the calculated values of K_c^{AD} and ϵ_λ^{AD} are very sensitive to errors in the directly measured quantities, and in particular to concentration errors.[4, 18] This does not mean that errors in optical density measurements are not significant. Considerable differences in experimentally determined values of K_c^{AD} may arise through small errors in optical density due to mismatched cells, inaccurate compensation for absorption by the solvent and excess solute, and poor instrument calibration.

The effect of solvent competing with the excess solute component has already been discussed (Section 6.B.1, p. 127).

The effect of various concentration scales on the evaluation of ϵ_λ^{AD} was pointed out by Scott[19] in 1955. The differences due to the different scales become very significant for values of $K_c^{AD} < \sim 1 \text{ l mol}^{-1}$. For determinations based on the Benesi–Hildebrand equation [equation (6.13)] and similar relationships, the simple dilute solution relationships $K_m^{AD} = K_c^{AD} d_s$ and $K_x^{AD} = K_c^{AD} 1000 \, d_s/M_s$ are not valid for such low values of K_c^{AD}.[114] Instead, the relationships (6.4) and (6.6) (p. 127) must be used. Hanna and Trotter[114] have shown that although a given system can only be ideal on one scale, nevertheless, on any concentration scale, the Benesi–Hildebrand type of plot will be linear. For example, if the system is ideal on a mole-fraction scale, but the donor concentration is expressed as a molarity, then the Benesi–Hildebrand equation [equation (6.13)] becomes:

$$\frac{[A]_0}{A} = \frac{1000}{M_s} \cdot \frac{1}{K_x^{AD}} \cdot \frac{1}{\epsilon_\lambda^{AD}} \cdot \frac{1}{[D]_0} + \frac{1}{\epsilon_\lambda^{AD}} \left(1 + \frac{1}{K_x^{AD}} - \frac{1}{K_x^{AD}} \cdot \frac{M_D d_s}{M_s d_D} \right) \quad (6.69)$$

If the system is ideal on a molal scale, but the donor concentration is expressed as a molarity, then:

$$\frac{[A]_0}{A} = \frac{d_s}{K_m^{AD}} \cdot \frac{1}{\epsilon_\lambda^{AD}} \cdot \frac{1}{[D]_0} + \frac{1}{\epsilon_\lambda^{AD}} \left(1 - \frac{M_D d_s}{1000 \, d_D} \cdot \frac{1}{K_m^{AD}} \right) \quad (6.70)$$

For a system which is ideal on a molal scale, but for which the donor concentration is expressed as a mole-fraction:

$$\frac{[A]_0}{A} = \frac{M_s}{1000K_m^{AD}} \cdot \frac{1}{\epsilon_\lambda^{AD}} \cdot \frac{1}{X_D} + \frac{1}{\epsilon_\lambda^{AD}}\left(1 - \frac{M_s}{1000K_m^{AD}}\right) \qquad (6.71)$$

If the simple Benesi–Hildebrand relationships corresponding to equations (6.69), (6.70) and (6.71) are used, then the determinations of ϵ_λ^{AD} will be scale-dependent and this will therefore also affect the evaluations of equilibrium constant. The effect is large for very small values of association constant,[114, 115] but is relatively small for values of $K_c^{AD} > \sim 1$ l mol^{-1},[15, 116] (Table 6.6). Some workers have argued that the molar scale is the correct scale to use in such cases.[117]

TABLE 6.6. Association constants for 2,3,5,6-tetrafluoro-p-benzoquinone–methylbenzene complexes in carbon tetrachloride at 33·5°C on the kg solution per mol scale (K_r^{AD}), mole fraction scale (K_X^{AD}), and molal scale (K_m^{AD}).*

Donor	K_r^{AD}	K_X^{AD}	K_m^{AD}	K_X^{AD}/K_r^{AD}†	K_m^{AD}/K_r^{AD}‡
Toluene	0·9₆	5·₉	1·1	6·2	1·1
p-Xylene	1·5	9·₃	1·7	6·2	1·1
Mesitylene	2·2	14·₄	2·4	6·6	1·0₉
Durene	4·9	30·₁	4·7	6·1	0·9₆
Pentamethylbenzene	7·9	49·₆	8·0	6·3	1·0₁
Hexamethylbenzene	15·4	100·₆	16·1	6·5	1·0₅

* Ref. 116.
† For ideal solutions the ratio is 6·5.
‡ For ideal solutions the ratio is 1·0.

However, there remain gross anomalies in the variously determined values of K^{AD} and ϵ_λ^{AD} which cannot be accounted for in terms of a single 1:1 complex which, in the equilibrium mixture, obeys Beer's law. Some of these anomalies are discussed in the following section.

6.C.2. Observed Anomalies in K_c^{AD} and ϵ_λ^{AD} Values Obtained Experimentally

In Chapter 3 it was pointed out that, for a set of complexes of a given acceptor with series of structurally related donors, in which no significant steric effects occur, ϵ_λ^{AD} may be expected to increase as the intermolecular interaction increases.[118–121] Based on the simple assumption that a single 1:1 complex is formed, optical experimental data suggest that in many cases the converse trend often occurs[26, 121–123] (Table 3.7, p. 74). For some systems the same assumptions would imply that the value of K^{AD} is zero and the value of ϵ_λ^{AD} is infinite.[26]

For a given system, optical methods yield values of K^{AD} and ϵ_λ^{AD}

6*

which are often dependent on the concentration conditions: $[D]_0 \gg [A]_0$, $[D]_0 \ll [A]_0$ or $[D]_0 = [A]_0$.[28, 124–127] Some examples are given in Table 6.7. In two systems at least, this concentration dependence is significant over the range of concentration ratio $[D]_0/[A]_0$ from unity to *ca.* 10:1. Above this ratio, the changes in the apparent value of K^{AD} are less than

TABLE 6.7. Comparison of association constants derived from optical and from n.m.r. data, together with the molar absorptivity calculated from the optical data (all at 33·5°C).*

Solvent	Method	Relative conc.	$K_c^{AD}/$ l mol^{-1}	$\epsilon \times 10^{-3}/$ l mol^{-1} cm^{-1}	$K_c^{AD} \epsilon \times 10^{-3}/$ l^2 mol^{-2} cm^{-1}
\multicolumn{6}{c}{N,N-dimethylaniline–1,3,5-trinitrobenzene}
CHCl$_3$	Opt.	$[A]_0 \gg [D]_0$	1·0$_8$	0·9$_6$	1·0$_3$
	Opt.	$[A]_0 \ll [D]_0$	0·6$_8$	1·5$_9$	1·0$_8$
	Opt.	$[A]_0 = [D]_0$	0·4$_0$	2·6$_6$	1·0$_5$
	n.m.r.	$[A]_0 \ll [D]_0$	0·4$_1$	—	
\multicolumn{6}{c}{N,N,N′,N′-tetramethyl-p-phenylenediamine–1,3,5-trinitrobenzene}
CH$_2$Cl–CH$_2$Cl	Opt.	$[A]_0 \gg [D]_0$	1·0$_7$	1·1$_7$	2·3$_1$
	Opt.	$[A]_0 \ll [D]_0$	1·4$_5$	1·6$_0$	2·3$_2$
	Opt.	$[A]_0 = [D]_0$	1·0$_4$	2·2$_0$	2·3$_1$
	n.m.r.	$[A]_0 \ll [D]_0$	1·0$_1$		
\multicolumn{6}{c}{Hexamethylbenzene–fluoranil}
CH$_2$Cl–CH$_2$Cl	Opt.	$[A]_0 \ll [D]_0$	3·3$_0$	3·4$_7$	11·4$_3$
	Opt.	$[A]_0 = [D]_0$	2·5$_5$	4·3$_7$	11·1$_5$
	n.m.r.	$[A]_0 \ll [D]_0$	2·6$_6$	—	
CCl$_4$	Opt.	$[A]_0 \ll [D]_0$	16·5	2·9$_9$	49·4
	Opt.	$[A]_0 = [D]_0$	11·5	4·2$_9$	49·4
	n.m.r.	$[A]_0 \ll [D]_0$	9·7		

* Refs. 124 and 126.

the experimental error[126] (see Figs. 6.8 and 6.9). A feature of experimentally determined values of K^{AD} and ϵ_λ^{AD} from such data is that the *product* $(K_{AD} \cdot \epsilon_\lambda^{AD})$ is independent of the concentration conditions (Table 6.6). By contrast, a considerable number of determinations of K^{AD} based on n.m.r. chemical shift data yield K^{AD} values which are independent of the two conditions $[D]_0 \gg [A]_0$ or $[D]_0 \ll [A]_0$, see p. 144 and Table 6.2. It has so far not proved possible to determine K^{AD} satis-

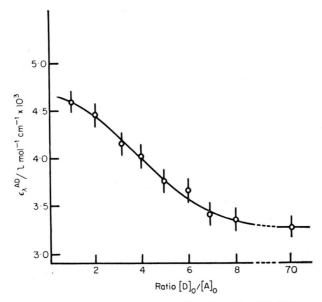

FIG. 6.8. Variation in apparent values of ϵ_λ^{AD} as the ratio $[D]_0/[A]_0 = n$ is altered. Calculated from optical data for the system fluoranil–hexamethylbenzene in carbon tetrachloride at 486 nm, 33·3°C. (From data in Ref. 126.)

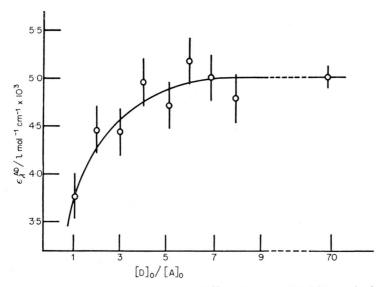

FIG. 6.9. Variation in apparent values of ϵ_λ^{AD} as the ratio $[D]_0/[A]_0 = n$ is altered. Calculated from optical data for the system tetracyanoethylene–hexamethylbenzene in 1,2-dichloroethane at 540 nm, 6·2°C. (From data in Ref. 126.)

factorily for weak complexes from chemical shift measurements under the condition $[D]_0 = [A]_0$ because of the relatively small chemical shift differences. Reasonable agreement is obtained between values of K^{AD} obtained from n.m.r. data on solutions in which one component is in large excess, and values of K^{AD} obtained from optical data when the latter are measured under the experimental condition $[D]_0 = [A]_0$, see Table 6.7.[124, 126]

An apparent wavelength-dependence of the experimentally determined values of K^{AD} is observed in some systems.[18, 127–130] The largest variations have been observed in systems where there is more than one intermolecular charge-transfer band. This anomalous behaviour only occurs in solutions where one component is in large excess.[128] For equimolar mixtures of the components of the same system, this variation is not observed. The basic assumptions upon which methods for estimating K^{AD} and ϵ_λ^{AD} are based have been questioned and modified in several respects by various groups of workers in an endeavour to explain some or all of these anomalies.

Recently, several systems have been observed where the variation of the n.m.r. chemical shift of non-equivalent nuclei with concentration of the second component yields values of association constant which are dependent on the nucleus measured[61, 62] (Table 6.3). Similar observations have been made on the values of enthalpy of complex formation obtained from the temperature-dependence of such shifts.[60] Whilst in principle some explanation could be found for the latter systems in terms of a temperature-dependence of the chemical shift for the pure complex, obviously another explanation must be sought which could apply to systems measured at a single temperature. Any such explanation would then, *per se*, also explain the anomalous enthalpy results.

It should be emphasized that a complete explanation of the behaviour of any one system may require more than one extra postulate. In this respect, none of the following modifications is exclusive of the rest.

6.C.3. Deviations from Ideality

The effect of using different concentration scales for an assumed ideal system has already been discussed (p. 156). There is also the more general possibility that one or more of the solute species does not behave ideally on any scale. In such a case the measured quotient of concentrations will not be equal to the thermodynamic equilibrium constant. For example, if γ_A, γ_D and γ_{AD} are the activity coefficients for the molar concentrations of A, D and AD at equilibrium, then:

$$K_c^{AD}(\text{therm}) = \frac{[AD]}{[D][A]} \cdot \frac{\gamma_{AD}}{\gamma_A \gamma_D} \qquad (6.72)$$

Corresponding relationships may be written for association constants on other concentration scales. Only if the term $(\gamma_{AD}/\gamma_A\gamma_D)$ is unity, will the quotient $[AD]/[D][A]$ be equal to K_c^{AD}(therm). Such behaviour may obtain, though it is difficult to demonstrate experimentally. In principle it could be used to explain the concentration dependence of the calculated values of K^{AD} based on direct *optical* determinations. However, it cannot be the complete explanation since non-optical methods yield values of K^{AD} which are constant under the same range of concentration conditions (see p. 144). Furthermore, the high correlation of experimental points with a linear relationship between $\Delta/[D]_0$ and Δ (Fig. 6.6) implies that either $\gamma_{AD}/\gamma_A\gamma_D$ is unity or is a linear function of $[D]_0$ over the concentration used. This alternative condition is demonstrated by combining equation (6.52) with equation (6.72), whence

$$\frac{\Delta}{[D]_0} = -K_c^{AD}\left(\frac{\gamma_{AD}}{\gamma_A\gamma_D}\right)\Delta + \Delta_0 K_c^{AD}\left(\frac{\gamma_{AD}}{\gamma_A\gamma_D}\right) \qquad (6.73)$$

If we replace the term $\gamma_{AD}/\gamma_A\gamma_D$ by the expression $(1 + B[D]_0)$, where B is independent of $[D]_0$, but could vary with $[A]_0$, then equation (6.55) becomes

$$\frac{\Delta}{[D]_0} = -(K_c^{AD} + B)\Delta + K_c^{AD}\Delta_0 \qquad (6.74)$$

This would lead to an "incorrect" value $(K_c^{AD} + B)$ for the association constant and for Δ_0, although the product $K^{AD}\Delta_0$ would be "correct". The "incorrect" value of $(K_c^{AD} + B)$ corresponds to Guggenheim's [131] "sociation constant".

By the same arguments, the *optical* relationships corresponding to equation (6.13) and equation (6.15) for such non-ideal solutions on a molar concentration scale are:

$$\frac{A}{[D]_0} = \frac{1}{\epsilon_\lambda^{AD}K_c^{AD}}\cdot\frac{1}{[D]_0} + \frac{1}{\epsilon_\lambda^{AD}}\left(\frac{K_c^{AD}+B}{K_c^{AD}}\right) \qquad (6.75)$$

and

$$\frac{A}{[D]_0} = -(K_c^{AD}+B)A + K_c^{AD}\epsilon_\lambda^{AD}[A]_0 \qquad (6.76)$$

This would therefore also lead to an "incorrect" value for the association constant of $(K_c^{AD} + B)$ but a "correct" value for the product $(K_c^{AD}\epsilon_\lambda^{AD})$. Therefore, although both methods may lead to a "sociation constant" rather than a true thermodynamic association constant, the differences between the measured values of the formation constant, depending on whether the chemical shift method or the charge-transfer band optical method is used, *cannot* be explained in terms of deviations from ideality.

6.C.4. Isomeric 1:1 Complexes

The possibility of isomeric complexes in solution was first suggested by Orgel and Mulliken.[26] If AD^i is the ith 1:1 complex with an association constant $K_c^{AD^i}$ and an extinction coefficient $\epsilon_\lambda^{AD^i}$ at a wavelength λ, let us make the following definitions:

$$[AD]' = \sum_i [AD^i] \tag{6.77}$$

$$K_c^{AD'} = \sum_i K_c^{AD^i} \tag{6.78}$$

whence, for a 1-cm path-length of solution, the absorbance due to complex species will be:

$$A = \sum_i [AD^i]\,\epsilon_\lambda^{AD^i} = [AD]'\,\epsilon_\lambda^{AD'} \tag{6.79}$$

where $\epsilon_\lambda^{AD'}$ is defined by equation (6.79)
Now:

$$K_c^{AD^i} = \frac{[AD^i]}{([D]_0 - [AD]')([A]_0 - [AD]')} \tag{6.80}$$

whence from equations (6.78) and (6.80)

$$K_c^{AD^i}/K_c^{AD'} = [AD^i]/[AD]' \tag{6.81}$$

and from equations (6.79) and (6.81):

$$\epsilon_\lambda^{AD'} = \left(\sum_i K_c^{AD^i}\,\epsilon_\lambda^{AD^i} \right)/K_c^{AD'} \tag{6.82}$$

and from equations (6.77), (6.78) and (6.80):

$$K_c^{AD'} = \frac{[AD]'}{([D]_0 - [AD]')([A]_0 - [AD]')} \tag{6.83}$$

For a system where isomeric 1:1 complexes are present, the forms of equations (6.79) and (6.83) are identical with those for the case of a single 1:1 complex. This should result in experimental values for the experimental equilibrium constant K_c^{exp} where:

$$K_c^{exp} = K_c^{AD'} = \sum_i K_c^{AD^i} \tag{6.84}$$

and experimental values for the extinction coefficient ϵ_λ^{exp} at the wavelength (λ) of measurement where:

$$\epsilon_\lambda^{exp} = \epsilon_\lambda^{AD'} = \frac{\sum_i K_c^{AD^i}\,\epsilon_\lambda^{AD^i}}{\sum_i K_c^{AD^i}} \tag{6.85}$$

The value of K_c^{exp} should be independent of the wavelength of measurement. Included in the term $\sum_i K_c^{AD^i}$ will be K_c^{AD} values for all 1:1 complexes, whether or not they absorb at the wavelength at which the optical measurements are made. It is therefore not possible to determine directly whether isomeric 1:1 complexes are present. However, if such isomers exist and they have different enthalpies of formation, then the apparent enthalpy change ($\Delta H^{\ominus}$) will be temperature-dependent. Although the term ϵ_λ^{exp} should also be temperature-dependent because of this effect, a certain decrease in ϵ_λ^{exp} with increase in temperature is in any case expected. A better measure of the presence of isomeric complexes would be the oscillator strength of the charge-transfer band determined over a range of temperature. Relatively few measurements have been made of K_c^{exp} at more than two or three temperatures. In such cases where measurements at a considerable number of temperatures have been made, no deviations from linearity of the plot of $\log K_c^{exp}$, or of oscillator strength, against the reciprocal of the absolute temperature, have been observed [30] (see Fig. 6.10). However, the temperature ranges used were not large.

If isomeric 1:1 complexes are present and have different molar absorptivities, then the observed anomalies in optically derived properties may be accounted for, provided that a further postulate is made: namely, that the ratio of the isomers formed is dependent on the solute concentrations in the extreme cases where one component is in large and varying excess. [130] Such a system is equivalent to a single hypothetical complex, the molar absorptivity of which is dependent on the solvent–solute environment (see Section 6.C.8).

The effect of the presence of isomeric 1:1 complexes on the evaluation of the equilibrium constant by the n.m.r. method described in Section 6.B.4 should be analogous to the result obtained for the optical method. Thus for the case of several 1:1 complexes, equation (6.50) becomes:

$$\delta = P_A \delta_A + \sum_i P_{AD^i} \delta_{AD^i} \tag{6.86}$$

where δ is the measured shift of a nucleus in component A, where P_{AD^i} is the fraction of component A, in the form of the 1:1 complex AD^i. Equation (6.86) may be written:

$$\delta = P_A \delta_A + P_{AD'} \delta_{AD'} \tag{6.87}$$

where

$$P_{AD'} = \sum_i P_{AD^i} \tag{6.88}$$

and

$$\delta_{AD'} = \frac{\sum_i K_c^{AD^i} \delta_{AD^i}}{K_c^{AD'}} \tag{6.89}$$

and

$$K_c^{AD'} = \sum_i K_c^{AD^i} \tag{6.84}$$

The form of equation (6.87) is identical with that for the case where only one complex is formed, namely, equation (6.50). Consequently, association constants evaluated by the type of n.m.r. method described in

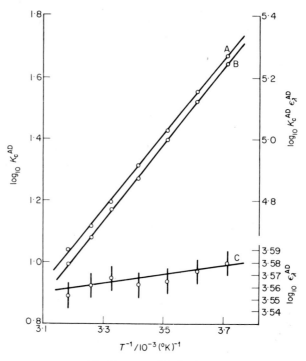

[R. Foster and I. B. C. Matheson, Ref. 30]

FIG. 6.10. Plots of (A) $\log_{10} K_c^{AD}$; (B) $\log_{10} K_c^{AD} \epsilon_\lambda^{AD}$; (C) $\log_{10} \epsilon_\lambda^{AD}$ against the reciprocal of the absolute temperature for the complex tetracyanoethylene–hexamethylbenzene in 1,2-dichloroethane.

Section 6.B.4. but in which two or more isomeric $1:1$ complexes are actually formed, will yield a value of association constant $K_c^{AD'}$, which is the sum of all the $1:1$ association constants [equation (6.84)], and a shift value which is a weighted average of the shifts of all the complexes: that is $\delta_{AD'}$ defined by equation (6.89), or the corresponding $\Delta_0^{AD'}$ value where:

$$\Delta_0^{AD'} = \frac{\sum_i K_c^{AD^i} \Delta_0^{AD^i}}{K_c^{AD'}} \tag{6.90}$$

An important result of this analysis is that a single value for the experimental value of the association constant (namely $\sum_i K_c^{AD^i}$) should be obtained, irrespective of the particular nuclei measured. If different values of association constant are obtained (see, for example, Table 6.3), then this means that the δ_{AD^i} values (or $\Delta_0^{AD^i}$ values) are effectively concentration-dependent. This could be the result of actual changes in the relative shifts for the various nuclei in the pure complexes, or of changes in the proportion of isomers (which may be dependent on the concentrations of the various species, particularly the concentration of the second component present in large and varying excess). In either case, when treated as a single complex, the behaviour will be tantamount to a complex with a concentration-dependent $\delta_{AD'}$ (or $\Delta_0^{AD'}$) term. This conclusion closely parallels the conclusion that some anomalies in optical determinations of association constant are accountable for in terms of deviations from Beer's law (see Section 6.C.8). It appears to be contrary to the suggestion of Ronayne and Williams[132] that observations similar to those in Table 6.7 can be accounted for by supposing that complexing at one site in one component gives rise to changes in chemical shifts of two non-equivalent nuclei in that component, which are different from the changes which occur when complexing takes place at a second site. Although their argument was in relation to temperature-variation in a two-component system in which the complex could be described as a dipole–induced-dipole interaction, the argument would have been applicable to charge-transfer complexes.

6.C.5. Contact Charge-Transfer

Mulliken and Orgel[26] have suggested that intermolecular charge transfer absorption may occur not only by complexed AD pairs but also during random encounters whenever the D and A molecules are sufficiently close to one another. This has been termed "contact charge-transfer". For a system in which only contact charge-transfer occurs, a Benesi–Hildebrand or similar plot might be expected to yield a value of $K_c^{AD} = 0$ and $\epsilon_\lambda^{AD} = \infty$, thus accounting for the experimental observations on such systems as iodine–cyclohexane.

Prue[133] has argued that even if there were only contact charge-transfer, the random collisions would lead to a finite value for K_c^{exp}. This suggestion appears to be in conflict with some experimental results where values of K_c^{exp}, far smaller than an anticipated "random" constant, have been obtained. On the basis of a simple lattice model, Scott[134] has suggested that in fact the experimental methods measure the association in excess of any random effect.

For systems in which a complex AD is formed as well as contact charge-transfer, Mulliken and Orgel [26] had assumed that every free donor–acceptor pair is capable of contact charge-transfer interaction. It may be shown for such systems that the Benesi–Hildebrand relationship yields an approximately correct value of K_c^{AD}, but that the value of ϵ_λ^{exp} is related to the extinction coefficient of the complex (ϵ_λ^{AD}), thus:

$$\epsilon_\lambda^{exp} = \epsilon_\lambda^{AD} + \frac{\alpha\epsilon'}{K_c^{AD}} \tag{6.91}$$

where α is the number of possible contact sites for the species in excess around any molecule of the second species, ϵ' is the extinction coefficient for the contact charge-transfer process based on the potential contact concentration. This postulate has been used to explain the apparent decrease in extinction coefficient values for a series of complexes of increasing stability, involving a common acceptor (Table 3.7).*

6.C.6. Specific Solvation of Solute Species

Murrell, Carter and Rosch [122, 135] have suggested that the donor, acceptor and complex occur in solution with well-defined solvent shells so that the equilibrium governing complex-formation should be written:

$$AS_n + DS_m \rightleftharpoons \overline{AD}\,S_p + qS \tag{6.92}$$

where n, m, and p are the numbers of solvent molecules (S) associated with the species, A, D and AD respectively, and $n + m = p + q$. Such a possibility had been considered earlier by Ross et al. [136] It may be shown that the experimentally determined values of association constant (K_c^{exp}) and molar absorptivity ϵ_λ^{exp}, based on the assumption that the simple equilibrium $A + D \rightleftharpoons AD$ occurs rather than the equilibrium involving solvated species, are such that:

$$K_c^{exp} = K_c^t - \frac{q(m+1)}{[S]_0} \tag{6.93}$$

and

$$\epsilon_\lambda^{exp} = \epsilon_\lambda^t \left(\frac{K_c^t}{K_c^{exp}} \right) \tag{6.94}$$

where K_c^t is the equilibrium constant for the equilibrium (6.92), ϵ_λ^t is the corresponding value for the molar absorptivity of $\overline{AD}\,S_p$, and $[S]_0$ is the total solvent concentration when $[D]_0 = 0$. If the term $q(m+1)/[S]_0$ in equation (6.93) has a value of 1–$3\,l\,mol^{-1}$, values of K_c^{AD} of the order of $10\,l\,mol^{-1}$ may be in error to the extent of 10% or more, whilst values of K_c^{AD} of the order of $1\,l\,mol^{-1}$ may be more than 100% in error. There will

* However, see pp. 76 and 167.

be corresponding errors if ϵ_λ^{exp} is equated with ϵ_λ^{AD}. Reasonable values of n, m and p can lead to molar absorptivities for a series of complexes with a common acceptor, which increase as the stabilities of the various complexes increase. More recently, a simpler version of equation (6.92) in which n = 0 has been proposed.[114] For sufficiently weak interactions between donor and acceptor in solvents of high solvating power, the theory predicts that K_c^{exp} could become negative, when

$$\frac{q(m+1)}{[S]_0} > K_c^t$$

in equation (6.93). It may also be noted that by equation (6.94) the product $(K_c^{exp} . \epsilon_\lambda^{exp})$ is equal to $K_c^{AD} \epsilon_\lambda^{AD}$ on this theory. The result of such specific solvation on the experimental estimation of molar absorptivity could account for the anomalies in the magnitudes of ϵ_λ^{AD} referred to in Section 6.C.1 and Table 6.5. It seems likely that such specific solvation will occur with strongly interacting solvents such as ethers and alcohols. Whether it is of importance when more weakly interacting solvents are used is yet to be determined. As expressed in equation (6.93), it cannot account for the *differences* in the experimentally determined values of K_c^{AD} obtained from optical and from n.m.r. data referred to in Section 6.C.1. This follows from the fact that, under the same concentration conditions, the effect will be independent of the physical method used.

6.C.7. Presence of Termolecular Complexes

Several groups of workers have suggested that anomalous results obtained from optical data might be accounted for by the presence of termolecular (or higher) complexes such as AD_2, A_2D along with the complex AD in equilibrium with unassociated D and A molecules.[7, 18, 28, 127, 135, 137] A similar interpretation has been given for Ag^+ ion–olefin interactions based on solubility behaviour.

For simplicity, let us consider a system in which $[D]_0 \gg [A]_0$ for which it is reasonable to assume that concentrations of any termolecular species A_2D will be insignificant compared with concentrations of the species AD_2. If the equilibria are written:

$$A + D \rightleftharpoons AD \quad K_c^{AD} = [AD]/[D][A] \simeq [AD]/[D]_0([A]_0 - [AD] - [AD_2])$$
$$(6.95)$$

$$AD + D \rightleftharpoons AD_2 \quad K_c^{AD_2} = [AD_2]/[D][AD] \simeq [AD_2]/[D]_0[AD] \quad (6.96)$$

from equations (6.95) and (6.96):

$$[AD] = \frac{K_c^{AD}[D]_0[A]_0}{1 + K_c^{AD}[D]_0 + K_c^{AD} K_c^{AD_2}[D]_0^2} \qquad (6.97)$$

$$[AD_2] = \frac{K_c^{AD} K_c^{AD_2}[D]_0^2[A]_0}{1 + K_c^{AD}[D]_0 + K_c^{AD} K_c^{AD_2}[D]_0^2} \qquad (6.98)$$

For optical determinations, let us assume that the absorption at the wavelength of measurement (λ) is due only to the species AD and AD$_2$ and that the measured absorbances A are for a 1-cm path-length of solution, then:

$$A = \epsilon_\lambda^{AD}[AD] + \epsilon_\lambda^{AD_2}[AD_2] \qquad (6.99)$$

where ϵ_λ^{AD} and $\epsilon_\lambda^{AD_2}$ are the molar absorptions of AD and AD$_2$ respectively at λ. From a combination of equations (6.97), (6.98) and (6.99) we have:

$$A = \frac{\epsilon_\lambda^{AD} K_c^{AD}[A]_0[D]_0 + \epsilon_\lambda^{AD_2} K_c^{AD} K_c^{AD_2}[A]_0[D]_0^2}{1 + K_c^{AD}[D]_0 + K_c^{AD} K_c^{AD_2}[D]_0^2} \qquad (6.100)$$

This may be arranged, for example, into the form of the Benesi–Hildebrand equation [equation (6.13)]:

$$\frac{[A]_0}{A} = \frac{1 + [D]_0 K_c^{AD_2}}{\left(1 + \dfrac{\epsilon_\lambda^{AD_2}}{\epsilon_\lambda^{AD}} [D]_0 K_c^{AD_2}\right)} \cdot \frac{1}{\epsilon_\lambda^{AD}}$$

$$+ \frac{1}{\left(1 + \dfrac{\epsilon_\lambda^{AD_2}}{\epsilon_\lambda^{AD}} [D]_0 K_c^{AD_2}\right)} \cdot \frac{1}{[D]_0 \epsilon_\lambda^{AD} K_c^{AD}} \qquad (6.101)$$

or the Scott equation [equation (6.14)]:

$$\frac{[A]_0[D]_0}{A} = \frac{(1 + [D]_0 K_c^{AD_2})}{\left(1 + \dfrac{\epsilon_\lambda^{AD_2}}{\epsilon_\lambda^{AD}} [D]_0 K_c^{AD_2}\right)} \cdot \frac{[D]_0}{\epsilon_\lambda^{AD}}$$

$$+ \frac{1}{\left(1 + \dfrac{\epsilon_\lambda^{AD_2}}{\epsilon_\lambda^{AD}} [D]_0 K_c^{AD_2}\right)} \cdot \frac{1}{\epsilon_\lambda^{AD} K_c^{AD}} \qquad (6.102)$$

or in the form of equation (6.15):

$$\frac{A}{[D]_0} = -A K_c^{AD}(1 + [D]_0 K_c^{AD_2}) + \epsilon_\lambda^{AD} K_c^{AD}[A]_0 \left(1 + \frac{\epsilon_\lambda^{AD_2}}{\epsilon_\lambda^{AD}} [D]_0 K_c^{AD_2}\right) \qquad (6.103)$$

If the appropriate functions are plotted using any of the equations (6.13), (6.14) and (6.15) on the assumption that only one complex with a stoichiometry 1:1 is present, whereas in fact significant quantities of a complex with a different stoichiometry are also present, then apparent association constant values which are dependent on the concentration $[D]_0$ will be obtained. Calculations using synthetic data show that the association constant and molecular absorptivity, based on the assumption of a single simple $1:1$ interaction, may be increased or decreased by contributions from termolecular complexes.[137]

The presence of the terms ϵ_λ^{AD} and $\epsilon_\lambda^{AD_2}$ in equations (6.101), (6.102) and (6.103) implies that the apparent values of K_c^{AD} are wavelength-dependent. This has been observed in some systems [18, 127–130] (see p. 160) and appears to lend support for the postulate of termolecular complexes, since the wavelength-dependence of the apparent value of the association constant only occurs when one component is in large excess, i.e. when termolecular complex formation is favoured. Furthermore, it has recently been shown that the absorption spectra for the naphthalene–tetracyanoethylene system in carbon tetrachloride and in chloroform can be closely simulated, using synthetic data, by assuming the presence of termolecular complexes along with $1:1$ complexes.[7, 137]

However, if it is assumed that under the condition $[D]_0 = [A]_0$ the observed association constants may be equated to K_c^{AD}, and if these values are then used to calculate the magnitude of $K_c^{AD_2}$, the latter values are relatively large,[138] whereas they might be expected to be considerably smaller than the corresponding K_c^{AD} values. This conclusion is also supported by the use of synthetic data to calculate the absorbance from assumed values of the various parameters. The variation of the calculated absorbance with change in total concentration of one of the initial species is such that straight lines are obtained in Benesi–Hildebrand plots even when relatively large values of $K_c^{AD_2}$ and $\epsilon_\lambda^{AD_2}$ are used.[18] When reasonably small values of $K_c^{AD_2}$ and $\epsilon_\lambda^{AD_2}$ are used, the effect on the experimental estimate of K_c^{AD} is small and cannot account for the large differences such as are noted in Table 6.7.

The arguments which have been applied to optically determined values of K_c^{AD} have their analogy in determinations which use other physical methods. In the n.m.r. determination of association constants, if it is assumed that the complex species AD and AD_2 are present in a series of solutions in which $[D]_0 \gg [A]_0$, then the observed chemical shift of a nucleus in the acceptor moiety, relative to some standard, will be:

$$\delta = P_A \delta_A + P_{AD} \delta_{AD} + P_{AD_2} \delta_{AD_2} \qquad (6.104)$$

where P_A, P_{AD}, P_{AD_2} are the relative molecular populations of the species A, AD and AD_2 respectively, and δ_A, δ_{AD}, δ_{AD_2} are the chemical shifts of these species if each were the sole solute [cf. equation (6.50)].

If $\varDelta = (\delta_A - \delta)$, $\varDelta_0^{AD} = (\delta_A - \delta_{AD})$ and $\varDelta_0^{AD_2} = (\delta_A - \delta_{AD_2})$, it may be shown that:

$$\frac{\varDelta}{[D]_0} = -K_c^{AD}(1 + [D]_0 K_c^{AD_2}) \varDelta + \left(1 + \frac{\varDelta_0^{AD_2}}{\varDelta_0^{AD}} [D]_0 K_c^{AD_2}\right) K_c^{AD} \varDelta_0^{AD} \quad (6.105)$$

which may be compared with equation (6.55). Now the experimental error in these determinations is such that an upper limit can be placed on $K_c^{AD_2}$. It amounts to 5% of K_c^{AD} for values of $K_c^{AD} \simeq 10 \ l \ mol^{-1}$ when $\varDelta_0^{AD} = \varDelta_0^{AD_2}$. (It is under this latter condition that the deviation of a plot of equation (6.105) from linearity is a minimum.) However, for smaller values of K_c^{AD} significant amounts of termolecular species could be present and yet plots using equations (6.54) or (6.55) could be linear within experimental error. This would account for at least some of the reported anomalies.

Job plots[139-144] for the type of A–D systems under discussion give good 1:1 curves. However, such plots are not sensitive to small additions of a 2:1 complex. A more satisfactory piece of evidence is that, in many[145] though not all,[146] iodine-donor systems, the electronic spectra show remarkably sharp isobestic points indicative of the presence of only two absorbing species, namely the free iodine and the complexed iodine. This of course argues only for these particular systems.

6.C.8. The Effect of Self-association of the Component Species

Tanner and Bruice[147] have shown that large errors arise in the calculated values of association constants when the reactants themselves undergo self-association. For a system for which $[D]_0 \gg [A]_0$ in a Benesi–Hildebrand treatment where there is a self-association of D to form an n-mer D_n, if it is assumed that the n-mer does not interact with A, then we may write:

$$K_c^n = [D_n]/[D]^n \quad (6.106)$$

and

$$[D]_0 = [D] + nK_c^n[D]^n \quad (6.107)$$

If, for simplicity, we consider the case where, at the wavelength measurement (λ), only A and the charge-transfer complex absorb, and this complex is a 1:1 adduct AD, then:

$$A' = \epsilon_\lambda^A([A]_0 - [AD]) + \epsilon_\lambda^{AD}[AD] \quad (6.17)$$

If an apparent association constant K_c' is defined as:

$$K_c' = \frac{[AD]}{([A]_0 - [AD])[D]_0} \quad (6.108)$$

when a combination of equations (6.17) and (6.108) gives:

$$\frac{[A]_0(\epsilon_\lambda^A - \epsilon_\lambda^{AD})}{A' - \epsilon_\lambda^A[A]_0} = \frac{1}{K_c'} \cdot \frac{1}{[D]_0} + 1 \qquad (6.109)$$

from a plot of the left-hand side of equation (6.109) against $1/[D]_0$, K_c' may be obtained. K_c' is in fact related to the true association constant K_c^{AD} by the relationship:

$$K_c' = K_c^{AD} \cdot \frac{1}{(1 + nK_c^n[D^{n-1}])} \qquad (6.110)$$

It is seen therefore that K_c' is dependent on $[D]_0$. Synthetic data show that the curvature of a Benesi–Hildebrand plot is not large except for very large values of K_c^n. However, the values of association constant obtained from such a plot, even when K_c^n is relatively small, are significantly less than K_c^{AD}. This is apparent from equation (6.110).

The corresponding case, where there is self-association between molecules of A, is also discussed by Tanner and Bruice. An obvious condition which minimizes such effects is to use very dilute solutions. For systems where both component species may self-associate, the use of solutions in which $[D]_0 \approx [A]_0$ is advantageous, because both components will then be present in relatively low concentration.

Such self-complexing cannot, however, explain the differences between the results obtained from n.m.r. chemical shift data, and the values obtained from optical data under the same concentration conditions (Section 6.C.2 and Table 6.6), because the effect of self-association, as described by equation (6.110), will apply irrespective of the particular physical method used. Significant self-association of commonly used donor and acceptor species in solution has been demonstrated experimentally.[148]

6.C.9. Apparent Deviations from Beer's Law by the Complex Species

The anomalies referred to in Section 6.C.2 derive from measurements which contain one component in large excess and are usually observed when *optical* methods are used.[124] This suggests that a false assumption has been made concerning the optical properties of such systems, rather than that some gross deviation from the simple mass-action relationship involving the formation of a 1:1 complex occurs.

Virtually all optical methods assume Beer's law to be obeyed for the various solute species and in particular by the complex. In principle, a deviation from Beer's law by the absorption band of the complex species could account for the differences between optically determined K_c^{AD} values depending on the relative concentrations of D and A; and, by contrast, the constancy of the value obtained for K_c^{AD} when non-optical methods are used.

It has been suggested that the intensity of the charge-transfer band is sensitive to the solvent environment, which, under conditions which yield anomalous K_c^{AD} values, may well vary from 0·5M to 0·05M or less with respect to the component in excess. Although the extinction co-efficient of the complex may be concentration-dependent, Benesi–Hildebrand or similar plots will yield effectively straight lines from which experimental values of association constant, K_c^{exp} and extinction coefficient of the complex ϵ_λ^{exp} may be obtained. These are related to the true values K_c^{AD} and ϵ_{AD} by the relationship:

$$K_c^{exp}.\epsilon_\lambda^{exp} = K_c^{AD}.\epsilon_\lambda^{AD} \tag{6.111}$$

For systems with donor in excess, experimental results fit the relationship:[149]

$$\epsilon_\lambda^{exp} = \epsilon_\lambda^{AD}(1 + \alpha[D]_0^{1/2}) \tag{6.112}$$

where α is a constant for a given system at a given temperature.

The constancy of the product $(K_c^{exp}.\epsilon_\lambda^{exp})$ has often been observed for systems under varying concentration conditions, in which normal analytical procedures yield concentration-dependent values for the *separated* terms K_c^{exp} and ϵ_λ^{exp}.[27, 122, 124] However, it does not provide positive evidence in support of Beer's law deviations since a similar prediction follows from the postulate of termolecular complexes, and from specific solvation (see above).

If it is assumed that optical methods provide a correct value for the product $(K^{AD}.\epsilon_\lambda^{AD})$, and the n.m.r. methods yield correct values for K^{AD}, then ϵ_λ^{AD} may be obtained from the quotient $(K^{AD}.\epsilon_\lambda^{AD})_{optical}/(K^{AD})_{nmr}$. Such a set of ϵ_λ^{AD} values have been obtained for a series of polymethyl-benzene-1,3,5-trinitrobenzene complexes.[56] They show the theoretically predicted increase in ϵ_λ^{AD} with increasing donor strength (Table 3.8) in contrast to the values of ϵ_λ^{AD} derived purely from optical measurements.

The effect of large concentrations of D, expressed in equation (6.112) will have its counterpart in solutions where $[D]_0 \ll [A]_0$. Consequently the optimum conditions to minimize this effect will be when $[D]_0 = [A]_0$. It is under such conditions that the anomalies disappear (see Section 6.C.1). These include the wavelength-dependence of K_c^{exp} in some systems which show two charge-transfer bands. Although observations are also consistent with the postulate that the anomalies are the result of termolecular complexes, this latter argument does not account for the fact that many of the anomalies vanish when non-optical methods of estimating the concentration of the complex are used.

However, a major criticism of the suggestion that Beer's law is not obeyed is that the postulated deviations have to be remarkably large and may reach $\sim 50\%$ of the value of ϵ_λ^{AD}. Furthermore, if such deviations are the result of changes in the solvent shell of the complex, then ϵ_λ^{AD} might be expected to be sensitive to changes in the nature of the diluting solvent. In fact ϵ_λ^{AD} is virtually independent of the solvent.

The effect of contact charge-transfer (Section 6.C.4), if it were to occur concurrently with an optically absorbing charge-transfer complex, could itself be considered to be a deviation from Beer's law by the complex species at wavelengths where there was absorption by both mechanisms. In the original description of contact charge-transfer, Orgel and Mulliken[26] postulated that only uncomplexed A and D molecules could be involved. This leads to the prediction that K_c^{exp} will be equal to K_c^{AD} and only the extinction coefficient value is affected (Section 6.C.4). In order to explain the observed variation of K_c^{AD} in terms of contact charge-transfer, it would be necessary to modify Orgel and Mulliken's postulate; for example, by supposing that contact charge-transfer could occur between AD and the species in excess.[58]

As already mentioned in Section 6.C.3, if the product of the association in fact consists of a mixture of two or more isomeric 1 : 1 complexes, then a variation of the ratio of isomers with concentration of solute could account, at least partially, for the apparent deviations from Beer's law, provided that the different isomers have different molar absorptivities. This assumption is not unreasonable. However, the relatively rare occurrence of analogous anomalies in association constants derived from n.m.r. data would lead to the conclusion that the chemical shift values for the various isomers of a D–A pair do not differ greatly.

It is very apparent that a complete solution to these problems is still to be found.

6.D. Methods for Evaluating Enthalpies and Entropies of Complex Formation

The numerical values of enthalpy ($\Delta H^\ominus$) and entropy ($\Delta S^\ominus$) of complex formation are dependent on the concentration scale used. If $\Delta H^{\ominus c}$, $\Delta H^{\ominus X}$ and $\Delta H^{\ominus m}$ are the enthalpies of formation referred to the standard states, unit molar concentration, pure component and unit molal concentration respectively, corresponding to K_c^{AD}, K_X^{AD} and K_m^{AD}, then:[150]

$$\Delta H^{\ominus X} = \Delta H^{\ominus m} = \Delta H^{\ominus c} - \alpha R T^2 \tag{6.113}$$

where α is the coefficient of expansion of the solvent.

Estimates of $\Delta H^{\ominus}$ and $\Delta S^{\ominus}$ of many organic charge-transfer complexes have been obtained from van't Hoff plots of the variation of $\ln K^{\mathrm{AD}}$ with T^{-1}, where T is the absolute temperature. The temperature range is usually small, viz. $\sim -10°$ to $+50°\mathrm{C}$. For optical determinations an alternative procedure may be used to obtain $\Delta H^{\ominus}$, namely to plot $\ln(K^{\mathrm{AD}}\epsilon_{\lambda}^{\mathrm{AD}})$ against T^{-1}.[30] This yields virtually the same value of $\Delta H^{\ominus}$ as a strict van't Hoff plot since $\epsilon_{\lambda}^{\mathrm{AD}}$ is nearly temperature-independent (Fig. 6.9). The advantage is, that whereas the product $K^{\mathrm{AD}}\epsilon_{\lambda}^{\mathrm{AD}}$ may be obtained readily, there is serious doubt as to the correct separation of the terms K^{AD} and $\epsilon_{\lambda}^{\mathrm{AD}}$ (see previous sections).

The same principle may be applied to chemical shift measurements. If P is the fraction of the solute species which is complexed, then:

$$\frac{P}{1-P} \propto \exp(\Delta S^{\ominus}/\mathrm{R}) \cdot \exp(-\Delta H^{\ominus}/\mathrm{R}T) \qquad (6.114)$$

At a given temperature $P = \Delta/\Delta_0$ [see equation (6.52)], hence a plot of $\ln \Delta$ against T^{-1} will give $-\Delta H^{\ominus}$ directly, providing that Δ_0 is temperature-independent.[151] Recent work has shown that for many charge-transfer complexes proton Δ_0 values are in fact temperature-independent.[152, 153] This does not appear to be the case for $^{19}\mathrm{F}$ Δ_0 values.[152, 153]

An alternative method for determining $\Delta H^{\ominus}$ is by direct calorimetry. It has been shown (p. 150) that:[88]

$$\frac{1}{K_{\mathrm{c}}^{\mathrm{AD}}} = \frac{1000\,\Delta H}{v\,\Delta H^{\ominus c}} + \frac{[\mathrm{D}]_0[\mathrm{A}]_0\,v\,\Delta H^{\ominus c}}{1000\,\Delta H} - ([\mathrm{A}]_0 + [\mathrm{D}]_0) \qquad (6.60)$$

where ΔH is the experimentally measured enthalpy of reaction for a volume v ml of solution. Although the method has only been applied to organo-metallic Lewis acid complexes of electron donors, the method could be applied to purely organic systems.

It has been pointed out that if more than one complex is present and the enthalpies of formation are not identical, then the apparent enthalpy of formation $\Delta H^{\ominus}(\exp)$ will be related to the various complexes by the expression:

$$\Delta H^{\ominus}(\exp) = \frac{\sum_{\mathrm{i}} K_{\mathrm{c}}^{\mathrm{i}}\,\Delta H_{\mathrm{i}}^{\ominus}}{\sum_{\mathrm{i}} K_{\mathrm{c}}^{\mathrm{i}}} \qquad (6.115)$$

where $K_{\mathrm{c}}^{\mathrm{i}}$ and $\Delta H_{\mathrm{i}}^{\ominus}$ are the association constant and enthalpy of formation of the ith complex. In practice, the small temperature range normally used would require there to be large differences in $\Delta H^{\ominus}$ for the effect to be detectable in, say, a plot of $\log K_{\mathrm{c}}$ against T^{-1}.

A least-squares treatment of experimental data for the determination of the thermodynamic constants $\Delta H^{\ominus}$ and $\Delta S^{\ominus}$ has been described by Wentworth et al.[5]

For a more detailed discussion of the evaluation of thermodynamic functions from equilibrium constants, the reader is referred to a paper by Clarke and Glew.[154]

REFERENCES

1. C. P. Nash, *J. phys. Chem.*, *Ithaca* **64**, 950 (1960).
2. M. Tamres, *J. phys. Chem.*, *Ithaca* **65**, 654 (1961).
3. P. A. D. de Maine and R. D. Seawright, "Digital Computer Programs for Physical Chemistry," Vols. I and II, Macmillan, New York (1963).
4. K. Conrow, G. D. Johnson and R. E. Bowen, *J. Am. chem. Soc.* **86**, 1025 (1964).
5. W. E. Wentworth, W. Hirsch and E. Chen, *J. phys. Chem.*, *Ithaca* **71**, 218 (1967).
6. D. R. Rosseinsky and H. Kellawi, *J. chem. Soc.* (A), 1207 (1969).
7. W. L. Stone, Thesis, Marshall University (1968).
8. M. I. Foreman, unpublished work.
9. E. Grunwald and W. C. Coburn, Jr., *J. Am. chem. Soc.* **80**, 1322 (1958).
10. P. D. Groves, P. J. Huck and J. Homer, *Chemy Ind.* 915 (1967).
11. F. J. C. Rossotti and H. S. Rossotti, "The Determination of Stability Constants and Other Equilibrium Constants in Solution," McGraw-Hill, New York (1961).
12. R. E. Merrifield and W. D. Phillips, *J. Am. chem. Soc.* **80**, 2778 (1958).
13. R. S. Drago, T. F. Bolles and R. J. Niedzielski, *J. Am. chem. Soc.* **88**, 2717 (1966).
14. M. Brandon, M. Tamres and S. Searles, Jr., *J. Am. chem. Soc.* **82**, 2129 (1960).
15. M. Tamres and M. Brandon, *J. Am. chem. Soc.* **82**, 2134 (1960).
16. Cf. H. J. G. Hayman, *J. chem. Phys.* **37**, 2290 (1962).
17. H. A. Benesi and J. H. Hildebrand, *J. Am. chem. Soc.* **71**, 2703 (1949).
18. G. D. Johnson and R. E. Bowen, *J. Am. chem. Soc.* **87**, 1655 (1965).
19. R. L. Scott, *Proc. Int. Conf. on Co-ordination Compounds*, Amsterdam (1955), p. 265; *Recl Trav. chim. Pays-Bas Belg.* **75**, 787 (1956).
20. R. Foster, D. Ll. Hammick and A. A. Wardley, *J. chem. Soc.* 3817 (1953).
21. G. Cilento and D. L. Sanioto, *Z. phys. Chem.* (Leipzig) **223**, 333 (1963).
22. J. A. A. Ketelaar, C. van de Stolpe, A. Goudsmit and W. Dzcubas, *Recl Trav. chim. Pays-Bas Belg.* **71**, 1104 (1952).
23. R. Foster, *Nature, Lond.* **173**, 222 (1954).
24. J. M. Corkill, R. Foster and D. Ll. Hammick, *J. chem. Soc.* 1202 (1955).
25. P. A. D. de Maine, *Spectrochim. Acta* **12**, 1051 (1959).
26. L. E. Orgel and R. S. Mulliken, *J. Am. chem. Soc.* **79**, 4839 (1957).
27. W. B. Person, *J. Am. chem. Soc.* **87**, 167 (1965).
28. S. D. Ross and M. M. Labes, *J. Am. chem. Soc.* **79**, 76 (1957).
29. R. Foster, *J. chem. Soc.* 5098 (1957).
30. R. Foster and I. B. C. Matheson, *Spectrochim. Acta* **23A**, 2037 (1967).
31. N. J. Rose and R. S. Drago, *J. Am. chem. Soc.* **81**, 6138, 6141 (1959).
32. S. Nagakura, *J. Am. chem. Soc.* **76**, 3070 (1954).
33. W. Liptay, *Z. Elektrochem.* **65**, 375 (1961).

34. R. G. Satterfield, *Ber. Bunsenges. physik. Chem.* **69**, 88 (1965); cf. W. Liptay, *Ber. Bunsenges. physik. Chem.* **69**, 89 (1965).
35. W. E. Deming, "Statistical Adjustment of Data," Wiley, New York (1943).
36. J. Lauransan, P. Pineau and J. Lascombe, *J. Chim. phys.* **63**, 635 (1966).
37. H. Yamada and K. Kozima, *J. Am. chem. Soc.* **82**, 1543 (1960).
38. J. Morcillo and E. Gallego, *An. R. Soc. esp. Fís. Quím.* **56B**, 263 (1960).
39. A. I. Popov, R. E. Humphrey and W. B. Person, *J. Am. chem. Soc.* **82**, 1850 (1960).
40. R. E. Kagarise, *Spectrochim. Acta* **19**, 629 (1963).
41. A. G. Maki and E. K. Plyler, *J. phys. Chem., Ithaca* **66**, 766 (1962).
42. R. F. Lake and H. W. Thompson, *Proc. R. Soc.* **A297**, 440 (1967).
43. G. Michel and G. Duyckaerts, *Spectrochim. Acta* **21**, 279 (1965).
44. G. Leclere and G. Duyckaerts, *Spectrochim. Acta* **22**, 403 (1966).
45. R. Foster and C. A. Fyfe, *in* "Progress in Nuclear Magnetic Resonance Spectroscopy," Vol. 4, eds. J. W. Emsley, J. Feeney and L. H. Sutcliffe, Pergamon, Oxford (1969), Ch. 1.
46. C. J. Creswell and A. L. Allred, *J. phys. Chem., Ithaca* **66**, 1469 (1962).
47. J. Homer and P. J. Huck, *J. chem. Soc.* (A), 277 (1968).
48. M. Nakano, N. I. Nakano and T. Higuchi, *J. phys. Chem., Ithaca* **71**, 3954 (1967).
49. E. T. Strom, W. L. Orr, B. S. Snowden, Jr. and D. E. Woessner, *J. phys. Chem., Ithaca* **71**, 4017 (1967).
50. R. Mathur, E. D. Becker, R. B. Bradley and N. C. Li, *J. phys. Chem., Ithaca* **67**, 2190 (1963).
51. F. Takahashi and N. C. Li, *J. phys. Chem., Ithaca* **69**, 1622 (1965).
52. F. Takahashi and N. C. Li, *J. Am. chem. Soc.* **88**, 1117 (1966).
53. P. J. Berkeley, Jr. and M. W. Hanna, *J. phys. Chem., Ithaca* **67**, 846 (1963).
54. M. W. Hanna and A. L. Ashbaugh, *J. phys. Chem., Ithaca* **68**, 811 (1964).
55. A. A. Sandoval and M. W. Hanna, *J. phys. Chem., Ithaca* **70**, 1203 (1966).
56. R. Foster and C. A. Fyfe, *Trans. Faraday Soc.* **61**, 1626 (1965).
57. R. Foster and C. A. Fyfe, *J. chem. Soc.* (B), 926 (1966).
58. R. Foster and I. B. C. Matheson, unpublished work.
59. R. Foster and C. A. Fyfe, *Chem. Commun.* 642 (1965).
60. R. E. Klink and J. B. Stothers, *Can. J. Chem.* **44**, 37 (1966).
61. R. Foster and D. R. Twiselton, unpublished work.
62. R. Foster and M. I. Foreman, unpublished work.
63. N. M. D. Brown, R. Foster and C. A. Fyfe, *J. chem. Soc.* (B), 406 (1967).
64. R. Foster and C. A. Fyfe, *Nature, Lond.* 213, 591 (1967).
65. T. S. Moore, F. Shepherd and F. Goodall, *J. chem. Soc.* 1447 (1931).
66. H. D. Anderson and D. Ll. Hammick, *J. chem. Soc.* 1089 (1950).
67. R. Foster, D. Ll. Hammick and S. F. Pearce, *J. chem. Soc.* 244 (1959).
68. R. L. Denyer, A. Gilchrist, J. A. Pegg, J. Smith, T. E. Tomlinson and L. E. Sutton, *J. chem. Soc.* 3889 (1955).
69. L. J. Andrews and R. M. Keefer, *J. Am. chem. Soc.* **71**, 3644 (1949).
70. L. J. Andrews and R. M. Keefer, *J. Am. chem. Soc.* **72**, 3113 (1950).
71. L. J. Andrews and R. M. Keefer, *J. Am. chem. Soc.* **72**, 5034 (1950).
72. H. C. Brown and J. D. Brady, *J. Am. chem. Soc.* **74**, 3570 (1952).
73. H. C. Brown and J. J. Melchiore, *J. Am. chem. Soc.* **87**, 5269 (1965).
74. M. A. Muhs and F. T. Weiss, *J. Am. chem. Soc.* **84**, 4697 (1962).
75. W. E. Falconer and R. J. Cvetanović, *Analyt. Chem.* **34**, 1064 (1962).
76. E. Gil-Av and J. Herling, *J. phys. Chem., Ithaca* **66**, 1208 (1962).

77. R. J. Cvetanović, F. J. Duncan and W. E. Falconer, *Can. J. Chem.* **41**, 2095 (1963).
78. R. J. Cvetanović, F. J. Duncan and W. E. Falconer, *Can. J. Chem.* **42**, 2410 (1964).
79. R. J. Cvetanović, F. J. Duncan, W. E. Falconer and R. S. Irwin, *J. Am. chem. Soc.* **87**, 1827 (1965).
80. R. J. Cvetanović, F. J. Duncan, W. E. Falconer and W. A. Sunder, *J. Am. chem. Soc.* **88**, 1602 (1966).
81. A. R. Cooper, C. W. P. Crowne and P. G. Farrell, *Trans. Faraday Soc.* **62**, 2725 (1966).
82. A. R. Cooper, C. W. P. Crowne and P. G. Farrell, *Trans. Faraday Soc.* **63**, 447 (1967).
83. M. E. Peover, *Proc. chem. Soc.* 167 (1963).
84. M. E. Peover, *Trans. Faraday Soc.* **60**, 417 (1964).
85. H. M. N. H. Irving, *in* "Advances in Polarography" (Proceedings of the 2nd International Congress of Polarography),Vol. 1, ed. I. S. Longmuir, Pergamon, Oxford, London, New York, Paris (1960), p. 42.
86. D. R. Twiselton, unpublished work.
87. R. D. Holm, W. R. Carper and J. A. Blancher, *J. phys. Chem., Ithaca* **71**, 3960 (1967).
88. T. F. Bolles and R. S. Drago, *J. Am. chem. Soc.* **87**, 5015 (1965).
89. T. F. Bolles and R. S. Drago, *J. Am. chem. Soc.* **88**, 3921 (1966).
90. T. D. Epley and R. S. Drago, *J. Am. chem. Soc.* **89**, 5770 (1967).
91. B. Nelander, *Acta chem. scand.* **20**, 2289 (1966).
92. A. V. Few and J. W. Smith, *J. chem. Soc.* 2781 (1949).
93. R. J. Bishop and L. E. Sutton, *J. chem. Soc.* 6100 (1964).
94. D. Ll. Hammick, A Norris and L. E. Sutton, *J. chem. Soc.* 1755 (1938).
95. J. J. Neumayer, *Analytica chim. Acta* **20**, 519 (1959).
96. D. E. Burge, *J. phys. Chem., Ithaca* **67**, 2590 (1963).
97. J. van Dam, *Recl Trav. chim. Pays-Bas Belg.* **83**, 129 (1964).
98. W. I. Higuchi, M. A. Schwartz, E. G. Rippie and T. Higuchi, *J. phys. Chem., Ithaca*, **63**, 996 (1959).
99. J. R. Larry, *Diss. Abstr.* **27**, 2316B (1967).
100. A. S. Bailey and J. M. Evans, *J. chem. Soc.* (C), 2105 (1967).
101. S. D. Ross and I. Kuntz, *J. Am. chem. Soc.* **76**, 3000 (1954).
102. R. M. Keefer, J. H. Blake, III and L. J. Andrews, *J. Am. chem. Soc.* **76**, 3062 (1954).
103. A. K. Colter and S. S. Wang, *J. Am. chem. Soc.* **85**, 114 (1963).
104. A. K. Colter, S. S. Wang, G. H. Mergerle and P. S. Ossip, *J. Am. chem. Soc.* **86**, 3106 (1964).
105. A. K. Colter and L. M. Clemens, *J. Am. chem. Soc.* **87**, 847 (1965).
106. A. K. Colter, F. F. Guzik and S. H. Hui, *J. Am. chem. Soc.* **88**, 5754 (1966).
107. F. M. Menger and M. L. Bender, *J. Am. chem. Soc.* **88**, 131 (1966).
108. R. Foster and I. Horman, *J. chem. Soc.* (B), 1049 (1966).
109. R. Foster and B. Dodson, unpublished work.
110. Z. Rappoport, *J. chem. Soc.* 4498 (1963); Z. Rappoport and A. Horowitz, *J. chem. Soc.* 1348 (1964).
111. R. Foster and P. Hanson, *Tetrahedron* **21**, 255 (1965).
112. E. F. Caldin, "Fast Reactions in Solution," Blackwell, Oxford (1964), Ch. 4, 5; E. F. Caldin and J. E. Crooks, *J. scient. Instrum.* **44**, 449 (1967); K. Bergmann,

M. Eigen and L. De Maeyer, *Ber. Bunsenges. physik. Chem.* **67**, 819 (1963);
K. Bergmann, *Ber. Bunsenges. physik. Chem.* **67**, 826 (1963).

113. P. R. Hammond, *J. chem. Soc.* 479 (1964).

114. P. J. Trotter and M. W. Hanna, *J. Am. chem. Soc.* **88**, 3724 (1966).

115. T. G. Beaumont and K. M. C. Davis, *J. chem. Soc.* (B), 1131 (1967).

116. R. Foster, C. A. Fyfe and J. W. Morris, unpublished work.

117. I. D. Kuntz, Jr., F. P. Gasparro, M. D. Johnston, Jr. and R. P. Taylor, *J. Am. chem. Soc.* **90**, 4778 (1968).

118. R. S. Mulliken, *J. Am. chem. Soc.* **74**, 811 (1952).

119. H. Murakami, *Bull. chem. Soc. Japan* **28**, 577 (1955).

120. H. Murakami, *Bull. chem. Soc. Japan* **28**, 581 (1955).

121. S. P. McGlynn, *Chem. Rev.* **58**, 1113 (1958).

122. S. Carter, J. N. Murrell and E. J. Rosch, *J. chem. Soc.* 2048 (1965).

123. J. N. Murrell, *J. Am. chem. Soc.* **81**, 5037 (1959).

124. P. H. Emslie, R. Foster, C. A. Fyfe and I. Horman, *Tetrahedron* **21**, 2843 (1965).

125. P. H. Emslie, R. Foster and R. Pickles, *Can. J. Chem.* **44**, 9 (1966).

126. C. A. Fyfe and I. B. C. Matheson, unpublished work.

127. M. J. S. Dewar and C. C. Thompson, Jr., *Tetrahedron*, **Suppl. 7**, 97 (1966).

128. R. Foster and I. Horman, *J. chem. Soc.* (B), 171 (1966).

129. H. Kuroda, T. Amano, I. Ikemoto and H. Akamatu, *J. Am. chem. Soc.* **89**, 6056 (1967).

130. T. Matsuo and O. Higuchi, *Bull. chem. Soc. Japan* **41**, 518 (1968).

131. E. A. Guggenheim, *Trans. Faraday Soc.* **56**, 1159 (1960).

132. J. Ronayne and D. H. Williams, *J. chem. Soc.* (B), 540 (1967).

133. J. E. Prue, *J. chem. Soc.* 7534 (1965).

134. R. L. Scott, personal communication.

135. S. Carter, *J. chem. Soc.* (A), 404 (1968).

136. S. D. Ross, D. J. Kelley and M. M. Labes, *J. Am. chem. Soc.* **78**, 3625 (1956).

137. C. C. Thompson, Jr., unpublished work.

138. R. Foster, unpublished work.

139. P. Job, *Compt. Rend.* **180**, 928 (1925); *Annls Chim.* [10], **9**, 113 (1928); *Annls Chim.* [11], **6**, 97 (1936).

140. R. K. Gould and W. C. Vosburgh, *J. Am. chem. Soc.* **64**, 1630 (1942); W. C. Vosburgh and G. R. Cooper, *J. Am. chem. Soc.* **63**, 437 (1941).

141. P. Hagemuller, *Compt. Rend.* **230**, 2190 (1950).

142. M. M. Jones and K. K. Innes, *J. phys. Chem., Ithaca* **62**, 1005 (1958).

143. M. M. Jones, *J. Am. chem. Soc.* **81**, 4485 (1959).

144. E. Asmus, *Z. analyt. Chem.* **183**, 321, 401 (1961).

145. P. Klaeboe, *J. Am. chem. Soc.* **84**, 3458 (1962); S. Nagakura, *J. Am. chem. Soc.* **80**, 520 (1958); H. Tsubomura, *J. Am. chem. Soc.* **82**, 40 (1960).

146. R. S. Drago, R. L. Carlson, N. J. Rose and D. A. Wenz, *J. Am. chem. Soc.* **83**, 3572 (1961).

147. D. W. Tanner and T. C. Bruice, *J. phys. Chem., Ithaca* **70**, 3816 (1966).

148. B. E. Mann, Thesis, Oxford University (1968).

149. R. Foster and I. B. C. Matheson, unpublished work.

150. J. A. A. Ketelaar, C. van de Stolpe and H. R. Gersmann, *Recl Trav. chim. Pays-Bas Belg.* **70**, 499 (1951).

151. R. J. Abraham, *Molec. Phys.* **4**, 369 (1961).

152. R. Foster, C. A. Fyfe and M. I. Foreman, *Chem. Commun.* 913 (1967).

153. R. Foster, C. A. Fyfe and M. I. Foreman, unpublished work.

154. E. C. W. Clarke and D. N. Glew, *Trans. Faraday Soc.* **62**, 539 (1966).

Chapter 7

Determination of the Position of Equilibrium—Results

7.A. General

Determinations of association constant have been made on a large range of organic charge-transfer complexes under various conditions. None of the methods currently used yields results in which, for all systems, high confidence can be placed. The comparison of association constant values obtained from spectroscopic data of several systems containing π-donors and π-acceptors under different concentration conditions with association constants derived from n.m.r. chemical shift measurements (Chapter 6) suggests that, for these complexes, possibly more weight should be given to association constant values obtained from optical data under the condition $[D]_0 = [A]_0$ and to values derived from n.m.r. chemical shift data, rather than from optical density measurements of the charge-transfer band on solutions in which one solute species is present in large excess. In fact, this last category accounts for the vast majority of all association constant determinations of organic charge-transfer complexes. Apart from complications due to the possible presence of termolecular complexes and of isomeric complexes, specific solvation, and such like problems, the experimental determination of the optical absorption due to the complex may be made difficult by the absorptions of the component molecules themselves. This is particularly evident for weak complexes where only a small fraction of either component is complexed.[1, 2] Even for interactions which are not particularly weak, some of the reported association constants obtained from optical measurements on solutions of comparable concentration are disconcertingly disparate. As an example, the association constants reported by different groups of workers for hexamethylbenzene–1,3,5-trinitrobenzene in various solvents are compared in Table 7.1. A comparison of values of K^{AD} for various tetracyanoethylene complexes obtained by various workers is included in Table 7.11. These various inconsistencies reduce confidence in published thermodynamic data. However, amongst some groups of complexes there can perhaps be more optimism. The agreement between values of association constants obtained from absorbance measurements of the charge-transfer bands of various complexes between iodine and n-donors,[3] and from absorbance measurements of the shifted iodine transition in the same complexes[4]

TABLE 7.1. A comparison of literature values for the association constant (K_c^{AD}) of 1,3,5-trinitrobenzene–hexamethylbenzene at 20°C obtained from optical measurements under the condition $[D]_0 \gg [A]_0$.

Solvent	$K_c^{AD}/l\ mol^{-1}$		
Cyclohexane	$17 \cdot 50 \pm 0 \cdot 20$*		$13 \cdot 5 \pm 0 \cdot 4$†
Carbon tetrachloride	$4 \cdot 87 \pm 0 \cdot 08$*	$7 \cdot 10 \pm 0 \cdot 05$‡	$5 \cdot 7\ \ \pm 0 \cdot 3$†
Chloroform	$0 \cdot 81$§	$0 \cdot 92 \pm 0 \cdot 10$*	$0 \cdot 76 \pm 0 \cdot 05$†

* Ref. 33.
† Ref. 70.
‡ G. Briegleb and J. Czekalla, Z. Elektrochem. **59**, 184 (1955).
§ Ref. 72.

(Table 6.1, p. 128) suggests that, in some of these systems at least, a degree of reliance can be attached to the association constant values. Evidence of compliance with Beer's law and for the absence of ter-molecular complexes is shown by some,[5] though certainly not all,[6] iodine complexes in solution by the presence of sharp isosbestic points in the near-ultraviolet–visible absorption spectra.

Whilst published values may not in all cases represent accurate measures of the absolute values of association constant, the estimated association constants under set conditions for a given series of complexes may provide a reasonable measure of their relative values. Care should be taken before combining the results of different groups of workers.

Much of the earlier work on systems involving diatomic halogen acceptors has been reviewed by Andrews and Keefer.[7] A detailed discussion of thermodynamic constants for a wide range of complexes reported in the literature up to 1961 has been given by Briegleb.[8] The amount of data which has now been published makes it virtually impossible to give a comprehensive review here. Appendix Table 2 lists

references to some of the very many papers concerned with experimentally determined association constants and other thermodynamic
constants for various charge-transfer systems.

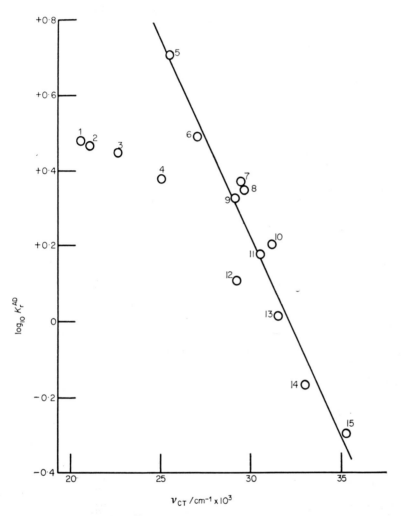

FIG. 7.1. Values of $\log_{10} K_r^{AD}$ for various donor–1,3,5-trinitrobenzene complexes,
calculated from n.m.r. shift data, plotted against ν_{CT} for the various complexes,
carbon tetrachloride solution, 33·5°C. The donors are: (1) N,N-dimethylaniline;
(2) 4-bromoaniline; (3) N-methylaniline; (4) aniline; (5) hexamethylbenzene; (6)
pentamethylbenzene; (7) 1,2,3,4-tetramethylbenzene; (8) 1,2,3,5-tetramethyl-
benzene; (9) 1,2,4,5-tetramethylbenzene; (10) 1,2,4-trimethylbenzene; (11) 1,2,3-
trimethylbenzene; (12) 1,3,5-trimethylbenzene; (13) m-xylene; (14) toluene;
(15) benzene. (Data from Refs. 49 and 50.)

7

In the following sections the influence of various factors on the magnitude of the association constant for complex formation is discussed.

It is important to bear in mind that, as already pointed out in earlier chapters, the contribution of charge-transfer forces to the stability of the ground state of the complex, though sometimes significant, is rarely the overwhelming factor. This is particularly evident amongst complexes of π-donors and is demonstrated by the general lack of correlation between association constant and the energy of the lowest intermolecular charge-transfer transition (Fig. 7.1). Nevertheless, within a closely related group of donors, some correlation can sometimes be observed (Fig. 7.1).

7.B. Association Constants

7.B.1. Effect of Solvent

In this section we are concerned with the effect of solvent on the position of equilibrium between the component species and the complex, and not with processes such as the ionization of complexes by solvating ionizing solvents, which are described in Chapter 11.

From published data, the position of equilibrium appears to be very dependent on the nature of the solvent (see Table 7.2). Several determinations of association constants for electron donor–electron acceptor systems in the gaseous phase have been made [9-18] (Table 7.3). These results should provide a near-ideal standard by which to compare the effect of various solvents in the liquid phase. The experimental difficulties are such that the probable errors in the calculated constants are considerably greater in the gaseous phase than in the liquid phase. Nevertheless, there appears to be little doubt that, for virtually all of the complexes studied, the association constant in the gaseous phase is larger, often considerably larger, than that observed in an "inert" solvent in the liquid phase. The corresponding values of the maximum molar absorptivity (extinction coefficient, ϵ_{max}^{AD}) show the opposite trend. One exception is the sulphur dioxide–trimethylamine complex,[17] for which the values of K_c^{AD} and of ϵ_{max}^{AD} are remarkably similar in the gas phase compared with the values obtained from solutions in heptane (see Table 7.3). It has been pointed out (Chapter 6) that, although the product $(K^{AD}\epsilon^{AD})$ can be obtained with relative ease, difficulties and ambiguities arise when attempts are made to separate these two factors. However, this situation cannot provide the explanation of the recorded differences between K^{AD} and ϵ^{AD} values for the liquid and gaseous states since the product $(K^{AD}\epsilon^{AD})$ is significantly different for a given complex at a given temperature in the two states.

For a thermodynamic function $X^{\ominus}$, the relation between the change

TABLE 7.2. The effect of solvent on the optically determined association constant (K_c^{AD}) of some charge-transfer complexes.

Tetrachlorophthalic anhydride–hexamethylbenzene[*]

Solvent	K_c^{AD}/l mol^{-1}	D.C.[†]
n-Hexane	34	1·89
Carbon tetrachloride	14·0	2·24
Dibutyl ether	13·0	3·06[‡]
Benzotrifluoride	6·4	9·18[§]
Fluorobenzene	2·7	5·42[‡]
Cyclohexanone	2·4	18·3
Benzene	2·3	2·28

1,3,5-Trinitrobenzene–N,N-dimethylaniline[‖] *at* 19·7 $\pm$ 1·2°C

Solvent	K_c^{AD}/l mol^{-1}	D.C.[†]
Cyclohexane	9·6	2·02
n-Hexane	8·2	1·89
n-Heptane	8·2	1·92
Decalin[¶]	7·2	2·18[**]
Carbon tetrachloride	3·4	2·24
Chloroform	1·3	4·81
1,1,2,2-Tetrachloroethane	0·2	8·2
1,4-Dioxane	0·15	2·21[‡]

1,3,5-Trinitrobenzene–naphthalene[††] *at* 20°C

Solvent	K_c^{AD}/l mol^{-1}	D.C.[†]
n-Heptane	9·58	1·92
Cyclohexane	9·15	2·02
n-Hexane	7·82	1·89
Carbon tetrachloride	5·16	2·24
Carbon disulphide	3·25	2·64
Chloroform	1·82	4·81

[*] J. Czekalla and K.-O. Meyer, *Z. phys. Chem. Frankf. Ausg.* **27**, 185 (1961).
[†] D.C. = dielectric constant at 20°C unless otherwise stated.
[‡] At 25°C.
[§] At 30°C.
[‖] Ref. 19.
[¶] Mixture of *cis* and *trans*.
[**] Average value.
[††] Ref. 32.

TABLE 7.3. Association constants and other data from measurements of charge-transfer complexes in the gaseous phase with comparative measurements in the liquid phase.

Complex	Phase	ν_{CT}/cm^{-1}	ϵ_{max}^{AD}/mol^{-1} cm^{-1} l	Temp./°C	K_c^{AD}/l mol^{-1}	$-\Delta H^{\ominus}$/kcal mol^{-1}	$-\Delta S^{\ominus}$/cal mol^{-1} deg^{-1}	Ref.*
Diethyl ether–iodine	Gas	42,700	2100 ± 400	25	6.4 ± 1.2	3.2 ± 1		9
	CCl$_4$	40,200	4700	20	0.81	4.3†		a
	n-Heptane	39,700	5650	25	0.86	4.2†		4
	n-Heptane			21.5 ± 0.5	0.72			b
	CS$_2$			30	0.43			c
Benzene–iodine	Gas	37,300	1650 ± 100	25	4.5 ± 0.6	2.0 ± 0.1		9
	Gas	33,700			3.4	2.44		16
	CCl$_4$	33,800	15,400	25	0.157			d
	CCl$_4$	33,800	16,700		0.158	1.3		e
	CCl$_4$	34,200	14,700	Room	0.149			f
	CCl$_4$		16,400	25	0.15			54
	CCl$_4$				0.145	1.32†		76
Dimethyl sulphide–iodine	Gas	34,970	5000 ± 2500	25	220 ± 140	7.4 ± 0.5	13.6 ± 2.4	10
	Gas‡	34,500	11,200 ± 1900	25	226	8.4 ± 0.4	15.0 ± 1.1	13, 15
			(8770)	25	(285)	(8.4 ± 0.4)	(14.8 ± 1.0)	13, 15
Diethyl sulphide–iodine	Gas	34,480	3500 ± 1000	25	750 ± 250	8.3 ± 0.5	14.2 ± 1.9	10
	n-Heptane‡	33,000	27,600	25	180	8.9 ± 0.6	19.4 ± 2.0	15
			(24,800)	25	(200)	8.3 ± 0.2	(17.6 ± 0.5)	g
Tetrahydrothiophene–iodine	Gas	34,130	1750 ± 400	25	2400 ± 400	9.0 ± 0.5	14.2 ± 1.6	10
	n-Heptane	32,730	27,000	25	252 ± 5	7.8	15	10

			K_c^{AD}	Temp (°C)	K_c	$-\Delta H^{\ominus}$	$-\Delta S^{\ominus}$	Refs*
o-Xylene–tetracyanoethylene	Gas	25,380	1100 ± 350	25	350 ± 120	8.7 ± 0.5	17 ± 2.0	10
	CH₂Cl₂	23,530 26,880	800 ± 250§	22	275	7.2 ± 1	13.1 ± 1.6	11
p-Xylene–tetracyanoethylene	Gas	21,700	2650	25	280 ± 100	8.1 ± 0.5	15.5 ± 2	10
	CH₂Cl₂	24,100	2770	22	0.49	3.37†	7.38‖	20
Mesitylene–tetracyanoethylene	Gas	23,640	1250 ± 375	25	1020 ± 350	9.9 ± 0.5	18.7 ± 2.0	10
	CH₂Cl₂	21,700	3120	22	1.11	4.52†	9.61‖	20
Durene–tetracyanoethylene	Gas	21,140 23,260	400 ± 150§	25	11,800 ± 5900	10.8 ± 0.8	17.3 ± 3.1	10
	CH₂Cl₂	20,800	2075	22	3.4	5.07†	9.23‖	20
Diethyl ether–carbonyl cyanide	Gas	38,200	65		45			12, 14
	n-Hexane	34,800	420		3.6			12, 14
Benzene–carbonyl cyanide	Gas	34,600	125		14			12, 14
	n-Hexane	32,400	950		0.4			12, 14
Trimethylamine–sulphur dioxide	Gas	36,200	5700 ± 350¶	25	340	9.7 ± 0.4	21.0 ± 1.0	17
	n-Heptane	36,600	5400 ± 56	25	2550 ± 50	11.0 ± 0.3	21.4 ± 0.9	17

* Refs.: [a] P. A. D. de Maine, J. chem. Phys. 26, 1192 (1957). [b] J. S. Ham, J. chem. Phys. 20, 1170 (1952). [c] H. Yamada and K. Kozima, J. Am. chem. Soc. 82, 1543 (1960). [d] H. A. Benesi and J. H. Hildebrand, J. Am. chem. Soc. 71, 2703 (1949). [e] J. A. A. Ketelaar, J. Phys. Radium, Paris 15, 197 (1954). [f] M. Tamres, D. R. Virzi and S. Searles, J. Am. chem. Soc. 75, 4358 (1953). [g] M. Tamres and S. Searles, J. phys. Chem., Ithaca 66, 1099 (1962).

† Values of $\Delta H^{\ominus}$x.

‡ The experimental values of K_c^{AD} in the gaseous phase are based on estimates of the reactants by absorption in n-heptane. The values for the gaseous phase quoted in parentheses correspond to those experimental values for K_c^{AD} in solution which are quoted in parentheses.

§ Refers to higher energy bands.

‖ Values of $\Delta S^{\ominus}$x.

¶ At 40°C.

in the gaseous phase $\Delta X^{\ominus}$(g) and the change in the liquid solution phase $\Delta X^{\ominus}$(soln) on the complex formation:

$$
\begin{array}{ccccc}
\text{D(g)} & + & \text{A(g)} & \xrightarrow{\;\Delta X^{\circ}\text{(g)}\;} & \text{AD(g)} \\
\downarrow^{\text{solv.}}{\scriptstyle\Delta X^{\circ}_{\text{D}}\text{(soln)}} & & \downarrow^{\text{solv.}}{\scriptstyle\Delta X^{\circ}_{\text{A}}\text{(soln)}} & & \downarrow{\scriptstyle\Delta X^{\circ}_{\text{AD}}\text{(soln)}} \\
\text{D(soln)} & + & \text{A(soln)} & \xrightarrow{\;\Delta X^{\circ}\text{(soln)}\;} & \text{AD(soln)}
\end{array}
\qquad (7.1)
$$

will be:

$$
\Delta X^{\ominus}\text{(g)} = \Delta X^{\ominus}\text{(soln)} - [\Delta X^{\ominus}_{\text{AD}}\text{(soln)} - \Delta X^{\ominus}_{\text{A}}\text{(soln)} - \Delta X^{\ominus}_{\text{D}}\text{(soln)}]
$$

$$(7.2)$$

In the case of the free energy change $\Delta G^{\ominus}(= -RT \ln K)$, the difference between $\Delta G^{\ominus}$(g) and $\Delta G^{\ominus}$(soln), reflected by the often large difference in association constants for the two phases, indicates that the sum of the solvation energies of the uncomplexed-donor and uncomplexed-acceptor molecules is not identical with the solvation energy of the complex.

The observed differences in the functions $\Delta X^{\ominus}$(g) and $\Delta X^{\ominus}$(soln) in equation (7.1) often represent small differences between relatively large quantities, and the experimental values must be treated with circumspection. However, there is little doubt that the solvent can affect considerably the magnitudes of the thermodynamic constants which describe the equilibrium of a complex with its components.

For equilibria in solution, there often appears to be competition by the solvent molecule (S) for either the donor or the acceptor species (or both).[19-22] The effect of this on the apparent value of the equilibrium:

$$
\text{A} + \text{D} \rightleftharpoons \text{AD} \qquad (7.3)
$$

has already been discussed in Chapter 6. For example, in dilute solutions where the solvent shows donor properties in competition with the solute D molecules, by equation (6.7) (p. 127) we have:

$$
K_c^{\text{AD}}(\text{corr}) = K_c^{\text{AD}}(\text{obs})\{1 + K_c^{\text{AS}}[\text{S}]\} \qquad (7.4)
$$

where K_c^{AD}(obs) is the observed value of the association constant, K_c^{AD}(corr) the association constant for the complex AD had there been no solvent competition, and K_c^{AS} the association constant for the complex AS. The solvent does not have to compete as either an electron donor or an electron acceptor: competitive interaction by some other mechanism such as hydrogen bonding would be equally effective. Even in cases where K_c^{AS} has a low value, the effect may be significant since the concentration of the solvent is large relative to either $[D]_0$ or $[A]_0$. This competition is probably the major solvent effect for solvents such as dioxane and ethers which are known to be effective n-donors; chloroform and dichloromethane which have been shown to hydrogen-bond to aromatic π-donor molecules; and even carbon tetrachloride, which appears to behave as an electron acceptor.[20, 23–29]* The gross differences in association constants for some tetracyanoethylene complexes in different solvents have been accounted for in this way.[20] Good agreement for association constant values and for enthalpy values has similarly been obtained, after correction for solvent competition, for the system dimethylacetamide–iodine in benzene solution compared with the system in carbon tetrachloride.[22] It may be chance that, for these two complexes, the solvent differences can be more or less completely accounted for by solvent competition. However, one can at least conclude that such effects are significant. The paucity and probable unreliability of the necessary K^{AS} (or K^{DS}) values make this type of calculation possible only in a few systems.

Correction for solvent competition must be applied before any possible correlation between association constant (or $\Delta H^{\ominus}$, $\Delta S^{\ominus}$) and some solvent property such as dielectric constant or polarizability can be tested. This has not yet been attempted. For example, it is irrelevant that the uncorrected values of K^{AD} do not show any correlation with the dielectric constant of the solvent (Table 7.2).

The possibility[30] of specific solvation of the species A, D and AD (Chapter 6) provides an explanation for the decreased intensity of the charge-transfer absorption in the vapour phase relative to the absorption in solution. However, there are features of the present specific solvation model which are difficult to explain; for example, why the effects observed in optical experiments are not also observed in chemical shift measurements; for such appears to be the case (see Chapter 6).

The suggestion of Trotter,[31] that in the liquid phase the complexes are under a very large mechanical pressure, of the order 10^3–10^4 atm. (see Chapter 6), leads to the expectation that the association constants should be smaller in the gaseous phase. This would follow if there is a decrease in the volume of the interacting molecules when the complex is formed. Of course, it could be that the compression is so great that considerable repulsive forces are encountered, thus decreasing the stability in the ground state. In fact, even when allowance for competition is made, there are as yet virtually no cases where, for a given complex, K^{AD} is larger for the liquid phase than for the gaseous phase.

The gross effects of various relatively inert solvents on the *observed*

* See also Chapter 10, Section B, which contains further references.

values of K^{AD} for associations between neutral donors and neutral acceptors are generally agreed upon,[19, 32, 33] namely, that the poorest

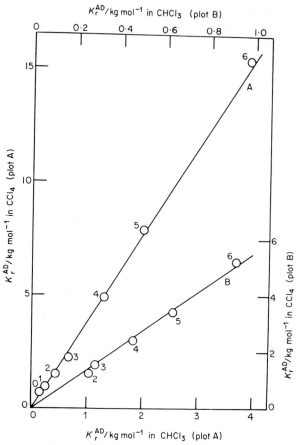

[N. M. D. Brown, R. Foster and C. A. Fyfe, Ref. 49]

FIG. 7.2. Plots of K_r^{AD} for complexes of a series of donors with (A) fluoranil; (B) 1,4-dicyano-2,3,5,6-tetrafluorobenzene in carbon tetrachloride solution, against K_r^{AD} for the same complexes in chloroform solution, all at 33·5°C. Donors: (0) benzene; (1) toluene; (2) p-xylene; (3) mesitylene; (4) durene; (5) pentamethylbenzene; (6) hexamethylbenzene.

ionizing solvents give rise to the highest values of K_c^{AD}: paraffins > carbon tetrachloride > carbon disulphide > chloroform > dichloromethane (Table 7.2). For some series of complexes of structurally related donors with a common acceptor, the ratio of the value of K^{AD} for a given complex in two solvents is remarkably independent of the particular complex (see

Fig. 7.2). When a wide range of donor types is taken, or the acceptor moiety is altered, this simple relationship fails.

7.B.2. Effect of Temperature

The association constant for a given system decreases as the temperature increases. Both in liquid solution and in the gaseous phase the behaviour has been simply interpreted in terms of temperature-independent enthalpy and entropy changes in the normal van't Hoff relationship. The temperature ranges used have tended to be wider in the case of the gaseous phase equilibria than for studies in the liquid phase. However, large temperature ranges in the liquid phase can introduce new factors— for example, changes in the structure of the solvent. The evaluations of $\Delta H^{\ominus}$ and $\Delta S^{\ominus}$ for complex formation from the temperature variation of K^{AD} are discussed in Section 7.C.

7.B.3. Effect of Pressure

Gibson and Loeffler[34] were the first to report on the intensification of the visible spectrum of a mixture of an electron donor and an electron acceptor when the system is compressed.

The effect of pressure on various iodine–electron donor equilibria in n-hexane has been measured.[35] The absorbance increases by about 50% as the pressure is increased from 1 to 10^3 atm. This increase is only partly accounted for by the increased concentrations of the various solute species due to solvent compressibility. Ham[35] suggested that the difference was due to an increase in K^{AD} rather than in ϵ^{AD}.

More recently, the equilibria between tetracyanoethylene and various aromatic hydrocarbons in dichloromethane at pressures up to 10^4 atm. have been studied.[36] Considerable intensity increases are observed for the benzene–tetracyanoethylene complex after large corrections have been applied for material compression and expansion of the pressure

TABLE 7.4. The effect of pressure on the association constant (K_x^{AD}), the uncorrected oscillator strength (f) and the corrected oscillator strength (f_n) of the benzene-tetracyanoethylene complex in dichloromethane at 30°C.*

Pressure/atm.	K_x^{AD}	f_n	f
1	8·7	0·026	0·033
1000	8·1	0·030	0·039
2000	7·5	0·035	0·046
4000	7·1$_5$	0·042	0·057

* Ref. 36.

vessel. This increase is the net effect of changes in the oscillator strength of the complex (see Chapter 3), changes in the refractive index of the solvent and the possible shift in the equilibrium constant. Estimates of some of these quantities are given in Table 7.4. The molar absorptivity, the uncorrected oscillator strength, and the oscillator strength corrected for the refractive index effect using the Debye theory of dielectric polarization,[37] all increase as the pressure increases. By contrast K_x^{AD} apparently decreases slightly. This result is somewhat unexpected. Gott and Maisch[36] suggest that this may be the result of increased solvent-competition as the pressure is increased.

7.B.4. Effect of the Donor

The values obtained for the association constants of charge-transfer complexes are highly dependent on the nature of the donor and acceptor moieties. For complexes between σ-acceptors, such as iodine, and n-donors, where the intermolecular overlap may be considerable, values of K_c^{AD} greater than 10^4 l mol^{-1} have been observed although other complexes within this group are very much less stable. Some selected values are given in Table 7.5 (see also references in Appendix Table 2, p. 396). Amongst the many observations reported, it is noted that the ethers and amides are very much weaker n-donors than their thio-analogues[38] (Table 7.5). In some instances n-donors and π-donors are contrasted by very large differences in their degree of association with σ-acceptors compared with π-acceptors. For example, the complex of the n-donor methyl disulphide with the π-acceptor tetracyanoethylene is very much weaker[39] than the complex with the σ-acceptor iodine[40] in comparable solvents. However, when the corresponding complexes of the π-donor hexamethylbenzene are compared, the tetracyanoethylene complex[20] is very much stronger than the iodine complex.[7]

This division between n-donors and π-donors is not always sharp. Thus, some compounds behave as n-donors towards some acceptors and as π-donors towards other acceptors. For example, N,N-dimethylaniline appears to interact with iodine via its n-electrons,[41] whilst with π-acceptors it has the properties of a π-donor. The possibility of the simultaneous existence of two types of complex involving both n- and π-electron binding has been proposed.[42] The K^{AD} values for some series of complexes with a common acceptor component have been related to various measures of the electron-donating ability of the electron donor. A linear correlation has been demonstrated between $\log K^{AD}$ and the Taft σ^* constant of the donor for series of complexes of substituted acetonitriles with various diatomic halogen acceptors (Fig. 7.3).[43] Likewise, a good correlation of $\log K^{AD}$ with Hammett σ values of the

TABLE 7.5. A selection of experimentally determined values of thermodynamic constants for various n-donor–σ-acceptor complexes.*

Donor	Acceptor	Solvent	Temp. for K_c^{AD}/°C	K_c^{AD}/l mol^{-1}	$-\Delta H^{\ominus c}$/kcal mol^{-1}	$-\Delta S^{\ominus c}$/cal mol^{-1} deg^{-1}	Ref.†
Ammonia	I$_2$	n-Heptane	20	67	4·8	8·0	77
Methylamine	I$_2$	n-Heptane	20	530	7·1	12·3	77
Ethylamine	I$_2$	n-Heptane	20	720	7·4	12·3	77
n-Butylamine	I$_2$	n-Heptane	20	1230	8·4	14·8	77
Dimethylamine	I$_2$	n-Heptane	20	6800	9·8	15·9	77
Diethylamine	I$_2$	n-Heptane	20	7120	9·7	18·4	77
Piperidine	I$_2$	n-Heptane	20	9400	10·3	16·1	77
Trimethylamine	I$_2$	n-Heptane	20	12,100	12·1	22·6	77
Triethylamine	I$_2$	n-Heptane	20	6320	12·0	23·5	a
Tri-n-propylamine	I$_2$	n-Heptane	20	1390	12·1	26·6	77
Tri-n-butylamine	I$_2$	n-Heptane	20	1600	12·3	27·8	77
N,N-Dimethylaniline	I$_2$	n-Heptane	27	18·8	8·2 ± 0·3	21·6 ± 0·6	41
N,N-Dimethylacetamide	I$_2$	CCl$_4$	25	6·8 ± 0·2	3·9 ± 0·1	9·3 ± 0·6	b
N,N-Dimethylthio-acetamide	I$_2$	CCl$_4$	25	1190 ± 40	9·5 ± 1·5		38
1,4-Dioxane	I$_2$	CCl$_4$	25	1·00	3·3 ± 0·3	11·0 ± 1·0	c
1,4-Dithiane	I$_2$	CCl$_4$	25	76·9	6·2 ± 0·3	12·1 ± 1·0	c
Pyridine	I$_2$	CCl$_4$	25	97			d
Pyridine	ICl	CCl$_4$	25	4·8 × 10^5			d
Pyridine	IBr	CCl$_4$	25	1·3 × 10^4			d
N,N-Dimethylacetamide	Br$_2$	CCl$_4$	25	1·8 ± 0·4	1·6 ± 0·3	4 ± 1	e

* For a more comprehensive set of values see references in Appendix Table 2.

† Refs.: a S. Nagakura, J. Am. chem. Soc. 80, 520 (1958). b R. S. Drago, D. A. Wenz and R. L. Carlson, J. Am. chem. Soc. 84, 1106 (1962); cf. R. S. Drago, R. L. Carlson, N. J. Rose and D. A. Wenz, J. Am. chem. Soc. 83, 3572 (1961). c J. D. McCullough and I. C. Zimmermann, J. phys. Chem., Ithaca 65, 888 (1961). d A. I. Popov and R. H. Rygg, J. Am. chem. Soc. 79, 4622 (1957). e R. S. Drago and D. A. Wenz, J. Am. chem. Soc. 84, 526 (1962).

donor substituents for a series of complexes between iodine and substituted benzamides,[44] and substituted thioanisoles [45] acting as n-donors, has been reported (Fig. 7.4).

It has been more difficult to discern correlations of published values of K^{AD} with donor properties amongst complexes involving π-donors. This is due in part at least to the uncertainty of many of the experimental evaluations of K^{AD} (see above, also Chapter 6). This criticism may well apply to the data treated by Charton [46] who has used an extended form of the Hammett relationship to correlate the K^{AD} values of complexes of a

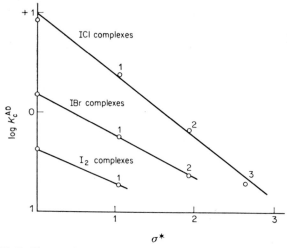

[W. B. Person, W. C. Golton and A. I. Popov, Ref. 43]

FIG. 7.3. Relationship between $\log K_c^{AD}$ and Taft σ^* values for nitrile–halogen complexes, carbon tetrachloride solution, 25°C. (1) acetonitrile; (2) chloroacetonitrile; (3) dichloroacetonitrile.

range of π-donors with various acceptors, and to the correlations of $\log K^{AD}$ discussed in detail by Briegleb.[8] A similar correlation has also been described by Rosenberg et al.[47]

It may be that n.m.r. chemical shift measurements provide more satisfactory data for the evaluation of K^{AD} (Chapter 6). It is certainly the case that the values of K^{AD} obtained by this type of measurement (Table 7.6) show regularities within series of complexes, which are not always apparent in the values derived from optical data on solutions which have the customary concentration condition $[D]_0 \gg [A]_0$. For example, the plots of $\log K^{AD}$ for a series of methylbenzene complexes with one electron acceptor, against the $\log K^{AD}$ values for the complexes of the same series of donors with other acceptors, are linear (Fig. 7.5). It is difficult to find comparable values from optical data; however, the irregular plot shown in Fig. 7.6 is typical of the lower correlation often observed in K^{AD} values obtained from optical data.

Amongst the K^{AD} values obtained from n.m.r. chemical shift data, the association constants of a series of complexes with a common acceptor

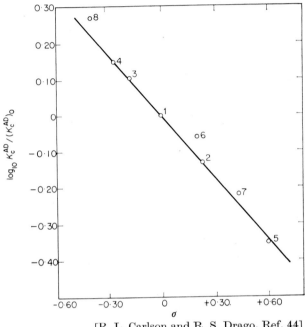

[R. L. Carlson and R. S. Drago, Ref. 44]

FIG. 7.4. Plot of $\log K_c^{AD}/(K_c^{AD})_0$ against Hammett σ values for complexes of iodine with: (1) N,N-dimethylbenzamide; (2) p-chloro-N,N-dimethylbenzamide; (3) p-methyl-N,N-dimethylbenzamide; (4) p-methoxy-N,N-dimethylbenzamide; (5) 3,4-dichloro-N,N-dimethylbenzamide; (6) o-chloro-N,N-dimethylbenzamide; (7) 2,4-dichloro-N,N-dimethylbenzamide; (8) o-methoxy-N,N-dimethylbenzamide, $(K_c^{AD})_0$ refers to N,N-dimethylbenzamide.

species increase as the Lewis basicity of the donor increases.[48-51] However, no *linear* relationship between $\log K^{AD}$ and the ionization potential of the donor (or ν_{max}) is observed. This is to be expected since the dative structure in the ground state corresponds to only a fraction, often a minor fraction, of the total ground-state binding energy[52] (see Chapter 2). A reasonably linear correlation between $\log K^{AD}$ and Hammett σ values for the donor substituent is observed for the series of complexes of 1,3,5-trinitrobenzene with various mono-substituted benzenes. The correlation breaks down when various electron-withdrawing groups are

TABLE 7.6. Selected values of K_r^{AD} for charge-transfer complexes measured in carbon tetrachloride at 33·5°C, together with the chemical shift of the measured nucleus in the pure complex in solution relative to the chemical shift in the pure acceptor (Δ_0), as determined by an n.m.r. method.*

Donor	1,3,5-Trinitrobenzene† K_r^{AD}/kg mol^{-1}	Δ_0/Hz§	1,4-Dinitrobenzene† K_r^{AD}/kg mol^{-1}	Δ_0/Hz§	Fluoranil‡ K_r^{AD}/kg mol^{-1}	Δ_0/Hz‖	1,4-Dicyano-2,3,5,6-tetrafluorobenzene‡ K_r^{AD}/kg mol^{-1}	Δ_0/Hz‖
Benzene	0.5_0	75	0.2_7	85	0.7_0	99	0.8_0	82
Toluene	0.6_7	72	0.2_7	90	0.9_6	127		
Ethylbenzene	0.7_1	62			1.1_0	120		
n-Propylbenzene	0.5_8	63			0.8_9	106		
i-Propylbenzene	0.6_1	60						
n-Butylbenzene	0.6_2	74			0.9_4	105		
t-Butylbenzene	0.4_5	72						
n-Pentylbenzene	0.6_1	72			0.7_3	145		
n-Hexylbenzene	0.5_7	70			0.9_9	106		
1,2-Dimethylbenzene (o-xylene)	1.0_9	66	0.4_2	81	1.4_6	149		
1,3-Dimethylbenzene (m-xylene)	1.0_3	65	0.4_1	76	1.5_3	161		
1,4-Dimethylbenzene (p-xylene)	0.9_2	68	0.3_5	86	1.5	161	1.2	100
1,2,3-Trimethylbenzene (hemimellitene)	1.5_0	65	0.4_9	83	2.6_6	158		
1,2,4-Trimethylbenzene (pseudocumene)	1.5_8	61	0.4_9	88	2.5_9	175		
1,3,5-Trimethylbenzene (mesitylene)	1.2_7	61	0.4_1	79	2.2	172	1.5	111
1,2,3,4-Tetramethylbenzene (prehnitene)	2.3_3	64	0.6_5	79	4.5_3	192		
1,2,3,5-Tetramethylbenzene (isodurene)	2.2_2	67	0.6_6	71	4.5_4	193		
1,2,4,5-Tetramethylbenzene (durene)	2.1_1	62	0.6_2	84	4.9	208	2.4	136
Pentamethylbenzene	3.0_9	66	0.7_5	90	7.9	229	3.4	154
Hexamethylbenzene	5.1_2	65	1.0_1	91	15.4	261	5.2	168

* For a full description of the method, see p. 140, also Ref. 48. Association constants expressed in kg. solution per mol.
† Refs. 48–50.
‡ Refs. 49, 50 and 53.
§ Proton chemical shifts measured at 60·004 MHz.
‖ ^{19}F chemical shifts measured at 56·462 MHz.

substituents in the donor molecule. The stabilities of these complexes are in fact larger than that of unsubstituted benzene (Fig. 7.7). It seems probable that in these complexes dipole–dipole and other polarization

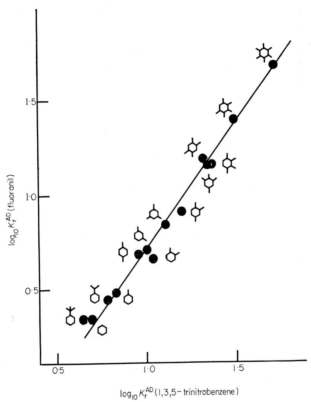

FIG. 7.5. Plots of $\log_{10} K_r^{AD}$ for a series of alkylbenzene–1,3,5-trinitrobenzene complexes against $\log_{10} K_r^{AD}$ for the corresponding fluoranil complexes calculated from n.m.r. chemical shift measurements: carbon tetrachloride solution 33·5°C. (K_r^{AD} values from Refs. 49, 50, 51 and 53.)

forces completely overwhelm any effects due to charge-transfer interaction.[53]

In none of the examples so far quoted is there any very significant steric hindrance of the approach of the two components of the complex. The effects of such steric hindrance can be large. By comparison with hexamethylbenzene, hexaethylbenzene forms complexes of low stability[20, 53–56] (Table 7.7) although the position of the intermolecular charge-transfer band of the hexaethylbenzene complex is at lower

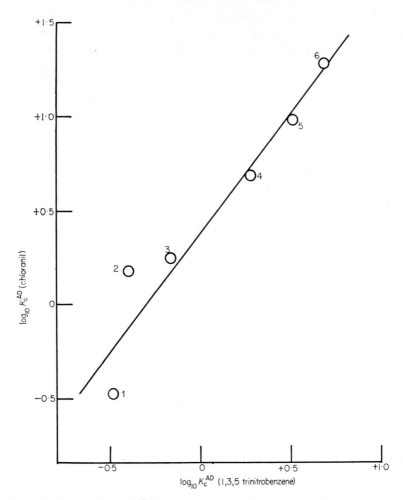

Fig. 7.6. Plots of $\log_{10} K_c^{AD}$ for a series of alkylbenzene–1,3,5-trinitrobenzene complexes against $\log_{10} K_c^{AD}$ for the corresponding chloranil complexes calculated from optical absorption measurements: carbon tetrachloride solution, 25°C. Donors: (1) toluene; (2) m-xylene; (3) mesitylene; (4) durene; (5) pentamethylbenzene; (6) hexamethylbenzene. (K_c^{AD} values from Refs. 48–51.)

energies in comparable complexes. A similar, though smaller, decrease in K^{AD} is observed in the 1,3,5-trinitrobenzene [53] and iodine monochloride [55] complexes of 1,3,5-triethylbenzene, compared with the mesitylene complexes. The decrease in K^{AD} in the series of complexes of 1,3,5-trinitrobenzene with toluene, ethylbenzene, iso-propylbenzene and t-butylbenzene [53] and in other alkylbenzenes [53, 57] may also be accounted for in terms of a primary steric effect. The measured values of K_r^{AD} for various methylpyridine complexes of 1,3,5-trinitrobenzene[53] (Table 7.8) suggest that the pyridines act as n-donors rather than π-donors towards this acceptor so that the presence of substituents *ortho* to the heterocyclic nitrogen atom causes a primary steric effect which adversely affects K_r^{AD}.

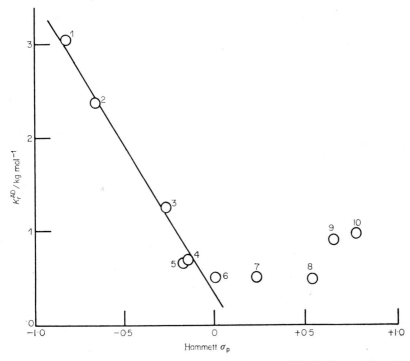

[R. Foster and J. W. Morris, Ref. 53]

FIG. 7.7. Plot of K_r^{AD} for the 1,3,5-trinitrobenzene complexes of monosubstituted benzenes (C_6H_5X, where X = (1) —NMe$_2$; (2) —NH$_2$; (3) —OMe; (4) —CH$_2$CH$_3$; (5) —CH$_3$; (6) —H; (7) —Br; (8) —CF$_3$; (9) —CN; (10) —NO$_2$) in carbon tetrachloride solution at $33 \cdot 5°C$, against the Hammett σ_p constants for the respective substituents.

However, there are inconsistencies: for example, K^{AD} for 3,5-dimethylpyridine is smaller than the value for 3-methylpyridine (Table 7.8). The pronounced alternation of K_r^{AD} with increasing chain-length of n-alkylbenzene complexed with 1,3,5-trinitrobenzene, after the first few members of the series (Fig. 7.8), may be the result of steric packing requirements.[53] A similar, though less pronounced, effect in the association

TABLE 7.7. Association constants (K_c^{AD}) for various hexaethylbenzene complexes (HEB) compared with the corresponding hexamethylbenzene complexes (HMB).

Acceptor	Solvent	Temp./ °C	HEB K_c^{AD}/l mol^{-1}	HMB K_c^{AD}/l mol^{-1}	Method*	Ref.
Iodine	CCl$_4$	25	0·13	1·35	Optical	54
Iodine monochloride	CCl$_4$	25	1·24	22·7	Optical	54, 55
Chloranil	Cyclohexane	20	1·3	28·9	Optical	56
Tetracyano-ethylene	CH$_2$Cl$_2$	22	0·32	16·9	Optical	20
1,3,5-Trinitro-benzene	CCl$_4$	33·5	<0·05	3·2	N.m.r.	53

* See Chapter 6.

TABLE 7.8. Association constants (K_r^{AD}) for some 1,3,5-trinitrobenzene–methylpyridine complexes at 33·5°C, together with the chemical shift of the acceptor proton in the pure complex, relative to the chemical shift of the same proton in the free acceptor (Δ_0),* determined by an n.m.r. method.†

Donor	Solvent Carbon tetrachloride K_r^{AD}/kg mol^{-1}‡	Δ^0/Hz§	Chloroform K_r^{AD}/kg mol^{-1}‡	Δ^0/Hz§
Pyridine	1·4$_7$	7	0·2$_8$	14
2-Methylpyridine	0·9$_6$	18	<0·3	—
3-Methylpyridine	1·9$_6$	11	0·2$_4$	25
4-Methylpyridine	2·4$_1$	10	0·3$_4$	18
2,6-Dimethylpyridine	0·7$_7$	31	<0·1	—
3,5-Dimethylpyridine	1·7$_9$	16	0·2$_8$	28
2,3-Dimethylpyridine	1·4$_1$	23	0·1$_7$	47
2,4-Dimethylpyridine	1·4$_1$	21	0·2$_8$	31
2,4,6-Trimethylpyridine	1·0$_4$	38	0·2$_5$	47

* J. W. Morris, unpublished work.
† Ref. 48, see also Chapter 6.
‡ K_r^{AD} measured in kg solution per mol.
§ Proton chemical shifts measured at 60·004 MHz.

constants of the n-alkylbenzene picrates based on distribution measurements had been described much earlier by Anderson and Hammick.[58]

The complexes of stilbenes with 1,3,5-trinitrobenzene may provide examples of a somewhat different type of primary steric effect. The stability constant for trans-stilbene is, within experimental error, twice

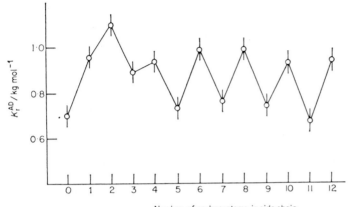

[R. Foster and J. W. Morris, Ref. 53]

FIG. 7.8. Plot of K_r^{AD} for a series of fluoranil–n-alkylbenzene complexes against the number of carbon atoms in the side chain of the donor, calculated from n.m.r. data, carbon tetrachloride solution, 33·5°C.

TABLE 7.9. Association constants (K_r^{AD}) of some 1,3,5-trinitro-benzene–stilbene and related complexes in carbon tetrachloride at 33·5°C, together with the chemical shift of the measured nucleus in the pure complex in solution, relative to the shift in the pure acceptor (Δ^0) as determined by an n.m.r. method.*

Donor	K_r^{AD}/kg mol^{-1}	Δ_0/Hz†
cis-Stilbene	1·2$_5$	72
α-Ethyl-*cis*-stilbene	0·9$_1$	51
α-Propyl-*cis*-stilbene	0·9$_0$	51
α-Methyl-*cis*-stilbene	1·2$_4$	53
4-Methyl-*cis*-stilbene	3·0$_8$	48
trans-Stilbene	2·3$_0$	74
α-Methyl-*trans*-stilbene	1·2$_9$	70
α-Cyano-*trans*-stilbene	2·1$_4$	36
4-Methyl-*trans*-stilbene	3·1$_1$	69
2-Methyl-*trans*-stilbene	2·5$_3$	51
αα′-Dichloro-*trans*-stilbene	0·2$_9$	131
Styrene	1·2$_0$	53
Diphenylmethane	1·0$_1$	72
Triphenylmethane	0·9$_2$	65
Dibenzyl	1·3$_7$	58
Tolan	2·0$_0$	57
Phenylacetylene	0·9$_5$	48
Indene	1·5$_6$	80
2,3-Diphenylbutane	0·2$_9$	72

* Ref. 59.
† Proton chemical shifts measured at 60·004 MHz.

that for *cis*-stilbene [59] (Table 7.9). If the acceptor is located over one of the benzene rings, then *trans*-stilbene provides statistically twice the number of sites for interaction compared with *cis*-stilbene, in which the orientation of the two rings within the molecule naturally screens one side of each benzene ring from the approach of the acceptor molecule. A

TABLE 7.10. Association constants (K^{AD}) of some 1,3,5-trinitrobenzene–alkyl-aniline complexes at 33·5°C, together with the chemical shift of the measured nucleus in the pure complex in solution, relative to the chemical shift in the pure acceptor (Δ^0) as determined by an n.m.r. method.*

Donor	Solvent Carbon tetrachloride $K_r^{AD}/\text{kg mol}^{-1}$†	Δ^0/Hz‡	Chloroform $K_r^{AD}/\text{kg mol}^{-1}$†	Δ^0/Hz‡
Aniline	$2\cdot3_8$	53	$0\cdot5_6$	64
N-Methylaniline	$2\cdot7_9$	59	$0\cdot6_7$	71
N-Ethylaniline	$3\cdot1_8$	56	$0\cdot7_7$	66
N-Propylaniline	$2\cdot8_0$	58		
N,N-Dimethylaniline	$3\cdot0_5$	61	$0\cdot7_8$	75
N,N-Diethylaniline	$2\cdot6_4$	60	$0\cdot7_4$	66
2-Methylaniline	$3\cdot0_0$	48		
N,N-Dimethyl-2-methylaniline	$1\cdot6_2$	53		
N,N-Diethyl-2-methylaniline	$1\cdot4_2$	56		
3-Methylaniline	$3\cdot2_0$	53		
4-Methylaniline	$3\cdot2_0$	54		
2-Ethylaniline	$2\cdot9_0$	56		
4-Ethylaniline	$3\cdot2_2$	53		
2,4-Dimethylaniline	$4\cdot1_0$	57		
N,N-Dimethyl-2,4-dimethylaniline	$2\cdot1_0$	56		
N,N-Diethyl-2,4-dimethylaniline	$3\cdot8_4$	64		
2,6-Dimethylaniline	$3\cdot5_4$	61		
N,N-Dimethyl-2,6-dimethylaniline	$3\cdot2_1$	59		
N,N-Diethyl-2,6-dimethylaniline	$2\cdot8_2$	57		
N,N-Dimethyl-4-bromoaniline	$2\cdot9_2$	49		

* R. Foster and J. W. Morris, unpublished work, see also Chapter 6.
† K_r^{AD} measured in kg solution per mol.
‡ Proton chemical shifts measured at 60·004 MHz.

combination of steric and electronic effects can be used to rationalize the association constants for the other stilbenes listed in Table 7.9. The concept of the dependence of K^{AD} on the statistical number of donor sites within the molecule (though yielding an average 1:1 stoichiometry) could explain why the K^{AD} values for the 1,3,5-trinitrobenzene complexes of dibenzyl, tolan, and *trans*-stilbene are each twice the values for the corresponding "half" molecules, namely, toluene, phenylacetyl-ene and styrene (Table 7.9). By contrast, the association constants for the tetracyanoethylene complexes of *cis*- and *trans*-stilbene are nearly equal,[60] as are the corresponding iodine complexes.[61] It is possible that in these latter cases the acceptor molecules are sited over the central carbon–carbon double bond rather than over one or other ring, although there is no proof that such is the structure. Laarhoven and Nivard have, in fact, suggested that two isomeric complexes are formed between tetracyanoethylene and the given stilbene donor.

Secondary steric effects, in which the resonance delocalization is reduced by intramolecular steric hindrance, appears to be operative in the 2,6-dialkyl-N,N-dialkylaniline complex of 1,3,5-trinitrobenzene [53] (Table 7.10) and is in harmony with the earlier observations of the steric hindrance of these donor molecules, based on ultraviolet spectroscopy.[62]

7.B.5. Effect of the Acceptor

The *relative* values of the equilibrium constants for a series of donors with a given acceptor are dependent on the particular acceptor. Similarly, the effect of the solvent on the relative values of K^{AD} is dependent on the particular complex chosen.[49–51, 63] Because of these differential effects it is impossible to arrange acceptors in an unambiguous order of strength based on values of K^{AD}. Nevertheless, gross differences may be seen. Amongst the diatomic halogen acceptors for which complexes with *n*-donors have been measured, iodine monochloride is by far the strongest in the series $ICl \gg IBr > I_2 > Br_2$. The greater stability of iodine atom complexes compared with the corresponding iodine molecule complexes [64] reflects the higher electron affinity of the iodine atom.

Amongst the π-acceptors, comparison of the association constants of hexamethylbenzene complexes in relatively inert solvents, such as carbon tetrachloride, indicates that 2,3-dichloro-5,6-dicyano-*p*-benzoquinone (DDQ) is one of the strongest, readily available acceptors. The published values of K^{AD} for various complexes involving this acceptor are remarkably large.[65–67] Experimentally, considerable care has to be taken with a compound with such a high redox potential; optical measurements can be seriously affected by traces of water, for example. Further determinations of K^{AD} on such systems appear to be desirable. There is little doubt that complexes of tetracyano-*p*-benzoquinone [68, 69] would have yet higher association constants. Complexes of tetracyanoethylene show relatively high stabilities (Table 7.11).* The association constants of a range of acceptors with a common donor species under closely similar conditions are listed in Table 7.12. It should be re-emphasized that the order of the "strength" of the acceptor, as

* For further references to these and other complexes see Appendix Table 2, p. 396.

TABLE 7.11. Measured association constants (K_c^{AD}) of some tetracyanoethylene complexes determined by optical methods.

Donor	Solvent	Temp./°C	K_c^{AD}/l mol^{-1}	ϵ_{max}^{AD}	ν_{CT}	Ref.*
Benzene	CCl$_4$	20	1.04†	2330	26,100	a
Benzene	CHCl$_3$	25	0.28 ± 0.05	3294 ± 52	25,600	52
Benzene	CH$_2$Cl$_2$	22	0.128†	3570	26,000	20
Benzene	CH$_2$Cl$_2$	30	0.55†	1175		36
Benzene	CHCl$_3$	22	0.25 ± 0.01	2900	25,600	b
Toluene	CH$_2$Cl$_2$	22	0.237†	3330	24,600	20
o-Xylene	CH$_2$Cl$_2$	22	0.446†	3860	23,300	20
m-Xylene	CH$_2$Cl$_2$	22	0.384†	3300	22,700	20
p-Xylene	CH$_2$Cl$_2$	22	0.489†	2650	21,700	20
Mesitylene	CH$_2$Cl$_2$	22	1.11†	3120	21,700	20
Durene	CCl$_4$	20	14.2†	3320	19,900	a
Durene	CH$_2$Cl$_2$	22	3.4†	2075	20,800	20
Pentamethylbenzene	CH$_2$Cl$_2$	22	7.88†	3275	19,200	20
Hexamethylbenzene	CCl$_4$	20	148	4780	18,800	a
Hexamethylbenzene	CH$_2$Cl$_2$	22	16.8†	4390	18,300	20
Naphthalene	CCl$_4$	25	~4.35–5.4‡	1635	18,120	c
Naphthalene	CCl$_4$	20	3.29†			a
Naphthalene	CHCl$_3$	25	1.05 ± 0.09 / 1.01 ± 0.11 §	2001 ± 107§ / 1725 ± 78	17,900 § / 23,100	52
Naphthalene	CHCl$_3$	22	1.01 ± 0.04	1600‖	18,200 / 23,300	b
Naphthalene	CH$_2$Cl$_2$	22	0.750†	1240‖	18,200 / 23,300	20
Phenanthrene	CCl$_4$	20	6.94†	2040	19,000	a
Phenanthrene	CHCl$_3$	25	2.16 ± 0.18	1637 ± 65	18,500	52
Phenanthrene	CHCl$_3$	22	1.99 ± 0.12	1750	18,700	b
Pyrene	CHCl$_3$	25	2.79 ± 0.56	1664 ± 29 / 1221 ± 24	13,700 / 20,000	52
Pyrene	CH$_2$Cl$_2$	25	1.93 ± 0.15	1140	13,600	d
Pyrene	CH$_2$Cl$_2$	22	1.89†	1137‖	13,800 / 20,200	52
Benzo[a]pyrene (3,4-benzpyrene)	CHCl$_3$	25	5.10 ± 0.65	1819 ± 361	12,100	20
Indene	CHCl$_3$	22	1.63 ± 0.04	1350‖	18,500 / 23,300 / 17,500	b

Compound	Solvent	Temp.	K	ε	$\tilde{\nu}$	Ref.
Biphenylene	CH_2Cl_2	22	5·1 ± 0·5†	500 ± 50	14,700	[f]
Acenaphthylene	$CHCl_3$	25	2·84 ± 0·18	2180 ± 117	19,400	[52]
Azulene	$CHCl_3$	25	14·60 ± 0·95		13,600	[52]
Triphenylene	CCl_4	20	9·17†	2290	18,000	[a]
Triphenylene	$CHCl_3$	25	3·52 ± 0·41	{ 1994 ± 45 / 841 ± 116	{ 17,600 / 22,700	[52]
Biphenyl	CCl_4	20	1·86†	1360	20,000	[a]
Biphenyl	$CHCl_3$	25	0·67 ± 0·29	{ 5298 ± 1786 / 1499 ± 110	{ 19,800 / 25,400	[52]
Biphenyl	$CHCl_3$	22	0·73 ± 0·02	1050	19,800	[b]
Biphenyl	CH_2Cl_2	22	0·262†	1450	20,000	[20]
o-Terphenyl	$CHCl_3$	25	0·15 ± 0·06	{ 2726 ± 1446 / 5615 ± 2430	{ 19,200 / 24,300	[52]
m-Terphenyl	$CHCl_3$	25	0·64 ± 0·16	{ 2065 ± 67 / 1605 ± 90	{ 19,600 / 25,300	[52]
m-Terphenyl	CH_2Cl_2	22	0·352†	1470	20,000	[20]
p-Terphenyl	$CHCl_3$	25	0·85 ± 0·30	{ 5163 ± 3500 / 1707 ± 257	{ 17,600 / 25,900	[52]
p-Terphenyl	CH_2Cl_2	22	0·730†	830	17,700	[20]
Indole	$CHCl_3$	22	4·8 ± 0·2	3000	18,400	[b]
Indole	CH_2Cl_2	25	2·8		17,900	[g]
Carbazole	$CHCl_3$	22	5·12 ± 0·13	2900	16,500	[b]
N,N-Dimethylaniline	$CHCl_3$	32·5	15·0		14,800	[h]
N,N-N',N'-Tetramethyl-p-phenylenediamine	Et_2O	20	$1·9 \times 10^3 \pm 400$	3600 ± 360‖	{ 10,400 / 23,500	[i]
Methyl disulphide	CH_2Cl_2	25	0·16	2150	23,500	[j]
Ferrocene	cyclo C_6H_{12}		4·9†	465¶	{ 9300 / 11,100	[k]

* Refs.: [a] G. Briegleb, J. Czekalla and G. Reuss, *Z. phys. Chem. Frankf. Ausg.* **30**, 333 (1961). [b] A. R. Cooper, C. W. P. Crowne and P. G. Farrell, *Trans. Faraday Soc.* **62**, 18 (1966). [c] G. D. Johnson and R. E. Bowen, *J. Am. chem. Soc.* **87**, 1655 (1965). [d] R. Foster and I. Horman, *J. chem. Soc.* (B), 171 (1966). [e] P. H. Emslie, R. Foster and R. Pickles, *Can. J. Chem.* **44**, 9 (1966). [f] D. G. Farnum, E. R. Atkinson and W. C. Lothrop, *J. org. Chem.* **26**, 3204 (1961). [g] R. Foster and P. Hanson, *Tetrahedron* **21**, 255 (1965). [h] Z. Rappoport, *J. chem. Soc.* 4498 (1963). [i] W. Liptay, G. Briegleb and K. Schindler, *Z. Elektrochem.* **66**, 331 (1962). [j] W. M. Moreau and K. Weiss, *J. Am. chem. Soc.* **88**, 204 (1966). [k] M. Rosenblum, R. W. Fish and C. Bennett, *J. Am. chem. Soc.* **86**, 5166 (1964).

† Calculated from values of K_x^{AD} on the assumption that the relationship $K_c^{AD} = K_x^{AD}$ ($10^{-3} \times$ molar volume of solvent in ml) holds.
‡ Very dependent on wavelength: the authors suggest that termolecular species may be present (see Chapter 6).
§ Average values for low and high naphthalene concentrations.
‖ Value for lower energy band.
¶ Average value.

measured by K^{AD}, will show some variations as the common donor species or the common solvent is changed (see pp. 182 and 190).

Within chemically related groups of acceptors the stabilities of complexes with a common donor species appear to be in the expected order; for example:

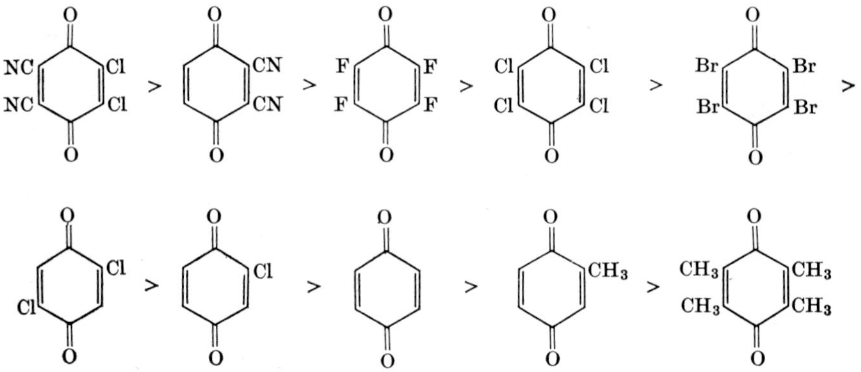

Amongst the nitrobenzenes the order of stability with a common donor species appears to be 1,2,3,5-tetranitrobenzene > 1,3,5-trinitrobenzene > 1,2,4-trinitrobenzene > 1,2,3-trinitrobenzene > 1,4-dinitrobenzene > 1,3-dinitrobenzene > 1,2-dinitrobenzene > nitrobenzene. The order of stabilities of the three trinitrobenzenes is reasonably interpreted in terms of a steric hindrance to planarity of adjacent nitro groups.[70] This will have two effects. The increase in the effective "thickness" of the acceptor molecule will result in a primary steric effect impeding the approach of the donor molecule. Secondly, in the non-planar acceptor molecule, there will be a reduction in the delocalization with a consequent decrease in the Lewis acidity of the acceptor. A similiar argument may also account for the low values of association constants for complexes of 1,2-dinitrobenzene. This contrasts with complexes of the dicyanobenzenes,[71] the linear substituents of which cause no steric interference in the ortho position one to the other: here there is evidence that the 1,2-isomer forms stronger complexes than the 1,3-isomer. A very good example of an ortho steric-effect increasing the effective bulk of an acceptor molecule is dipicryl which forms much weaker complexes with acceptors than does 1,3,5-trinitrobenzene.[72] The decrease in the association constants of hexamethylbenzene with picryl chloride, picryl bromide and picryl iodide compared with the corresponding 1,3,5-trinitrobenzene complex[73] (Table 7.12) may likewise be the result of reduced delocalization in the acceptor and increased intermolecular steric repulsion. A similar comparison between 1,3,5-trinitro-benzene and 1,3,5-trichloro-2,4,6-trinitrobenzene was made many years ago.[74]

TABLE 7.12. Association constants K_r^{AD} of hexamethylbenzene with various acceptors either from n.m.r. chemical shift data, or from optical measurements under the condition $[D]_0 = [A]_0$, or from kinetic data, all at 33·5°C.

Solvent	CCl$_4$		CHCl$_3$		CH$_2$ClCH$_2$Cl	
Acceptor	K_r^{AD}/kg mol^{-1}	Ref.*	K_r^{AD}/kg mol^{-1}	Ref.*	K_r^{AD}/kg mol^{-1}	Ref.*
2,3-Dicyano-5-chloro-p-benzoquinone					4·5	a
Tetracyanoethylene					16·5	75
Fluoranil	15·4	49	3·9	49	3·6	49
2,3-Dicyano-p-benzoquinone					2·8	b
1,4-Dicyano-2,3,5,6-tetra-fluorobenzene	5·2	49	0·9$_2$	49	0·7$_2$	49
7,7,8,8-Tetracyanoquino-dimethane					0·7$_0$	b
1,2,3,5-Tetranitrobenzene					0·6$_6$	b
2,5-Dichloro-p-benzoquinone					0·6$_2$	b
1,3,5-Trinitrobenzene	5·1	48	0·8$_6$	49	0·5$_9$	49
1,2,4,5-Tetracyanobenzene					0·4$_2$	b
Picryl chloride			0·2$_2$			c
Picryl bromide			0·2$_3$			c
Picryl iodide			0·2$_1$			c
p-Benzoquinone					0·1$_5$	b
1,4-Dinitrobenzene	1·0	48			0·1$_5$	49
1,3,5-Tricyanobenzene					0·1$_1$	b

* a B. Dodson, unpublished work. b R. Foster and C. A. Fyfe, *Trans. Faraday Soc.* **62,** 1400 (1966). c J. W. Morris, unpublished work.

7.C. Enthalpies and Entropies of Formation

Compared with the very large amount of data which has been published on equilibrium constants of charge-transfer complexes, relatively few attempts have been made to evaluate the enthalpy change ($\Delta H^{\ominus}$) and the entropy change ($\Delta S^{\ominus}$) involved. As measured, these terms will represent the net changes in enthalpy and entropy respectively for the solvated species [cf. equations (7.1) and (7.2)]. Indications of effects of solvation on $\Delta H^{\ominus}$ and $\Delta S^{\ominus}$ are available for one or two systems for which measurements have been made in the gas phase as well as in solution.[9–11, 13, 15, 17, 18] In the case of the relatively stable diethyl-sulphide–iodine complex,[15] it appears that the enthalpy changes in the gas phase [$\Delta H^{\ominus}(g)$], and in solution [$\Delta H^{\ominus}(\text{soln})$], are comparable. The reason for an observed difference between $\Delta G^{\ominus}(g)$ and $\Delta G^{\ominus}(\text{soln})$ is the difference in the entropy of solvation of the two components compared

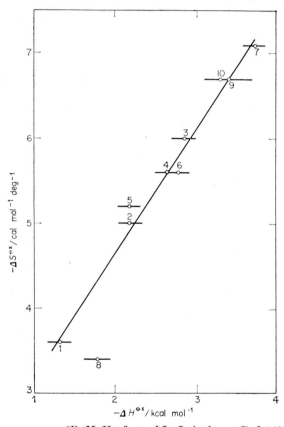

[R. M. Keefer and L. J. Andrews, Ref. 76]

FIG. 7.9. Plot of $-\Delta S^{\ominus x}$ against $-\Delta H^{\ominus x}$ for a series of iodine–donor complexes in carbon tetrachloride solution measured optically. Donors: (1) benzene; (2) *p*-xylene; (3) mesitylene; (4) 1,3,5-triethylbenzene; (5) 1,3,5-tri-*t*-butylbenzene; (6) durene; (7) hexamethylbenzene; (8) hexaethylbenzene; (9) *t*-butyl alcohol; (10) dioxane.

with the entropy of solvation of the complex. For the weaker iodine complexes of benzene and of diethyl ether,[10] the enthalpy as well as the entropy of solvation of the components compared with those of the complex appear to contribute to the observed free energy changes.

Some determinations of $\Delta H^{\ominus}$ and $\Delta S^{\ominus}$ are based on values of K^{AD} which have already been criticized (see above, also Chapter 6). Since the major difficulty in optical determinations often appears to be in the separation of the terms in the product $(K^{AD}\epsilon^{AD})$, and since ϵ^{AD} usually has only a relatively small temperature coefficient, it may be possible in favourable cases[75] to obtain estimates of $\Delta H^{\ominus}$ from the temperature variation of the total term $K^{AD}\epsilon^{AD}$ (Chapter 6).

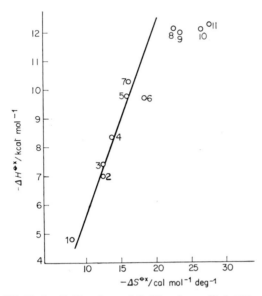

[H. Yada, J. Tanaka and S. Nagakura, Ref. 77]

Fig. 7.10. Plot of $-\Delta H^{\ominus x}$ against $-\Delta S^{\ominus x}$ for a series of iodine–aliphatic amine complexes in n-heptane. Amines: (1) ammonia; (2) methylamine; (3) ethylamine; (4) n-butylamine; (5) dimethylamine; (6) diethylamine; (7) piperidine; (8) trimethylamine; (9) triethylamine; (10) tri-n-propylamine; (11) tri-n-butylamine.

Andrews and Keefer[76] evaluated $\Delta H^{\ominus}$ and $\Delta S^{\ominus}$ for a series of alkylbenzene–iodine complexes from optically determined values of K^{AD}. A near-linear correlation between $\Delta H^{\ominus}$ and $\Delta S^{\ominus}$ is observed (Fig. 7.9). The only complex which shows a significant deviation is the sterically hindered hexaethylbenzene complex. There is a similar linear relationship between $\Delta H^{\ominus}$ and $\Delta S^{\ominus}$ for the strong iodine–aliphatic primary and secondary amine complexes. The relationship breaks down for the aliphatic tertiary amine complexes[77] (Fig. 7.10).

Briegleb[8] has shown that, for some groups of complexes having common acceptor species, there is no apparent correlation between the values of $\Delta H^{\ominus}$ and $\Delta S^{\ominus}$ (Fig. 7.11). This may well be the case if wide groups of structurally unrelated donors are compared. However, there must be doubt as to the correctness of $\Delta H^{\ominus}$ and $\Delta S^{\ominus}$ values based on

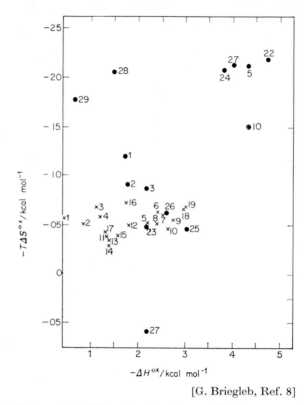

[G. Briegleb, Ref. 8]

FIG. 7.11. Values of $-T\Delta S^{\ominus x}$ plotted against $-\Delta H^{\ominus x}$ for 1,3,5-trinitrobenzene complexes with various donors: (1) benzene; (2) toluene; (3) m-xylene; (4) mesitylene; (5) naphthalene; (6) 1-methylnaphthalene; (7) 2-methylnaphthalene; (8) acenaphthene; (9) anthracene; (10) phenanthrene; (11) aniline; (12) N,N-dimethylaniline; (13) o-toluidine; (14) m-toluidine; (15) p-toluidine; (16) p-chloroaniline; (17) diphenylamine; (18) 1-naphthylamine; (19) 2-naphthylamine; (21) durene; (22) hexamethylbenzene; (23) styrene; (24) stilbene; (25) diphenylbutadiene; (26) diphenylhexatriene; (27) diphenylhexadiene; (28) trimethylethylene; (29) cyclohexene. Those in carbon tetrachloride are marked ●; those in chloroform are marked X. All are derived from optical measurements.

uncertain values of K^{AD}. Furthermore, in most cases the requisite measurements have been made at only a very few temperatures, often only two. The accumulation of errors consequently leads to values of $\Delta H^{\ominus}$ and $\Delta S^{\ominus}$ which can only be quoted within wide error-limits, even when the actual measurements have been made with extreme care. (See Fig. 7.13 below.)

Experimental methods which can yield more certain evaluations of $\Delta H^{\ominus}$ and $\Delta S^{\ominus}$ are therefore of consequence. For example, the enthalpies of formation of iodine complexes of various disulphides and of 1,2-dithiane have been determined together with their association constants by a calorimetric method.[40] No significant differences were noted in the

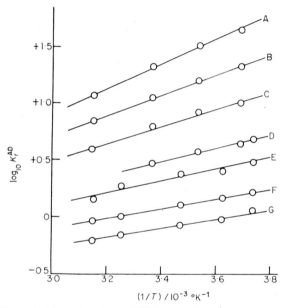

FIG. 7.12. Plot of $1/T$ against $\log_{10} K_r^{\mathrm{AD}}$ for a series of fluoranil–methylbenzene complexes in carbon tetrachloride solution. K_r^{AD} values calculated from n.m.r. data. The donors are (A) hexamethylbenzene; (B) pentamethylbenzene; (C) durene; (D) mesitylene; (E) p-xylene; (F) toluene; (G) benzene. (Data from Ref. 78.)

$\Delta H^{\ominus}$ values for the various dialkyl disulphide iodine complexes apart from the di-*tert*-butyl disulphide complex.

The evaluation of $\Delta H^{\ominus}$ and $\Delta S^{\ominus}$ for a range of charge-transfer complexes has been made recently, using the classical van't Hoff relationship but in which the association constants are obtained by an n.m.r. chemical shift method (Chapter 6) using series of measurements made at a considerable number of temperatures.[78, 79] Typical plots are shown in Fig. 7.12. Some values of $\Delta H^{\ominus}$ and $\Delta S^{\ominus}$ are given in Table 7.13. A plot of $\Delta H^{\ominus}$ against the ionization potential of the donor for a series of complexes of structurally related donors with a common acceptor shows a

TABLE 7.13. Some $\Delta H^{\ominus}$ and $\Delta S^{\ominus}$ values for charge-transfer complex formation derived from n.m.r. chemical shift data (solvent carbon tetrachloride).*

Acceptor Donor	1,3,5-Trinitrobenzene†		1,4-Dinitrobenzene‡		Fluoranil†		1,4-Dicyano-2,3,5,6-tetrafluorobenzene‡	
	$-\Delta H^{\ominus}$ kcal mol⁻¹	$-\Delta S^{\ominus}$/cal mol⁻¹ deg⁻¹	$-\Delta H^{\ominus}$ kcal mol⁻¹	$-\Delta S^{\ominus}$/cal mol⁻¹ deg⁻¹	$-\Delta H^{\ominus}$ kcal mol⁻¹	$-\Delta S^{\ominus}$/cal mol⁻¹ deg⁻¹	$-\Delta H^{\ominus}$ kcal mol⁻¹	$-\Delta S^{\ominus}$/cal mol⁻¹ deg⁻¹
Benzene	1·9	7·8	1·0	6·2	2·0	7·4		
Toluene	2·2	7·9			2·3	7·5		
p-Xylene	2·5	8·2	1·5	6·4	2·7	8·0		
Mesitylene	2·8	8·8	1·9	7·3	3·0	8·1		
Durene	3·1	8·5	2·3	8·1	3·9	9·6	2·4	6·2
Pentamethylbenzene	3·4	8·6			4·4	10·1	3·5	9·0
Hexamethylbenzene	3·7	8·8	3·0	9·7	5·4	12·2	4·3	10·8

* Derived from K_r^{AD} values expressed in kg solution per mol units, these constants are effectively equivalent to those referred to a standard state of a 1 molal solution.

† Ref. 78.

‡ Ref. 79.

regular relationship when $\Delta H^{\ominus}$ is evaluated from n.m.r. chemical-shift data (Fig. 7.13). By comparison, thermodynamic constants for the same systems derived from optical data show a much poorer correlation[33] (Fig. 7.13). The relationship between values of $\Delta H^{\ominus}$ and $\Delta S^{\ominus}$ for such

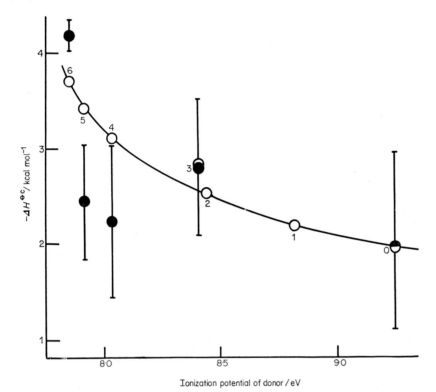

Ionization potential of donor / eV

FIG. 7.13. Plots of $-\Delta H^{\ominus c}$ obtained from optical data, ●, and from n.m.r. data, ○, for a series of 1,3,5-trinitrobenzene complexes in carbon tetrachloride, plotted against the ionization potentials of the donors: (0) benzene; (1) toluene; (2) p-xylene; (3) mesitylene; (4) durene; (5) pentamethylbenzene; (6) hexamethyl-benzene. The vertical lines are measures of the errors in the optical determinations. [Data from Ref. 78, and C. C. Thompson, Jr. and P. A. D. de Maine, *J. phys. Chem., Ithaca* **69**, 2766 (1965).]

complexes obtained from n.m.r. chemical-shift data is smooth (Fig. 7.14). Although the plots of $\Delta H^{\ominus}$ against $\log K^{AD}$ (or $\Delta G^{\ominus}$) for a set of complexes of a common acceptor species with a series of structurally related donors derived from n.m.r. data are near-linear (Fig. 7.15), there is no simple correlation between $\Delta H^{\ominus}$ and $\log K^{AD}$ for complexes un-restricted in respect of the donor or acceptor species.

Drago and Wayland[80, 81] have correlated the enthalpies of a variety of charge-transfer complexes in poorly solvating solvents, using a double-scale equation of the form

$$-\Delta H^{\ominus} = E_A \, E_D + C_A \, C_D \qquad (7.5)$$

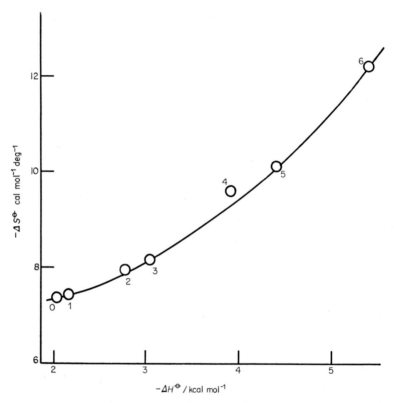

Fɪɢ. 7.14. Plot of $-\Delta S^{\ominus}$ against $-\Delta H^{\ominus}$ for a series of fluoranil complexes in carbon tetrachloride solution, based on K_r^{AD} values calculated from n.m.r. data. Donors: (0) benzene; (1) toluene; (2) p-xylene; (3) mesitylene: (4) durene; (5) pentamethylbenzene; (6) hexamethylbenzene. (From data in Ref. 78.)

where E_A and C_A are two constants assigned to the acceptor. They are considered to represent the electrostatic and covalent binding properties of the acceptor. E_D and C_D are the corresponding constants for the donor. An advantage of using enthalpies, rather than free energies of complex formation, is that the former function is nearly independent of the particular solvent, providing that it is not a strongly solvating solvent.

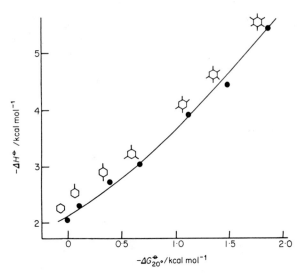

FIG. 7.15. Plot of $-\Delta H^{\ominus}$ against $-\Delta G^{\ominus}$ for a series of fluoranil–alkylbenzene complexes in carbon tetrachloride solution, based on K_r^{AD} values calculated from n.m.r. data. (From data in Ref. 78.)

REFERENCES

1. J. Peters and W. B. Person, *J. Am. chem. Soc.* **86**, 10 (1964).
2. W. B. Person, *J. Am. chem. Soc.* **87**, 167 (1965).
3. M. Tamres and M. Brandon, *J. Am. chem. Soc.* **82**, 2134 (1960).
4. M. Brandon, M. Tamres and S. Searles, Jr., *J. Am. chem. Soc.* **82**, 2129 (1960).
5. P. Klaeboe, *J. Am. chem. Soc.* **84**, 3458 (1962); P. Klaeboe, *Acta chem. scand.* **18**, 999 (1964); H. Yada, J. Tanaka and S. Nagakura, *Bull. chem. Soc. Japan* **33**, 1660 (1960).
6. R. S. Drago, R. L. Carlson, N. J. Rose and D. A. Wenz, *J. Am. chem. Soc.* **83**, 3572 (1961); H. Tsubomura and R. P. Lang, *J. Am. chem. Soc.* **83**, 2085 (1961).
7. L. J. Andrews and R. M. Keefer, *in* "Advances in Inorganic Chemistry and Radiochemistry," Vol. III, eds. H. J. Emeléus and A. G. Sharpe, Academic Press, New York (1961), p. 91.
8. G. Briegleb, "Elektronen-Donator-Acceptor-Komplexe," Springer-Verlag, Berlin (1961).
9. F. T. Lang and R. L. Strong, *J. Am. chem. Soc.* **87**, 2345 (1965).
10. M. Kroll, *J. Am. chem. Soc.* **90**, 1097 (1968).
11. M. Kroll and M. L. Ginter, *J. phys. Chem., Ithaca* **69**, 3671 (1965).
12. J. Prochorow, *J. chem. Phys.* **43**, 3394 (1965).
13. J. M. Goodenow and M. Tamres, *J. chem. Phys.* **43**, 3393 (1965).
14. J. Prochorow and A. Tramer, *J. chem. Phys.* **44**, 4545 (1966).
15. M. Tamres and J. M. Goodenow, *J. phys. Chem., Ithaca* **71**, 1982 (1967).
16. D. Atack and O. K. Rice, *J. phys. Chem., Ithaca* **58**, 1017 (1954).

8

17. S. D. Christian and J. Grundnes, *Nature, Lond.* **214**, 1111 (1967).
18. E. I. Ginns and R. L. Strong, *J. phys. Chem., Ithaca* **71**, 3059 (1967).
19. R. Foster and D. Ll. Hammick, *J. chem. Soc.* 2685 (1954).
20. R. E. Merrifield and W. D. Phillips, *J. Am. chem. Soc.* **80**, 2778 (1958).
21. M. Tamres, *J. phys. Chem., Ithaca* **65**, 654 (1961).
22. R. S. Drago, T. F. Bolles and R. J. Niedzielski, *J. Am. chem. Soc.* **88**, 2717 (1966).
23. D. P. Stevenson and G. M. Coppinger, *J. Am. chem. Soc.* **84**, 149 (1962).
24. F. Dörr and G. Buttgereit, *Ber. Bunsenges. physik. Chem.* **67**, 867 (1963).
25. R. Anderson and J. M. Prausnitz, *J. chem. Phys.* **39**, 1225 (1963).
26. R. Anderson and J. M. Prausnitz, *J. chem. Phys.* **40**, 3443 (1964).
27. R. F. Weimer and J. M. Prausnitz, *J. chem. Phys.* **42**, 3643 (1965).
28. K. M. C. Davis and M. F. Farmer, *J. chem. Soc.* (B), 28 (1967).
29. D. R. Rosseinsky and H. Kellawi, Chemical Society Anniversary Meeting, *Exeter* (1967); *J. chem. Soc.* (A),1207 (1969).
30. S. Carter, J. N. Murrell and E. J. Rosch, *J. chem. Soc.* 2048 (1965).
31. P. J. Trotter, *J. Am. chem. Soc.* **88**, 5721 (1966).
32. C. C. Thompson, Jr. and P. A. D. de Maine, *J. Am. chem. Soc.* **85**, 3096 (1963).
33. C. C. Thompson, Jr. and P. A. D. de Maine, *J. phys. Chem., Ithaca* **69**, 2766 (1965).
34. R. E. Gibson and O. H. Loeffler, *J. Am. chem. Soc.* **62**, 1324 (1940).
35. J. S. Ham, *J. Am. chem. Soc.* **76**, 3881 (1954).
36. J. R. Gott and W. G. Maisch, *J. chem. Phys.* **39**, 2229 (1963).
37. W. Kauzmann, "Quantum Chemistry," Academic Press, New York (1957), p. 581.
38. R. J. Niedzielski, R. S. Drago and R. L. Middaugh, *J. Am. chem. Soc.* **86**, 1694 (1964).
39. W. M. Moreau and K. Weiss, *J. Am. chem. Soc.* **88**, 204 (1966).
40. B. Nelander, *Acta chem. scand.* **20**, 2289 (1966).
41. H. Tsubomura, *J. Am. chem. Soc.* **82**, 40 (1960).
42. B. B. Wayland and R. S. Drago, *J. Am. chem. Soc.* **86**, 5240 (1964).
43. W. B. Person, W. C. Golton and A. I. Popov, *J. Am. chem. Soc.* **85**, 891 (1963).
44. R. L. Carlson and R. S. Drago, *J. Am. chem. Soc.* **85**, 505 (1963).
45. J. van der Veen and W. Stevens, *Recl Trav. chim. Pays-Bas Belg.* **82**, 287 (1963).
46. M. Charton, *J. org. Chem.* **31**, 2991, 2996 (1966).
47. H. M. Rosenberg, E. C. Eimutis and D. Hale, *Can. J. Chem.* **45**, 2859 (1967).
48. R. Foster and C. A. Fyfe, *Trans. Faraday Soc.* **61**, 1626 (1965).
49. N. M. D. Brown, R. Foster and C. A. Fyfe, *J. chem. Soc.* (B), 406 (1967).
50. R. Foster and D. R. Twiselton, unpublished work.
51. R. Foster and M. I. Foreman, unpublished work.
52. M. J. S. Dewar and C. C. Thompson, Jr., *Tetrahedron* **Supp.** 7, 97 (1966).
53. R. Foster and J. W. Morris, unpublished work.
54. L. J. Andrews and R. M. Keefer, *J. Am. chem. Soc.* **74**, 4500 (1952).
55. N. Ogimachi, L. J. Andrews and R. M. Keefer, *J. Am. chem. Soc.* **77**, 4202 (1955).
56. R. Foster, D. Ll. Hammick and B. N. Parsons, *J. chem. Soc.* 555 (1956).
57. R. E. Lovins, L. J. Andrews and R. M. Keefer, *J. phys. Chem., Ithaca* **68**, 2553 (1964).
58. H. D. Anderson and D. Ll. Hammick, *J. chem. Soc.* 1089 (1950).
59. R. Foster, I. Horman and J. W. Morris, unpublished work.
60. W. H. Laarhoven and R. J. F. Nivard, *Recl Trav. chim. Pays-Bas Belg.* **84**, 1478 (1965).

61. S. Yamashita, *Bull. Chem. Soc. Japan* **32**, 1212 (1959).
62. B. M. Wepster, *in* "Steric Effects in Conjugated Systems," Chemical Society Symposium, ed. G. W. Gray, Butterworth, London (1958), p. 82.
63. A. S. Bailey, B. R. Henn and J. M. Langdon, *Tetrahedron* **19**, 161 (1963).
64. R. L. Strong and J. Pérano, *J. Am. chem. Soc.* **83**, 2843 (1961); *ibid.*, **89**, 2535 (1967).
65. P. R. Hammond, *J. chem. Soc.* 3113 (1963).
66. R. Foster and I. Horman, *J. chem. Soc.* (B), 1049 (1966).
67. R. D. Srivastava and G. Prasad, *Spectrochim. Acta* **22**, 1869 (1966).
68. K. Wallenfels and G. Bachmann, *Angew. Chem.* **73**, 142 (1961).
69. K. Wallenfels, G. Bachmann, D. Hofmann and R. Kern, *Tetrahedron* **21**, 2239 (1965).
70. R. Foster, *J. chem. Soc.* 1075 (1960).
71. R. Foster and T. J. Thomson, *Trans. Faraday Soc.* **59**, 2287 (1963).
72. C. E. Castro, L. J. Andrews and R. M. Keefer, *J. Am. chem. Soc.* **80**, 2322 (1958).
73. J. W. Morris, unpublished work.
74. D. Ll. Hammick and A. Hellicar, *J. chem. Soc.* 761 (1938).
75. R. Foster and I. B. C. Matheson, *Spectrochim. Acta* **23A**, 2037 (1967).
76. R. M. Keefer and L. J. Andrews, *J. Am. chem. Soc.* **77**, 2164 (1955).
77. H. Yada, J. Tanaka and S. Nagakura, *Bull. chem. Soc. Japan* **33**, 1660 (1960).
78. R. Foster, C. A. Fyfe and M. I. Foreman, *Chem. Commun.* 913 (1967).
79. R. Foster, C. A. Fyfe and M. I. Foreman, unpublished work.
80. R. S. Drago and B. B. Wayland, *J. Am. chem. Soc.* **87**, 3571 (1965).
81. R. S. Drago, *Chem. in Britain*, **3**, 516 (1967).

Chapter 8

Crystal Structures

8.A. Introduction

There has been considerable discussion concerning the relative orientation of the donor and acceptor moieties in charge-transfer complexes in *solution*. Many conflicting data have been reported and considerable uncertainty remains as to the exact geometry of the complexes. An example of this is the problem of the orientation of the iodine molecule in the iodine–benzene complex (see Chapter 3). By contrast, detailed crystallographic structures have now been determined for a number of organic charge-transfer complexes in the solid phase. A major difference between solid-phase structures and complexes in solution is that, although a given complex may have the same stoichiometry, say 1:1, in both phases, whilst the complex will probably exist as discrete pairs containing single donor and acceptor molecules in solution, in the solid phase the structures usually consist of infinite stacks of alternate donor and acceptor molecules. In such cases no obvious pairing exists but rather each donor has two nearest acceptor partners and each acceptor has two nearest donor partners. This type of structure is particularly evident in complexes of π-donors. Furthermore, there may be significant interactions with yet other molecules within the lattice. Consequently, care must be taken in applying properties of the solid-phase complexes to the discrete complex in a diluting liquid solvent.

8.B. Complexes Formed by n-Donors

Crystallographic studies have been made on various solids containing
n-donors complexed with acceptors, often halogens.[1] Although in most
systems the donor and acceptor functions are in separate molecular
species, for example the halogen complexes of ethers, amines, sulphides
and selenides (see Table 8.1), some crystal structures have been studied in
which there is apparently an intermolecular charge-transfer interaction
between a donor site in one molecule and an acceptor site in a second
molecule of the same species. Evidence for such interactions is given in
Section 8.B.2.

8.B.1. Complexes of n-Donors with Separate Acceptor Species

In these complexes, the intermolecular charge-transfer bond is between
two specific atoms, the n-donor atom and an acceptor atom. This latter
atom is usually a halogen which utilizes an expanded shell (probably
$sp^3d_z^2$). Many of the complexes are unstable and X-ray measurements
have often had to be made at low temperatures. In the complexes with
diatomic halogen acceptors, a linear orientation of the halogen molecule
with the donor atom is generally observed (Table 8.1). A similar observation has also been made for complexes containing acceptors other than
molecular halogens, though they usually involve a halogen acceptor *atom*
(see below).

Attempts to prepare the 1:1 iodine–pyridine charge-transfer complex*
in a crystalline form have not been successful. Hassel and Rømming[3]
have prepared the crystalline pyridine–iodine monochloride complex,
however, and have shown that this has a linear arrangement of the
nitrogen atom of the pyridine ring with the acceptor molecule (Fig. 8.1).
Reid and Mulliken[4] had proposed a structure for the corresponding
pyridine–iodine complex, in which one iodine atom was linked to the
nitrogen atom and situated in the plane of the pyridine ring, whilst the
second iodine atom was in a plane perpendicular to the plane of the
pyridine ring (Fig. 8.2). Other iodine complexes containing an intermolecular I···N association[5–7] show a linear I—I···N orientation similar
to the Cl—I···N arrangement in the pyridine–iodine monochloride
complex. Hassel and his co-workers have also shown that a linear arrangement occurs with halogen–ether[2, 8–11] and halogen–ketone[12] complexes

---

* The 1:1 charge-transfer complex between iodine and pyridine should not be confused
with the ionic compound which has the stoichiometry $2I_2$.pyridine. The structure of this
latter product has been determined by Hassel and Hope [*Acta chem. scand.* **15**, 407 (1961)],
who have shown it to consist of nearly planar Py_2I^+ ions, the remaining iodine atoms
forming a network built up of centrosymmetric triiodide ions with an I–I distance of
2·93 Å and of iodine molecules with an I–I distance equal to 2·74 Å.

TABLE 8.1. Some charge-transfer complexes of n-donor molecules for which the crystal structures have been determined.

Donor (D)	Acceptor (A)	Ratio D:A	Atoms X, Y, involved in charge-transfer bond					Acceptor Y–Z* bond length		Ref.†
			X–Y	Length X–Y (Å)	Sum of covalent radii X,Y (Å)	Sum of Van der Waals radii X,Y (Å)	Angle* XYZ	In complex (Å)	In free acceptor (Å)	
Trimethylamine	Iodine	1:1	N—I	2·27	2·03	3·65	179°	2·83	2·67	5, 6
4-Picoline	Iodine	1:1	N—I	2·31	2·03	3·65	Approx. linear	2·83	2·67	7
Phenazine	Iodine	1:1	N—I	2·92	2·03	3·65	180°	2·73	2·67	a
Pyridine	Iodine monochloride	1:1	N—I	2·26	2·03	3·65	Approx. linear	2·51	2·32	3
Trimethylamine	Iodine monochloride	1:1	N—I	2·30	2·03	3·65	180°	2·52	2·32	6, b
Pyridine	Iodine monobromide	1:1	N—I	2·26	2·03	3·65	Approx. linear	2·66	2·47	c
Pyridine	Iodine cyanide	1:1	N—I	2·57	2·03	3·65	180°			c
Quinoline	Iodoform	3:1	N—I	3·05	2·03	3·65	177°			14
Dioxane	Iodine	1:1	O—I	2·81	1·99	3·55				8
Dioxane	Iodine monochloride	1:2	O—I	2·57	1·99	3·55	Approx. linear	2·33	2·32	2, 9
Dioxane	Iodoform	1:1	O—I	2·94	1·99	3·55				8
Cyclohexan-1,4-dione	Diiodoacetylene	1:1	O—I	2·94	1·99	3·55	165, 166°	1·98	2·03	33
Benzyl sulphide	Iodine	1:1	S—I	2·78	2·37	4·00	179°	2·82	2·67	2, d
1,4-Dithiane	Iodine	1:2	S—I	2·87	2·37	4·00	178°	2·79	2·67	21, 22
1,4-Dithiane	Diiodoacetylene	1:1	S—I	3·27	2·37	4·00	174, 175°	2·79	2·67	8, 32
1,4-Dithiane	Iodoform	1:1	S—I	3·32	2·37	4·00	175°	2·03, 2·20	2·12	15

Acceptor	Donor	Ratio	X—Y				Angle			Refs[†]
Sulphur (S$_8$)	Iodoform	3:1	S—I	3·50	2·37	4·00	178°	2·10	2·12	16
1,4-Diselenane	Iodine	1:2	Se—I	2·83	2·50	4·15	180°	2·87	2·67	21,23
Tetrahydro-selenophene	Iodine	1:1	Se—I	2·76	2·50	4·15	179°	2·91	2·67	26
1,4-Diselenane	Diiodoacetylene	1:1	Se—I	3·34	2·50	4·15	173°			8,32
1,4-Diselenane	Iodoform	1:2	Se—I	3·47[‡]	2·50	4·15	179°	2·05	2·12	8,17
1,4-Diselenane	Tetraiodoethylene	1:1	Se—I	3·43	2·50	4·15			2·67	30
1,4-Oxaselenane	Iodine	1:1	Se—I	2·76	2·50	4·15	175°	2·96	2·88	25
Acetonitrile	Bromine	2:1	N—Br	2·84	1·84	3·45	179·4°	2·328		20
Hexamethylene-tetramine	Bromine	1:2	N—Br	2·16	1·84	3·45	180°	2·43	2·28	1, e
Acetone	Bromine	1:1	O—Br	2·82	1·80	3·35	180°	2·28	2·28	12
Dioxane	Bromine	1:1	O—Br	2·71	1·80	3·35	180°	2·31	2·28	10
Methanol	Bromine	2:1	O—Br	2·80	1·80	3·35	175°	2·28	2·28	19
Dioxane	Oxalyl bromide	1:1	O—Br	3·21	1·80	3·35	165°	1·96		18,39
Dioxane	Chlorine	1:1	O—Cl	2·67	1·65	3·20	178°	2·02	1·99	11
Dioxane	Oxalyl chloride	1:1	O—Cl	3·18	1·65	3·20	180°	1·70	1·72	18
Dioxane	Dinitrogen tetroxide	1:1	O—N	2·76			76°	1·75§	1·64§	35,36

* Where Z is the atom within the acceptor molecule to which the donor is attached, in cases where there are two Y–Z or two X–Y distances, the data refer to the shorter distances.

† Refs.: [a] T. Uchida, *Bull. chem. Soc. Japan* **40**, 2244 (1967). [b] O. Hassel and H. Hope, *Acta chem. scand.* **14**, 391 (1966). [c] T. Dahl, O. Hassel and K. Sky, *Acta chem. scand.* **21**, 592 (1967). [d] C. Rømming, *Acta chem. scand.* **14**, 2145 (1960). [e] G. Eia and O. Hassel, *Acta chem. scand.* **10**, 139 (1956).

‡ Second X–Y distance is 3·51 Å.

§ N–N distance.

(Table 8.1), for which structures such as those shown in Figs. 8.3 and 8.4, respectively, had originally been proposed on theoretical grounds by Mulliken.[13]

In general, with these complexes the intermolecular distance is less than the van der Waals distance and the intramolecular bond involving the acceptor atom within the acceptor molecule is increased (Table 8.1). In many of the complexes of molecular halogens, where the interaction is localized between a specific atom in the donor and a specific atom in the acceptor, although the length of the intermolecular bond is sensitive to the nature of these atoms, it is only slightly affected by the

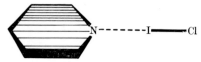

[After O. Hassel and C. Rømming, Ref. 3]

FIG. 8.1. Structure of the pyridine–iodine monochloride complex.

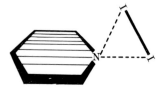

[After C. Reid and R. S. Mulliken, Ref. 4]

FIG. 8.2. Proposed structure of the pyridine–iodine complex.

nature of the *total* structures of the donor and acceptor.[8] The lengthening of the bond in a given acceptor appears to be only slightly dependent on the strength of the donor (Table 8.1). However, with the very weak acceptors such as iodoform,[14–17] and the oxalyl halides,[18] the intermolecular distance more nearly approaches the van der Waals distance (Table 8.1).

In complexes such as iodine–4-picoline[7] and iodine–trimethylamine,[5,6] only one of the iodine atoms is directly concerned in charge-transfer bonding. By contrast, the structure of the iodine–dioxane complex[8] involves endless chains of dioxane and iodine molecules in which both halogen atoms within a molecule participate equally in intermolecular bonding with separate donor molecules, the four atoms immediately concerned being collinear. This type of structure, illustrated in Fig. 8.5, and described by Hassel as containing "halogen-molecule bridges," is also shown by the corresponding chlorine [11] and bromine [10] complexes of dioxane. The observed donor–acceptor separation is only slightly larger in the bromine complex than it is in the chlorine complex (Table 8.1), despite the considerably larger van der Waals radius of bromine; this suggests that the effect of the difference of size of bromine and chlorine is nearly compensated by the increase in the intermolecular interaction.

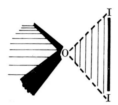

[After R. S. Mulliken, Ref. 13]

FIG. 8.3. Proposed structure of an ether–iodine complex.

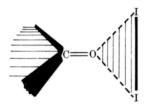

[After R. S. Mulliken, Ref. 13]

FIG. 8.4. Proposed structure of a ketone–iodine complex.

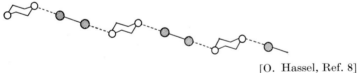

[O. Hassel, Ref. 8]

FIG. 8.5. Structure of the dioxane–iodine complex.

Halogen-molecule bridges have also been observed [12] in the bromine–acetone complex which contains endless chains involving the basic structure illustrated in Fig. 8.6. They also appear to be involved in the structure of the 2:1 complex between methanol and bromine, [19] in which each oxygen atom is surrounded tetrahedrally by a methyl group, two hydrogen-bonded bridges and one "bromine-molecule bridge" (Fig. 8.7). Although such linkages occur usually between oxygen-atom acceptors, an example of a "halogen-molecule bridge" between two nitrogen donor

8*

atoms is observed in the weak complex between bromine and aceto-nitrile.[20] This has an effectively linear structure in which two acetonitrile molecules are linked through a bromine molecule by their nitrogen atoms (Fig. 8.8). When the homopolar acceptor molecule is replaced by the unsymmetrical acceptor iodine monochloride, the product with dioxane,

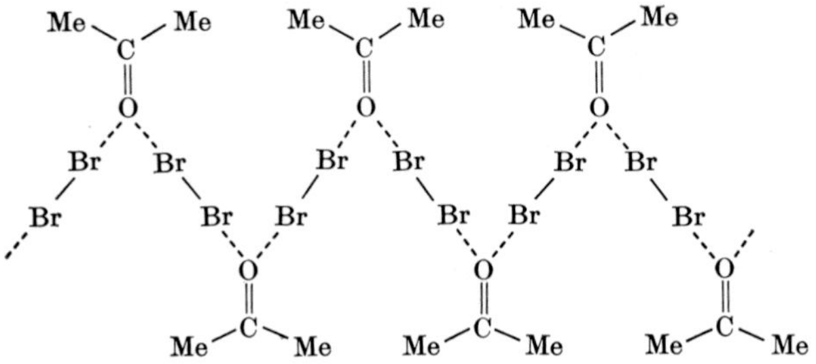

Me—C—Me   Me—C—Me   Me—C—Me
      ‖   ‖   ‖
      O   O   O

Br   Br   Br   Br   Br   Br

Br   Br   Br   Br   Br   Br

         O   O   O
         ‖   ‖   ‖
Me—C—Me   Me—C—Me   Me—C—Me

[After O. Hassel and K. O. Strømme, Ref. 12]

FIG. 8.6. Basic structure of the acetone–bromine complex.

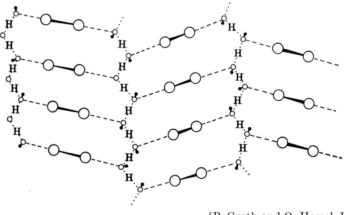

[P. Groth and O. Hassel, Ref. 19]

FIG. 8.7. Basic structure of the methanol–bromine complex.

which is readily obtained, has a stoichiometry of $2:1$. Here the oxygen atoms of the dioxane are each linked to iodine atoms on separate acceptor molecules without the formation of endless chains.[2, 9] The intermolecular Cl–Cl distance (3·38 Å) is the normal van der Waals separation.

The structures of the $2:1$ iodine complexes of 1,4-dithiane,[21, 22] 1,4-diselenane[21, 23] and 1,4-selenothiane[24] have been determined. Various projections are given in Figs. 8.9 and 8.10. It has been noted that,

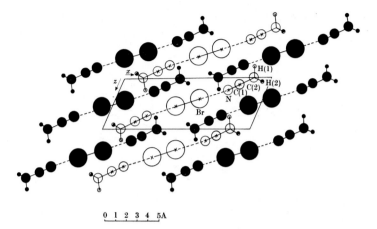

0 1 2 3 4 5A

[K.-M. Marstokk and K. O. Strømme, Ref. 20]

FIG. 8.8. Lattice structure of the acetonitrile–bromine 2:1 complex as seen along the y-axis. White circles represent atoms in molecules at $y = 0$. Black circles represent atoms in molecules at $y = \pm\frac{1}{2}$. Intermolecular N---Br bonding is indicated by broken lines. Covalent bonds are represented by full lines.

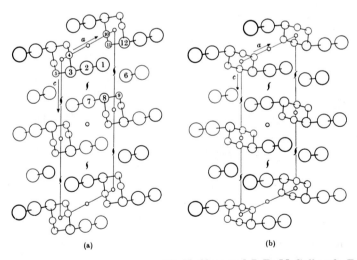

(a)   (b)

[G. Y. Chao and J. D. McCullough, Ref. 23]

FIG. 8.9. (a) Projection of the structure of the diselenane–iodine 1:2 complex down the b-axis. (b) Projection of the structure of the dithiane–iodine 1:2 complex down the b-axis.

whereas the bonding to the acceptor atom in the dioxane, the dithiane and the selenothiane complexes corresponds to an equatorial direction with respect to the ring of the donor (Fig. 8.11), the iodine is bonded to

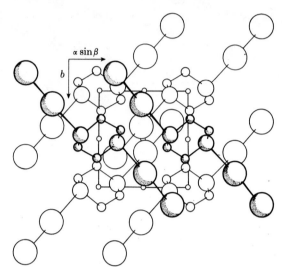

[G. Y. Chao and J. D. McCullough, Ref. 22]

FIG. 8.10. Projection of the structure of the dithiane–iodine 1 : 2 complex down the c-axis.

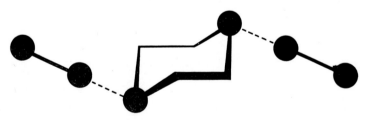

FIG. 8.11. Equatorial orientation of the acceptor in the dithiane– and selenothiane–iodine 1 : 2 complexes.

selenium in an axial position in diselenane (Fig. 8.12). Iodine forms a 1 : 1 complex with 1,4-oxaselenane;[25] the oxygen atom is not involved in intermolecular bonding. The iodine is linked in the axial position to the selenium as in 1,4-diselenane (Fig. 8.13).

An axial orientation has been observed for the tetrahydroselenophene complex,[26] although the intermolecular arrangement is different in each of these selenium-containing complexes.[24] The secondary bonds between an Se atom in one molecule of 1,4-diselenane and an iodine atom are relatively weak (length 3·89 Å) and give rise to a structure in which the molecules are linked together about alternate 2_1 screw axes. The secondary Se—I bonds in the tetrahydroselenophene complex are considerably stronger (length 3·64 Å) and give a near-linear chain of

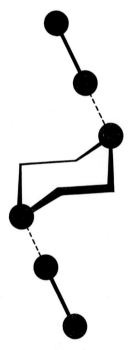

FIG. 8.12. Axial orientation of the acceptor in the diselenane–iodine 2:1 complex.

alternate donor–acceptor molecules. The interaction of iodine with 1,4-oxaselenane is of intermediate character. Zigzag chains of alternate donor and acceptor molecules result. In all cases there is an effectively linear orientation of the iodine molecule to the selenium atom with which it is primarily associated (Table 8.1). The intermolecular Se—I distances in some of these complexes are only little more than the normal covalent distance. They may legitimately be described as molecular complexes,[26] although the structures are tending towards the bonding found in *compounds* of the type R_2SeBr_2 and R_2SeCl_2.[27–29]

The ability of the atoms in diatomic halogen molecules to participate in charge-transfer bonding is also shown by halogen atoms in other types of acceptor molecule. For example, 1,4-diselenane forms endless chains of alternate donor and acceptor molecules with tetraiodoethene.[30] The donor atom–halogen atom carbon arrangement is nearly linear. The second iodine atom of the acceptor molecule is linked to a selenium atom of a donor molecule in a neighbouring chain. The similarity of the

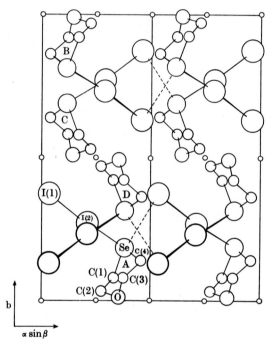

[H. Maddox and J. D. McCullough, Ref. 25]

FIG. 8.13. Structure of the 1:1 complex 1,4-oxaselenane–iodine, projection down the c-axis.

crystal structure of the 1:1 pyrazine–tetrabromoethene complex with that of tetrabromoethene itself has been noted by Dahl and Hassel.[31]

1,4-Dithiane and 1,4-diselenane both form 1:1 complexes with diiodoacetylene,[32] the structures of which consist of chains of alternate donor and acceptor molecules extending through the crystals. In the dithiane complex the S—I intermolecular bonds are roughly equatorial to the dithiane ring, whilst in the diselenane complex the Se—I intermolecular bonds are approximately axial to the ring (Fig. 8.14). These intermolecular bonds, which are relatively long compared with the

equivalent bonds in the corresponding molecular iodine complexes
(Table 8.1), provide an indication of the poorer electron-acceptor
properties of diiodoacetylene.

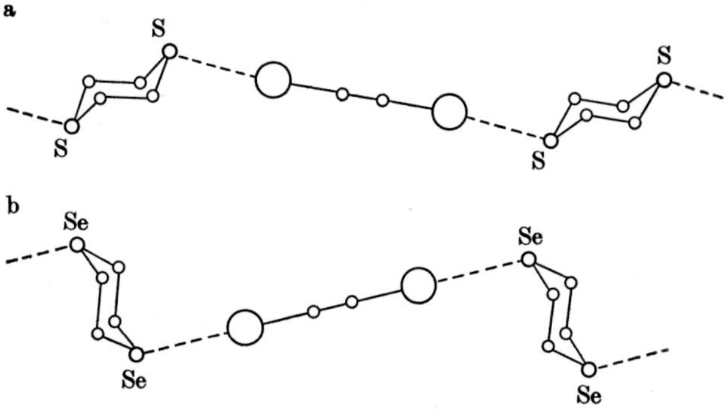

[O. Holmesland and C. Rømming, Ref. 32]

FIG. 8.14. Structure of the (a) 1,4-dithiane–diiodoacetylene, and (b) 1,4-
diselenane–diiodoacetylene chains.

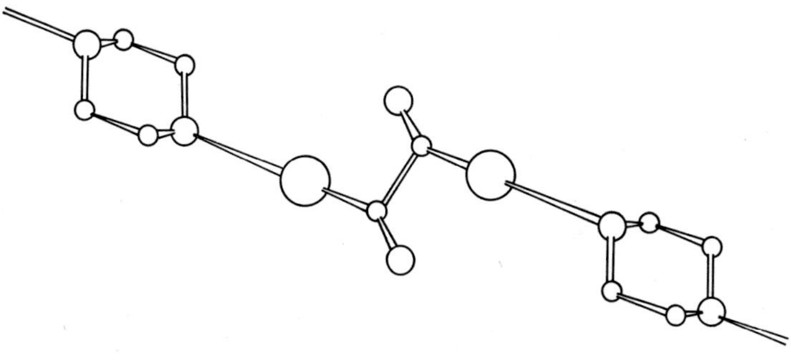

[E. Damm, O. Hassel and C. Rømming, Ref. 18]

FIG. 8.15. Schematic drawing illustrating the chains of alternating donor and
acceptor molecules in oxalyl halide–dioxane complexes.

The cyclohexan-1,4-dione–diiodoacetylene complex shows a disordered structure.[33] The dione molecules are in a twisted-boat conformation. The angles made by the intermolecular O–I direction with the
linear diiodoacetylene for the two oxygen positions have been measured
as 165° and 166°. This observation, together with the relatively long

O–I intermolecular distance (2·94, 2·95 Å), and the fact that the angle between the two carbonyl bonds in the donor molecule (155·6°) is approximately that found in the dione itself (154°) compared with the angle observed in the mercuric chloride adduct[34] (176°), suggests that the charge-transfer bonding of the dione to the diiodoacetylene is

[T. Bjorvatten and O. Hassel, Ref. 15]

FIG. 8.16. Structure of the iodoform–dithiane chains. Open circles indicate sulphur, large closed circles iodine atoms, and small closed circles carbon atoms.

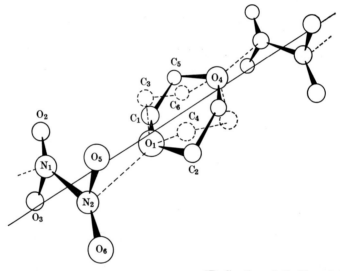

[P. Groth and O. Hassel, Ref. 35]

FIG. 8.17. Schematic drawing of the dioxane–dinitrogen tetroxide complex.

relatively weak. The structures of the complexes of dioxane with oxalyl chloride and with oxalyl bromide are isomorphous, and consist of endless chains of alternating donor and acceptor molecules joined by charge-transfer bonds connecting oxygen and halogen atoms[18] (Fig. 8.15). In the weaker oxalyl chloride complex the O–hal distance is only slightly less than the van der Waals distance, whereas for the oxalyl bromide this distance is significantly shorter. Chains of alternating donor and acceptor molecules also occur in the 1:1 dithiane–iodoform complex, where each iodoform molecule is linked to two neighbouring dithiane molecules and vice versa[15] (Fig. 8.16). An intermolecular I–S distance, noticeably shorter than the van der Waals distance, is observed (Table 8.1). By contrast, iodoform forms 1:3 complexes with quinoline[14] and with sulphur[16] (S_8), in which each iodine atom of the acceptor is involved in direct charge-transfer bonding. However, in all these complexes the C—I bond of the acceptor and the charge-transfer bond are nearly colinear.

The 1:1 complex between dioxane and dinitrogen tetroxide[35, 36] has a continuous chain structure in which the acceptor has a planar conformation and each molecule is linked to two dioxane molecules via weak N···O bonds of length 2·76 Å (Fig. 8.17).

8.B.2. Intermolecular Charge-Transfer Complexing of n-Donor Sites with Acceptor Moieties in the same Molecular Species

The possibility that the effects of intermolecular charge-transfer bonding between electron-donating and electron-accepting atoms of a single molecular species in the solid phase might be observable, was

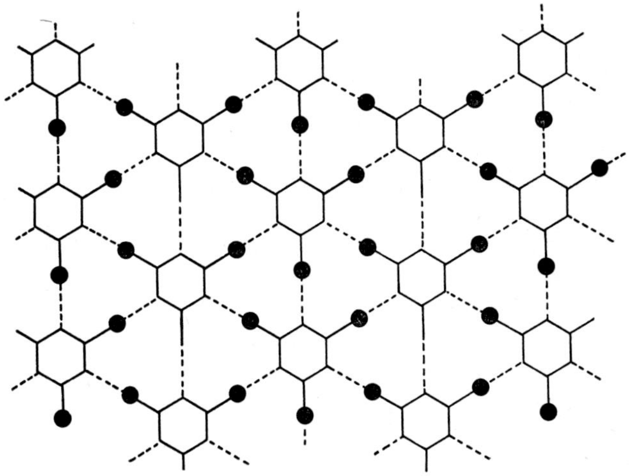

[After W. Hoppe, H. U. Lenné and G. Morandi, Ref. 38]

FIG. 8.18. Bonding within a layer of a crystal of cyanuric chloride.

anticipated by Mulliken,[13] and by Hassel,[37] who has suggested that the Cl–N distance in cyanuric chloride,[38] which is somewhat smaller than the expected van der Waals distance, is the result of charge-transfer interaction. This suggestion readily accounts for the planar structure and the linear C—Cl···N arrangement (Fig. 8.18).

The details of the crystal structure of oxalyl bromide[39] indicate that charge-transfer forces may be significant. The crystal is built up of non-planar sheets in which each molecule appears to be linked to four neighbours through O—Br charge-transfer bonds. The length of these bonds, 3·27 Å, is measurably less than the van der Waals distance of 3·35 Å. The angle C—Br···O is 169°.

The somewhat shortened intermolecular Te–Cl distance in solid dimethyltellurium dichloride is possibly indicative of a charge-transfer interaction,[40] as are details of the crystal structure of di-*p*-chloro-diphenyltellurium diiodide,[41] which include relatively long intra-molecular Te–I distances (2·95 Å), and short intermolecular I–I distances (3·85 Å).

In the original analysis of crystalline *p*-nitroaniline, a structure was claimed which contained an abnormally short distance between the oxygen atom of a nitro group and a carbon atom of an adjacent molecule of *p*-nitroaniline.[42] It was suggested that this is a result of a charge-transfer interaction involving these specific atoms.[43] However, it has now been shown that this structure is incorrect and, in fact, the lattice contains no abnormally short distances.[44]

8.C. Complexes of π-Donors

Most solid charge-transfer complexes involving π-electron donors have a feature in common with many charge-transfer complexes of *n*-electron donors, namely infinite chains of alternate donors and acceptor molecules in which the donor–acceptor distance is less than the van der Waals distance. In the case of π-donors this shortening is considerably less than in cases of a strong transfer-interaction between an *n*-electron donor and a specific acceptor-atom.

Structural determinations of the benzene–bromine complex[45] and of the isomorphous benzene–chlorine complex were made by Hassel and Strømme.[46] They reported that the crystals were monoclinic, that the structure consisted of alternate benzene and halogen molecules in infinite stacks, that the ring planes of the benzene molecules were parallel, that each halogen molecule was symmetrically situated between successive benzene molecules, and that the line of the halogen molecules was coincident with the six-fold axis of the benzene rings within a given chain. No lengthening of the halogen–halogen intra-molecular distance on complex formation was observed. The shorter perpendicular distances from ring plane to halogen atom were determined as 3·36 Å for the bromine complex, and 3·28 Å for the chlorine complex, compared with 3·65 Å and 3·50 Å respectively for the van der

Waals distances. Alternate chains were reported to be staggered so that the halogens are surrounded, edge-on, by donor molecules. These results were of considerable interest in view of the possible hypothetical structures discussed earlier by Mulliken[47] for isolated halogen–benzene complexes in solution. Symmetry requirements had lent support for either the resting structure (R) or the oblique structure (O) in preference to the axial structure (A) (Fig. 8.19) which is the basic unit in Hassel's experimentally determined structure. Structure A (and structure O) are in harmony with the original interpretation of the infrared absorption of such complexes in solution[48, 49] (see Chapter 4). Any argument, based on the difference in structure of the complex in the solid phase compared with the structure in solution, to explain these conflicting views, is not tenable since Person and his co-workers[50] have shown

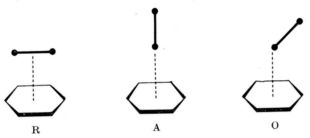

R   A   O

[After R. S. Mulliken, *J. Am. chem. Soc.* **74**, 811 (1952)]

FIG. 8.19. Hypothetical structures for halogen–benzene complexes. R = resting, A = axial, O = oblique.

that in the solid bromine–benzene complex the Br–Br stretching vibration has about the same infrared absorption intensity and frequency as the complex in solution. This and other details of the infrared spectrum can only be explained if neither component molecule of the complex is located at a site possessing a centre of symmetry.[50] This contrary picture had led to a reappraisal of the crystal structure described by Hassel and Strømme.[45, 46] If, in fact, the crystals are rhombohedral rather than monoclinic, with both benzenes and bromines located at sites of C_{3v} symmetry, only a minor adjustment of the crystal structure is required in order to fit the diffraction data, and to accommodate the infrared absorption properties. The change in the structure is effectively to displace benzene molecules from their central positions between pairs of bromine molecules in the stack so that they are closer to one molecule than to the other in the pair. Either a random array of the short neighbour-distances, or the effect of vibrational motion could cause an average distance to be measured. Hassel and Strømme [45, 46] did in fact observe large temperature factors in their original measurements.

Evidence for electron donor–acceptor interaction in solid crystal lattices between halogen acceptor atoms linked to carbon, with π-donor

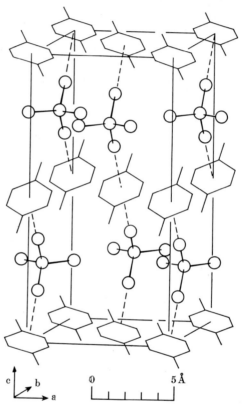

[F. J. Strieter and D. H. Templeton, Ref. 51]

FIG. 8.20. Crystal structure of the carbon tetrabromide–p-xylene complex.

molecules, is typified by the carbon tetrabromide–p-xylene complex, in which each acceptor molecule is associated with two donor molecules.[51] The bromine atom nearest to a xylene molecule, and the carbon atom to which it is attached, lie close to the two-fold axis perpendicular to that particular xylene molecule (Fig. 8.20). The distance of the bromine atom along this axis from the plane of the donor (3·34 Å) is the same, within experimental error, as the corresponding distance in the bromine–benzene complex.

The generally observed structure of a parallel arrangement of molecules in infinite stacks, amongst the planar or near-planar π-donor–π-acceptor complexes, cannot be claimed to be a characteristic only of charge-transfer complexes. Many single species, which have planar molecules, but for which there is no evidence of charge transfer, crystallize with such a structure. However, the *alternate* arrangement of the two molecular species within the stacks does appear to be typical though, of course, not exclusive, of many charge-transfer complexes. This type of stacking is not shown by the superficially analogous complex between hexabromobenzene and 1,2,4,5-tetrabromobenzene,[52] for which there is *no* evidence of charge-transfer complexing. In this case the molecular structure consists of stacks of molecules in a quasi-hexagonal array but containing only one molecular component in each stack.

The structures of a number of complexes between π-donors and π-acceptors have been determined, some details of which are summarized in Table 8.2. The early determinations by Powell and his co-workers[53, 54] showed that the interplanar distances between donor and acceptor molecules in complexes such as p-iodoaniline–1,3,5-trinitrobenzene are only slightly less than the van der Waals separation. This observation was vital to the development of ideas concerning the nature of bonding in this type of complex (see Chapter 1).

TABLE 8.2. Some complexes of π-donor molecules for which the crystal structures have been determined*

Donor (D)	Acceptor (A)	Ratio D:A	Mean layer separation (Å)†	Ref.
Naphthalene	1,3,5-Trinitrobenzene	1:1	3·35	72
Anthracene	1,3,5-Trinitrobenzene	1:1	3·28	66
Azulene	1,3,5-Trinitrobenzene	1:1	{ 3·33 / 3·38	64 / 81
Acepleiadylene	1,3,5-Trinitrobenzene	1:1	3·26	63
p-Iodoaniline	1,3,5-Trinitrobenzene	1:1		53
Indole	1,3,5-Trinitrobenzene	1:1	3·29	65
Skatole	1,3,5-Trinitrobenzene	1:1	3·30	65
2,4,6-Tri(dimethylamino)-1,3,5-triazine	1,3,5-Trinitrobenzene	1:1	3·36	74
N,N,N′,N′-Tetramethyl-p-phenylenediamine	Chloranil	1:1	{ 3·26 / 3·284	72 / 73
Hexamethylbenzene	Chloranil	1:1		68–70
8-Hydroxyquinoline	Chloranil	2:1		86
Bis-8-hydroxyquinolinato-palladium	Chloranil	1:1		85
N,N,N′,N′-Tetramethyl-p-phenylenediamine	Bromanil	1:1	3·31	72

TABLE 8.2.—*continued*

Donor (D)	Acceptor (A)	Ratio D : A	Mean layer separation (Å)†	Ref.
Perylene	Fluoroanil	1 : 1	3·23	72
p-Chlorophenol	p-Benzoquinone	2 : 1	3·20	56
p-Bromophenol	p-Benzoquinone	2 : 1		56
p-Chlorophenol	p-Benzoquinone	1 : 1	3·31	57
p-Bromophenol	p-Benzoquinone	1 : 1		57
Quinol	p-Benzoquinone	1 : 1	3·16	58, 59
Phenol	p-Benzoquinone	1 : 1	3·33	55
N,N,N′,N′-Tetramethyl-p-phenylenediamine	7,7,8,8-Tetracyano-quinodimethane	1 : 1	3·27	76
Bis-(8-hydroxy-quinolinato)copper(II)	7,7,8,8-Tetracyano-quinodimethane	1 : 1	∼3·2	67
Naphthalene	Tetracyanoethylene	1 : 1	3·30	71
Perylene	Tetracyanoethylene	1 : 1	3·23	81
Pyrene	Tetracyanoethylene	1 : 1	3·32	80
Naphthalene	1,2,4,5-Tetracyano-benzene	1 : 1	3·43	75
Bis-8-hydroxy-quinolinato-palladium	1,2,4,5-Tetracyano-benzene	1 : 1		89
Anthracene	Pyromellitic dianhydride	1 : 1	3·23	82
Perylene	Pyromellitic dianhydride	1 : 1	3·33	82
13,14-Dithiatricyclo-[8,2,1⁴, ⁷]tetradeca-4,6,10,12-tetraene	Benzotrifuroxan	1 : 1		79
Bis-8-hydroxy-quinolinato-copper(II)	Benzotrifuroxan	1 : 1		88
Bis-8-hydroxy-quinolinato-copper(II)	Picryl azide	1 : 1		87
p-Xylene	Carbon tetrabromide	1 : 1		51
Benzene	Bromine	1 : 1	3·36	6, 45
Benzene	Chlorine	1 : 1	3·28	46

* Less detailed structures for various complexes of pyromellitic dianhydride, benzotrifuroxan, tetracyanoethylene and hexafluorobenzene have also been reported [J. C. A. Boeyens and F. H. Herbstein, *J. phys. Chem., Ithaca* **69**, 2153 (1965)].

† In the case of planar or near planar π-acceptor molecules; for other acceptors the distance given is the perpendicular distance from the plane of the donor molecule to the nearest acceptor atom.

Recent measurements on a variety of complexes have shown that the average layer separation of the two components within a stack is only slightly dependent on the particular donor–acceptor pair. Apart from complexes where there is some participation from hydrogen bonding, all such distances which have been measured are within the range 3·23–3·38 Å (Table 8.2).

8.C.I. Structures in which there is also Participation by Hydrogen Bonding

Complexes between p-benzoquinone and π-electron donors, which are also potential hydrogen donors in hydrogen-bond complex formation, have been studied particularly by Wallwork and his co-workers.[55–57]

In the normal monoclinic form of quinhydrone both constituent molecules may be deformed to some extent.[58] The quinone and hydroquinone molecules are linked alternately through O—H⋯H hydrogen bonds to form a zigzag chain which extends through the crystal. Planes of molecules of the same species are all parallel within a given chain, the angle between the quinone and hydroquinones being about $2\frac{1}{2}°$.

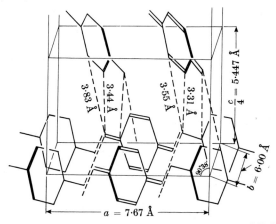

[H. Matsuda, K. Osaki and I. Nitta, Ref. 58]

FIG. 8.21. Perspective view of a part of the structure of quinhydrone.

These chains are stacked side by side to form a molecular sheet. The sheets themselves are piled into a type of layer lattice (Fig. 8.21). The overlap between neighbouring molecules within a column is illustrated in Fig. 8.22. In a triclinic modification of quinhydrone, the only difference in structure is that the direction of the molecular chain in successive sheets varies from the [120] direction to the [1$\bar{2}$0] direction according to the glide plane operation.[59]

In the 2:1 complex of phenol and p-benzoquinone, each quinone molecule is sandwiched between two phenol molecules which are parallel to the quinone.[55] Groups of three such molecules are stacked discontinuously in columns. Hydrogen bonds link the phenolic hydroxyl groups with quinone–oxygen atoms in adjacent columns. The stoichiometry on the one hand, and the parallel arrangement of neighbouring

electron donor and acceptor molecules on the other, suggest a compromise between hydrogen bonding and charge-transfer bonding
respectively. The line joining the centres of the rings within a triad makes

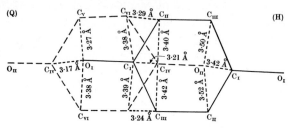

[H. Matsuda, K. Osaki and I. Nitta, Ref. 58]

FIG. 8.22. Perpendicular projection of two neighbouring molecules within a
column in quinhydrone.

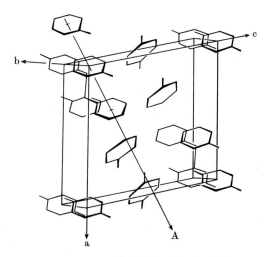

[T. T. Harding and S. C. Wallwork, Ref. 55]

FIG. 8.23. The phenol–p-benzoquinone 2:1 complex. An orthographic drawing
showing the contents of one unit cell with an additional phenol molecule, to
illustrate the sandwiching of the quinone at 0, 0, 0 by phenols at 1/3, 0, 1/6 and 2/3,
0, 5/6. Such groups of three molecules are stacked in columns parallel to the [201]
axis, marked A.

an angle of about 30° to the normal of the rings. The resulting sideways
displacement of each ring has been measured[55] as 1·98 Å (Fig. 8.23).

4-Chlorophenol forms two complexes with p-benzoquinone, namely a
complex with a 2:1 stoichiometry,[56] similar in structure to that of the

complex of unsubstituted phenol with p-benzoquinone, and a 1:1 complex.[57] 4-Bromophenol behaves similarly, the two bromophenol complexes formed being isomorphous with the 4-chlorophenol complexes.[56, 57] It would appear that in the 2:1 complexes hydrogen

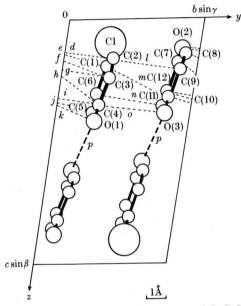

[G. G. Shipley and S. C. Wallwork, Ref. 57]

FIG. 8.24. The structure of the chlorophenol–p-benzoquinone 1:1 complex projected along the a-axis showing principal intermolecular contacts:

(d)	C(2)---O(2)	3·37 Å	(e)	C(1)---O(2)	3·24 Å
(f)	C(3)---C(8)	3·31	(g)	C(3)---C(7)	3·38
(h)	C(4)---C(7)	3·36	(i)	C(4)---C(12)	3·40
(j)	C(5)---C(12)	3·31	(k)	O(1)---C(10)	3·36
(l)	C(9)---C(2)	3·45	(m)	C(10)---C(1)	3·36
(n)	C(11)---C(6)	3·42	(o)	O(3)---C(4)	3·44
(p)	O(1)---O(3)	2·70 (H-bond)			

bonding is more dominant than in the 1:1 complexes. In the 1:1 4-chlorophenol–p-benzoquinone complex, which was studied in detail,[57] the phenol and quinone molecules are stacked alternately in infinite columns. The columns are themselves linked in pairs by hydrogen bonds. The donor and acceptor molecules are tilted to about 30° in the same fashion as in the 2:1 complexes. As a consequence, in all these complexes, the C—O or C=O groups in successive molecules are approximately over the centres of the rings of adjacent molecules (Fig. 8.24).

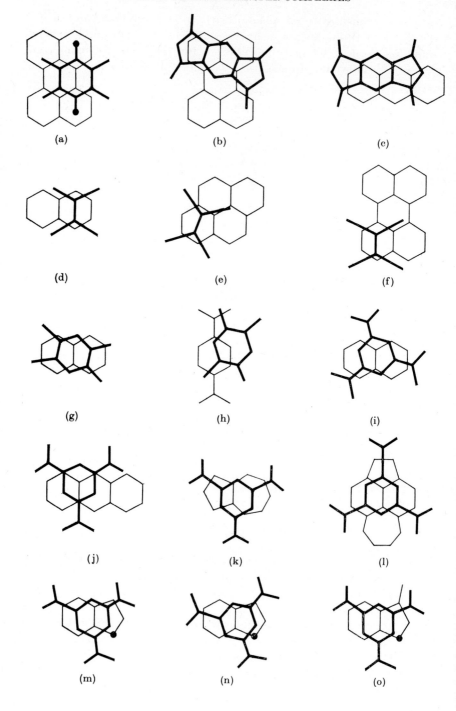

(a)   (b)   (c)

(d)   (e)   (f)

(g)   (h)   (i)

(j)   (k)   (l)

(m)   (n)   (o)

8.C.2. Structures in which there is no Hydrogen Bonding

Structure determinations on pairs of solid complexes derived from the same electron-donor–acceptor pair but with different stoichiometry and in which there is no possibility of hydrogen bonding would be of interest. Although such systems have been described, containing for example pyrene and pyromellitic dianhydride[60] or hexamethylbenzene and tetracyanoethylene,[61] detailed X-ray crystal structures have not been reported. The infrared absorption characteristics of the latter complex (see Chapter 4) suggest that the extra donor molecules are incorporated into the complex stacks, rather than forming separate stacks of pure hydrocarbon alongside stacks of alternating donor and acceptor molecules.[61] Pyromellitic dianhydride appears to be somewhat exceptional in the variable stoichiometry of its hydrocarbon complexes.[62]

Infinite stacks of alternate donor and acceptor molecules appear to be a common feature of all solid 1:1 charge-transfer complexes in which there is no participation through hydrogen bonding. In some cases, deviations from planarity of otherwise planar component molecules have been observed, although they are not large. Significant twisting of the nitro groups in 1,3,5-trinitrobenzene in, for example, complexes with acepleiadylene,[63] azulene,[64] skatole,[65] indole[65] and anthracene,[66] has been noted. The normally planar 7,7,8,8-tetracyanoquinodimethane molecule in the crystalline complex with bis-(8-hydroxyquinolinato)-copper(II) is likewise non-planar.[67] Examples of buckling of the donor molecule have also been observed. However, the reported[68] large distortion of the component molecules in crystals of the hexamethylbenzene–chloranil complex appears to be incorrect.[69, 70]

The stacking angle varies from complex to complex. In some, such as naphthalene–tetracyanoethylene,[71] it is about 30°, whilst in others, such as the complexes of N,N,N',N'-tetramethyl-p-phenylenediamine with

---

Fig. 8.25. Overlap of the donor and acceptor components in various solid crystalline complexes viewed approximately normal to their mean planes. In each case the acceptor molecule is distinguished by the heavier lines. (a) perylene–fluoranil, after Ref. 78; (b) perylene–pyromellitic dianhydride, after Ref. 82; (c) anthracene–pyromellitic dianhydride, after Ref. 82; (d) naphthalene–tetracyanoethylene, after Ref. 71; (e) pyrene–tetracyanoethylene, after Ref. 80; (f) perylene–tetracyanoethylene, after Ref. 81; (g) naphthalene–1,2,4,5-tetracyanobenzene, after Ref. 75; (h) N,N,N',N'-tetramethyl-p-phenylenediamine–1,2,4,5-tetracyanobenzene, after Ref. 84; (i) naphthalene–1,3,5-trinitrobenzene, after Ref. 72; (j) anthracene–1,3,5-trinitrobenzene, after Ref. 72; (k) azulene–1,3,5-trinitrobenzene, after Refs. 64 and 83; (l) acepleiadylene–1,3,5-trinitrobenzene, after Ref. 63; (m) indole–1,3,5-trinitrobenzene, after Ref. 65; (n) indole–1,3,5-trinitrobenzene, second arrangement, after Ref. 65; (o) skatole–1,3,5-trinitrobenzene, after Ref. 65.

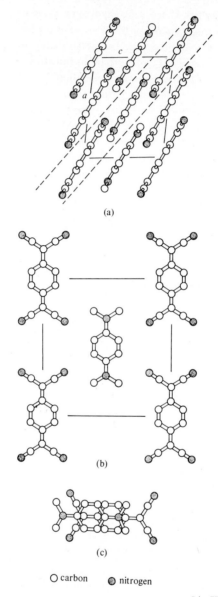

(a)

(b)

(c)

○ carbon   ◉ nitrogen

[A. W. Hanson, Ref. 76]

FIG. 8.26. (a) The structure of the 7,7,8,8-tetracyanoquinodimethane–N,N,N′,N′-tetramethyl-*p*-phenylenediamine complex viewed along the *b*-axis. The heavily outlined molecules lie at $y = \frac{1}{2}$, and the rest at $y = 0$. The dotted lines define a sheet of molecules. (b) A sheet of molecules in plan. (c) Overlapping molecules viewed normal to their plane.

chloranil [72, 73] and with bromanil,[72] the 2,4,6-tri(dimethylamino)-1,3,5-triazine–1,3,5-trinitrobenzene complex,[74] and the naphthalene–1,2,4,5-tetracyanobenzene complex,[75] it is zero. In these latter cases, this means

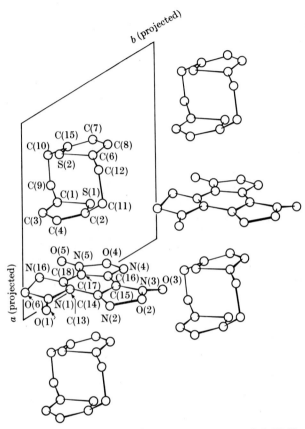

[B. Kamenar and C. K. Prout, Ref. 79]

FIG. 8.27. Part of two of the plane-to-plane stacks in the crystal of the 13,14-dithiatricyclo[8,2,1,1^{4,7}]tetradeca-4,6,10,12-tetraene–benzotrifuroxan complex projected down the c-axis.

that the donor and acceptor molecules are vertically above one another. Orientations of one molecule, relative to adjacent molecules of the second species in the same stack for various complexes, are depicted in Fig. 8.25. The dihedral angle between the planes of donor–acceptor pairs appears usually to be small, rarely more than a few degrees. For many complexes the dihedral angle has not been determined. In other cases, where there is buckling of one or both of the components, it is difficult to describe such a dihedral angle.

The "overlap and orientation principle" (Chapter 2, p. 21) would lead one to expect that for maximum charge-transfer interaction the planes of the two components should lie parallel, one directly above the other, if the bonding is via the delocalized π-orbitals in both components. This

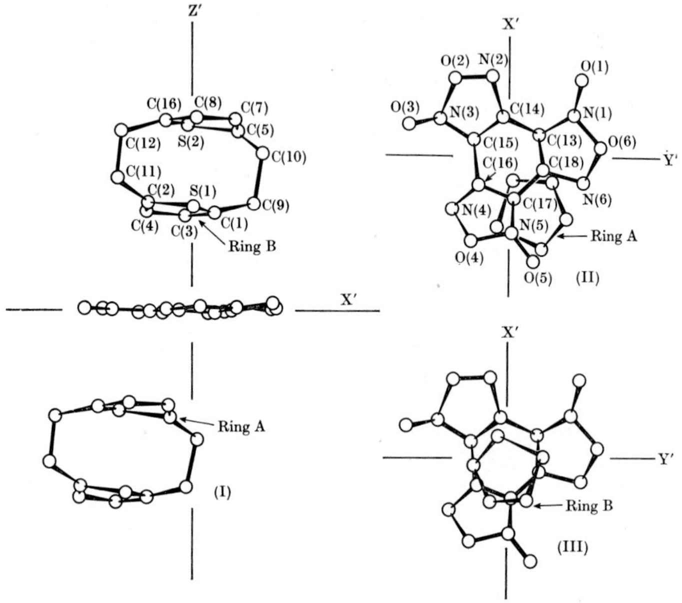

[B. Kamenar and C. K. Prout, Ref. 79]

FIG. 8.28. The 13,14 - dithiatricyclo[8,2,1,1^{4,7}]tetradeca - 4, 6, 10, 12 - tetraene–benzotrifuroxan complex: (I) two donor molecules are projected perpendicular to the least-squares best plane of the benzene nucleus of the benzotrifuroxan molecule. The relationships between the two thiophene rings, A and B, of the donor molecules and the benzotrifuroxan are shown in (II) and (III) respectively.

could be the case for complexes such as the N,N,N',N'-tetramethyl-p-phenylenediamine complexes of chloranil and bromanil, where, perhaps as Wallwork[72] suggests, the charge-transfer forces overwhelm any other forces operative in crystal lattices which might otherwise lead to an alternative orientation. However, care must be taken in drawing conclusions from an observed orientation.[74] Thus, in the 2,4,6-tri(dimethylamine)-1,3,5-triazine–1,3,5-trinitrobenzene complex,[74] both molecules are similarly orientated with respect to the crystal axes, and the stacking angle is zero. Nevertheless, from the nature of the donor molecule, a very strong interaction is not anticipated. Indeed, some confirmation of the relative weakness of the interaction is obtained from the energy of the charge-transfer transition and from the rather large mean perpendicular separation of the planes of the molecules (3·36 Å), although it is recognized that this latter is a rather insensitive measure of binding energy for π–π complexes. It may be argued that the shapes of these two molecules

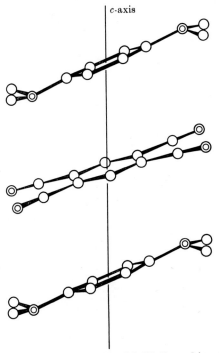

[Y. Ohashi, H. Iwasaki and Y. Saito, Ref. 84]

FIG. 8.29. Structure of the N,N,N′,N′-tetramethyl-p-phenylenediamine–1,2,4,5-tetracyanobenzene complex viewed along the [1$\bar{1}$0] plane.

provide a good fit for each other. In the case of the 7,7,8,8-tetracyanoquinodimethane – N, N, N′, N′ - tetramethyl - p - phenylenediamine complex,[76] where there is also a strong resemblance between the shapes of the donor and acceptor molecules, the stacking angle is large, as is shown by the non-coincident overlap (Fig. 8.26).

Prout and Wallwork[77] have suggested that, in the case of some quinone complexes, the displacement of the molecular centres of the two components may be the result of a specific interaction between the carbonyl group of the acceptor and the aromatic ring of the donor. This

obviously would apply, not only to systems in which there is also hydrogen bonding, where it had been suggested that the observed configuration
might be the result of a compromise between charge-transfer forces and
hydrogen bonding (p. 235), but also to systems where there can be no

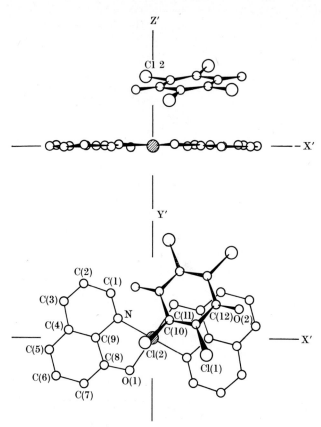

[B. Kamenar, C. K. Prout and J. D. Wright, Ref. 85]

Fig. 8.30. A chloranil molecule in the bis-8-hydroxyquinolinatopalladium(II)–
chloranil complex projected parallel to and perpendicular to, the least-squares best
plane of the donor molecule.

hydrogen bonding, for example, in the perylene–fluoranil complex.[78]
However, this argument might lead one to expect a stacking angle of
about 30° for all p-benzoquinone complexes (including substituted
p-benzoquinones), which is not what is observed.

The crystal structure of the 1:1 complex between benzotrifuroxan
and 13,14-dithiatricyclo[8,2,1,1$^{4, 7}$]tetradeca-4,6,10,12-tetraene has been

determined.[79] The crystals are formed from plane-to-plane stacks of alternate donor and acceptor molecules (Fig. 8.27). There is considerable distortion of the planar systems. The dihedral angle between the mean plane of the acceptor and the plane of the thiophene ring of each donor adjacent to the acceptor is $9\frac{1}{2}°$. These two rings (A and B in Fig. 8.28) are differently related to the acceptor molecule.

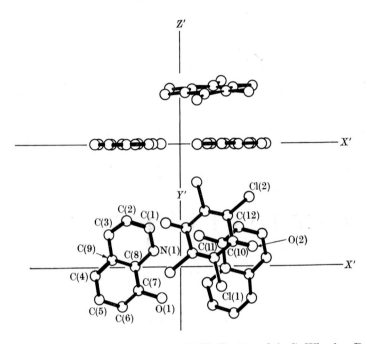

[C. K. Prout and A. G. Wheeler, Ref. 86]

FIG. 8.31. The molecular arrangement in the crystal of the 8-hydroxyquinoline–chloranil molecule projected parallel to and perpendicular to the least-squares best plane of the 8-hydroxyquinoline dimer.

Other examples of unusual orientations are shown by the pyrene[80] and the perylene[81] complexes of tetracyanoethylene, the anthracene– and the perylene–pyromellitic dianhydride complex,[82] and the 1,3,5-trinitrobenzene complexes of naphthalene,[72] anthracene,[66, 72] azulene,[64, 83] acepleiadylene,[63] indole[65] and skatole[65] (Fig. 8.25). In the last two complexes, Hanson[65] has suggested that there may be some specific interaction between the nitrogen atom in the donor molecule and an unsubstituted position in the acceptor molecule. A localized interaction may also be the explanation of the unusual orientation of the

9

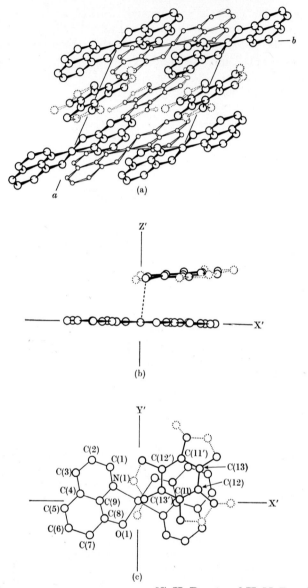

[C. K. Prout and H. M. Powell, Ref. 88]

Fig. 8.32. The bis-(8-hydroxyquinolinato)copper(II)–benzotrifuroxan complex: (a) crystal structure projected perpendicular to the (100) planes of the crystal; in (b) and (c) respectively the benzotrifuroxan molecule is projected on to, and perpendicular to, the plane of the bis-(8-hydroxyquinolinato)copper(II) molecule. In all cases the oxygen atom sites (broken lines) are to be regarded as undetermined.

donor with respect to the acceptor in crystals of the N,N,N′,N′-tetra-methyl-p-phenylenediamine – 1, 2, 4, 5 - tetracyanobenzene complex [84] (Figs. 8.25 and 8.29). In all these cases the actual position of the donor with respect to the acceptor will be a compromise of the preferred

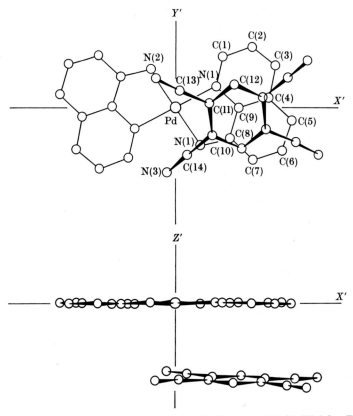

[B. Kamenar, C. K. Prout and J. D. Wright, Ref. 89]

FIG. 8.33. The 1,2,4,5-tetracyanobenzene–bis-(8-hydroxyquinolinato)palladium-(II) complex showing a projection parallel to, and perpendicular to, the least-squares best plane of the donor molecule.

orientation for maximum charge-transfer overlap, specific interactions including local polarizing–polarization interactions, together with a minimization of repulsions and a maximal occupation of space within the crystal.

The structures of charge-transfer complexes of π-acceptors with metal 8-hydroxyquinolinates are of particular interest. The 1:1 complex between bis-8-hydroxyquinolinato-palladium(II) and chloranil shows a structure of approximately parallel donor and acceptor molecules[85] (Fig. 8.30). It has been suggested that specific interaction between the chlorine atoms and the palladium atoms affects the relative orientation of the two components; which is not what would be anticipated from the overlap and orientation principle. The structure of this complex is remarkably similar to the 2:1 complex between 8-hydroxyquinoline and

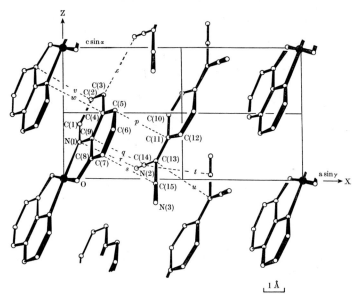

[R. M. Williams and S. C. Wallwork, Ref. 67]

Fig. 8.34. The structure of the bis-(8-hydroxyquinolinato)copper(II)–7,7,8,8-tetracyanoquinodimethane complex projected along the b-axis, showing short intermolecular contacts. The distances (Å) indicated are $p = 3 \cdot 236$, $q = 3 \cdot 267$, $r = 3 \cdot 303$, $s = 3 \cdot 314$, $t = 3 \cdot 387$, $u = 3 \cdot 451$, $v = 3 \cdot 457$, $w = 3 \cdot 410$, $x = 3 \cdot 467$, all $\pm$ ca. $0 \cdot 007$ Å.

chloranil.[86] In the latter complex the donor exists as a hydrogen-bonded dimer which corresponds to the chelated palladium system in the previous complex (Fig. 8.31). Evidence for charge-transfer complexing between bis-8-hydroxyquinolinato-copper(II) and picryl azide in the solid 1:2 complex has also been given.[87] Here the relative donor–acceptor orientation is in harmony with the overlap and orientation principle. By contrast, the 1:1 complex of bis-8-hydroxyquinolinato-copper(II) with benzotrifuroxan[88] shows features more akin to the chloranil–bis-8-hydroxyquinolinato-palladium(II) complex, namely a dihedral angle of about 10° between the mean planes of the donor and acceptor molecules and a relative orientation of the two components well removed from optimal charge-transfer interaction, due almost certainly to specific copper–nitrogen interaction (Fig. 8.32). The 1 : 1 complex of bis-8-hydroxyquinolinato-palladium(II) and 1,2,4,5-tetracyanobenzene[89] (Fig. 8.33) has a general structure intermediate between that of the corresponding 1 : 1 chloranil complex and that of the 2 : 1 bis-8-hydroxyquinolinato-copper(II)–picryl azide complex both of which are described above. The crystal structure of the complex between 7,7,8,8-tetracyanoquinodimethane and bis-(8-hydroxyquinolinato)-copper(II) has been the subject of a very careful study by Williams and Wallwork.[67] The general features of the structure are indicated in Fig. 8.34. The molecules are in the usual plane-to-plane arrangement of alternate components. These are orientated in such a manner that the double bond adjacent to one dicyanomethylene group of the acceptor lies over the 5 : 8 position of a donor molecule, whilst the other double bond of the acceptor is similar placed with respect to the benzenoid ring of the centrosymmetrically related donor molecule. (For recent reviews see Refs. 90 and 91.)

REFERENCES

1. O. Hassel and C. Rømming, *Quart. Rev. (London)* **16**, 1 (1962).
2. O. Hassel, *Proc. chem. Soc.* 250 (1957).
3. O. Hassel and C. Rømming, *Acta chem. scand.* **10**, 696 (1956).
4. C. Reid and R. S. Mulliken, *J. Am. chem. Soc.* **76**, 3869 (1954).
5. K. O. Strømme, *Acta chem. scand.* **13**, 268 (1959).
6. O. Hassel, *Molec. Phys.* **1**, 241 (1958).
7. O. Hassel, C. Rømming and T. Tufte, *Acta chem. scand.* **15**, 967 (1961).
8. O. Hassel, *Acta chem. scand.* **19**, 2259 (1965).
9. O. Hassel and J. Hvoslef, *Acta chem. scand.* **10**, 138 (1956).
10. O. Hassel and J. Hvoslef, *Acta chem. scand.* **8**, 873 (1954).
11. O. Hassel and K. O. Strømme, *Acta chem. scand.* **13**, 1775 (1959).
12. O. Hassel and K. O. Strømme, *Nature, Lond.* **182**, 1155 (1958); *Acta chem. scand.* **13**, 275 (1959).
13. R. S. Mulliken, *J. Am. chem. Soc.* **72**, 600 (1950).
14. T. Bjorvatten and O. Hassel, *Acta chem. scand.* **13**, 1261 (1959); *Acta chem. scand.* **16**, 249 (1962).
15. T. Bjorvatten and O. Hassel, *Acta chem. scand.* **15**, 1429 (1961).
16. T. Bjorvatten, *Acta chem. scand.* **16**, 749 (1962).
17. T. Bjorvatten, *Acta chem. scand.* **17**, 2292 (1963).
18. E. Damm, O. Hassel and C. Rømming, *Acta chem. scand.* **19**, 1159 (1965).
19. P. Groth and O. Hassel, *Molec. Phys.* **6**, 543 (1963); *Acta chem. scand.* **18**, 402 (1964).
20. O. Hassel, *Svensk kem. Tidskr.* **72**, 88 (1960); K.-M. Marstokk and K. O. Strømme, *Acta crystallogr.* **B24**, 713 (1968).

21. J. D. McCullough, G. Y. Chao and D. E. Zuccaro, *Acta crystallogr.* **12**, 815 (1959).
22. G. Y. Chao and J. D. McCullough, *Acta crystallogr.* **13**, 727 (1960).
23. G. Y. Chao and J. D. McCullough, *Acta crystallogr.* **14**, 940 (1961).
24. H. Hope and J. D. McCullough, *Acta crystallogr.* **15**, 806 (1962).
25. H. Maddox and J. D. McCullough, *Inorg. Chem.* **5**, 522 (1966).
26. H. Hope and J. D. McCullough, *Acta crystallogr.* **17**, 712 (1964).
27. J. D. McCullough and R. E. Marsh, *Acta crystallogr.* **3**, 41 (1950).
28. J. D. McCullough and G. Hamburger, *J. Am. chem. Soc.* **63**, 803 (1941).
29. J. D. McCullough and G. Hamburger, *J. Am. chem. Soc.* **64**, 508 (1942).
30. T. Dahl and O. Hassel, *Acta chem. scand.* **19**, 2000 (1965).
31. T. Dahl and O. Hassel, *Acta chem. scand.* **20**, 2009 (1966).
32. O. Holmesland and C. Rømming, *Acta chem. scand.* **20**, 2601 (1966).
33. P. Groth and O. Hassel, *Acta chem. scand.* **19**, 1733 (1965).
34. P. Groth and O. Hassel, *Acta chem. scand.* **18**, 1327 (1964).
35. P. Groth and O. Hassel, *Acta chem. scand.* **19**, 120 (1965).
36. P. Groth and O. Hassel, *Proc. chem. Soc.* 379 (1962).
37. O. Hassel, *Tidsskr. Kjemi Bergv. Metall.* **21**, 60 (1961).
38. W. Hoppe, H. U. Lenné and G. Morandi, *Z. Kristallogr.* **108**, 321 (1957).
39. P. Groth and O. Hassel, *Proc. chem. Soc.* 343 (1961).
40. G. D. Christofferson, R. A. Sparks and J. D. McCullough, *Acta crystallogr.* **11**, 782 (1958).
41. G. Y. Chao and J. D. McCullough, *Acta crystallogr.* **15**, 887 (1962).
42. S. C. Abrahams and J. M. Robertson, *Acta crystallogr.* **1**, 252 (1948).
43. S. C. Abrahams, *J. Am. chem. Soc.* **74**, 2692 (1952).
44. J. Donohue and K. N. Trueblood, *Acta crystallogr.* **9**, 960 (1956).
45. O. Hassel and K. O. Strømme, *Acta chem. scand.* **12**, 1146 (1958).
46. O. Hassel and K. O. Strømme, *Acta chem. scand.* **13**, 1781 (1959).
47. R. S. Mulliken, *J. Am. chem. Soc.* **74**, 811 (1952).
48. J. Collin and L. D'Or, *J. chem. Phys.* **23**, 397 (1955).
49. L. D'Or, R. Alewaeters and J. Collin, *Recl Trav. chim. Pays-Bas Belg.* **75**, 862 (1956).
50. W. B. Person, C. F. Cook and H. B. Friedrich, *J. chem. Phys.* **46**, 2521 (1967).
51. F. J. Strieter and D. H. Templeton, *J. chem. Phys.* **37**, 161 (1962).
52. G. Gafner and F. H. Herbstein, *J. chem. Soc.* 5290 (1964).
53. H. M. Powell, G. Huse and P. W. Cooke, *J. chem. Soc.* 153 (1943).
54. H. M. Powell and G. Huse, *J. chem. Soc.* 435 (1943).
55. T. T. Harding and S. C. Wallwork, *Acta crystallogr.* **6**, 791 (1953).
56. G. G. Shipley and S. C. Wallwork, *Acta crystallogr.* **22**, 585 (1967).
57. G. G. Shipley and S. C. Wallwork, *Acta crystallogr.* **22**, 593 (1967).
58. H. Matsuda, K. Osaki and I. Nitta, *Bull. chem. Soc. Japan* **31**, 611 (1958).
59. T. Sakurai, *Acta crystallogr.* **19**, 320 (1965).
60. I. Ilmet and L. Kopp, *J. phys. Chem., Ithaca* **70**, 3371 (1966).
61. B. Hall and J. P. Devlin, *J. phys. Chem., Ithaca* **71**, 465 (1967).
62. Y. Nakayama, Y. Ichikawa and T. Matsuo, *Bull. chem. Soc. Japan* **38**, 1674 (1965).
63. A. W. Hanson, *Acta crystallogr.* **21**, 97 (1966).
64. A. W. Hanson, *Acta crystallogr.* **19**, 19 (1965).
65. A. W. Hanson, *Acta crystallogr.* **17**, 559 (1964).
66. D. S. Brown, S. C. Wallwork and A. Wilson, *Acta crystallogr.* **17**, 168 (1964).
67. R. M. Williams and S. C. Wallwork, *Acta crystallogr.* **23**, 448 (1967).

68. T. T. Harding and S. C. Wallwork, *Acta crystallogr.* **8**, 787 (1955).
69. N. D. Jones and R. E. Marsh, *Acta crystallogr.* **15**, 809 (1962).
70. S. C. Wallwork and T. T. Harding, *Acta crystallogr.* **15**, 810 (1962).
71. R. M. Williams and S. C. Wallwork, *Acta crystallogr.* **22**, 899 (1967).
72. S. C. Wallwork, *J. chem. Soc.* 494 (1961).
73. J. L. de Boer and A. Vos, *Acta crystallogr.* **B24**, 720 (1968).
74. R. M. Williams and S. C. Wallwork, *Acta crystallogr.* **21**, 406 (1966).
75. S. Kumakura, F. Iwasaki and Y. Saito, *Bull. chem. Soc. Japan* **40**, 1826 (1967).
76. A. W. Hanson, *Acta crystallogr.* **19**, 610 (1965).
77. C. K. Prout and S. C. Wallwork, *Acta crystallogr.* **21**, 449 (1966).
78. A. W. Hanson, *Acta crystallogr.* **16**, 1147 (1963).
79. B. Kamenar and C. K. Prout, *J. chem. Soc.* 4838 (1965).
80. H. Kuroda, I. Ikemoto and H. Akamatu, *Bull. chem. Soc. Japan* **39**, 547 (1966).
81. I. Ikemoto and H. Kuroda, *Bull. chem. Soc. Japan* **40**, 2009 (1967).
82. J. C. A. Boeyens and F. H. Herbstein, *J. phys. Chem., Ithaca* **69**, 2160 (1965).
83. D. S. Brown and S. C. Wallwork, *Acta crystallogr.* **19**, 149 (1965).
84. Y. Ohashi, H. Iwasaki and Y. Saito, *Bull. chem. Soc. Japan* **40**, 1789 (1967).
85. B. Kamenar, C. K. Prout and J. D. Wright, *J. chem. Soc.* 4851 (1965).
86. C. K. Prout and A. G. Wheeler, *J. chem. Soc.* (A), 469 (1967).
87. A. S. Bailey and C. K. Prout, *J. chem. Soc.* 4867 (1965).
88. C. K. Prout and H. M. Powell, *J. chem. Soc.* 4882 (1965).
89. B. Kamenar, C. K. Prout and J. D. Wright, *J. chem. Soc.* (A), 661 (1966).
90. C. K. Prout and J. D. Wright, *Angew. Chem.* **80**, 688 (1968).
91. H. A. Bent, *Chem. Rev.* **68**, 587 (1968).

Chapter 9

Electrical and Magnetic Properties

9.A. Dipole Moments

The dipole moment of a charge-transfer complex is of particular interest in that it should be directly related to the dative contribution to the ground state of the complex. It has been observed that non-polar acceptor molecules have apparently non-zero dipole moments when dissolved in symmetrical electron-donor solvents: for example, iodine in benzene (0·6D), in p-xylene (0·9D), and in dioxane (1·3D).[1, 2] This seems to provide convincing evidence for a stable complex with sufficient polarity to orientate itself in an applied electric field. Further support has come from similar systems in which the non-polar donor and acceptor species are dissolved in a diluting "inert" solvent,[3] and from systems in which, although at least one component has a dipole moment, the moment of the complex is not equal to the sum of the moments of the component molecules.[4-6] Obviously, in this last case, there may be less certainty since the relative orientation of donor and acceptor must be determined, or assumed, if both components have dipoles.

For a complex (AD) formed between a non-polar donor and a non-polar acceptor, the ground state of the complex (ψ_N) may be represented as:

$$\psi_N = a\psi_0(A, D) + b\psi_1(A^- - D^+) \tag{9.1}$$

where for weak interactions $a \gg b$ [see equation (2.1), Chapter 2]. If the "no bond" structure (ψ_0) has a dipole moment $\mu_0 = 0$, then the ground-state dipole moment μ_N will be related to the dipole moment (μ_1) of the dative structure ψ_1 by the relationship:

$$\mu_N = \mu_1(b^2 + abS) \tag{9.2}$$

where a and b are the coefficients in equation (9.1) ($a \gg b$), and S is the overlap integral of the "no bond" and dative structures ($S = \int \psi_1 \psi_0 \, d\tau$) (see Chapter 2).

Normalizing conditions (Chapter 2) require that:

$$a^2 + 2abS + b^2 = 1 \tag{9.3}$$

Correspondingly, for the coefficients of the wave equation for the excited state ψ_E of the complex AD:

$$\psi_E = a^{\mp} \psi_1(A - D^+) - b^{\mp} \psi_0(A, D) \tag{9.4}$$

and the coefficients are related by:

$$a^{\mp 2} - 2a^{\mp} b^{\mp} S + b^{\mp 2} = 1 \tag{9.5}$$

For this simple case of a complex between non-polar components, the value of μ_1 has been approximated by the product of the intermolecular separation of the two components in the ground state (r_{AD}), multiplied by unit electronic charge. Weiss[7] has suggested that μ_1 should be equated to $(e.r_{AD})/D$, where D, the effective dielectric constant, could have a value considerably greater than unity. In general, workers do not appear to have made any allowance for such a possible dielectric screening. If a knowledge of the degree of association of the components under the conditions of the measurement is available, μ_N may be calculated from suitable experimental polarization data. From this quantity and estimates of μ_1 and S, the coefficients a, b, $a^{\mp}$ and $b^{\mp}$ in equations (9.1) and (9.4) may be evaluated by using equations (9.2), (9.3) and (9.5) (Table 9.1). By and large, workers have ignored the possible polarization in the "no bond" structure of the complex. Such polarizations could in fact make significant contributions to the observed dipole.[8, 9] Hence, if no allowance is made, the percentage dative structure will be overestimated. Weiss[7] has pointed out that in solvents of low dielectric constant, dipolar systems may dimerize to form more complex aggregates with a very much smaller dipole moment. The values of the coefficients quoted in Table 9.1 do not take into account such effects.

It is clear from the foregoing remarks that, quite apart from experimental difficulties, these estimates can, at best, only be approximate. In the case of some systems such as iodine with aliphatic amines, the presence of traces of water makes the occurrence of ionic reactions more likely,[10] so that an ion pair, rather than a charge-transfer complex, may

9*

TABLE 9.1. Dipole moments (μ_N) of some charge-transfer complexes together with estimated values of the coefficients a and b for equation (9.1) and the percentage dative structure in ground state.

Donor	Acceptor	Solvent	μ_N (Debye)	a	b	% Dative structure	Ref.*
Ammonia	Iodine	Dioxane†	6.1	0.74	0.49	35	13
Ethylamine	Iodine	Benzene†	6.2	0.74	0.49	35	13
Isopropylamine	Iodine	Benzene†	6.2	0.74	0.49	36	13
Diethylamine	Iodine	Benzene†	6.2	0.74	0.49	34	13
Pyridine	Iodine	Heptane	4.5	0.865	0.50	25	a
Benzene	Iodine	Cyclohexane	1.8	0.93	0.286	8.2	3
Durene	1,3,5-Trinitrobenzene	Carbon tetrachloride	0.55	0.975	0.145	2.1	5
Hexamethylbenzene	1,3,5-Trinitrobenzene	Carbon tetrachloride	0.87	0.962	0.193	3.8	5
Naphthalene	1,3,5-Trinitrobenzene	Carbon tetrachloride	0.69	0.969	0.168	2.8	5
trans-Stilbene	1,3,5-Trinitrobenzene	Carbon tetrachloride	0.82	0.964	0.186	3.5	5
Durene	Chloranil	Carbon tetrachloride	0.90	0.960	0.199	4.1	b
Naphthalene	Chloranil	Carbon tetrachloride	0.90	0.960	0.199	4.1	b
Durene	Tetracyanoethylene	Carbon tetrachloride	1.26	0.946	0.243	6.2	b
Hexamethylbenzene	Tetracyanoethylene	Carbon tetrachloride	1.35	0.943	0.253	6.7	b
Naphthalene	Tetracyanoethylene	Carbon tetrachloride	1.28	0.945	0.245	6.3	b

* Refs.: [a] C. Reid and R. S. Mulliken, *J. Am. chem. Soc.* **76**, 3869 (1954). [b] G. Briegleb, J. Czekalla and G. Reuss, *Z. phys. Chem. Frankf. Ausg.* **30**, 333 (1961).

† μ_N shown to be concentration dependent, increasing as the concentration of the amine is increased, so that a decreases whilst b and the percentage dative structure increases.

be observed. For tertiary aliphatic amines with iodine in solvents such as dioxane, this type of reaction may occur even when extremely anhydrous conditions are applied.[11] This probably accounts for the very large dipole moments which are measured (10–12D). However, when the components are mixed at a low temperature (−40 to −50°C) in toluene solution, values of 6·5D for the trimethylamine–iodine complex and 6·9D for the triethylamine–iodine complex are obtained.[12] Absence of ionization was confirmed by the inability to detect I_3^- spectroscopically. The dipole moments of several other aliphatic amine–iodine complexes have been determined recently.[13] The dipole moments of these systems increase with increasing amine concentration. This is accounted for by the increased contribution of the dative function [an increase in b in equation (9.1)] as the dielectric constant of the solution increases[13] (see Table 9.1).

The possibility of estimating the degree of association of electron donors with electron acceptors by measuring the concentration-dependence of polarization has been discussed in Chapter 6.

9.B. Electrical Conductivity and Paramagnetism of Solid Complexes*

9.B.I. General

A large number of solid charge-transfer complexes, formed from non-radical electron-donor and electron-acceptor species, are dia-magnetic[15–18] and are very poor electrical conductors as would be expected on the basis of Mulliken's[19] valence-bond description (Chapter 2). Such complexes are usually composed of components, one at least of which is a relatively weak interactant. By contrast, there are solid complexes, involving strong donor and strong acceptor species, which are paramagnetic, or have diamagnetic susceptibilities which are significantly less than the sum of the diamagnetic susceptibilities of their components.[20] In the latter type of complex the diminution has been accounted for in terms of a paramagnetic contribution to the total susceptibility of the complex.

For many of the solid complexes, formed between strong electron donors and strong electron acceptors, there are some indications of ionic character in the ground state. Probably the most direct evidence for this is obtained from infrared spectra. For weak interactions, the infrared spectrum of the complex appears to be effectively the superposition of the

* For a recent discussion of general aspects of organic semiconductors, see Ref. 14, also J. Kommandeur, "Physics and Chemistry of the Organic Solid State," Vol. 2, eds. D. Fox, M. M. Labes and A. Weissberger, Wiley, New York, London and Sydney (1965), Ch. 1; O. H. LeBlanc, Jr., op. cit., Vol. 3, Chapter 3.

spectra of the neutral donor (D) and acceptor (A) components. With strongly interacting species there is no such resemblance, but rather a close correspondence to the sum of the spectra of the negative ion (A$^-$) and the positive ion (D$^+$). Similar, though not such distinctive correspondence has been observed in the near-ultraviolet–visible absorption spectra[21–24] and in specular reflection spectra[25] (Chapter 3). In general, the weakly interacting complexes have very much lower melting points than those of the strongly interacting complexes.[16] Notwithstanding the ultraviolet–visible and infrared spectral evidence, although the "strong" complexes generally show some paramagnetic behaviour and have high electrical conductivities, the detailed behaviour is not consistent with a simple ionic lattice of A$^-$ and D$^+$ paramagnetic ions (see below). Indeed, the electrical and magnetic properties of these "strong" charge-transfer complexes may be compared with those of the chemically and structurally distinct organic salts formed from the anion radical of 7,7,8,8-tetracyanoquinodimethane (TCNQ) and various cations,[27] having the general formula M$^+$ (TCNQ)$_n^-$, where n can have the values 1, 1·5 or 2. The cations of these salts are diamagnetic whilst (TCNQ)$_n^-$ is paramagnetic.[28–34] Of particular interest is the quinolinium salt of (TCNQ)$_2^-$. This shows a paramagnetism and an electrical conductivity which are temperature-independent.[33, 34] Such behaviour has been discussed in terms of Pauli spin paramagnetism as applied to semi-metals.

The division of solid complexes into "strong" and "weak" types gives rise to some anomalies. Thus, the crystal structure of the supposed "strong" complex N,N,N',N'-tetramethyl-p-phenylenediamine–chloranil is similar to that observed for many "weak" complexes.[35, 36] The basic structure consists of stacks of alternate donor and acceptor moieties with their planes parallel, arranged with a zero stacking angle. Although this structure is perhaps less usual than one in which there is a finite stacking angle, there are examples of weak complexes where this angle is zero. The interplanar distance for this complex is not remarkably different from that for other, weaker complexes;[32] also the bond lengths and angles of the acceptor and donor moieties in this complex correspond closely to those in the pure neutral components.[36] Furthermore, the spin concentration is surprisingly low. Pott and Kommandeur[37, 38] suggest that the pure complex has a crystal lattice built up of neutral donor and neutral acceptor molecules together with ca. 20% bivalent positive tetramethyl-p-phenylenediamine ions and bivalent negative chloranil ions (a so-called "molionic" lattice). However, there is evidence for the presence of the monovalent ions A$^-$ and D$^+$ in crystals grown from solutions containing an excess of N,N,N',N'-tetramethyl-p-phenylene-diamine[37, 38] (see below).

There are considerable experimental problems involved in electrical conductivity and electron spin resonance (e.s.r.) measurements. The effect of trace impurities can be serious. In cases where non-stoichiometric solids are obtained, the properties of the product can be very dependent on the constitution of the solid. [39-46]

Demonstrations [36, 37, 47, 48] that the electrical and magnetic properties of a complex are, or are not, independent of the method of preparation are therefore of consequence. In some cases the product has been shown to be polymorphic. [37, 38, 49, 50] There is good evidence that, on occasion, solid-state reactions proceed after isolation of the initial product. For example, Inokuchi and his co-workers [51] have observed both increases in spin concentration and changes in hyperfine structure in samples of N,N,N',N'-tetramethyl-p-phenylenediamine–chloranil adduct over a period of many days. Changes in the electrical and magnetic properties of this complex have also been noted by Pott and Kommandeur. [37, 38] Chemical changes of this system in the solid phase are readily demonstrated by X-ray diffraction, [36] and by the fact that the freshly prepared complex dissolves in water to form the appropriate free D^+ and A^- ions, whereas week-old specimens are virtually insoluble in water.

The importance of local charge-transfer centres, formed in the growth of molecular crystals, has been emphasized by Blyumenfeld and Benderskii. [52]

It is unfortunate that relatively little work has been carried out on single crystals [37, 38, 53] (see Appendix Table 1), and remarkably few studies have been made on complexes where the detailed crystal structure is known.

9.B.2. Electrical Conductivity of Solid Molecular Complexes [14]

The majority of charge-transfer complexes have resistivities (ρ) which show an exponential temperature-dependence:

$$\rho = \rho_0 \exp(E_a/2kT) \qquad (9.6)$$

where ρ_0 and E_a are constants, the latter being the experimentally determined activation energy as defined by equation (9.6). Such complexes have been described as semiconductors. [39, 54] On the band model for conduction the energy gap ϵ, which is the difference between the highest point in the valence band and the lowest point in the conduction band, has been related to the activation energy (E_a) by $\epsilon = E_a$. Some workers quote the activation energy for conduction as $E_a'(=E_a/2)$, so that care must be taken to determine whether quoted energies refer to

E_a or E_a'.* The application of the relationship $\epsilon = E_a$ to the organic molecular complexes under consideration, though widely made, is questionable. The band model, upon which intrinsic semiconduction is based, assumes a perfect periodic lattice. The fact that the resistivity is an exponential function of temperature [equation (9.6)] is no evidence that the mechanism of carrier generation is necessarily intrinsic. Such behaviour may arise in various other ways, for example: (a) through surface generation of charge carriers and injection; (b) by small polaron formation with hopping transport (see below), thus yielding a ρ_0 which itself has an activation energy; (c) there may be extrinsic semiconduction. The various quinone–electron donor charge-transfer complexes provide examples of the effect of the structure of the complex on the conductive properties of the complex [47, 55, 56] (Appendix Table 1). Virtually all the complexes which have resistivities of less than 10^6 ohm cm have infrared spectra which indicate "strong" interactions. A parallel correlation has been observed between the visible spectra of the solid complexes of 1,6-diaminopyrene with chloranil, bromanil and 2,3-dichloro-5,6-dicyano-p-benzoquinone, all of which indicate a significant ionic contribution and all of which have resistivities $\ll 10^4$ ohm cm, compared with the iodanil complex, which shows very little ionic character in its visible spectrum and has a resistivity of 10^6 ohm cm.[55]

The effectiveness of the molecular halogens as electron acceptors, which has been noted in the thermodynamic properties of charge-transfer complexes in solution (Chapter 7), and manifests itself in the crystal structures of their solid charge-transfer complexes (Chapter 8), is also reflected in the semiconductivities of the halogen complexes (Appendix Table 1). Reference has already been made to non-stoichiometric complexes.[39–46] These include the iodine complexes of perylene[40, 44] (I), of violanthrene[41, 44, 46] (II), and of aromatic diamines.[42] The variation in resistivity as a function of iodine content is shown in Fig. 9.1.

The complexes of iodine with pyrene (III) and with perylene (I) show remarkably high conductivities (Appendix Table 1). The possibility that this conductivity is the result of ionic conduction, by products of chemical reaction, is countered by the observation that there is no significant change in conductivity even after the total charge passed has exceeded 10,000-fold the amount allowed by Faraday's laws.[57, 58] These particular measurements are of importance since they yield thermal

---

* The factor two arises from the mechanism of intrinsic semiconduction, which is due to electrons in the valence band acquiring sufficient energy to reach the conduction band. For every electron so excited, a positive hole is left behind in the valence band. Both the electron and the hole are charge carriers.

I   II   III

activation energies which are identical with the corresponding activation energies for spin concentration [58, 59] (see Section 9.B.5). The generally high conductivities of the molecular halogen complexes are a reflection of the strong overlap in these complexes.

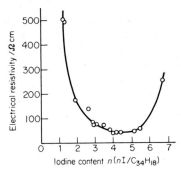

[H. Akamatu, H. Inokuchi and Y. Matsunaga, Ref. 44]

FIG. 9.1. Electrical resistivity of violanthrene–iodine complex as a function of the composition.

Complexes between 1,3,5-trinitrobenzene and electron donors, whether they be aromatic hydrocarbons of intermediate or low ionization potential or aromatic amines, all behave as "weak" complexes with high resistivities, usually $>10^{16}$ ohm cm at room temperature, and high activation energies [60] (Appendix Table 1). Relatively high resistivities are also observed for most complexes of tetracyanoethylene (Appendix Table 1).[61-63] The powerful electron-donating ability of violanthrene (II) is reflected in the considerably lower resistivity of its complexes with both these acceptors.[60, 61] The conductivity of the complex of tetracyanoethylene with tetrathiotetracene (IV) is exceptionally low.[62]

IV

Attempts to measure Hall voltages in several different charge-transfer complexes have in general not been successful.[53, 57, 64] This may well be the result of low mobilities of the carriers in the complexes.

In some complexes, for example those between p-phenylenediamine and chloranil[64, 65] or 2,3-dichloro-5,6-dicyano-p-benzoquinone,[26] positive Seebeck coefficients were obtained; this has been taken as an indication of the predominance of hole carriers, although it is not necessarily the case if the mobilities are low. In other cases, for example in the 2,3-dichloro-5,6-dicyano-p-benzoquinone complexes of pyrene[26] and perylene,[26] negative Seebeck coefficients have been observed. The major contribution to conduction has been assumed to be by electrons, although here again this cannot be simply argued.

The basic arrangement of the donor and acceptor moieties in most crystal lattices, namely stacks of alternate donor and acceptor, suggests that the semiconductivity in solid charge-transfer complexes might be anisotropic. Some evidence of this has been obtained in simple crystal determinations,[53, 66] although in most cases, unfortunately, the crystal structure has not been determined so that the various conductivities with respect to the molecular axes remains undetermined. However, for the perylene–fluoranil complex, sufficient is known concerning the structure of the complex to demonstrate that the conductivity along a stack (i.e. perpendicular to the molecular planes) is about three times the conductivity in an orthogonal direction.[66]

Amongst the "weak" aromatic hydrocarbon–tetracyanoethylene complexes there is a good linear correlation between $h\nu_{max}^{AD}$ and the energy gap ϵ ($=E_a$) for the naphthalene, acenaphthene, pyrene, perylene, anthanthrene and phenanthrene complexes.[61] For each of these the difference in energy ($\epsilon - h\nu_{max}^{AD}$) is only about 0·1 eV. For the tetracyanoethylene complexes of hexamethylbenzene, pentamethylbenzene and azulene, $h\nu_{max}^{AD}$ is significantly greater than ϵ[61] (Fig. 9.2). The energy difference ($\epsilon - h\nu_{max}^{AD}$) is considerable ($\sim0·5$ eV) for a series of 1,3,5-trinitrobenzene–polycyclic aromatic hydrocarbon complexes[60] (Fig. 9.3). Since ϵ can be less than $h\nu_{max}^{AD}$ and even less than the high-energy edge of the charge-transfer band, it would appear that the first excited "charge-transfer" state of these complexes is not the conducting state (see below).

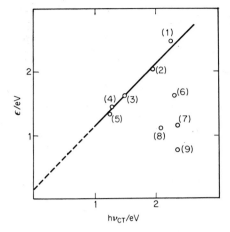

[H. Kuroda, M. Kobayashi, M. Kinoshita and S. Takemoto, Ref. 61]

FIG. 9.2. Plot of the energy gap (ϵ) against the energy of the lowest charge-transfer transition ($h\nu_{CT}$) for tetracyanoethylene complexes of naphthalene (1), acenaphthene (2), pyrene (3), perylene (4), anthanthrene (5), phenanthrene (6), hexamethylbenzene (7), pentamethylbenzene (8) and azulene (9).

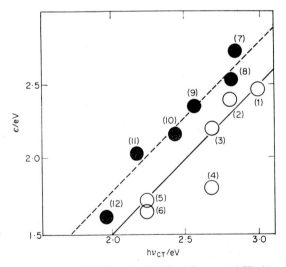

[H. Kuroda, K. Yoshihara and H. Akamatu, Ref. 60]

FIG. 9.3. Plot of the energy gap (ϵ) against the energy of the lowest charge-transfer transition ($h\nu_{CT}$) for 1,3,5-trinitrobenzene complexes of phenanthrene (1), chrysene (2), pyrene (3), anthracene (4), anthanthrene (5), perylene (6), p-chloro-aniline (7), aniline (8), 1-naphthylamine (9), N,N-dimethylaniline (10), p-phenyl-enediamine (11), N,N,N′,N′-tetramethyl-p-phenylenediamine (12), —— $\epsilon = h\nu_{CT} - 0.5$ [eV], - - - - $\epsilon = h\nu_{CT} - 0.2$ [eV].

As opposed to studies on homogeneous samples, studies have also been made on the change in conductivity of a pure donor or acceptor when a thin layer of a second component is deposited on the surface. The conductivity of film of phthalocyanine is potentiated $\times 10^7$ when treated in this way with o-chloranil.[67] Similar large increases in conductivity have been observed when a matrix of violanthrene is coated with one of a variety of acceptors including o-chloranil.[68] Other donor–acceptor systems have also been used[68] including anthracene–iodine.[69, 70] The

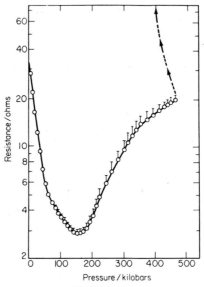

[W. H. Bentley and H. G. Drickamer, Ref. 74]

FIG. 9.4. Resistance vs. pressure for the complex perylene–$3I_2$, at 298°K.

converse experiment has also been demonstrated, for example o-chloranil treated with phthalocyanine,[68] and solid chloranil treated with various alkylamines and ammonia in the gaseous phase.[71]

The effects of high pressure on the electrical conductivity of various charge-transfer complexes have been investigated.[72–77] Initial increases in applied pressure are associated with increases in conductivity in all the cases reported. However, for some systems, when the pressure is sufficiently high, often, though not always, in excess of 100 kbar, the conductivity may decrease again[73, 74] (Fig. 9.4). This latter effect is irreversible and is associated with changes in the infrared and visible absorption spectra: this suggests that chemical reaction has occurred between donor and acceptor molecules.[73, 74] The effect of pressure will shorten somewhat the intermolecular distance and so increase the inter-molecular overlap of the appropriate donor and acceptor orbitals. Consistent with this view is the observation that the effect of pressure on the conductivity of the highly conducting complexes is not so great as it is on the more poorly conducting complexes for which the low initial orbital-overlap provides the possibility of a greater increase in overlap.[75] However, the compression of the complex will be relatively small. The main effect of the pressure may be on the thermal activation energy for conduction [E_a in equation (9.6)]. This has been measured for some complexes.[75, 76] The activation energies decrease with increasing pressure and for some tetrahalo-o-quinone–tetrathiotetracene (IV) complexes become almost zero at 35 kbar.[75] By comparison, the factor ρ_0 in equation (9.6) shows little change as the pressure increases.[75, 76]

It has been suggested[41] that a "hopping" model of electrical trans-port[78] might be applicable in some of these systems. If, initially, an electron is transferred from a donor molecule to a neighbouring acceptor molecule to yield a dative structure D^+A^-, the charges may separate subsequently through an orbital overlap of, say, D^+ with another neutral D molecule. By several such successive interactions the positive hole will become free of the negative charge and will give rise to hole conduction. The more effective the $D - D^+$ interaction the lower will be the activation energy for charge separation. The low mobility which is usually observed is in harmony with such a model.

9.B.3. Photo-Conductivity

The photo-conductivity of "weak" charge-transfer complexes is particularly amenable to study since the dark conductivity of these complexes is low and therefore does not swamp the photo-conductivity.

The spectral dependence of photo-conduction has been determined for the complexes of pyrene with tetracyanoethylene, with bromanil and with 1,3,5-trinitrobenzene[79] (Figs. 9.5–9.7). In each of these systems two photo-conductivity bands are observed, the one of lower energy is almost certainly due to intrinsic conduction. The fact that there is neither coincidence of the high-energy conduction band with the charge-transfer absorption band, nor even consistency as to whether or not the conductivity band is at higher energies than the charge-transfer band (Table 9.2), confirms the conclusion, based on dark current conductivity (p. 260), that the conduction state is not identical with the charge-transfer excited state. The difference between these states may be one of separation of the ions A^- and D^+. In the charge-transfer excited state, the complex will be present as a neighbouring ion-pair, whereas, in the conduction state, the ions will be separated. The relative magnitudes of

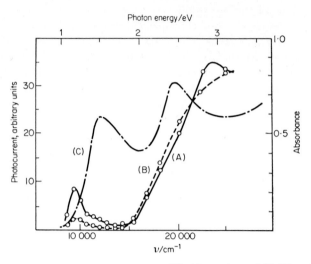

[H. Akamatu and H. Kuroda, Ref. 79]

FIG. 9.5. Spectral dependence of photo-conduction in pyrene–tetracyano-ethylene complex: (A) sandwich cell, (B) surface cell, (C) absorption spectrum.

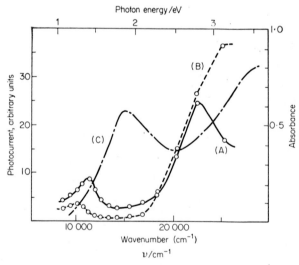

[H. Akamatu and H. Kuroda, Ref. 79]

FIG. 9.6. Spectral dependence of photo-conduction in pyrene–bromanil complex: (A) sandwich cell, (B) surface cell, (C) absorption spectrum.

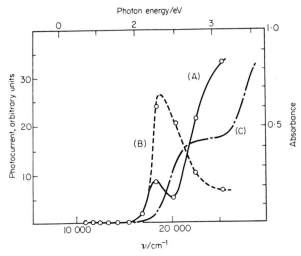

[H. Akamatu and H. Kuroda, Ref. 79]

FIG. 9.7. Spectral dependence of photo-conduction in pyrene–1,3,5-trinitro-benzene complex: (A) sandwich cell, (B) surface cell, (C) absorption spectrum.

the polarization energies of the two ions, the resonance energy, and Coulomb attraction of the ion pair, will determine which state is of higher energy. Apparently similar photo-conductive behaviour is shown by the complexes between chloranil and palladium(II)-8-hydroxyquinolinate and copper(II)-8-hydroxyquinolinate.[80]

A study has been made of the photo-conductivity of a donor molecule (N-isopropylcarbazole) in the absence, and in the presence, of an electron acceptor (picryl chloride), both as a 1:1 complex and as a 1% mixture of the acceptor in the donor.[81] The complex is of the "weak" rather than of the ionic type. In the photo-conductivity of the donor, the picryl chloride

TABLE 9.2. Threshold of photo-conduction.*

Complex	Threshold of photo-conduction (eV)	Energy gap (ϵ) (eV)	$h\nu_{CT}$ (eV)	Absorption edge (eV)
Pyrene–tetracyanoethylene	1·85	1·65	1·50	1·0
Pyrene–bromanil	2·08	1·88	1·86	1·3
Pyrene–1,3,5-trinitrobenzene	2·44	2·20	2·66	2·0

* Data from Ref. 79.

acts as a weak electron-trap, the hole mobility being little affected by the presence of the acceptor.

Although single crystals of perylene–fluoranil show photo-conduction, the mechanism is probably extrinsic.[66] The photo-current rise and decay times are relatively long (Fig. 9.8), which suggests that there are strong trapping processes.

Photo-conduction has been detected in "strong" charge-transfer complexes formed by depositing a layer of acceptor on a film of donor,

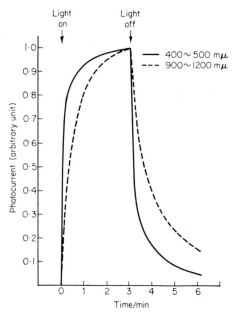

[H. Kokado, K. Hasegawa and W. G. Schneider, Ref. 66]

Fig. 9.8. Rise and decay of photo-current in perylene–fluoranil single crystal.

for example o-chloranil on metal-free phthalocyanine,[67] and o-chloranil on violanthrene (II).[68] In these structures, it is probable that the effect of light is to increase reversibly the total positive charge on the donor layer and the negative charge on the acceptor layer (see p. 263). The fact that this sometimes increases, and sometimes decreases, the unpaired electron spin concentration (Section 9.B.4) can be accounted for in terms of the proportions of singly charged paramagnetic ions A^- and doubly charged diamagnetic ions A^{--} which are formed in the process. The small increases in the photo-conductivity of chloranil crystals, caused by various aliphatic amine vapours of various wavelengths,[82] may possibly be accounted for in a similar fashion. Transient and steady-state photo-currents have been measured in appropriately illuminated single crystals of pyrene–tetracyanoethylene.[83] Estimates of hole mobilities as high as 30 cm^2/V sec at room temperature have been given. Such values appear to conflict with those predicted by the hopping model[78] for conduction.

9.B.4. Paramagnetism

In 1954 Kainer et al.[84] reported that the solid complexes of the Lewis bases p-phenylenediamine and of N,N,N',N'-tetramethyl-p-phenylene-diamine with certain quinone acceptors show a measurable e.s.r. absorption. They accounted for these observations[84, 85] by proposing that the ionic (dative) state of the complex is of lower energy than the "no bond" structure in such cases. The hypothetically separated components will be in doublet spin states, each component having one unpaired spin. At the normal distance of separation in the crystal lattice, these spins may interact to provide a singlet state and a triplet state of only slightly higher energy. It was suggested that the difference of energy is in fact sufficiently small that the triplet level has a significant thermal population. The spin populations in this type of complex appear to vary widely: estimates of 0·4% for the chloranil-p-phenylenediamine complex and 80% for the o-bromanil–N,N,N',N'-tetramethyl-p-phenylenedi-amine complex have been made.[84, 85] These complexes appear to obey Curie's law over the range 6–260°K.[84, 85] This would imply a constant concentration of radical species in this temperature range. These experimental results are consistent with a singlet–triplet model in which the separation between the two states is less than 0·0005 eV. Apart from the spectrum of p-phenylenediamine–bromanil complex, which appeared to consist of two unresolved lines, Bijl and his co-workers[84, 85] observed only a single line in the e.s.r. absorption of these complexes, which is in harmony with their triplet model. However, Matsunaga and McDowell[86] were soon to report that for the particular complex of chloranil with p-phenylenediamine, two peaks could be resolved in the e.s.r. spectrum. These were correlated to the ions A$^-$ and D$^+$, which, contrary to Bijl's interpretation, now appeared to be only weakly coupled, i.e. were in doublet states. A re-examination[87] of the p-phenylenediamine–chloranil system overturned the assignment of the double band to the two doublet species A$^-$ and D$^+$, in favour of the effect of an anisotropic g-tensor which gives rise to the two bands. The experimental results in this case were consistent with the hypothesis that the observed resonance is the result of a single unpaired electron on either the acceptor or the donor moiety (Fig.9.9). Benderskii et al.[88] have reported a line splitting in this complex

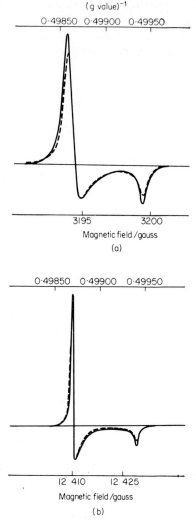

$(g \text{ value})^{-1}$

[M. E. Browne, A. Ottenberg and R. L. Brandon, Ref. 87]

FIG. 9.9. Electron paramagnetic resonance spectrum of powdered complex p-phenylenediamine–chloranil at room temperature at 8·9655 GHz (a) and at 34·820 GHz (b): —— experiment, ―――― theory.

at temperatures below −150°C, which they attribute to an electron spin–spin interaction. Pott and Kommandeur[37, 38] have made a similar observation, although in their case they observed a strong impurity signal as well.

These workers[37, 38] have shown that crystals of the N,N,N',N'-tetra-methyl-p-phenylenediamine–chloranil complex, grown from solutions containing an excess of the donor, yield two e.s.r. signals: one corresponding to a g-value for the monovalent cation derived from the donor, and the other with a g-value corresponding to the chloranil semiquinone. On the other hand, very pure crystals of the complex show no e.s.r. absorption near $g = 2 \cdot 003$, corresponding to the donor monopositive ion. It is therefore concluded that this ion is absent. In order to maintain electrical neutrality it is presumed that donor dipositive ions are present. From the spin concentration ($\sim 0 \cdot 8\%$ per DA pairs at 25°C) and crystallographic data[36] (see also Chapter 10) it is supposed that the chloranil ion concentration is small in what is a predominantly molecular lattice. The components of the g-tensor obtained are:

$$g_{xx} = 2 \cdot 0062 \pm 0 \cdot 0002$$
$$g_{yy} = 2 \cdot 0021 \pm 0 \cdot 0002 \qquad g \text{ average} = 2 \cdot 0052 \pm 0 \cdot 0002$$
$$g_{zz} = 2 \cdot 0072 \pm 0 \cdot 0002$$

Apart from this g-tensor anisotropy, a line-width anisotropy was also observed, which was different for different specimens (Fig. 9.10). It is thought that this difference might be the result of slightly differing orientations of the dimethylamino groups in the neighbourhood of a chloranil ion. When some specimens were cooled below 250°K, an anisotropic splitting of the e.s.r. line occurred (Fig. 9.11). This may be the result of chloranil ions not all having the same orientation in the crystal lattice.

The work by Pott and Kommandeur emphasizes the fact that the electrical and magnetic properties of complexes, formed between donors of low ionization potential and acceptors of high electron affinity, can be very dependent on the mode of preparation and purity of the product.

g-Tensor anisotropy has been given as the reason for the three-line e.s.r. spectrum of the N,N,N',N'-tetramethyl-p-phenylenediamine complex of 7,7,8,8-tetracyanoquinodimethane, which behaves as an ion-radical salt.[89] The relatively low spin-concentration in this case has led to the suggestion[89] that a triplet state is involved with an activation energy of 0·150 eV.

The assignment[90] of the five-line e.s.r. spectrum shown by systems containing triphenylamine and iodine to the triphenylamine cation radical has proved to be incorrect.[91] The absorption is, in fact, due to N,N,N',N'-tetraphenylbenzidine, which is probably formed from the triphenylamine radical.

A degree of ionic character, as indicated by infrared and other spectral

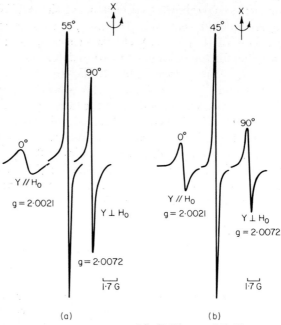

[G. T. Pott and J. Kommandeur, Ref. 38]

FIG. 9.10. Electron spin resonance signals of (a): N,N,N′,N′-tetramethyl-*p*-phenylenediamine–chloranil "type-B" crystal at three orientations of the *y*-axis with respect to the external magnetic field, (b) same for "type-A" crystals.

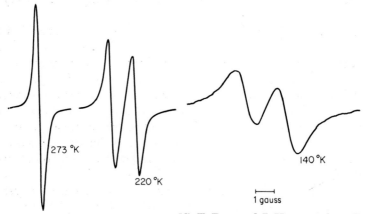

[G. T. Pott and J. Kommandeur, Ref. 38]

FIG. 9.11. Temperature dependence of the electron spin resonance absorption of a crystal of "type-B" N,N,N′,N′-tetramethyl-*p*-phenylenediamine–chloranil in the orientation of maximum splitting. The spectra were obtained with different spectrometer gain settings at the different temperatures.

evidence, does not appear to be a necessary prerequisite for para-magnetic behaviour. Thus, hydrocarbon–quinone complexes, such as perylene–o-chloranil, have infrared and electronic spectra which indicate that the interactions between donor and acceptor are weak; neverthe-less they show e.s.r. absorption.[68, 92] However, in these cases the un-paired electrons are located at imperfections in the solid. A similar cause has been suggested for the paramagnetism in complexes of 7,7,8,8-tetracyanoquinodimethane with hexamethylbenzene and with benzo[e]-pyrene,[93] and of complexes of tetracyanoethylene with various aromatic hydrocarbons.[94] The spin concentration in most cases is relatively low, and, by contrast with some of the ionic ground-state complexes, very dependent on the particular mode of preparation. The imperfections may be introduced irreversibly by heating otherwise diamagnetic complexes. On occasion, they can also be produced by crystallization. The imperfections appear to be caused by some reaction between the donor and acceptor molecules rather than by extraneous impurities.

There are various reports of the effect of illumination on the e.s.r. absorption of charge-transfer complexes.[67, 68, 95–101] Photo-induced transfer of an electron from donor to acceptor has been demonstrated in several instances by Lagercrantz and Yhland.[95–97] In the case of the tetracyanoethylene–tetrahydrofuran complex, Ilten and Calvin[98] have observed photo-induced e.s.r. absorption when the complex in an excess of tetrahydrofuran was irradiated in the charge-transfer band of the complex. The e.s.r. spectrum was characteristic of the tetracyano-ethylene mono-negative ion. The quantitative behaviour is consistent with the initial formation of an excited singlet state of the complex. A fraction of molecules in this state go to an excited triplet state. From both levels, relaxation to the ground state occurs. The primary source of the ions D^+ and A^- is from the excited triplet state. Lack of observation of an e.s.r. absorption corresponding to D^+ may be the result of signal broadening through chemical exchange. The complex between tetra-cyanoethylene and dimethylsulphoxide and other related systems have also been studied.[99, 100]

The e.s.r. spectrum of the complex between durene and 1,2,4,5-tetra-cyanobenzene in an ethanolic glass at 77°K, when irradiated in the region of the charge-transfer band, includes a signal which has been assigned to the charge-transfer triplet state of the complex.[101]

In the case of some lamellar donor–acceptor complexes, irradiation decreases, and in other cases increases, the spin concentration. A possible explanation,[67, 68] in terms of the formation of paramagnetic singly charged and diamagnetic doubly charged ions, has already been referred to in Section 9.B.3.

9.B.5. Relationship between Paramagnetism and Electrical Conductivity

From the foregoing sections, it is clear that a conducting solid charge-transfer complex may have a structure that is basically ionic (strong complex), or a structure in which the component molecules are only slightly perturbed (weak complex). Evidence has also been given for the dominance of hole conduction in some complexes and of electron conduction in others. It is therefore, perhaps, not surprising that the degree of correlation of the energy of activation for conduction, and for free-spin concentration, is very dependent on the particular complex.

TABLE 9.3. Activation energies for unpaired spin and for conduction for some charge-transfer complexes*†

Donor	Acceptor	$E_{a}^{\ddagger}$ (spin) (eV)	$E_{a}^{\ddagger}$ (conductivity) (eV)	Ref.
p-Phenylenediamine	Chloranil	0·26	0·86	48
N,N,N′,N′-Tetramethyl-p-phenylenediamine	Chloranil	0	0·54	85, 102
Diaminodurene	Chloranil	0·30, 0·32	0·50	28
Pyrene	Iodine§	0·28	0·28	57
Perylene	Iodine‖	0·038	0·038	57
N,N,N′,N′-Tetramethyl-p-phenylenediamine	TCNQ¶	0·150		89

 * After Ref. 48.
 † See also Appendix Table 1.
 ‡ Activation energy expressed in terms of $X = X_0 \exp (E_a/2kT)$.
 § Ratio D:A = 1:2.
 ‖ Ratio D:A = 2:3.
 ¶ TCNQ = 7,7,8,8-tetracyanoquinodimethane.

A selection of data is given in Table 9.3. Kommandeur and Singer[58, 59] have measured the free-spin concentrations of iodine complexes of pyrene and of perylene, the exponential temperature-dependences of which are in good agreement with the activation energies for electrical conduction[57, 58] (Table 9.3). There is little doubt that for these two complexes the unpaired spins are the charge-carriers. At low temperatures there is some evidence for the presence of small concentrations of trapped spins or free radicals.

The proposed zero-activation energy for the spin concentration of the N,N,N′,N′-tetramethyl-p-phenylenediamine–chloranil complex[85] compares with a value of 0·54 eV for the thermal energy of activation of this complex[102] (Table 9.3). A higher activation energy for conduction than for spin-concentration is also claimed for the chloranil–diaminodurene complex,[28] for the chloranil-p-phenylenediamine complex,[48] and others (see Table 9.3). It has been argued that E_a (spin) is the energy required to form adjacent ion radicals in the crystal lattice. The extra energy is needed for the stepwise cooperative transference of electrons, whereby the charges become sufficiently separated, so that the Coulomb attraction is effectively zero.

REFERENCES

1. F. Fairbrother, *Nature, Lond.* **160**, 87 (1947).
2. F. Fairbrother, *J. chem. Soc.* 1051 (1948).
3. G. Kortüm and H. Walz, *Z. Elektrochem.* **57**, 73 (1953).
4. R. C. Sahney, R. M. Beri, H. R. Sarna and M. Singh, *J. Indian chem. Soc.* **26**, 329 (1949).
5. G. Briegleb and J. Czekalla, *Z. Elektrochem.* **59**, 184 (1955).
6. J. Czekalla, *Z. Elektrochem.* **60**, 145 (1956).
7. J. J. Weiss, *Phil. Mag.* [8] **8**, 1169 (1963).
8. J. E. Anderson and C. P. Smyth, *J. Am. chem. Soc.* **85**, 2904 (1963).
9. R. J. W. Le Fèvre, D. V. Radford, G. L. D. Ritchie and P. J. Stiles, *Chem. Commun.* 1221 (1967).
10. H. Tsubomura and S. Nagakura, *J. chem. Phys.* **27**, 819 (1957).
11. K. Toyoda and W. B. Person, *J. Am. chem. Soc.* **88**, 1629 (1966).
12. A. J. Hamilton and L. E. Sutton, *Chem. Commun.* 460 (1968).
13. S. Kobinata and S. Nagakura, *J. Am. chem. Soc.* **88**, 3905 (1966).
14. F. Gutmann and L. E. Lyons, "Organic Semiconductors," Wiley, New York, London and Sydney (1967).
15. R. C. Sahney, S. L. Aggarwal and M. Singh, *J. Indian chem. Soc.* **23**, 335 (1946).
16. B. Puri, R. C. Sahney, M. Singh and S. Singh, *J. Indian chem. Soc.* **24**, 409 (1947).
17. H. Mikhail and F. G. Baddar, *J. chem. Soc.* 590 (1944).
18. M. Kondo, M. Kishita, M. Kimura and M. Kube, *Bull. chem. Soc. Japan* **29**, 305 (1956).
19. R. S. Mulliken, *J. Am. chem. Soc.* **74**, 811 (1952).
20. Y. Matsunaga, *Bull. chem. Soc. Japan* **28**, 475 (1955).
21. H. Kainer and A. Überle, *Chem. Ber.* **88**, 1147 (1955).
22. H. Kainer and W. Otting, *Chem. Ber.* **88**, 1921 (1955).
23. R. Foster and T. J. Thomson, *Trans. Faraday Soc.* **59**, 296 (1963).
24. R. Foster, *Photoelect. Spectrom. Grp Bull.* **15**, 413 (1963).
25. B. G. Anex and E. B. Hill, Jr., *J. Am. chem. Soc.* **88**, 3648 (1966).
26. A. Ottenberg, R. L. Brandon and M. E. Browne, *Nature, Lond.* **201**, 1119 (1964).
27. D. S. Acker, R. J. Harder, W. R. Hertler, W. Mahler, L. R. Melby, R. E. Benson and W. E. Mochel, *J. Am. chem. Soc.* **82**, 6408 (1960).
28. D. B. Chesnut and W. D. Phillips, *J. chem. Phys.* **35**, 1002 (1961).
29. R. G. Kepler, P. E. Bierstedt and R. E. Merrifield, *Phys. Rev. Lett.* **5**, 503 (1960).
30. D. B. Chesnut, H. Foster and W. D. Phillips, *J. chem. Phys.* **34**, 684 (1961).
31. D. B. Chesnut and P. Arthur, Jr., *J. chem. Phys.* **36**, 2969 (1962).

32. W. J. Siemons, P. E. Bierstedt and R. G. Kepler, *J. chem. Phys.* **39**, 3523 (1963).
33. R. G. Kepler, *J. chem. Phys.* **39**, 3528 (1963).
34. R. W. Tsien, C. M. Huggins and O. H. LeBlanc, Jr., *J. chem. Phys.* **45**, 4370 (1966).
35. S. C. Wallwork, *J. chem. Soc.* 494 (1961).
36. J. L. de Boer and A. Vos, *Acta crystallogr.* **B24**, 720 (1968).
37. G. T. Pott "Molionic Lattices," Thesis, University of Groningen (1966).
38. G. T. Pott and J. Kommandeur, *Molec. Phys.* **13**, 373 (1967).
39. H. Inokuchi and H. Akamatu, *in* "Solid State Physics," Vol. 12, eds. F. Seitzand and D. Turnbull, Academic Press, New York and London (1961), p. 140.
40. T. Uchida and H. Akamatu, *Bull. chem. Soc. Japan* **34**, 1015 (1961).
41. T. Uchida and H. Akamatu, *Bull. chem. Soc. Japan* **35**, 981 (1962).
42. S. Nishizaki and H. Kusakawa, *Bull. chem. Soc. Japan* **36**, 1681 (1963).
43. H. Kusakawa and S. Nishizaki, *Bull. chem. Soc. Japan* **38**, 313 (1965).
44. H. Akamatu, H. Inokuchi and Y. Matsunaga, *Bull. chem. Soc. Japan* **29**, 213 (1956).
45. A. Taniguchi, S. Kanda, T. Nogaito, S. Kusabayashi, H. Mikawa and K. Ito, *Bull. chem. Soc. Japan* **37**, 1386 (1964).
46. H. Akamatu, Y. Matsunaga and H. Kuroda, *Bull. chem. Soc. Japan* **30**, 618 (1957).
47. H. Scott, P. L. Kronick, P. Chairge and M. M. Labes, *J. phys. Chem., Ithaca* **69**, 1740 (1965).
48. A. Ottenberg, C. J. Hoffman and J. H. Osiecki, *J. chem. Phys.* **38**, 1898 (1963).
49. Y. Matsunaga, *J. chem. Phys.* **42**, 1982 (1965).
50. Y. Matsunaga, *Nature, Lond.* **211**, 183 (1966).
51. H. Inokuchi, K. Ikeda and H. Akamatu, *Bull. chem. Soc. Japan* **33**, 1622 (1960).
52. L. A. Blyumenfel'd and V. A. Benderskii, *Zh. strukt. Khim.* **4**, 405 (1963); *J. struct. Chem.* **4**, 370 (1963).
53. P. Kronick and M. M. Labes, *J. chem. Phys.* **35**, 2016 (1961).
54. Y. Okamoto and W. Brenner, "Organic Semiconductors," Reinhold, New York (1964), p. 63.
55. P. L. Kronick, H. Scott and M. M. Labes, *J. chem. Phys.* **40**, 890 (1964).
56. Y. Matsunaga, *Nature, Lond.* **205**, 72 (1965).
57. J. Kommandeur and F. R. Hall, *J. chem. Phys.* **34**, 129 (1961).
58. J. Kommandeur and L. S. Singer, *in* "Symposium on Electrical Conductivity in Organic Solids—Duke University, Durham" (1960), eds. H. Kallmann and N. Silver, Interscience, New York (1961), p. 325.
59. L. S. Singer and J. Kommandeur, *J. chem. Phys.* **34**, 133 (1961).
60. H. Kuroda, K. Yoshihara and H. Akamatu, *Bull. chem. Soc. Japan* **35**, 1604 (1962).
61. H. Kuroda, M. Kobayashi, M. Kinoshita and S. Takemoto, *J. chem. Phys.* **36**, 457 (1962).
62. Y. Matsunaga, *J. chem. Phys.* **42**, 2248 (1965).
63. H. Kusakawa and S. Nishizaki, *Bull. chem. Soc. Japan* **38**, 2201 (1965).
64. R. Sehr, M. M. Labes, M. Bose, H. Ur and F. Wilhelm, *in* "Symposium on Electrical Conductivity in Organic Solids—Duke University, Durham" (1960), eds. H. Kallmann and M. Silver, Interscience, New York (1961), p. 309.
65. M. M. Labes, R. Sehr and M. Bose, *J. chem. Phys.* **32**, 1570 (1960).

66. H. Kokado, K. Hasegawa and W. G. Schneider, *Can. J. Chem.* **42**, 1084 (1964).
67. D. R. Kearns, G. Tollin and M. Calvin, *J. chem. Phys.* **32**, 1020 (1960).
68. D. R. Kearns and M. Calvin, *J. Am. chem. Soc.* **83**, 2110 (1961).
69. M. M. Labes, O. N. Rudyj and P. L. Kronick, *J. Am. chem. Soc.* **84**, 499 (1962).
70. M. M. Labes and O. N. Rudyj, *J. Am. chem. Soc.* **85**, 2055 (1963).
71. P. J. Reucroft, O. N. Rudyj and M. M. Labes, *J. Am. chem. Soc.* **85**, 2059 (1963).
72. M. Schwarz, H. W. Davies and B. J. Dobriansky, *J. chem. Phys.* **40**, 3257 (1964).
73. R. B. Aust, G. A. Samara and H. G. Drickamer, *J. chem. Phys.* **41**, 2003 (1964).
74. W. H. Bentley and H. G. Drickamer, *J. chem. Phys.* **42**, 1573 (1965).
75. Y. Okamoto, S. Shah and Y. Matsunaga, *J. chem. Phys.* **43**, 1904 (1965).
76. T. N. Andersen, D. W. Wood, R. C. Livingston and H. Eyring, *J. chem. Phys.* **44**, 1259 (1966).
77. Y. Okamoto, *J. phys. Chem., Ithaca* **70**, 291 (1966).
78. J. Yamashita and T. Kurosawa, *Physics Chem. Solids* **5**, 34 (1958).
79. H. Akamatu and H. Kuroda, *J. chem. Phys.* **39**, 3364 (1963).
80. C. K. Prout, R. J. P. Williams and J. D. Wright, *J. chem. Soc.* (A), 747 (1966).
81. J. H. Sharp, *J. phys. Chem., Ithaca* **71**, 2587 (1967).
82. P. J. Reucroft, O. N. Rudyj, R. E. Salomon and M. M. Labes, *J. phys. Chem., Ithaca* **69**, 779 (1965).
83. M. C. Tobin and D. P. Spitzer, *J. chem. Phys.* **42**, 3652 (1965).
84. H. Kainer, D. Bijl and A. C. Rose-Innes, *Naturwissenschaften* **41**, 303 (1954).
85. D. Bijl, H. Kainer and A. C. Rose-Innes, *J. chem. Phys.* **30**, 765 (1959).
86. Y. Matsunaga and C. A. McDowell, *Nature, Lond.* **185**, 916 (1960).
87. M. E. Browne, A. Ottenberg and R. L. Brandon, *J. chem. Phys.* **41**, 3265 (1964).
88. V. A. Benderskii, I. B. Shevchenko and L. A. Blyumenfeld, *Optika Spektrosk.* **16**, 467 (1964); *Optics Spectrose., Wash.* **16**, 254 (1964).
89. M. Kinoshita and H. Akamatu, *Nature, Lond.* **207**, 291 (1965).
90. D. N. Stamires and J. Turkevich, *J. Am. chem. Soc.* **85**, 2557 (1963).
91. W. H. Bruning, R. F. Nelson, L. S. Marcoux and R. N. Adams, *J. phys. Chem., Ithaca* **71**, 3055 (1967).
92. J. W. Eastman, G. M. Androes and M. Calvin, *J. chem. Phys.* **36**, 1197 (1962).
93. W. Slough, *Trans. Faraday Soc.* **61**, 408 (1965).
94. W. Slough, *Trans. Faraday Soc.* **59**, 2445 (1963).
95. C. Lagercrantz and M. Yhland, *Acta chem. scand.* **16**, 1043 (1962).
96. C. Lagercrantz and M. Yhland, *Acta chem. scand.* **16**, 1799 (1962).
97. C. Lagercrantz and M. Yhland, *Acta chem. scand.* **16**, 1807 (1962).
98. D. F. Ilten and M. Calvin, *J. chem. Phys.* **42**, 3760 (1965).
99. F. E. Stewart, M. Eisner and W. R. Carper, *J. chem. Phys.* **44**, 2866 (1966).
100. F. E. Stewart and M. Eisner, *Molec. Phys.* **12**, 173 (1967).
101. H. Hayashi, S. Nagakura and S. Iwata, *Molec. Phys.* **13**, 489 (1967).
102. D. D. Eley, H. Inokuchi and M. R. Willis, *Discuss. Faraday Soc.* **28**, 54 (1959).

Chapter 10

Selected Charge-Transfer Complexes

In the previous chapters, various aspects of charge-transfer complexes have been discussed in such a way that the description of the properties of a given complex, or type of complex, is spread over several chapters. In this chapter, a few individual complexes, or closely related groups of complexes, have been selected for discussion, either because of their theoretical or experimental interest, or because of their unusual character.

10.A. The Pyridine–Iodine Complex

Many studies have been made of the pyridine–iodine system. The donor is of particular interest in that it is one of the simplest molecules which contain both n- and aromatic π-electrons. There are considerable experimental problems in making measurements on this system. Traces of impurities, particularly water, cause significant secondary ionic reactions. Although it may be impossible to eliminate these impurities, effectively stable solutions may be obtained if a solvent of low ionizing power such as n-heptane is used and the concentration of pyridine is kept small. Some of the reported values [1-6] for the association constant show reasonable agreement (Table 10.1). In a recent determination,[1] extreme attention was paid to the experimental conditions, and the results probably represent the most reliable value for K_c^{AD} and ϵ_{max}^{AD} at present. The relatively large value of K_c^{AD} is within the range of values for complexes of iodine with aliphatic amines which must act as n-donors, and is very much larger than the association constants for typical π-donors such as hexamethylbenzene. The high intensity of the charge-transfer band is of the same magnitude as those for the aliphatic–amine complexes and considerably larger than for the aromatic π-donor complexes of iodine. Furthermore, the energy of the charge-transfer band relative to the ionization potential of pyridine is appropriate to the correlation observed for n-donors and not for π-donors.[7] (See also Chapter 3.) These observations leave little doubt as to the n-donor nature of the pyridine in this complex.

TABLE 10.1. Experimental values for the association constant K_c^{AD} of the pyridine–iodine complexes in various solvents, obtained by different workers.*

Solvent	Temp./°C	K_c^{AD}/l mol^{-1}	Method	Ref.
n-Heptane	16·7	290†	Ultraviolet, charge-transfer	2
n-Heptane	25	185‡		
n-Heptane	25	132 ± 4·2§	Ultraviolet, charge-transfer	1
n-Heptane	25	138 ± 5·1	Ultraviolet $ca.$ 305 nm	1
n-Heptane	25	140 ± 1·3	Shifted visible I_2 band	1
Cyclohexane	~26	107 ± 25	Infrared band	3
Cyclohexane	18 ± 2	131 ± 30	Far infrared band	4
CCl$_4$	25	101	Ultraviolet, 290–550 nm	5
CCl$_4$	26	107 ± 25	Far infrared band	3
CHCl$_3$	~28	43·7	Shifted visible I_2 band	6

* Taken mainly from the table in Ref. 1.
† $\nu_{CT} = 42{,}500$ cm^{-1}, $\epsilon_{max}^{AD} = 50{,}000$.
‡ Interpolated by Bist and Person[1] from the data in Ref. 2.
§ $\nu_{CT} = 42{,}300$ cm^{-1}, $\epsilon_{max}^{AD} = 51{,}700$.

In terms of Mulliken's valence-bond description, the ionic character of the complex in solvents such as n-heptane appears to be relatively large, ~25% ionic based on dipole moment data,[8, 9] or 20% ionic based on infrared frequency-shift measurements.[10] As the ionizing power of the solvent is increased, the formation of conducting species occurs, possibly arising from an electron transfer in which the ion pair $(pyI)^+$ I^- is formed $(py = pyridine)$.[2] Further interaction of I^- with molecular iodine will yield the triiodide ion I_3^-.[9]

Although attempts to prepare the 1:1 pyridine–iodine charge-transfer complex as a solid crystalline material have so far not been successful, Hassel et $al.$ have determined the crystal structures of many related complexes including pyridine–iodine monochloride[11] and 4-picoline–iodine[12] (see Chapter 8). By analogy with these complexes, it seems reasonable to assume that, in the solid state at least, the geometry of the

10

complex is such that the nitrogen atom is colinear with the iodine molecule and the latter lies in the plane of the pyridine ring. Further support for this structure has been adduced from the dichroic behaviour of polyvinylpyridine–iodine films.[13] Calculations of the transition moment of the iodine stretching mode in the complex, based on such a linear model, give results which are in reasonable agreement with experimental observations,* namely 183, 184 cm^{-1}.[4, 14, 15] The absorption band corresponding to the intermolecular stretching mode has also been observed (94 cm^{-1}) using cyclohexane solutions.[4] A major contribution to the intensities of these two bands is thought to arise from a delocalization moment[4] (see Chapter 4).

10.B. Complexes of Electron Donors with Carbon Tetrachloride

Because of the wide use which is made of carbon tetrachloride as an "inert" solvent, it is obviously of importance to determine whether or not it can act as a component of a charge-transfer complex. Prausnitz and his co-workers[16, 17] have adduced evidence that carbon tetrachloride behaves as an electron acceptor and forms complexes of small association constant with various aromatic hydrocarbon donors in a diluting "inert" solvent (Table 10.2). A 1:1 solid complex of p-xylene and carbon tetrachloride has been obtained at low temperatures, although no evidence of charge-transfer complexing was obtained.[18] Support for complex formation between benzene and carbon tetrachloride also includes the results from measurements of volumes and heats of mixing,[19–23] from freezing-point diagrams,[19, 24–26] and from far infrared spectroscopic measurements.[27] For the hexamethylbenzene–carbon tetrachloride complex, a smaller value for the association constant has been obtained by Dörr and Buttgereit,[28] although Rosseinsky and Kellawi[29] report a value close to Prausnitz's (Table 10.2). Association constants of carbon tetrachloride with N,N-dimethylaniline and with N,N,N',N'-tetramethyl-p-phenylenediamine in n-hexane have also been reported[30] (Table 10.2). These results suggest that Prausnitz's value for the hexamethylbenzene complex may be too large.† The various determinations depend on spectroscopic measurements, made at relatively short wavelengths, where small band shifts, together with significant contributions by the absorptions of the component molecules, make the estimation of the optical absorption by the complex species open to error. Furthermore, the effect of possible specific solvation is very much

---

* See also Table 4.2, p. 100.

† However, the prediction of association constant values from known values for various donor–acceptor pairs can be misleading (see p. 201).

greater for systems in which the association constant is small[31] (see Chapter 6). In some of the systems studied, the quantitative estimates were further complicated by the abnormal concentration[17] and wavelength dependence[28] of the optical absorption, which, it has been suggested, could be the result of the presence of termolecular species.

Although the experimental data may be reasonably interpreted in terms of weak association, the actual quantitative estimates which have been obtained may require some modification. This conclusion also applies to the estimates of association constants for complexes between the donor p-xylene and various polar organic solvents.[32]

TABLE 10.2. Values of association constant (K_c^{AD}) obtained for various donor–carbon tetrachloride complexes in n-hexane

Donor	Temp./°C	K_c^{AD}/l mol^{-1}	$-\Delta H^{\ominus c}$/cal mol^{-1}	Ref.
Benzene	25	0·009 ± 0·004		16
Benzene		0·076 ± 0·057*		29
Benzene		0·043 ± 0·081†		29
m-Xylene		0·112 ± 0·028		29
p-Xylene		0·136 ± 0·018		29
Mesitylene	25	0·113 ± 0·044		16
Mesitylene		0·252 ± 0·036		29
Hexamethylbenzene	25	0·64 ± 0·12		17
Hexamethylbenzene	27	0·02 ± 0·01	540 ± 150	28
Hexamethylbenzene		0·550 ± 0·160		29
N,N-Dimethylaniline	25	0·06	500	30
N,N,N′,N′-Tetramethyl-p-phenylenediamine	25	0·13	1200	30

* From measurements near λ_{max}.
† From measurements at three wavelengths.

The effect of even very weak complexing, by the solvent species, with one or other, or both, the donor and acceptor solute species on the measured association constant for the complex between the two solute species, is discussed in Chapter 6.

Evidence from ultraviolet absorption spectra for the formation of charge-transfer complexes, or contact charge-transfer pairs between triethylamine and carbon tetrachloride and other halomethanes, has been given by Stevenson and Coppinger.[33] They suggest that the photochemical instability of aliphatic amines in carbon tetrachloride can be accounted for by such interactions.[34-38]

By contrast with complexes of carbon tetrachloride, alkyl halides, which contain at least one unsubstituted hydrogen atom, may complex with aromatic systems through a predominantly hydrogen-bonded interaction. Such systems are outside the scope of the present discussion.

10.C. Atomic-Iodine Complexes

A transient optical absorption has been observed in the flash photolysis of molecular-iodine–nitric oxide mixtures in the gaseous phase. This has been attributed to a complex between nitric oxide and an iodine atom, which may have the character of a charge-transfer complex.[39] Complexes have been likewise detected in the liquid phase, in which the solvent

TABLE 10.3. A comparison of the positions of the charge-transfer bands of a series of iodine atom complexes and of the corresponding iodine molecule complexes.

Donor	Wavelength of maximum absorption/nm		Ref.*
	I complex	I_2 complex	
Benzene	495	292†	42
Toluene	515	302†	42
o-Xylene	570	316	42
o-Xylene	590		50
p-Xylene	520	304†	42
Mesitylene	590	332†	42
Cyclohexane	330	242‡	45

* With respect to iodine-atom complex.
† In carbon tetrachloride solution, L. J. Andrews and R. M. Keefer, *J. Am. chem. Soc.* **74**, 4500 (1952).
‡ S. H. Hastings, J. L. Franklin, J. C. Schiller and F. A. Matsen, *J. Am. chem. Soc.* **75**, 2900 (1953).

appears to act as the donor species.[40–44] The positions of the absorption bands of many of these complexes correlate closely with the positions of the charge-transfer bands for the corresponding molecular-iodine complexes (Table 10.3), even to the extent that both the iodine-atom and the iodine-molecule complexes of p-xylene absorb at shorter wavelengths than might be expected from the ionization potential of p-xylene.* The transient absorption ($\lambda_{max} = 330$ nm), observed in the pulse radiolysis of solutions of iodine in cyclohexane, has been assigned to an iodine-atom–cyclohexane charge-transfer complex.[45] However, by comparison with other iodine-atom complexes (Table 10.3), the energy of

* The apparently anomalous position of the charge-transfer band, which occurs in various p-xylene complexes, may be the result of the overlapping of two charge-transfer bands [see L. E. Orgel. *J. chem. Phys.* **23**, 1352 (1955); also Chapter 3.]

this transition is remarkably low. The short-lived species observed in the pulse radiolysis of solutions of aromatic compounds in carbon tetrachloride [46, 47] have been identified as charge-transfer complexes of chlorine atoms with the aromatic solute. Comparison of the association constant of the iodine-atom–hexamethylbenzene complex [48] ($K_c^{AD} = 2\cdot7$ l mol^{-1} in carbon tetrachloride at room temperature) with the association constant of the corresponding molecular-iodine complex [49] ($K_c^{AD} = 1\cdot7$ l mol^{-1}) indicates that the iodine atom forms a more stable complex than molecular iodine in this case. A similar result is also obtained for the o-xylene complexes [50] (Table 10.4). The greater stability of the iodine-atom complex, compared with the iodine-molecule complex, can be accounted for in the greater electron affinity of the iodine atom ($3\cdot075 \pm 0\cdot005$ eV) [51] compared with molecular-iodine ($1\cdot7 \pm 0\cdot5$ eV),[52]

TABLE 10.4. A comparison of association constants, spectroscopic and thermodynamic properties of the iodine-atom and iodine-molecule complexes of o-xylene in carbon tetrachloride.*

	I complex*	I$_2$ complex†
K_x^{AD} (at 25°C)	7·4	2·96
λ_{max}^{AD}/nm	590	316
$-\Delta H^{\ominus x}$/kcal mol^{-1}	4·4	2·0
$-\Delta S^{\ominus x}$/kcal mol^{-1} deg^{-1}	10·9	4·9
ϵ_{max}^{AD}/cm^{-1} mol^{-1} l	3400	12,500

* Ref. 50.
† J. A. A. Ketelaar, *J. Phys. Radium, Paris* **15**, 197 (1954).

which would also account for the relative positions of the charge-transfer bands of the two complexes described in Table 10.3.

The possible participation of iodine-atom complexes in the gas-phase recombination of iodine atoms in the presence of a third body, such as benzene or a methylated benzene, has been proposed by Porter and his co-workers.[39, 53] Agreement between the degree of complex formation in the liquid phase and the kinetics of recombination in the gaseous phase is considered to be reasonable, although the exact quantitative effect which the solvent species has on such equilibria still remains to be determined (see Chapter 7).

10.D. Hexafluorobenzene Complexes

Patrick and Prosser[54] have reported the formation of complexes between hexafluorobenzene and various aromatic hydrocarbons: for example, a 1 : 1 solid complex by mixing benzene with hexafluorobenzene. More recently Swinton and his co-workers have made a careful study of the phase diagrams,[55] excess volumes of mixing,[56] and dipole moments,[56] of various binary hexafluorobenzene–hydrocarbon systems. Scott and his co-workers[57] have measured the heats of mixing of benzene with various polyfluorobenzenes. Their results were interpreted as indicating a decrease in complex formation in the series hexafluorobenzene, pentafluorobenzene, 1,2,4,5-tetrafluorobenzene, which is the expected order for charge-transfer complexes in which the fluorobenzenes are acting as electron acceptors. An increase in stability of the complex, as

TABLE 10.5. Association constants, spectroscopic and thermodynamic properties of various hexafluorobenzene–amine complexes in n-hexane*

Donor	ν_{CT}/cm^{-1}	$\epsilon_{max}^{AD}\dagger/cm^{-1}$ $mol^{-1}\,l$	$K_c^{AD}\ddagger/l$ mol^{-1}	$K_c^{AD}\S/l$ mol^{-1}
N,N,N′,N′-Tetramethyl-p-phenylenediamine	28,300	1393	0·53	0·47
p-Phenylenediamine	29,000			
N,N′-Diphenyl-p-phenylenediamine	30,100			
N,N-Diethylaniline	31,300	1086	0·40	0·37
N-Phenyl-p-phenylenediamine	31,400			
N,N-Dimethylaniline	31,600	736	0·42	0·46
Triethylamine	37,700‖			

* From Ref. 60.
† Values of ϵ are dependent on the concentration-scale of K^{AD} (see Chapter 6), the values quoted correspond with the values of K_c^{AD} quoted in the following column.
‡ Optically determined values of K_c^{AD} at 25°C.
§ N.m.r. determined values of K_c^{AD} at 25°C.
‖ Extrapolated value.

hexafluorobenzene is partnered with molecules of increasing donor-capabilities, has also been observed.[56] From the general crystallographic features of the equimolecular complexes of hexafluorobenzene with anthracene, pyrene and perylene, Boeyens and Herbstein[58] have considered these complexes to be of the charge-transfer class.

However, as has been pointed out by Swinton and co-workers,[56] the evidence for significant charge-transfer interaction in most cases is not overwhelming. The values of the measured dipole moment of hexafluorobenzene in various aromatic solvents are unusually low for charge-transfer complexes.[56] It has also been noted[59] that there is no detectable shift of the ^{19}F nuclear magnetic resonance absorption of hexafluorobenzene in the presence of a varying large excess of electron donors, such as occurs when other fluorine-containing electron acceptors such as fluoranil or 1,3-difluoro-2,4,6-trinitrobenzene are used. By itself, this observation does not necessarily imply the absence of complexes in solution. If the degree of charge transfer is small, the ^{19}F chemical shift in the complex, relative to the shift in the free-hexafluorobenzene, may be small (see Chapter 6).

Although attempts to detect an intermolecular charge-transfer absorption band in various hexafluorobenzene–aromatic hydrocarbon mixtures have failed,[54, 59] electronic absorption bands in mixtures of hexafluorobenzene with various n-donors, extra to the absorption of the components, have been observed.[60, 61] These have been assigned as intermolecular charge-transfer transitions (Table 10.5). Estimates of association constant have been obtained both from optical and from n.m.r. measurements. The results (Table 10.5) indicate the relatively weak complexing in these systems.

Measurements[62] of broad-line n.m.r. of the solid complexes of hexafluorobenzene with benzene, with mesitylene and with durene are referred to in Chapter 5. The results have been rationalized in terms of molecular rotation at room temperature.

10.E. Tetranitromethane Complexes

The well-known colourations which are observed when tetranitromethane is added to solutions of certain unsaturated compounds, particularly alkenes, and to a lesser extent aromatic hydrocarbons, have sometimes been generally classified as charge-transfer complexes.[63, 64] However, whilst there is some evidence for charge-transfer complex formation between some aromatic hydrocarbons and tetranitromethane, it would appear that the colourations observed with alkenes are more likely due to species such as I, formed by the attack of a nitronium ion on

I

the double bond.[65, 66] This would account for the new absorption maxima appearing at lower energies for the olefin interactions (usually 20,000–22,000 cm^{-1}), than for the majority of aromatic hydrocarbon interactions, most of which show absorptions with $\nu_{max}^{AD} > 25,000$ cm^{-1},[67] despite the fact that this latter group of hydrocarbons has considerably lower ionization potentials than have the alkenes. It could also account for the lower intensities of the optical absorptions in the case of the aromatic hydrocarbon interactions.[68] It should be remarked, however, that the positions of the charge-transfer bands, vis-à-vis the order of the ionization potentials of the donor molecules, provide no absolute criterion for the intermolecular charge-transfer nature of the observed transitions (see Chapter 3). Furthermore, Chaudhuri and Basu[66] found no correlation between these two functions amongst the aromatic donor complexes themselves, where there is some claim for charge-transfer interaction.

Quantitative measurements of equilibrium constants for the tetra-nitromethane–aromatic hydrocarbon donor systems present experimental problems. It is difficult to prepare and to store pure tetranitromethane. With aromatic compounds which are not fully alkylated, slow, irreversible reactions occur with the formation of the nitro-substituted compound. Even with the fully substituted aromatic donor, hexamethylbenzene, the solution mixtures are unstable.[68] Enthalpies of formation with the range 0·1–1·7 kcal mol^{-1} have been reported for various tetranitromethane–aromatic donor complexes,[69, 70] Early measurements on a series of polymethylbenzene complexes suggest that the stabilities increase as the degree of methylation is increased.[71] A value of $\Delta H^{\ominus c} = -2 \cdot 6$ kcal mol^{-1} has recently been obtained for the hexamethylbenzene complex.[68] However, in this case the measured association constant was very sensitive to the purity of the tetranitromethane. Probably the thermodynamic values quoted for all of the tetranitromethane complexes should be considered with circumspection.

The solvent shift of the ultraviolet absorption of tetranitromethane in cyclohexane, compared with the vapour-phase absorption, has been attributed to a solvent–solute contact charge-transfer absorption.[72]

10.F. Metallocene Complexes

Metallocenes such as ferrocene have relatively low ionization potentials.[73] It is therefore not surprising that, with acceptors of medium or strong electron-affinity, complete electron transfer often occurs with the formation of the corresponding metallocenium ion.[74–76] In this respect cobaltocene is more readily oxidized than ferrocene. However, with some electron acceptors, for example tetracyanoethylene (TCNE)[76–79] and various polynitrobenzene derivatives,[80–83] ferrocene does appear to be capable of forming charge-transfer complexes.

A ferrocene–tetracyanoethylene charge-transfer complex has been obtained as a green solid by grinding together the two components, by precipitation from a hot solution of the components in ethyl acetate, and by evaporation of a benzene solution.[77, 84] It has also been reported to be formed when a solution of the components in cyclohexane is irradiated in a Vycor vessel,[77] although other workers have not been successful in using this method.[76] The product was, at first, thought to be the ferricenium salt of TCNE$^-$.[84] The visible–near-infrared absorption spectrum of the solid does not possess an absorption maximum at 619 nm, characteristic of the ferrocenium ion,[85] but does show a broad absorption centred at ~1100 nm, with poorly defined maxima at 975 and 1150 nm.[77] The general appearance of this absorption is typical of an intermolecular charge-transfer band. The infrared vibrational spectrum of the complex is, to a first approximation, a superposition of the spectra of the components. These features, together with the relatively low melting point, ~115–119°C,[77] strongly suggest that this solid is basically a non-ionic charge-transfer complex, rather than the ferrocenium–TCNE$^-$ salt.[77] Electron transfer with the formation of these two ions does occur, however, when the solid is dissolved in solvents of only moderate ionizing power, e.g. chloroform or dichloroethane (see Chapter 11), owing presumably to stabilization of the ions by solvation. This accounts for the observations made by Webster et al.[84] The charge-transfer complex can exist in solution, providing the ionizing power of the solvent is very low. The association constant and the molar absorptivity are both relatively small (Table 10.6). Although there may be some doubt concerning the *separated* values of K_X^{AD} and ϵ, nevertheless the product $K_X^{AD}\epsilon$, which is probably a more reliable experimental term, is also low (see Chapter 6). Correspondingly low values of association constant have been obtained for the complexes with various polynitrobenzene acceptors.[80–82]

Two structures have been considered for this type of complex, namely II, in which the acceptor interacts with the filled π-orbital of a cyclopentadienyl ring, and III, in which non-bonding electrons of the metal

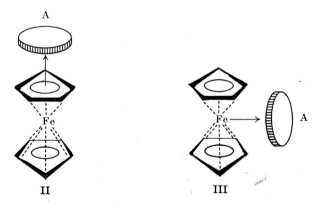

II   III

10*

TABLE 10.6. Association constants, spectroscopic and thermodynamic properties of some selected metallocene complexes.

Donor	Acceptor	Solvent	Temp./ °C	ν_{CT}/ cm^{-1}	K_x^{AD}	ϵ_{max}^{AD}/cm^{-1} mol^{-1} l	$-\Delta H^{\ominus}x$/ kcal mol^{-1}	$-\Delta S^{\ominus}x$/kcal mol^{-1} deg^{-1}	Ref.
Ferrocene	Tetracyano-ethylene	Cyclohexane	30	10,000	30·0	474	3·8	5·5	79
Ferrocene	Tetracyano-ethylene	Cyclohexane		{ 11,100 9,300	39	465*			77
Ferrocene	1,3,5-Trinitro-benzene	1,2-Dichloro-ethane	23	18,700	2·8 ± 0·3	630	1·8 ± 0·6	4 ± 2	81
1,1'Dimethyl-ferrocene	Tetracyano-ethylene	Cyclohexane	30	9,760	52·0	462	4·3	6·4	79
Ethylferrocene	1,3,5-Trinitro-benzene	1,2-Dichloro-ethane	23	18,500	3·2 ± 0·4				80
1,1'-Di-n-propyl-ferrocene	1,3,5-Trinitro-benzene	1,2-Dichloro-ethane	23	17,400	1·7 ± 0·2				80
Chloromercuri-ferrocene	1,3,5-Trinitro-benzene	1,2-Dichloro-ethane	23	19,400	0·75				80

* Average value for the two maxima.

TABLE 10.7. Types of solid product formed between metallocenes and various acceptors.*

Donor	$\mathscr{E}^{ox}/v$	Acceptor	$\mathscr{E}^{red}/v$	Product type
Ferrocene	+0·30†	p-Benzoquinone	−0·50‡	Charge-transfer complex
		Chloranil	+0·01‡	Charge-transfer complex
		Tetracyanoethylene	+0·15§	Charge-transfer complex
		DDQ‖	+0·51‡	Ionic
Cobaltocene	−1·16	Chloranil	+0·01	Ionic
		Tetracyanoethylene	+0·15	Ionic
		DDQ‖	+0·51	Ionic
Bis(tetrahydroindenyl)iron		Tetracyanoethylene	+0·15	Ionic
		DDQ‖	+0·51	Ionic

* From Ref. 76.
† In acetonitrile, from J. Tironflet, E. Laviran, R. Dabard and J. Komenda, *Bull. Soc. chim. Fr.* 857 (1963).
‡ In acetonitrile, for the reaction: quinone + e⁻ ⇌ semiquinone ion, from M. E. Peover, *J. chem. Soc.* 4540 (1962).
§ In acetonitrile, Ref. 84.
‖ Abbreviation for 2,3-dichloro-5,6-dicyano-p-benzoquinone.

are involved.[74, 77] The results of a single-crystal X-ray diffraction study[79] show that, in the solid phase for the ferrocene–tetracyano-ethylene complex at least, the geometry of the complex corresponds to II and not III. Further support for II comes from the Mossbauer experiments of Collins and Pettit,[78] and also from the observation that 1,1'-dimethylferrocene forms more stable complexes than unsubstituted ferrocene.[79] This latter result, corresponding to the behaviour of benzenoid systems, is more reasonably interpreted in terms of an interaction of the aromatic π-electrons, as in II, than the involvement of the metal atom. Prior to the crystallographic study of the tetracyano-ethylene complex, it had been argued that the relatively small values for the association constants and molar absorptivities, and the positions of the intermolecular charge-transfer bands for various ferrocene–electron acceptor complexes might be better interpreted in terms of structure III.[77, 83] Although it could be argued that this may be the case for complexes other than that of tetracyanoethylene, it now seems unlikely, since the tetracyanoethylene complex is amongst those which shows such an anomalous behaviour.[77] By comparison with ferrocene, cobaltocene and bis(tetrahydroindenyl)iron both form solid ionic salts with tetra-cyanoethylene[76] (Table 10.7). This reflects the greater ease of oxidation of the two latter metallocenes.

10.G. Paracyclophane Complexes

Various aspects of charge-transfer complexing by paracyclophanes with the general structures IV and V have been studied by Cram and his

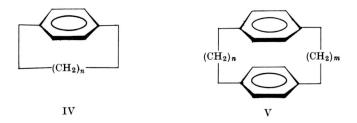

IV   V

co-workers.[86–89] Both of these types of paracyclophane form 1:1 complexes with tetracyanoethylene.[86] The energy of the charge-transfer band (or in cases where a particular complex exhibits two charge-transfer bands, then the band of lower energy) in every case is less than the energy of the transition for the tetracyanoethylene complex of the corresponding non-paracyclophane hydrocarbon p-xylene (Table 10.8). This observa-

TABLE 10.8. Energies of the lowest charge-transfer bands for a series of paracyclophane–tetracyanoethylene complexes in dichloromethane solution compared with the corresponding p-xylene complex.*

Donor†	ν_{CT}/cm^{-1}	Donor†	ν_{CT}/cm^{-1}
IV [3.3]	16,700	V [9]	20,300
IV [3.4]	18,600	V [12]	20,400
IV [1.7]	19,000	IV [6.6]	20,400
IV [2.2]	19,200	IV [1.12]	20,400
IV [3.6]	19,200	V [10]	20,600
IV [1.8]	19,400	IV [5.6]	20,600
IV [2.3]	19,600	IV [5.5]	20,600
IV [1.9]	19,600	IV [4.5]	20,600
IV [1.10]	19,800	IV [2.4]	20,700
IV [1.11]	20,000	IV [4.4]	21,000
IV [4.6]	20,300	p-Xylene	21,700‡

* From Ref. 86.

† The numbers in the square brackets refer to the values of m and n in structure VII or of n in structure VI (see text).

‡ Value from R. E. Merrifield and W. D. Phillips, *J. Am. chem. Soc.* **80**, 2778 (1958).

tion is in harmony with structures VI and VII for the paracyclophane complexes in which the transannular delocalization of charge in the

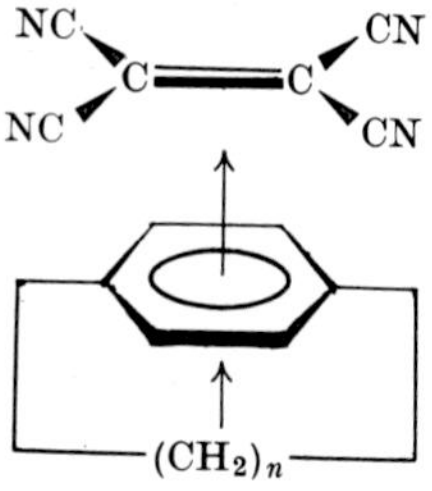

VI 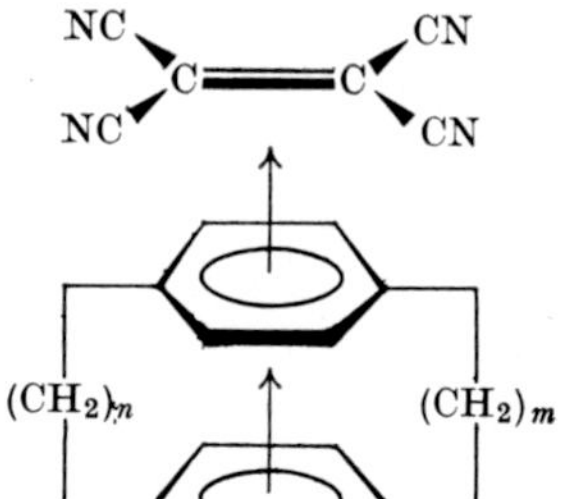   VII

paracyclophane increases the donor strength of the ring which is acting as the immediate donor partner to the acceptor. This increased donation is significant even when the backing effect is provided by a saturated hydrocarbon moiety as in VI. This effect appears to outweigh any possible decrease in donor strength of the paracyclophane, resulting from non-planarity of the benzene rings, and of the inhibition of π-inductive effects (or hyperconjugation) of the methylene groups attached to the benzene rings by their being bent out of the planes of the benzene rings. It is remarkable that this transannular delocalization is even observed in such paracyclophanes as V ($n = 1$, $m = 12$), in which the two benzene rings must be very far from coplanarity, and in the compounds IV ($n = 9$) and V ($n = 2$, $m = 2$), in each of which both benzene rings are warped. However, some inhibition of transannular delocalization of charge, due to the methylene bridges holding the benzene rings apart, is indicated by the fact that the complex of [6.6]paracyclophane with tetracyano-ethylene absorbs at a lower energy than does the [4.4]paracyclophane complex. The warping effect on the benzene rings and the decreased interaction over the benzyl methylene groups probably account for the

TABLE 10.9. Association constants and spectroscopic properties of some paracyclophane–tetracyanoethylene complexes in dichloromethane at 25°C.*

Donor	ν_{CT}/cm^{-1}	K_x^{AD}	$\epsilon/cm^{-1}\,mol^{-1}\,l$
4-Ethyl[2.2]PC†	18,500	52	1610
4-Acetyl[2.2]PC†	20,200	24·5	1450
4-Cyano[2.2]PC†	21,100	8	2000
p-Xylene‡	21,700	7·6	2650

* From Ref. 87.
† PC = paracyclophane.
‡ From R. E. Merrifield and W. D. Phillips, *J. Am. chem. Soc.* **80**, 2778 (1958).

higher energy of the charge-transfer band of the tetracyanoethylene complex of [2.2]paracyclophane. In complexes derived from [1.n]cyclo-phanes, the smaller the value of n the lower is the energy of the charge-transfer band. This is to be expected since the transannular delocalization increases as the two benzene rings become more nearly parallel.

The equilibrium constants of complexes of various 4-substituted-[2.2]-paracyclophanes with tetracyanoethylene have been measured[87] using the Benesi–Hildebrand method[90] (cf. Chapter 6). The values obtained are compared, in Table 10.9, with the association constants for the complex of tetracyanoethylene with p-xylene. The paracyclophanes containing electron-releasing groups form the more stable complexes VII, as would be expected. The complex of 4-ethyl-[2.2]paracyclophane is remarkably stable. The presence of the electron-withdrawing acetyl group will undoubtedly deactivate the ring to which it is attached. Since the complex involving 4-acetyl-[2.2]paracyclophane is significantly more stable than the tetracyanoethylene–p-xylene complex, it has been suggested that complexing is via the unsubstituted ring (VIII). The

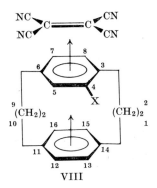

VIII

same explanation can be given for the 4-cyano[2.2]paracyclophane complex, although in this case the transannular electron-releasing ability of the non-complexed ring is about balanced by the electron-withdrawing effect of the cyano group (IX). It should be remarked that

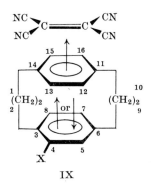

IX

the stability of complexes of various 1-X-substituted benzenes with 1,3,5-trinitrobenzene increases as certain electron-withdrawing groups are introduced into what is ostensibly the electron-donor molecule. Elsewhere this has been accounted for by suggesting that there is an increasing contribution to the stability of the complex from dipolar forces. However, the fact that, amongst the paracyclophanes, there is some correlation between the association constants and the charge-transfer band energies suggests that stabilization through dipolar forces may not have an overwhelming influence in the paracyclophane complexes.

Reference has been made in Chapter 3 to the behaviour of [2.2]-paracyclophanequinone (X), which has an electronic absorption band at

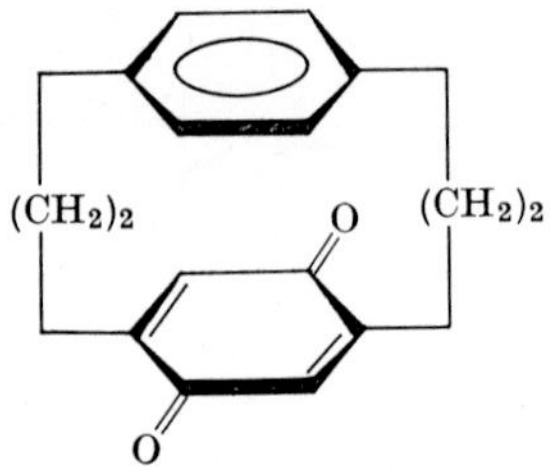

X

29,000 cm^{-1} ($\epsilon = 597$). It is suggested that this absorption is the result of an intramolecular charge-transfer transition across space.[88]

10.H. Complexes Involving Planar Ions

Reference has been made to the extensive studies made by Kosower *et al.*[91-100] on the charge-transfer interactions between the inorganic anions, particularly the iodide ion, and pyridinium and substituted-pyridinium cations, especially in regard to the sensitivity of their charge-transfer absorptions to the solvent environment (Chapter 3), and to the possible role of structures of this type on enzymatic oxidation-reduction processes (Chapter 12). The association constants of such complexes do not appear to be very sensitive to ring substitution[94] (Table 10.10).

TABLE 10.10. Association constants for various pyridinium iodide complexes.*

Pyridinium ion	K_c^{AD}/l mol^{-1}†
1-Methyl-	2·3 ± 0·3
1,2,6-Trimethyl-	1·6 ± 0·3
1,2,4,6-Tetramethyl-	1·8 ± 0·3

* From Ref. 94.
† At room temperature in aqueous solution at an ionic strength of 0·01.

The bis-pyridinium cation methylviologen (N,N'-dimethyl-4,4'-dipyridylium, XI), also known particularly in the form of the dichloride salt as paraquat, has been shown to form 1 : 1 charge-transfer complexes with various anionic and neutral donor species.[101-105] The association

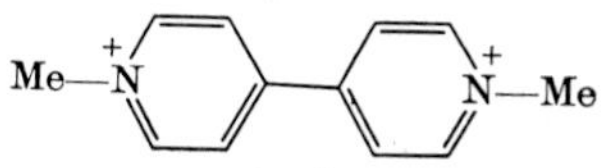

$$Me—\overset{+}{N}\underset{}{\bigcirc}—\bigcirc—\overset{+}{N}—Me$$

XI

constant for the complex with ferricyanide ion in water ($K_c^{AD} = 52 \pm 5$ 1 mol^{-1} at 23°C) shows that this is a relatively stable complex.[101] Only a very small concentration of the termolecular complex A_2D was detected under the conditions of the determinations. The association constants of complexes of **XI** with neutral donors are also relatively large. For example, at 20°C, $K_c^{AD} = 88$ 1 mol^{-1} for the N-phenyl-2-naphthylamine complex in methanol, compared with $K_c^{AD} = 2 \cdot 3$ 1 mol^{-1} for the corresponding chloranil complex in dichloromethane.[102] The possible role of charge-transfer complex formation in the herbicidal activity of this cation is referred to in Chapter 12.

There is evidence that a considerable number of other heterocyclic cationic systems can also act as electron acceptors in forming charge-transfer complexes with inorganic anions. The coloured nature of the pyrylium [106, 107] (**XII**), tropylium [108–110] (**XIII**), thiopyrylium [111] (**XIV**), 3-azapyrylium [112] (**XV**), 1,3-dithiolium [113, 114] (**XVI**), quinolinium [115, 116] (**XVII**), and acridinium [115] (**XVIII**) iodides, compared with the lack of

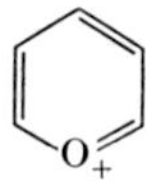

XII 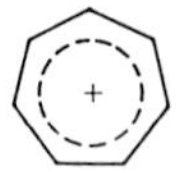   XIII 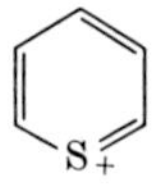   XIV 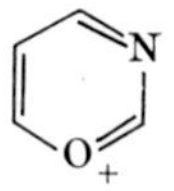   XV

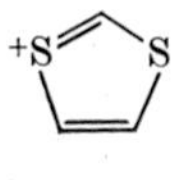

XVI 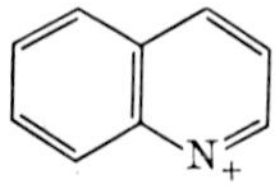   XVII 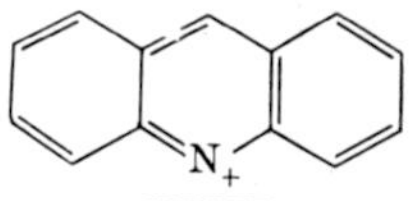   XVIII

visible absorption by most of their other salts, may be explained in terms of charge-transfer interaction between the two ions. Obviously, caution must be exercised before making such an assignment to ensure that the transition is not the result of some other process, such as irreversible oxidation or the formation of the triiodide ion.

The formation of charge-transfer complexes between neutral donors and aromatic cations has been reported.[117, 118] For example, complexes between tropylium and a series of aromatic hydrocarbons show resolved charge-transfer bands,[118] the energies of which are proportional to the energies of the bands for the corresponding hydrocarbon–2,3-dichloro-5,6-dicyano-p-benzoquinone complexes (Fig. 10.1). Only small solvent shifts were observed in the charge-transfer bands of two tropylium-hydrocarbon complexes.[119]

Of particular interest is the 1,2,3,4,5-pentacarbomethoxycyclopentadienyl anion XIX, which can act as a donor.[120] If the potassium salt

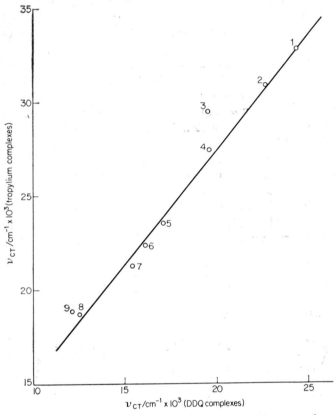

FIG. 10.1. Plot of ν_{CT} for a series of tropylium–hydrocarbon complexes against ν_{CT} for the corresponding 2,3-dichloro-5,6-dicyano-p-benzoquinone (DDQ) complexes, all in 1,2-dichloroethane solution. Donors: (1) benzene; (2) toluene; (3) p-xylene; (4) mesitylene; (5) phenanthrene; (6) naphthalene; (7) 2-methylnaphthalene; (8) pyrene; (9) anthracene. (Tropylium data from M. Feldman and S. Winstein, Ref. 118.)

of XIX is mixed with a solution of tropylium tetrafluoroborate in acetone, the salt XX may be isolated. This has an absorption at 470 nm, which is assigned as an interionic charge-transfer transition. Other salts which have been isolated, together with the position of their charge-transfer bands, are: 2,4,6-trimethylpyrylium (418 nm); 2,3,4,5,6-penta-

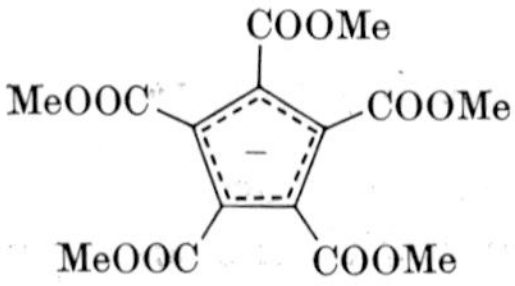

R

COOMe

MeOOC   COOMe

MeOOC   COOMe

XIX 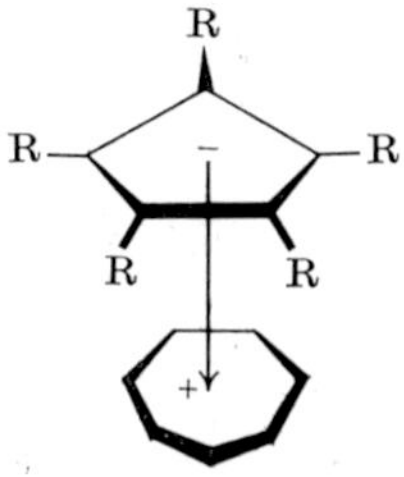   XX

phenylpyrylium (420 nm); N-methylpyridinium (350 nm) and N-methyl-quinolinium (402 nm). The ion-pair from XIX and triphenylcyclo-propenyl was obtained in acetonitrile solution but could not be isolated. Complexes between XIX and neutral acceptors such as 1,3,5-trinitro-benzene and 2,4,7-trinitrofluorenone may be obtained as their solid potassium salts.

Charge-transfer complex formation in solution by various relatively reactive carbonium ions with pyrene as donor has been demonstrated by Dauben and Wilson.[121] The low energies of the charge-transfer bands, for example $14 \cdot 1 \times 10^3$ cm^{-1} for the triphenylcarbonium complex and $12 \cdot 2 \times 10^3$ cm^{-1} for the tris-(p-chlorophenyl)carbonium complex, suggest that such ions are strong π-acceptors.

Charge-transfer complexes in the gas phase involving positively charged species may be formed in the field-ionization mass spectrometry of certain aromatic compounds and mixtures of aromatic compounds. In this type of mass spectrometry, fragmentation is negligible. The spectra of nitrobenzene and of pyridine show ions greater than the molecular ion (M^+), at m/c 246 and 158 respectively,[122] corresponding to the dimeric species M_2^+. In nitrobenzene the trimer, M_3^+, has been detected. In pyridine, peaks correspond to the doubly-charged trimer, M_3^{++}. Although, in the experiments of Job and Patterson,[122] aniline alone did not show any evidence of ions greater than the molecular ion, the spectrum of a mixture of nitrobenzene (X) and aniline (Y) showed peaks corresponding to the hetero-dimer XY^+, which were more abundant than those of the homo-dimer X_2^+. A hetero-dimer of pyridine and aniline was also observed. These positively charged dimer species have been assigned as charge-transfer complexes formed by the interaction of a molecular ion with a neutral molecule.[122] For example, in the presence of the nitrobenzene-positive ion, the nitrobenzene molecule may act as a donor, thus permitting the formation of the observed complex M_2^+. The ter-molecular species (M_3^+) derived from nitrobenzene may be represented as an AD_2 complex of an M^+ acceptor complexed to two neutral M molecules acting as donors. The M_3^{++} species observed in the spectrum of pyridine would correspond to an A_2D complex.

10.1. Complexes Involving Polymers

Ultraviolet absorption spectra of complexes formed between molecular bromine and poly-2-vinylpyridine, and with the co-polymer of styrene and 2-vinylpyridine, indicate the presence of intermolecular charge-transfer transitions.[123] A maximum of one bromine molecule per pyridine residue can be inserted. Similar evidence for the formation of complexes between pyridinium residues and chloride, bromide and iodide ions in acid solutions of poly-2-vinylpyridine, has also been obtained. Evidence for complex formation by other polymers including polynitrostyrene[123] (XXI), poly-α-vinylnaphthalene[123] (XXII), poly-acenaphthylene[123, 124] (XXIII), polyvinylmesitylene[123] (XXIV), poly-vinylanthracene[125] (XXV), and poly-N-vinylcarbazole[126] (XXVI), and

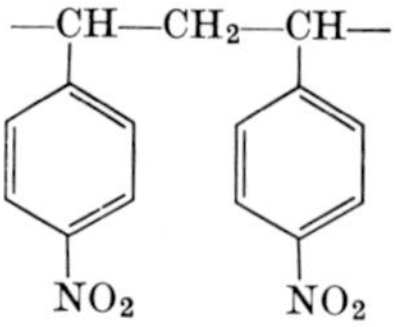

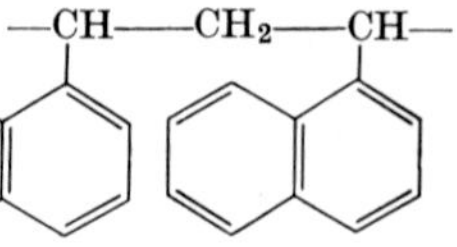

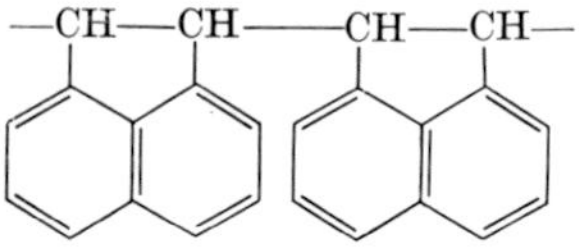

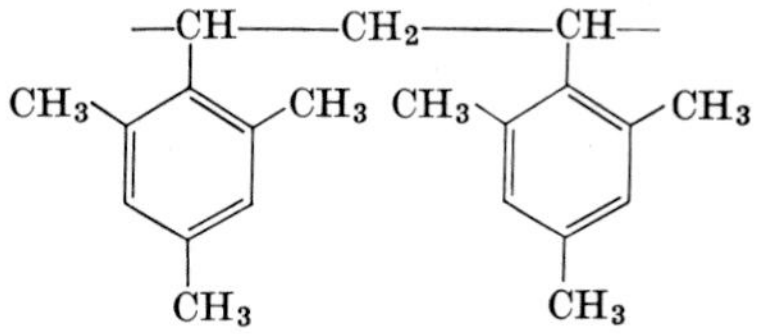

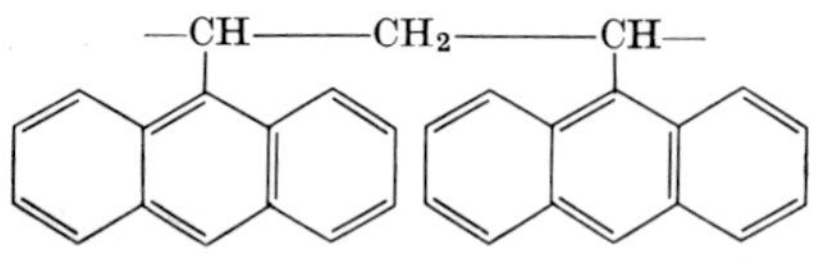

various electron acceptors including tetracyanoethylene, chloranil, 2, 3 - dichloro - 5, 6 - dicyano-p-benzoquinone, 7, 7, 8, 8-tetracyanoquino-dimethane and silver perchlorate has also been obtained. The conductivities of some of these complexes are relatively low and are not very sensitive to changes in the donor–acceptor ratio in the specimens.[123]

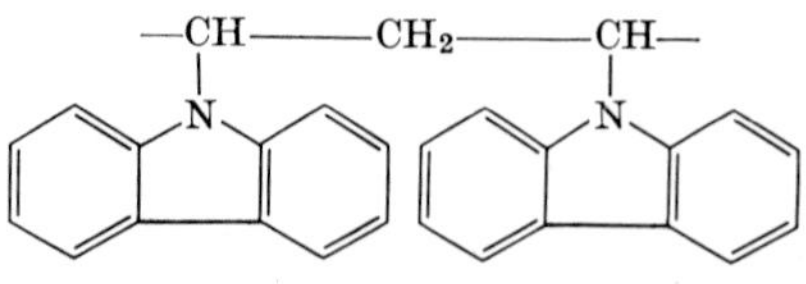

Seebeck coefficients indicate that, in many cases, the majority of the current carriers are positive.[123] Some discussion of the electrical properties of such complexes has been given in Chapter 9.

The anion-exchange properties of aromatic heterocyclic polymers[123] have been discussed. The use of tetranitrobenzylpolystyrene (XXVII) as

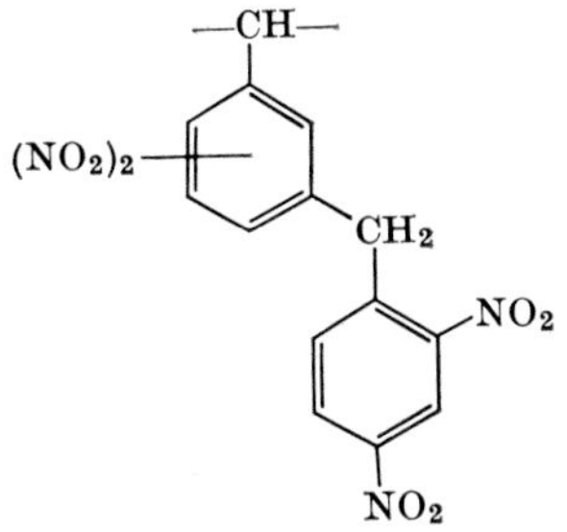

XXVII

the stationary phase in column chromatography of hydrocarbons has been proposed.[127]

The possible participation of charge-transfer complexes in various polymerization reactions has been suggested.[128-139] These include the co-polymerization reactions between 2,4,6-trinitrostyrene and 2-vinyl-pyridine, 4-vinylpyridine or 4-(dimethylamino)-styrene.[129] The products of these reactions have optical absorptions which may be the result of intramolecular, across-space, charge-transfer interactions.[129]

10.J. Self-Complexes

The possibility of self-complexation between two molecules of the same species to form a charge-transfer complex was considered by Mulliken[140] (see also Chapter 2). In practice, relatively few examples have been described. The majority of dimers involving strong donors or strong acceptors show none of the properties of charge-transfer complexes. In the main, such associations are probably polarization complexes.

The dimerization of molecular iodine has been studied by several groups of workers.[141-149] These studies began with the early observation that iodine possesses a weak absorption in the ultraviolet which does not obey Beer's law.[141] From measurements of the optical absorption it is concluded that the enthalpy of dimerization in solution is low, namely -1 to -2 kcal mol^{-1}.[147,148] However, quantitative measurements on this system are difficult to analyse because of the low concentration of dimer (see Chapter 6). Some of the estimates of K^{I_4} and ϵ^{I_4} may be optimistic. Nevertheless, the molar absorptivity of the species I_4 is undoubtedly large, possibly $\sim 16 \times 10^3 \, l \, mol^{-1} \, cm$ in carbon tetrachloride solution at $25.5°C$: the corresponding association constant may have a value $\sim 0.13 \, l \, mol^{-1}$.[148] There is a similar uncertainty concerning the evaluation of data from vapour-phase measurements, although here too the results indicate that the degree of association is small.[149] The charge-transfer band for the complex has been reported at 288 nm [148] and at 293·5 nm [147] in carbon tetrachloride, at $\sim$277·5 nm in hexane [147] and at 245 nm in the vapour phase.[149] However, the exact position is difficult to determine because of overlap with the band maximum of the monomer (280 nm in carbon tetrachloride, 270 nm in the vapour).

The dimerization of the free radical N-ethylphenazyl XXVIII in solution has been described by Hausser and Murrell.[150] An optical

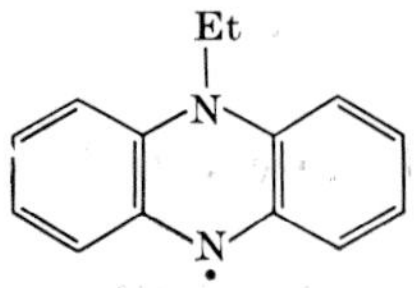

Et

XXVIII

absorption in the near infrared, which appears at low temperatures, intensifies as the paramagnetism decreases. This has been accounted for by the formation of a charge-transfer complex. An antisymmetrical combination of two charge-transfer configurations occurs, in which all the electrons are paired. A similar explanation has been given for the low-temperature behaviour of the cation of Wurster's blue perchlorate (the radical cation derived from N,N,N',N'-tetramethyl-p-phenylenedi-amine).[150]

Each of several α-aminophenyl-ω-nitrophenylalkanes XXIX in

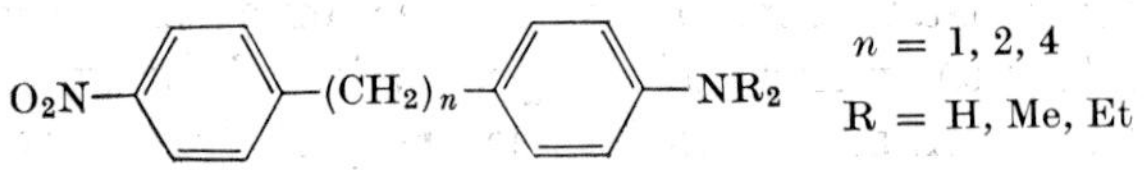

$$O_2N\!-\!\!\!\langle\ \rangle\!\!\!-\!(CH_2)_n\!-\!\!\!\langle\ \rangle\!\!\!-\!NR_2$$

$n = 1, 2, 4$
$R = H, Me, Et$

XXIX

solution has been shown to have an absorption band which does not obey Beer's law, and may be assigned as an intermolecular charge-transfer band.[151]* Solutions of 2,4,6-trinitro-p-terphenyl XXX in chloroform and in 1,2-dichloroethane, likewise, include in their spectra an absorption in the region 300–400 nm which, in concentrated solution,

* Cf. p. 78.

XXX

does not obey Beer's law.[152] The absorption band, which persists in dilute solution at 345 nm, has been assigned as an *intra*-molecular charge-transfer transition. Apparent molecular weight determinations indicate an association constant of 0.6 l mol^{-1} at temperatures somewhat above room temperature. Such systems may be contrasted with solutions of those molecules in which the donor moiety and the acceptor moiety of the same molecule interact with one another (see Chapter 3, p. 78).

REFERENCES

1. H. D. Bist and W. B. Person, *J. phys. Chem., Ithaca* **71**, 2750 (1967).
2. C. Reid and R. S. Mulliken, *J. Am. chem. Soc.* **76**, 3869 (1954).
3. A. G. Maki and E. K. Plyler, *J. phys. Chem., Ithaca* **66**, 766 (1962).
4. R. F. Lake and H. W. Thompson, *Proc. R. Soc.* **A297**, 440 (1967).
5. A. I. Popov and R. H. Rygg, *J. Am. chem. Soc.* **79**, 4622 (1957).
6. J. N. Chaudhuri and S. Basu, *Trans. Faraday Soc.* **55**, 898 (1959).
7. R. S. Mulliken and W. B. Person, *A. Rev. phys. Chem.* **13**, 107 (1962).
8. G. Kortüm and H. Walz, *Z. Elektrochem.* **57**, 73 (1953).
9. K. Toyoda and W. B. Person, *J. Am. chem. Soc.* **88**, 1629 (1966).
10. H. B. Friedrich and W. B. Person, *J. chem. Phys.* **44**, 2161 (1966).
11. O. Hassel and C. Rømming, *Acta chem. scand.* **10**, 696 (1956).
12. O. Hassel, C. Rømming and T. Tufte, *Acta chem. scand.* **15**, 967 (1961).
13. P. L. Kronick, *J. phys. Chem., Ithaca* **69**, 3178 (1965).
14. E. K. Plyler and R. S. Mulliken, *J. Am. chem. Soc.* **81**, 823 (1959).
15. S. G. W. Ginn and J. L. Wood, *Trans. Faraday Soc.* **62**, 777 (1966).
16. R. Anderson and J. M. Prausnitz, *J. chem. Phys.* **39**, 1225 (1963); *ibid.*, **40**, 3443 (1964).
17. R. F. Weimer and J. M. Prausnitz, *J. chem. Phys.* **42**, 3643 (1965).
18. H. O. Hooper, *J. chem. Phys.* **41**, 599 (1964).
19. J. R. Goates, R. J. Sullivan and J. B. Ott, *J. phys. Chem., Ithaca* **63**, 589 (1959).
20. G. H. Cheeseman and A. M. B. Whitaker, *Proc. R. Soc.* **A212**, 406 (1952).
21. R. M. A. Noordtzij, *Helv. chim. Acta* **39**, 637 (1956).
22. L. A. K. Staveley, W. I. Tupman and K. R. Hart, *Trans. Faraday Soc.* **51**, 323 (1955).
23. R. P. Rastogi, J. Nath and J. Misra, *J. phys. Chem., Ithaca* **71**, 1277 (1967).
24. W. F. Wyatt, *Trans. Faraday Soc.* **25**, 48 (1929).
25. R. P. Rastogi and R. K. Nigam, *Trans. Faraday Soc.* **55**, 2005 (1959).
26. J. B. Ott, J. R. Goates and A. H. Budge, *J. phys. Chem., Ithaca* **66**, 1387 (1962).
27. G. W. Chantry, H. A. Gebbie and H. N. Mirza, *Spectrochim. Acta* **23A**, 2749 (1967).

28. F. Dörr and G. Buttgereit, *Ber. Bunsenges. physik. Chem.* **67**, 867 (1963).
29. H. Kellawi and D. R. Rosseinsky, *J. chem. Soc.* (A), 1207 (1969).
30. K. M. C. Davis and M. F. Farmer, *J. chem. Soc.* (B), 28 (1967).
31. S. Carter, J. N. Murrell and E. J. Rosch, *J. chem. Soc.* 2048 (1965).
32. R. F. Weimer and J. M. Prausnitz, *Spectrochim. Acta* **22**, 77 (1966).
33. D. P. Stevenson and G. M. Coppinger, *J. Am. chem. Soc.* **84**, 149 (1962).
34. R. F. Collins, *Chemy Ind.* 704 (1957).
35. N. H. Cromwell, P. W. Foster and M. M. Wheeler, *Chemy Ind.* 228 (1959).
36. G. J. Beichl, J. E. Colwell and J. G. Miller, *Chemy Ind.* 203 (1960).
37. H. Williams, *Chemy Ind.* 900 (1960).
38. R. Foster, *Chemy Ind.* 1354 (1960).
39. G. Porter, Z. G. Szabó and M. G. Townsend, *Proc. R. Soc.* **A270**, 493 (1962).
40. N. K. Bridge, *J. chem. Phys.* **32**, 945 (1960).
41. S. J. Rand and R. L. Strong, *J. Am. chem. Soc.* **82**, 5 (1960).
42. R. L. Strong, S. J. Rand and J. A. Britt, *J. Am. chem. Soc.* **82**, 5053 (1960).
43. R. L. Strong, *J. phys. Chem., Ithaca* **66**, 2423 (1962).
44. T. A. Gover and G. Porter, *Proc. R. Soc.* **A262**, 476 (1961).
45. M. Ebert, J. P. Keene, E. J. Land and A. J. Swallow, *Proc. R. Soc.* **A287**, 1 (1965).
46. R. E. Bühler, T. Gäumann and M. Ebert, *in* "Pulse Radiolysis," eds. M. Ebert, J. P. Keene, A. J. Swallow and J. H. Baxendale, Academic Press, London (1965), p. 279.
47. R. E. Bühler and M. Ebert, *Nature, Lond.* **214**, 1220 (1967).
48. R. L. Strong and J. Pérano, *J. Am. chem. Soc.* **83**, 2843 (1961).
49. M. Tamres, D. R. Virzi and S. Searles, *J. Am. chem. Soc.* **75**, 4358 (1953).
50. R. L. Strong and J. Pérano, *J. Am. chem. Soc.* **89**, 2535 (1967).
51. B. Steiner, M. L. Seman and L. M. Branscomb, *J. chem. Phys.* **37**, 1200 (1962).
52. W. B. Person, *J. chem. Phys.* **38**, 109 (1963).
53. G. Porter and J. A. Smith, *Nature, Lond.* **184**, 446 (1959); *Proc. R. Soc.* **A261**, 28 (1961).
54. C. R. Patrick and G. S. Prosser, *Nature, Lond.* **187**, 1021 (1960).
55. W. A. Duncan and F. L. Swinton, *Trans. Faraday Soc.* **62**, 1082 (1966); W. J. Gaw and F. L. Swinton, *Nature, Lond.* **212**, 283 (1966).
56. W. A. Duncan, J. P. Sheridan and F. L. Swinton, *Trans. Faraday Soc.* **62**, 1090 (1966).
57. D. V. Fenby, I. A. McLure and R. L. Scott, *J. phys. Chem., Ithaca* **70**, 602 (1966).
58. J. C. A. Boeyens and F. H. Herbstein, *J. phys. Chem., Ithaca* **69**, 2153 (1965).
59. R. Foster and C. A. Fyfe, *Chem. Commun.* 642 (1965).
60. T. G. Beaumont and K. M. C. Davis, *J. chem. Soc.* (B), 1131 (1967).
61. P. R. Hammond, *J. chem. Soc.* (A), 145 (1968).
62. D. F. R. Gilson and C. A. McDowell, *Can. J. Chem.* **44**, 945 (1966).
63. L. F. Fieser and M. Fieser, "Advanced Organic Chemistry," Reinhold, New York (1961), p. 146.
64. C. Lagercrantz and M. Yhland, *Acta chem. scand.* **16**, 1807 (1962).
65. J. M. Patterson, *J. org. Chem.* **20**, 1277 (1955).
66. J. N. Chaudhuri and S. Basu, *J. chem. Soc.* 3085 (1959).
67. A. K. Macbeth, *J. chem. Soc.* **107**, 1824 (1915).
68. R. Foster and I. B. C. Matheson, unpublished work.
69. D. Ll. Hammick and R. P. Young, *J. chem. Soc.* 1463 (1936).

70. R. W. Maatman and M. T. Rogers, Paper presented to the Division of Petroleum Chemists, American Chemical Society, Cincinnati (1955).
71. T. T. Davies and D. Ll. Hammick, *J. chem. Soc.* 763 (1938).
72. D. F. Evans, *J. chem. Soc.* 4229 (1957).
73. L. Friedman, A. P. Irsa and G. Wilkinson, *J. Am. chem. Soc.* 77, 3689 (1955).
74. J. C. Goan, E. Berg and H. E. Podall, *J. org. Chem.* 29, 975 (1964).
75. L. R. Melby, R. J. Harder, W. R. Hertler, W. Mahler, R. E. Benson and W. E. Mochel, *J. Am. chem. Soc.* 84, 3374 (1962).
76. R. L. Brandon, J. H. Osiecki and A. Ottenberg, *J. org. Chem.* 31, 1214 (1966).
77. M. Rosenblum, R. W. Fish and C. Bennett, *J. Am. chem. Soc.* 86, 5166 (1964).
78. R. L. Collins and R. Pettit, *J. inorg. nucl. Chem.* 29, 503 (1967).
79. E. Adman, M. Rosenblum, S. Sullivan and T. N. Margulis, *J. Am. chem. Soc.* 89, 4540 (1967).
80. B. Hetnarski, *Bull. Acad. pol. Sci. Sér. Sci. chim.* 13, 557 (1965).
81. B. Hetnarski, *Bull. Acad. pol. Sci. Sér. Sci. chim.* 13, 515 (1965).
82. B. Hetnarski, *Bull. Acad. pol. Sci. Sér. Sci. chim.* 13, 523 (1965).
83. B. Hetnarski, *Bull. Acad. pol. Sci. Sér. Sci. chim.* 13, 563 (1965).
84. O. W. Webster, W. Mahler and R. E. Benson, *J. Am. chem. Soc.* 84, 3678 (1962).
85. G. Wilkinson, M. Rosenblum, M. C. Whiting and R. B. Woodward, *J. Am. chem. Soc.* 74, 2125 (1952).
86. D. J. Cram and R. H. Bauer, *J. Am. chem. Soc.* 81, 5971 (1959).
87. L. A. Singer and D. J. Cram, *J. Am. chem. Soc.* 85, 1080 (1963).
88. D. J. Cram and A. C. Day, *J. org. Chem.* 31, 1227 (1966).
89. L. J. Andrews and R. M. Keefer, "Molecular Complexes in Organic Chemistry," Holden-Day, San Francisco (1964), pp. 69–76.
90. H. A. Benesi and J. H. Hildebrand, *J. Am. chem. Soc.* 71, 2703 (1949).
91. E. M. Kosower, *J. Am. chem. Soc.* 77, 3883 (1955).
92. E. M. Kosower and P. E. Klinedinst, Jr., *J. Am. chem. Soc.* 78, 3493 (1956).
93. E. M. Kosower, *J. Am. chem. Soc.* 78, 3497 (1956).
94. E. M. Kosower and J. C. Burbach, *J. Am. chem. Soc.* 78, 5838 (1956).
95. E. M. Kosower, *J. Chim. phys.* 230 (1964).
96. E. M. Kosower, D. Hofmann and K. Wallenfels, *J. Am. chem. Soc.* 84, 2755 (1962).
97. E. M. Kosower, *J. Am. chem. Soc.* 80, 3253 (1958).
98. E. M. Kosower, *J. Am. chem. Soc.* 80, 3261 (1958).
99. E. M. Kosower, J. A. Skorcz, W. M. Schwarz, Jr. and J. W. Patton, *J. Am. chem. Soc.* 82, 2188 (1960).
100. E. M. Kosower and J. A. Skorcz, *J. Am. chem. Soc.* 82, 2195 (1960).
101. A. Nakahara and J. H. Wang, *J. phys. Chem., Ithaca* 67, 496 (1963).
102. A. Ledwith and D. H. Iles, *Chem. in Britain* 4, 266 (1968).
103. A. J. Macfarlane, Thesis, Oxford (1968).
104. R. Foster, unpublished work.
105. M. J. Strauss, unpublished work.
106. M. Feldman and S. Winstein, *Tetrahedron Lett.* 853 (1962).
107. A. T. Balaban, M. Mocanu and Z. Simon, *Tetrahedron* 20, 119 (1964).
108. W. E. Doering and H. Krauch, *Angew. Chem.* 68, 661 (1956).
109. K. M. Harmon, F. E. Cummings, D. A. Davis and D. J. Diestler, *J. Am. chem. Soc.* 84, 120 (1962).
110. K. M. Harmon, F. E. Cummings, D. A. Davis and D. J. Diestler, *J. Am. chem. Soc.* 84, 3349 (1962).

111. R. Wizinger and P. Ulrich, *Helv. chim. Acta* **39**, 207 (1956).
112. S. Hünig and K. Hübner, *Chem. Ber.* **95**, 937 (1962).
113. E. Klingsberg, *J. Am. chem. Soc.* **84**, 3410 (1962).
114. A. Lüttringhaus, M. Mohr and N. Engelhard, *Justus Liebigs Annln Chem.* **661**, 85 (1963).
115. S. F. Mason, *J. chem. Soc.* 2437 (1960).
116. G. Briegleb, W. Jung and W. Herre, *Z. phys. Chem. Frankf. Ausg.* **38**, 253 (1963).
117. G. Briegleb, W. Rüttiger and W. Jung, *Angew. Chem.* **75**, 671 (1963).
118. M. Feldman and S. Winstein, *J. Am. chem. Soc.* **83**, 3338 (1961).
119. M. Feldman and B. G. Graves, *J. phys. Chem., Ithaca* **70**, 955 (1966).
120. E. Le Goff and R. B. LaCount, *J. Am. chem. Soc.* **85**, 1354 (1963).
121. H. J. Dauben, Jr. and J. D. Wilson, *Chem. Commun.* 1629 (1968).
122. B. E. Job and W. R. Patterson, Symposium "Newer Methods in Structural Chemistry," Oxford (1966); United Trade Press, London (1967).
123. W. Slough, *Trans. Faraday Soc.* **55**, 1030, 1036 (1959); *ibid.*, **58**, 2360 (1962).
124. A. Inami, K. Morimoto and Y. Hayashi, *Bull. chem. Soc. Japan* **37**, 842 (1962).
125. H. Inowe, K. Noda, S. Takiuchi and E. Imoto, *J. chem. Soc. Japan, Ind. Chem. Sec.* (*Kogyo Kagaku Zasshi*) **65**, 1286 (1962).
126. A. Taniguchi, S. Kanda, T. Nogaito, S. Kusabayashi, H. Mikawa and K. Ito, *Bull. chem. Soc. Japan* **37**, 1386 (1964).
127. J. T. Ayres and C. K. Mann, *Analyt. Chem.* **36**, 2185 (1964).
128. H. Scott, G. A. Miller and M. M. Labes, *Tetrahedron Lett.* 1073 (1963).
129. N. C. Yang and Y. Gaoni, *J. Am. chem. Soc.* **86**, 5022 (1964).
130. G. Henrici-Olivé and S. Olivé, *Z. phys. Chem. Frankf. Ausg.* **47**, 286 (1965).
131. G. Henrici-Olivé and S. Olivé, *Z. phys. Chem. Frankf. Ausg.* **48**, 35 (1966).
132. G. Henrici-Olivé and S. Olivé, *Z. phys. Chem. Frankf. Ausg.* **48**, 51 (1966).
133. L. P. Ellinger, *Chemy Ind.* 1982 (1963).
134. L. P. Ellinger, *Polymer* **5**, 559 (1964).
135. C. E. H. Bawn, C. Fitzsimmons and A. Ledwith, *Proc. chem. Soc.* 391 (1964).
136. C. E. H. Bawn, A. Ledwith and A. Parry, *Chem. Commun.* 490 (1965).
137. C. E. H. Bawn, R. Carruthers and A. Ledwith, *Chem. Commun.* 522 (1965).
138. A. Ledwith, *J. appl. Chem., Lond.* **17**, 344 (1967).
139. S. Iwatsuki and Y. Yamashita, *Makromolek. Chem.* **89**, 205 (1965).
140. R. S. Mulliken, *J. Am. chem. Soc.* **74**, 811 (1952).
141. J. Gróh and S. Papp, *Z. phys. Chem.* **A149**, 153 (1930).
142. G. Kortüm and G. Friedhaim, *Z. Naturf.* **2A**, 20 (1947).
143. H. M. McConnell, *J. chem. Phys.* **22**, 760 (1954).
144. D. F. Evans, *J. chem. Phys.* **23**, 1424 (1955).
145. P. A. D. de Maine, *J. chem. Phys.* **24**, 1091 (1956).
146. P. A. D. de Maine, *Can. J. Chem.* **35**, 573 (1957).
147. M. M. de Maine, P. A. D. de Maine and G. E. McAlonie, *J. molec. Spectrosc.* **4**, 271 (1960).
148. R. M. Keefer and T. L. Allen, *J. chem. Phys.* **25**, 1059 (1956).
149. M. Tamres, W. K. Duerksen and J. M. Goodenow, *J. phys. Chem., Ithaca* **72**, 966 (1968).
150. K. H. Hausser and J. N. Murrell, *J. chem. Phys.* **27**, 500 (1957).
151. V. A. Gluschenkov, V. A. Ismail'skii and Y. S. Moshkovskii, *Dokl. Akad. Nauk SSSR* **153**, 1363 (1963); *Dokl. (Proc.) Acad. Sci. U.S.S.R. (Phys. Chem. Sec.)* **153**, 1125 (1963).
152. R. L. Hansen and J. J. Neumayer, *J. phys. Chem., Ithaca* **71**, 3047 (1967).

Chapter 11

Chemical Reactions which may involve
Charge-Transfer Complexes

There are many instances in the chemical literature where the participation of a charge-transfer complex in a chemical reaction has been postulated, although positive evidence for such processes is not plentiful. The immediate formation of a charge-transfer complex on mixing two reactants, and its slow disappearance as further irreversible chemical reaction proceeds is, in itself, no evidence for the participation of the charge-transfer complex as a reaction intermediate. Nevertheless, this assumption has often naïvely been made. The formation, and consequent decomposition, of the complex may be simply due to a side-equilibrium. Kinetic and other evidence for the participation of the charge-transfer complex within the reaction coordinate is often difficult to obtain.

Reviews of the subject have been given recently by Kosower[1] and by Andrews and Keefer.[2]

11.A. Ionization of Charge-Transfer Complexes in Solution

The ground states of most organic charge-transfer complexes appear to involve very little transfer of charge when the complexes exist in solution in a solvent of low ionizing power. In such solutions the complexed species appear to exist almost entirely as isolated complexed donor–acceptor pairs. By contrast, in the solid phase, it is often not realistic to describe the structure of a complex AD as a 1:1 complex, since often each donor has two neighbouring acceptors, and vice versa. The proximity of more than one molecule of the complementary species can cause the ground state of the complex in the solid phase to have considerably more ionic character than the ground state of an isolated donor–acceptor pair in an inert solvent. The same general result may arise by altering the character of the solvent. If a solvent of high ionizing power is chosen, ionization of the essentially non-ionic complex will occur. The driving force is the energy of solvation of the ions formed.[3, 4] This chemical reaction

$$AD + \text{solvent} \rightleftharpoons A^-_{\text{solv}} + D^+_{\text{solv}} \qquad (11.1)$$

is analogous to the ionization of hydrogen chloride in a polar solvent.

An example of a charge-transfer complex which shows this behaviour is the complex between N,N,N′,N′-tetramethyl-p-phenylenediamine (TMPD) and tetrachloro-p-benzoquinone (chloranil).[5-9] In cyclohexane solution a broad band with a maximum at $11,500 \text{ cm}^{-1}$ (870 nm), typical of a charge-transfer complex, is formed. In methanol solution this absorption is absent, but absorptions which are characteristic of the cation radical TMPD$^+$ (I) are observed, together with absorptions at $23,500, 22,100 \text{ cm}^{-1}$ (426, 452 nm), which have been assigned to the semi-quinone ion derived from chloranil (II) (Fig. 11.1). Further support for

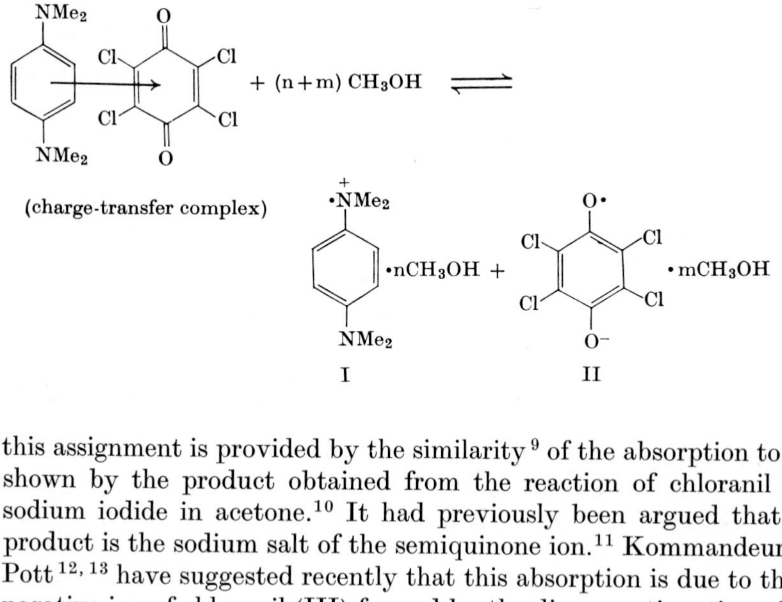

this assignment is provided by the similarity[9] of the absorption to that shown by the product obtained from the reaction of chloranil with sodium iodide in acetone.[10] It had previously been argued that this product is the sodium salt of the semiquinone ion.[11] Kommandeur and Pott[12, 13] have suggested recently that this absorption is due to the di-negative ion of chloranil (III) formed by the disproportionation of II*.

---

* Originating from studies of the solid complex between TMPD and chloranil, this assignment was based on measurements made on the solid sodium–chloranil compound and on electrolysed solutions of chloranil[12, 13] (see also Chapter 9).

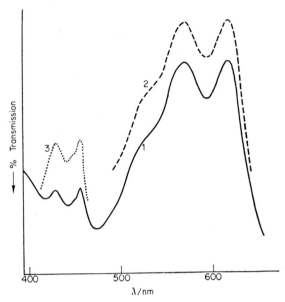

[R. Foster, Bulletin of the Spectrophotometric Group, No. 15, 413 (1963)]

Fig. 11.1. The absorption spectra of: (1) the solution obtained by dissolving the N,N,N′,N′-tetramethyl-*p*-phenylenediamine–chloranil complex in methanol; (2) a methanolic solution of TMPD$^+$ perchlorate; (3) a methanolic solution of the product obtained by the action of sodium iodide on a solution of chloranil in acetone.

Nevertheless, there seems little doubt that, initially at least, the TMPD–chloranil complex ionizes in solution to form A$^-$ and D$^+$.

$$\text{III}$$

The general type of behaviour exemplified by TMPD–chloranil is shown by many other complexes which involve relatively strong donors and acceptors.[9, 14] The spectra of mixtures of TMPD with each of a range of acceptors of increasing strength have been measured[9] (Table 11.1). In non-ionizing solvents only single broad maxima are observed, the frequencies of which are linearly related to the frequencies of the corresponding hexamethylbenzene complexes. This has been taken as evidence

TABLE 11.1. Absorption maxima of N,N,N',N'-tetramethyl-p-phenylenediamine (TMPD) acceptor systems (in nm).

Acceptor	Sol-vent*	Temp./ °C	TMPD complex				
			λ_1	λ_2	λ_3	λ_4	λ_5
2,4-Dinitrotoluene	C	Room					465
	M	Room					424
	A	Room					454
1,3-Dinitrobenzene	C	Room					500
	M	Room					454
	A	Room					480
2,4,6-Trinitro-m-xylene	C	Room					521
	M	Room					477
	A	Room					503
2,4,6-Trinitrotoluene	C	Room					558
	M	Room					510
	A	Room					544
2,4,6-Trinitrobiphenyl	C	Room					573
	M	Room					527
	A	Room					550
1,3,5-Trinitrobenzene	C	Room					614
	W	Room			569	617	
	M	Room			570	618	
	A	Room					583
Chloro-p-benzoquinone	C	Room					704
	W	Room			570†	617†	
	M	Room			569	617	
	M	−65°		433	569	617	
	A	Room			572†	617†	690
	A	−40°			572	617	692
2,3-Dichloro-p-benzoquinone	C	Room					761
	W	Room		438	568	616	
	M	Room			569	617	
	M	−65°		456	569	617	
	A	Room			571	620	750
	A	−40°		435	571	620	752
2,5-Dichloro-p-benzoquinone	C	Room					782
	W	Room		444	568	616	
	M	Room			568	616	
	M	−65°		451	568	616	
	A	Room			572	619	771
	A	−40°			572	619	771
2,6-Dibromo-p-benzoquinone	C	Room					779
	W	Room		449	568	616	
	M	Room		445	568	616	
	M	−65°		445	568	616	
	A	Room			572	619	769
	A	−40°		443	572	619	769

TABLE 11.1.—*continued*

Acceptor	Sol-vent*	Temp./°C	TMPD complex				
			λ_1	λ_2	λ_3	λ_4	λ_5
2,6-Dichloro-p-benzoquinone	C	Room					770
	W	Room	427	455	568	616	
	M	Room	427	453	568	616	
	M	−65°	427	453	568	616	
	A	Room	425	449	570	618	782
	A	−40°	425	449	570	618	782
Chloranil	C	Room					870
	W	Room	428	457	568	616	
	M	Room	428	452	568	616	
	M	−65°	426	452	568	616	
	A	Room	422	448	568	618	
	A	−40°	422	448	568	618	843
Tetrachloro-o-benzoquinone	C	Room					738‡
	W	Room			568	616	
	M	Room			568	616	
	M	−65°			568	616	
	A	Room			569	619	
	A	−40°			569	619	
Bromanil	C	Room					876
	W	Room	428	459	568	616	
	M	Room	430	455	568	616	
	M	−65°	430	455	568	616	
	A	Room	426	454	572	619	
	A	−40°	426	454	572	619	849
Iodanil	C	Room					882
	W	Room					
	M	Room			568†	618†	
	M	−65°			568†	618†	
	A	Room			571†	619†	
	A	−40°			571†	619†	
Tetracyanoethylene	C	Room					980‡
	W	Room					
	M	Room	Polybanded, 370–470				
	M	−65°	Polybanded, 370–470		568	618	
	A	Room	Polybanded, 370–470		570	619	
	A	−40°	Polybanded, 370–470		570	619	
Tetrabromo-o-benzoquinone	C	Room					737‡
	W	Room			568	616	
	M	Room			569	616	
	M	−65°			569	616	
	A	Room			569	619	
	A	−40°			569	619	

TABLE 11.1.—*continued*

Acceptor	Sol-vent*	Temp./ °C	TMPD complex				
			λ_1	λ_2	λ_3	λ_4	λ_5
7,7,8,8-Tetracyano-quinodimethane	C	Room					1163‡
	M	Room	744	844	570	617	
	M	−65°	744	843	570	617	
	A	Room	743, 760	844§	567	616	
2,3-Dichloro-5,6-dicyano-*p*-benzoquinone	W	Room		424	568	618	
	M	Room		396	568	618	
	M	−65°		400	568	618	
	A	Room		436	569	619	
	A	−40°		432	569	619	
TMPD+ as perchlorate salt	C	Room	Insoluble				
	W	Room			567	618	
	M	Room			568	618	
	A	Room			571	621	

* C = cyclohexane, W = water, M = methanol, A = acetonitrile.
† Bands less well resolved.
‡ Solutions in chloroform (too insoluble in cyclohexane).
§ Further maxima at 667, 681 and 683 nm.

in support of the assignment of these bands as intermolecular charge-transfer transitions. In methanolic solutions, mixtures of TMPD with the weaker acceptors—2,4-dinitrotoluene; 1,3-dinitrobenzene; 2,4,6-trinitro-*m*-xylene; 2,4,6-trinitrotoluene and 2,4,6-trinitrodiphenyl—also show a single band in the visible region in the anticipated position for an intermolecular charge-transfer transition. By contrast, mixtures of TMPD with 1,3,5-trinitrobenzene and with all the acceptors stronger than 1,3,5-trinitrobenzene (as measured by ν_{CT} in a non-ionizing solvent) show only absorptions characteristic of TMPD+ together with an absorption corresponding to an ion derived from the acceptor (cf. Fig. 11.1). For similar complexes involving donors weaker than TMPD, the onset of ionization in methanolic solution occurs for complexes of acceptors further along the scale of increasing strength.[14] For some complexes acetonitrile appears to be a solvent intermediate between cyclohexane and methanol, in that absorptions corresponding to the species A− (or A−−) and D+ are observed simultaneously with the inter-molecular charge-transfer absorption.[9, 14] Here the equilibrium repre-sented by equation (11.1) is fairly well-balanced compared with the solvolytic action of methanol or water. The equilibrium in acetonitrile can be shifted more in favour of the charge-transfer complex by reducing the temperature. Originally, a slow ionization was proposed for some of these systems.[9] The suggestion was based on the observation that the charge-transfer band disappeared in time, whilst the intensity of the TMPD$^+$ absorption increased. This conclusion is almost certainly incorrect;[15] the behaviour of these solutions can be completely accounted for by the slow, irreversible, further reaction of A$^-$ (or A^{--}) which upsets the equilibrium described by equation (11.1).

Another example of the solvation of a charge-transfer complex which causes ionization is given by the ferrocene–tetracyanoethylene system. In cyclohexane solution a charge-transfer complex is formed,[16] whereas in more polar solvents, such as acetonitrile, the ferricinium ion and the radical ion of tetracyanoethylene are formed.[17] Similar results have been obtained with nickelocene, ruthenocene and osmocene.[16] Liptay et al.[18] have made quantitative measurements on the system TMPD–tetracyano-ethylene in ethylene glycol dimethyl ether, in tetrahydrofuran and in benzotrifluoride. In these solvents of medium dielectric constant, they have estimated the association constants for the various equilibria between the free ions, ion-pairs and the neutral charge-transfer complex species. Tetracyanoethylene in dimethylsulphoxide solution immediately forms a charge-transfer complex.[19] With time, an e.s.r. absorption, attributable to the tetracyanoethylene radical anion, develops. No absorption from the corresponding cation of dimethylsulphoxide is observed. Its absence could be due to electron exchange with other molecules of dimethylsulphoxide in the bulk of the solvent. Recently, confirmation of the solvolytic nature of the ionization has been obtained by studying the reverse process.[20] The perchlorate salt of TMPD$^+$ has been dissolved together with the sodium salt derived from the chloranil semiquinone ion in acetonitrile. When the dielectric constant of the solvent is reduced by adding carbon tetrachloride, the absorptions due to the ionic species diminish and the broad charge-transfer band character-istic of the neutral complex develops. The spectral changes are reversed by further addition of acetonitrile. The formation of free radical ions by TMPD–chloranil system has been elegantly confirmed by two groups of workers.[7, 8] Whilst no e.s.r. absorption is observed for solutions of TMPD–chloranil in solvents of low ionizing power such as chloroform, strong absorption is obtained for acetonitrile solutions. Immediately after mixing, this consists of a singlet, assigned to the chloranil semiquinone ion, superimposed on a set of thirteen triplets attributed to TMPD$^+$ (Fig. 11.2). This absorption had been previously characterized in TMPD$^+$ ClO$_4^-$. The relative instability of the chloranil semiquinone ion compared with TMPD$^+$, which was remarked upon in discussing the

11

optical spectra, is also reflected in the collapse of the singlet absorption in the e.s.r. spectrum after a time (Fig. 11.2).

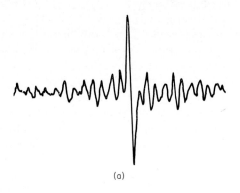

(a)

(b)

[J. W. Eastman, G. Engelsma and M. Calvin, Ref. 7]

FIG. 11.2. The derivative of the e.s.r. absorption obtained by dissolving equi-molecular amounts of N,N,N′,N′-tetramethyl-p-phenylenediamine and chloranil in acetonitrile; reaction time: (a) 25 min; (b) 1 h.

The electron donor tetrakis(p-dimethylaminophenyl)-ethylene (IV) has even greater potentialities for delocalizing a formal charge than has TMPD. Thus, whilst IV with chloranil shows only a broad band of the

IV

neutral charge-transfer complex in cyclohexane solution, and an absorption characteristic of the mono-positive ion in 1,2-dichloroethane (a solvent of medium dielectric constant), the optical absorption indicates that the original donor molecule exists mainly as the di-cation in methanolic solutions.[14] Earlier, the formation of the mono-cation of IV had been observed when the di-iodide salt of IV was dissolved in 1,2-dichloroethane, whereas in water it existed as the di-cation.[21, 22] The e.s.r. measurements on these solutions are in harmony with these conclusions. Similar results are obtained by dissolving the base IV and molecular iodine in 1,2-dichloroethane and subsequently diluting this solvent with methanol. If, on the other hand, the 1,2-dichloromethane solution is diluted with carbon tetrachloride, the charge-transfer band of the neutral complex is observed.

In a study of the unsymmetrical redox reaction between hydroquinone and chloranil, no evidence for electron transfer between the two neutral molecules was obtained.[23] There appears to be a two-stage electron-transfer process involving the hydroquinone negative ion and chloranil in alkaline solution, and electron transfer from hydroquinone to the protonated form of chloranil in acid solution. Although there is an immediate formation of the charge-transfer complex, it has not been possible to decide whether the complex is involved in the reaction co-ordinate.

II.B. Non-Photochemical Reactions which may proceed through Charge-Transfer Complexes

The products of the single electron-transfer reaction between TMPD and chloranil in polar solvents, described above, do undergo further irreversible reaction, although the chemistry involved has not, as yet, been studied in detail. However, the end-products of the somewhat similar donor–acceptor pair, N,N-dimethylaniline–chloranil have been shown to include crystal violet (V). This reaction has been studied in detail by Calvin and his co-workers.[7] Solutions of the acceptor in N,N-dimethylaniline, in the absence of a diluting solvent, were used. Under these conditions virtually all the chloranil is complexed and the broad absorption in the anticipated position at ~15,400 cm^{-1} (~650 nm) indicates the initial formation of a neutral charge-transfer complex. With time, the intensity of this band decreases, whilst a new absorption with a maximum of 24,500 cm^{-1} (410 nm) develops. Concurrently, the electrical conductivity of the solution rises rapidly and the solution becomes paramagnetic, the e.s.r. signal of which corresponds to the chloranil semiquinone ion. After a period of days, the absorption at

24,500 cm^{-1} diminishes and somewhat later the paramagnetism of the solution declines. The electrical conductivity of the solution is maintained, however. The solution finally contains crystal violet (V) as its cation. Although no e.s.r. absorption corresponding to the N,N-dimethylaniline cation is observed at any stage in the reaction, this absence does not rule out the existence of this radical. Indeed, one might not expect to see a signal since the exchange between the radical cation and the large excess of neutral N,N-dimethylaniline molecules present would give rise to such band broadening as to make the signal experimentally undetectable. Kosower[1] has proposed a detailed mechanism for the reaction in which electron-transfer to yield the ions VI and VII is followed by a hydrogen-atom transfer reaction (scheme 11.3).

It has been suggested that the reaction of chloranil with triethylamine proceeds via a charge-transfer complex and subsequent electron-transfer to give the two radical ions, which then react further. This reaction has been studied by Henbest and his co-workers,[24] who have shown that the end product is N,N-diethylaminovinyl-trichloro-p-benzoquinone (VIII). There is no direct evidence that a charge-transfer complex is in fact formed, let alone support for the hypothesis that the reaction proceeds via a charge-transfer complex. A solution of chloranil in triethylamine rapidly turns green and a green solid may be isolated from the solution by evaporation of the excess triethylamine.[25] This solid has an infrared absorption which is nearly identical with the sodium derivative obtained by the action of sodium iodide on chloranil, which had been considered to be the salt of the semiquinone ion but which Pott and Kommandeur[12, 13] have suggested is a mixture of the dianion and free chloranil.* The e.s.r. spectra of both the solid and the solution in excess triethylamine show a singlet which has been assigned to the chloranil radical ion. No signal has been observed which could be attributed to the counter-ion. If this ion were the amine radical $\cdot$NEt$_3^+$, the inability to observe a signal in solution could be explained in terms of band-broadening through a rapid electron-exchange reaction with other molecules of triethylamine. The absence of such a signal in the solid is more difficult to explain. Nevertheless the proposed mechanism[1, 24] (11.4) is a plausible pathway. Lucken and his co-workers[26] have studied the reaction of triphenylphosphine with chloranil. They suggest that IX may be formed from the charge-transfer complex; this is also oxidized by chloranil to VII and X.

The reactions of tetracyanoethylene with N-alkylanilines have been studied in detail by Rappoport and Horowitz.[27, 28] In chloroform solution, for example, tetracyanoethylene with N,N-dimethylaniline instantaneously forms a blue charge-transfer complex with a maximum

* See footnote on p. 304.

(11.3)

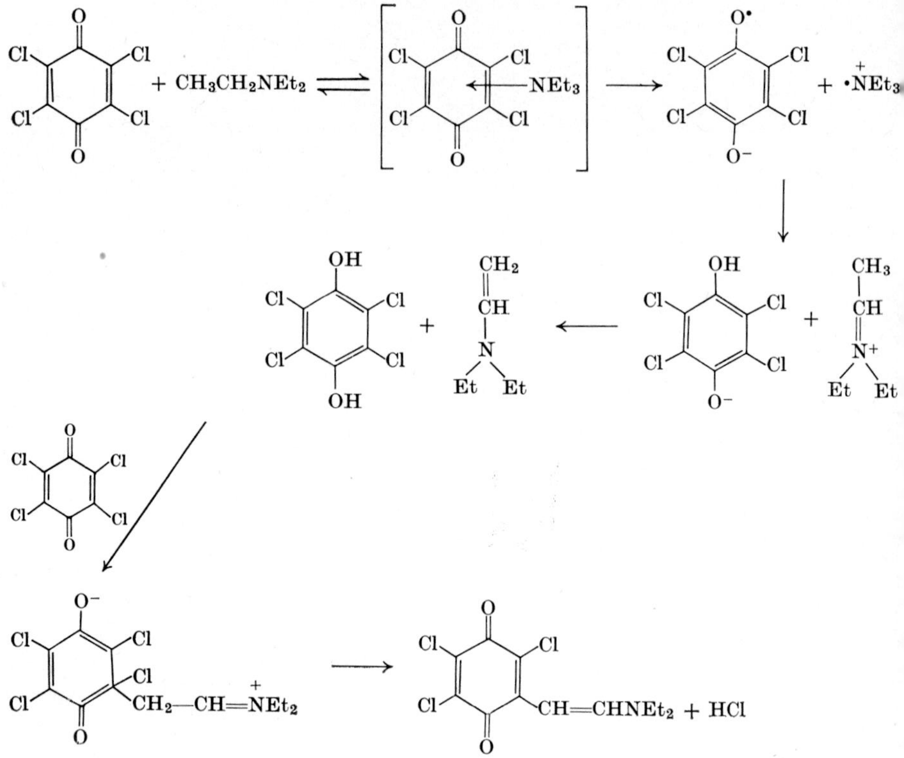

$$+ CH_3CH_2NEt_2 \rightleftharpoons \quad \longrightarrow \quad + \cdot NEt_3$$

VIII   (11.4)

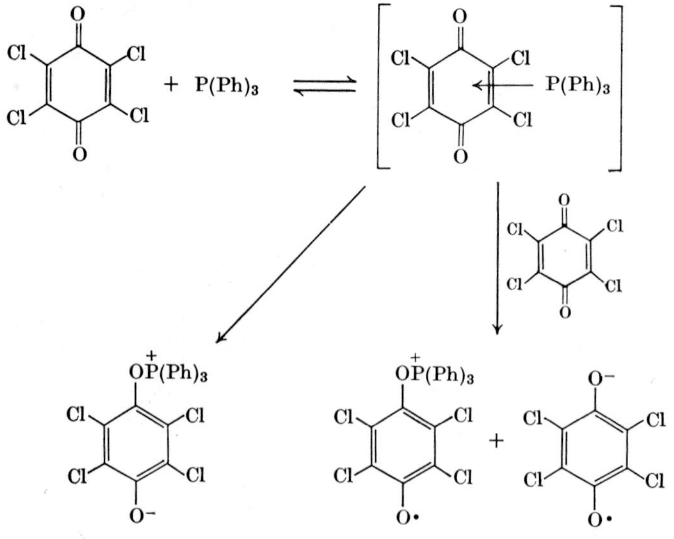

$$+ P(Ph)_3 \rightleftharpoons$$

IX   X   VII   (11.5) at 14,800 cm^{-1} (675 nm). The intensity of the band gradually diminishes as an intermediate develops which, in its turn, forms N,N-dimethyl-4-tricyanovinylaniline (XI) by a very slow reaction. A similar sequence is observed for the reaction of N-methylaniline with tetracyanoethylene although, in this case, the decomposition of the charge-transfer complex is fast and the formation of the final product is slower than in the reaction of N,N-dimethylaniline. Rappoport suggests that the intermediate formed is a σ-complex, for example XII from N,N-dimethylaniline. The kinetics were studied under conditions where there was a large excess of base. The various possible reaction schemes for the formation of the σ-complex include: (a) the mono-molecular transformation of the charge-transfer complex alone; (b) the bimolecular reaction of uncomplexed tetracyanoethylene with the aniline; (c) the reaction of all the tetra-cyanoethylene, both free and complexed, at the same rate in a bimolecu-lar reaction with the aniline; (d) the bimolecular reaction of the charge-transfer complex with a molecule of the free aniline; (e) the mono-molecular reaction (a) with the bimolecular reaction (b); and (f) the bimolecular reactions (b) and (d) proceeding concurrently but at different rates. The kinetic behaviour of the decomposition of the charge-transfer complex under the pseudo-first-order reaction conditions used was consistent with either (d) or (f) for the reaction involving N-methylaniline. Although the kinetics cannot distinguish between these two possibilities, it may be concluded that the reaction is proceeding *at least in part* via the charge-transfer complex. The kinetic data for the reaction of N,N-dimethylaniline with tetracyanoethylene are con-sistent with scheme (d). The suggested mechanism is given below. Rappoport's structure for the intermediate formed from the charge-transfer complex has been questioned by Kosower.[1] He has pointed out

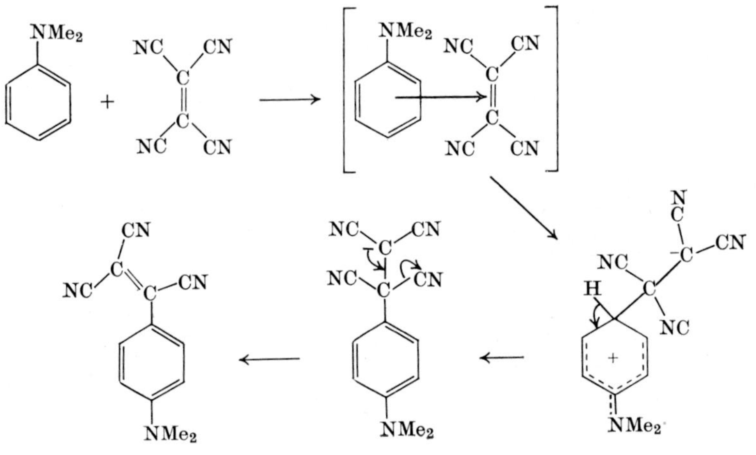

XI   XII   (11.6)

that such a σ-complex might be expected to show reversible decomposition to the charge-transfer complex whereas no reversibility is observed. Furthermore, this intermediate appears to be extractable into water, which Kosower suggests is unlikely for a species with the structure XII. The alternative structure proposed by Kosower is the bicyclo-[4.2.0]-octane derivative (XIII) formed via the σ-complex (XII). However, the properties of the intermediate, which has now been isolated as a

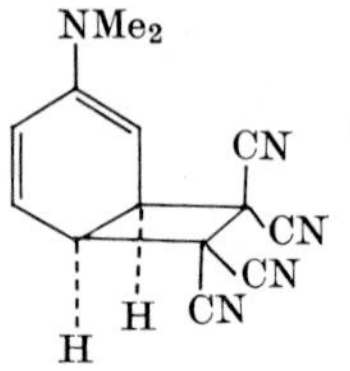

XIII

solid,[29] support Rappoport's proposal that the intermediate is the σ-complex (XII) (see also Isaacs [30]).*

The reaction of indole with tetracyanoethylene in dichloromethane follows a course exactly analogous to the reactions involving N-methylaniline.[31] The final product has been identified as 3-tricyanovinyl-indole.[32]

The 1,2-cyclo-addition products of the type proposed by Kosower for the intermediate in the N-alkylaniline–tetracyanoethylene reaction (XIII) are observed in the reaction between tetracyanoethylene and various alkenes.[33] On mixing the reactants, coloured solutions are immediately obtained. These colours, doubtless due to charge-transfer complex formation, fade with time. From the aged solutions, 1,1,2,2-tetracyano-3-X-cyclobutanes (XIV) may be isolated [X may be RO-, RS-, $R(R'CO)N$-, $C_6H_5SO_2N(R)$- or p-ROC_6H_4-].

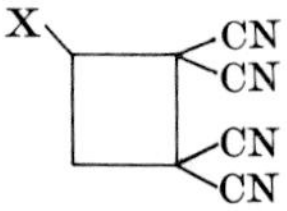

XIV

Dihydropyran gives the bicyclic adduct (XV) and 4-methylene dioxolane forms the spiro compound XVI. However, with 1-methoxy-1-3-butadiene, tetracyanoethylene gives the Diels-Alder product 1,1,2,2-tetracyano-3-methoxy-4-cyclohexene (XVII). It would appear that, when either cyclo-addition or Diels-Alder addition is possible, the latter

* Fresh doubt has been cast on the structure XII by recent observations [Z. Rappoport and E. Shohamy, *J. chem. Soc.* (B), 77 (1969)] on the reaction of tetracyanoethylene with 2,6-dimethylaniline.

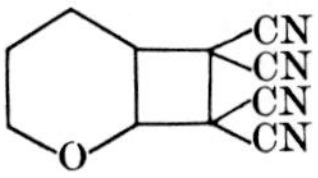

XV 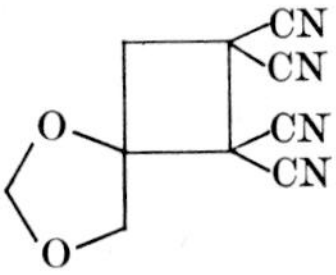   XVI 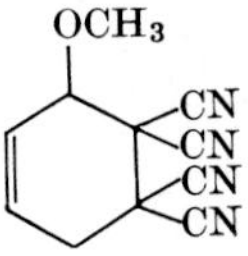   XVII

is preferred. In these and other addition reactions between electrophiles and tetracyanoethylene, there is as yet no direct evidence for the participation of a charge-transfer complex in the reaction sequence.

The electron acceptor 2,3-dichloro-5,6-dicyano-p-benzoquinone (DDQ) forms charge-transfer complexes with a variety of donors. In the course of a preliminary study of such interactions, it was noticed that, when a polymethylbenzene was used as the donor, the colour due to the complex slowly faded.[34] Condensation products of DDQ and the donor were isolated from the aged solutions. For example, from a mixture of DDQ and hexamethylbenzene in chloroform solution, the compounds XVIII and XIX were isolated. An analogous reaction has been reported[35] for 2,2-dimethylindane, in which the product XX is formed. The mechanisms of the reactions between DDQ and methylbenzenes have not been

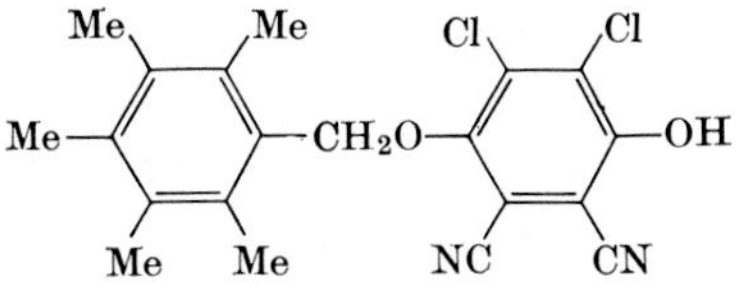

XVIII

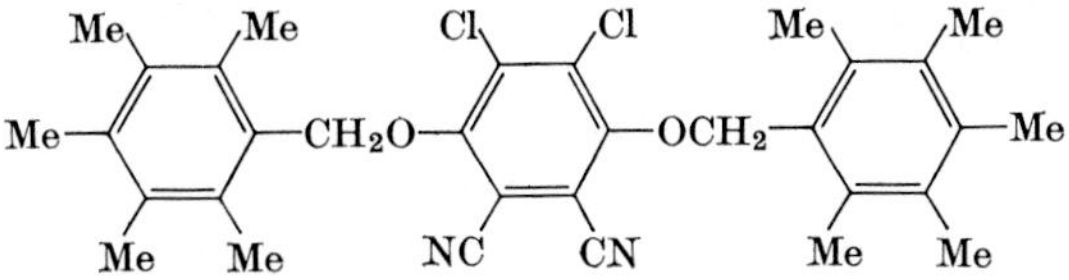

XIX

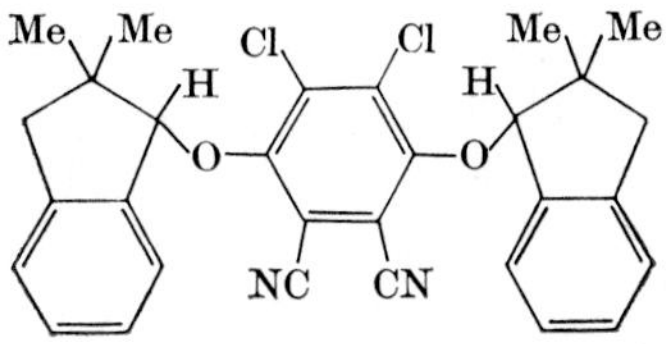

XX

11*

studied in such detail as the reactions of tetracyanoethylene with anilines. There is no doubt concerning either the initial formation of a charge-transfer complex or of the structures of the final products. The original data for the decomposition of the charge-transfer complex have been accounted for in terms of one of three of Rappoport's[27] proposed schemes (see p. 315), namely: the reaction of uncomplexed acceptor with donor; the monomolecular decomposition of the charge-transfer complex; or a combination of these two processes occurring simultaneously but at different rates. However, more recent experimental results on the same systems have given somewhat different, though less well-reproducible kinetic results.[36] These reactions of methylbenzenes with DDQ are still currently (1969) under investigation.[36]

Tropylium 2,3-dichloro-5,6-dicyano-1,4-hydroquinolate is rapidly formed by the action of DDQ on cycloheptatriene.[37] Analogous reactions have been observed with tetrachloro-o-benzoquinone and with chloranil. Reid and his co-workers,[37] have used this reaction to prepare a variety of tropylium salts by treatment with the appropriate acid in acetic acid solution. Under similar conditions, phenalene yields the phenalium ion XXI, whereas in the absence of strong acids the radical XXII is

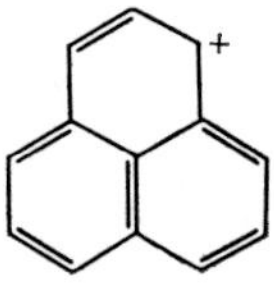

XXI 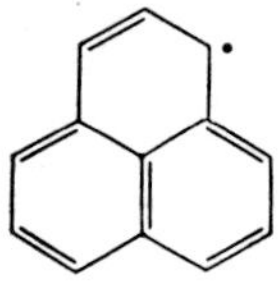   XXII

formed. In all these reactions the hydride-transfer may occur within the initially-formed charge-transfer complex, though at present there is no proof of such a mechanism.

The complex of anthracene with DDQ, when dissolved in solvents such as 1,2-dichloroethane, is unstable. Although no quantitative measurements have been made on this system, the product appears to involve an addition of the quinone across the 9,10-positions of anthracene to give a Diels-Alder-type product.[38] The reaction may bear some similarity to the reaction of anthracene with tetracyanoethylene, where again a Diels-Alder product is formed.[39-41] The present experimental data[41] do not permit a decision to be made about the possible participation of the charge-transfer complex. Nevertheless, in this latter reaction, the probable conformation of the complex (XXIII) is likely to be so similar to that of the Diels-Alder adduct XXIV, that it seems reasonable to propose a reaction proceeding via the complex. Details of possible

XXIII   XXIV

transition states between the charge-transfer complex and the adduct have been discussed by Brown and Cookson.[40] An interesting observation from this work is the evidence of lowering of the effective concentration of tetracyanoethylene by its complexing with the solvent.

There is ample evidence (see, for example, Chapter 6) for charge-transfer complex formation between iodine and amines, both aliphatic and aromatic. There is also no doubt that within these systems irreversible reactions can occur. Indeed, there is often considerable difficulty in minimizing these subsequent reactions in experiments designed to study the initial equilibrium of the free components with the charge-transfer complex, particularly when using aliphatic amines.[42, 43] Many of these further reactions have been studied to some extent. For example, fresh solutions of iodine in pyridine contain the 1 : 1 charge-transfer complex in equilibrium with the free components (Chapter 6). With time the electrical conductivities of the solutions become significant.[44] In these solutions the triiodide ion has been detected.[45, 46] The cation has been variously assigned the structures XXV and XXVI by different groups of

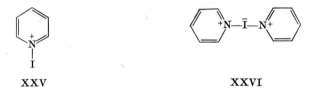

XXV   XXVI

workers.[46-50] The crystal structure of the solid isolated from such solutions contains XXVI with the linear structure indicated.[51] There appears to be no direct evidence to decide whether or not this type of ionization occurs through the charge-transfer complex or not. The kinetics of the formation of the quaternary iodonium triiodide salts from triphenylarsine and from triphenylstilbene have been studied by Rao and Bhat.[52] Here again, there is no complete proof that the reaction proceeds via the charge-transfer complex, however likely this may appear. The comment by Kosower[1] that such systems deserve additional study, is well made.

There is evidence that, although aromatic electrophilic substitution reactions may proceed via charge-transfer complexes, the formation of these complexes is not a rate-controlling step. Thus Brown and Brady[53] showed that, whilst there was a good correlation between the rate of halogenation of alkylbenzenes and the stabilities of the σ-complexes formed by these alkylbenzenes with $BF_3 + HF$, no such correlation existed if comparisons were made with the stabilities of weak charge-transfer complexes. This argued in favour of the rate-determining transition state being more closely represented by a σ-complex (XXVII)

 (The charge or otherwise of the initial electrophile will determine whether this species is charged).

XXVII

than by a charge-transfer complex. More recently, measurements by Olah and his co-workers[54] on the relative rates of nitration of alkylbenzenes by nitronium tetrafluoroborate suggested that, in this group of reactions, the charge-transfer complex was dominant in the rate-controlling step. The relative rates were obtained by competitive rate determinations. The differences in rate for the various substrates were found to be remarkably small: for example, the toluene/benzene ratio for nitration by nitronium tetrafluoroborate in sulpholane was 1·67 compared with 21–27 when a normal nitrating mixture was used. Tolgyesi[55] has pointed out that, for such fast reactions, the competitive method may yield incorrect values, since the reaction mixture may not remain homogeneous. The apparent correlation reported by Olah and his co-workers[54] may therefore be an artifact of their experimental method. Desolvation processes may also be important when sulpholane is used.

Dewar[56–58] has suggested that a "π-complex" might be formed in a rate-controlling step in aromatic electrophilic substitutions. Although the *general* structure which he attributes to his π-complex is identical with that which is described here as a charge-transfer complex, namely, an interaction between the lowest-vacant orbital of the acceptor species with the highest-filled molecular orbital of the aromatic substrate. Dewar considers that the energy of this interaction is much higher than the normally fairly weak binding which is usually assumed, and is effectively a *full covalent bond* involving these two orbitals. Dewar considers the energetics of "π-complex" formation involving electrophilic reagents with olefins to be of similar magnitude. In this respect his suggestions concerning the role of π-complexes in chemical reactions have been largely misinterpreted by other workers. Comprehensive accounts of his views have been given.[57, 58]

Andrews and Keefer[2] have considered the participation of the more conventional weak charge-transfer complexes in aromatic halogenation reactions. They have suggested that the charge-transfer complex may react with a third substance X to form the σ-complex (XXVIII). This substance X, which needs to be electrophilic in character, may be another molecule of the reagent, or a molecule of solvent. Arguments in favour of the scheme (11.7) have been given by Andrews and Keefer.[2]

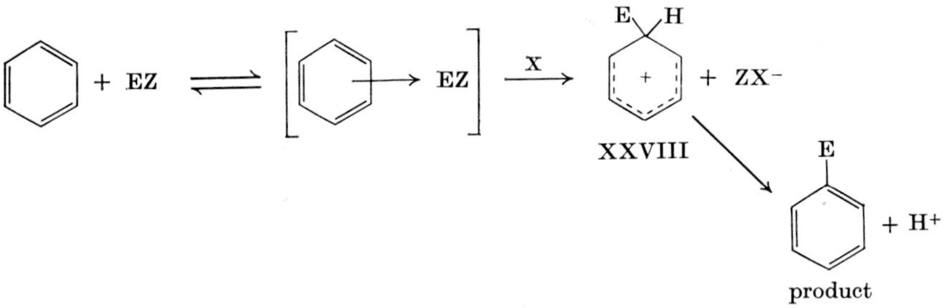

XXVIII

product

(11.7)

The reaction of a *non-polar* halogen with an aromatic substrate does not occur in a *non-polar* solvent unless a polar additive (X) is present.[59-62] When the polar halogen iodine monochloride is used, which reagent can act as the substance X, the halogenation will proceed in a non-polar solvent (carbon tetrachloride). In this case the reaction is third-order with respect to the halogen.[63] By contrast, this type of reaction is second-order with respect to halogen in acetic acid,[64] and first-order with respect to halogen in trifluoroacetic acid.[63] These observations are interpreted in terms of the increasing capacity of the solvent to assume the role of X in scheme (11.7). The third-order kinetics with respect to iodine monochloride are reasonably accounted for if it is assumed that a charge-transfer complex between iodine monochloride and the aromatic substrate is formed, together with a dimeric complex of two iodine monochloride molecules, and that the rate-determining step involves the interaction of these two complexes.[2] The experimental data show a good quantitative fit with this scheme, although they do not provide an unambiguous proof of the mechanism. Senkowski and Panson[65] have suggested that the iodination of N,N-dimethylaniline involves a charge-transfer complex (XXIX) which formed the ion (XXX) from which the iodonium ion is formed which attacks a molecule of N,N-dimethyl-aniline to form the σ-complex (XXXI) and thence N,N-dimethyl-*p*-iodoaniline.

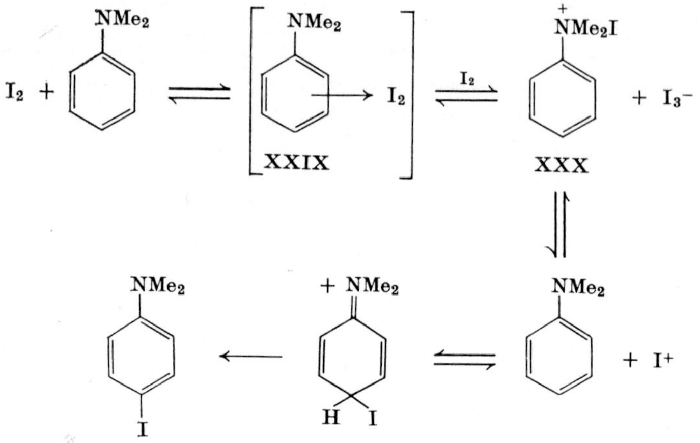

$$(11.8)$$

The possible participation of charge-transfer complexes in the halogenation of olefins has been considered by Dubois and Garnier.[66-69] Although they have shown the existence of transitory charge-transfer complexes (XXXII) evidence for their participation in the mechanism of halogenation, as represented, for example, in (11.9), is circumstantial.

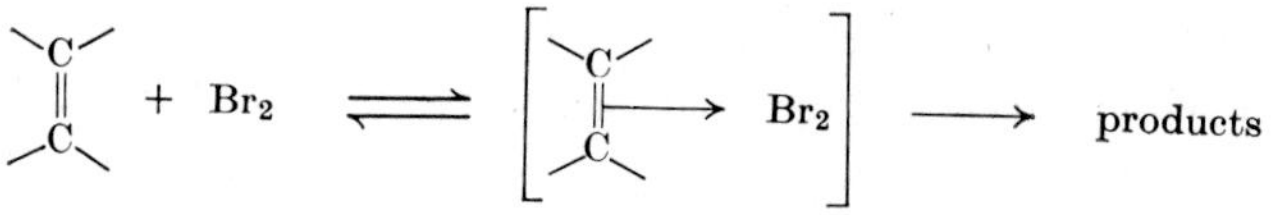

Aliphatic amines react with 1,3,5-trinitrobenzene even in solvents of only moderate ionizing ability to form σ-complexes.[70-76]* In many systems further reactions ensue.[77] In solvents of very low ionizing power, 1,3,5-trinitrobenzene does appear to form charge-transfer complexes with aliphatic amines when the amine is in low concentration.[78-80] At high amine concentrations, σ-complexes, either with a zwitterionic structure (XXXIII), or as a charged structure (XXXIV), may be formed. Excess base may solvate and so stabilize these σ-complexes. The optical spectra of such species are characteristic and differ considerably from intermolecular charge-transfer transitions, both in their band

* For reviews of σ-complexes formed from polynitroaromatics, see Refs. 74 and 75.

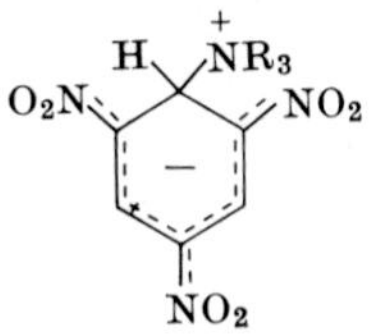

XXXIII 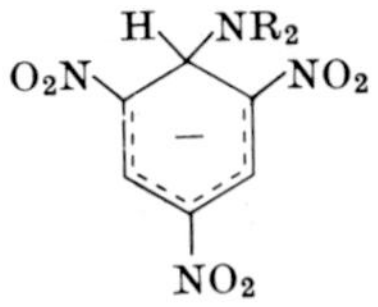   XXXIV

shape, and in their lack of linear dependence of band position on ionization potential of the amine component. The absorptions which have
been assigned as charge-transfer transitions in solvents of low ionizing
power are in the ultraviolet, at higher energies than the absorptions of the
σ-complexes, which occur in the visible region and give rise to the red
colour of these complexes. Although charge-transfer complexes may be
formed initially, it is not yet clear whether, if they are formed, they make
a significant contribution to the measured kinetics of the reaction at any
stage.

Ross and Kuntz[81] have interpreted the decrease in bimolecular rate
constants for the reaction between 2,4-dinitrochlorobenzene and aniline,
as the concentration of aniline was increased, in terms of charge-transfer
complex formation.

Several workers have demonstrated the effect of charge-transfer
complexing of one of the reactants by a species which is not itself a
reactant. Examples where the reactivity of the complexed reactant is
less than that of the free reactant have been reported; for example, the
competition of solvent acting as a donor in the tetracyanoethylene–
anthracene reaction,[40] which has already been mentioned. Similarly, it
has been shown that various carboxylate esters when complexed with the
3,5-dinitrobenzoate ion are unreactive towards hydroxyl ion.[82]

More impressive are cases in which the *complexed* reactant has an
enhanced reactivity. Thus Colter and Clemens[83] have observed that the
rates of racemization of (+)9,10-dihydro-3,4,5,6-dibenzophenanthrene
and of (+)1,1'-binaphthyl are accelerated by complex formation with
2,4,7-trinitrofluorenone. The enhancement of the rate of acetolysis of
2,4,7-trinitro-9-fluorenyl-p-toluenesulphonate (XXXV) by the presence

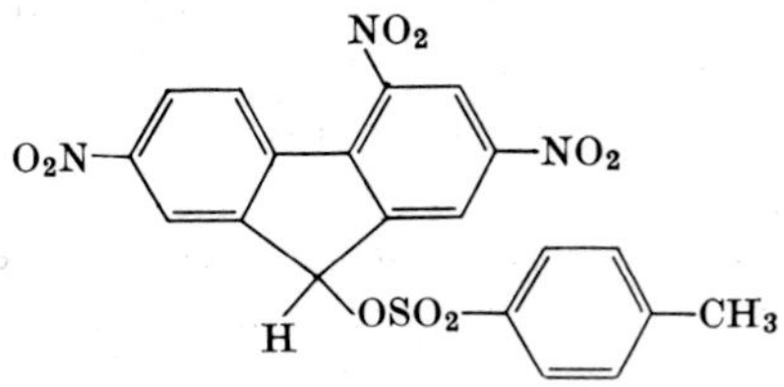

XXXV

of various aromatic donors has been measured.[84, 85] It was assumed that acetolysis of both the uncomplexed substrate (ROTs) and the complexed substrate (ROTs.D) occurs. The mechanism may be formulated as:

$$\text{ROTs} + \text{D} \quad \underset{\xleftarrow{\hspace{1cm}}}{\xrightarrow{\hspace{1cm}} K_T} \quad \text{ROTs.D}$$

$$\text{HOAc} \updownarrow K_u{}^* \qquad\qquad \text{HOAc} \updownarrow K_c{}^*$$

$$\text{(uncomplexed transition state)} + \text{D} \quad \underset{\xleftarrow{\hspace{1cm}}}{\xrightarrow{\hspace{1cm}} K^*} \quad \text{(complexed transition state)}$$

$$\text{HOTs} + \text{ROAc} + \text{D} \quad \underset{\xleftarrow{\hspace{1cm}}}{\xrightarrow{\hspace{1cm}} K_A} \quad \text{ROAc.D} + \text{HOTs}$$

$$(11.10)$$

The activation parameters for the phenanthrene-catalysed reaction in glacial acetic acid have been compared with the values for the un-catalysed reaction. The rates increase significantly as the phenanthrene concentration increases. Within the temperature range 56–85°C, the complexed substrate is twenty-one to twenty-seven times as reactive to acetolysis as the uncomplexed substrate. The experimental data indicate that this difference is due mainly to an entropy effect, the complexed transition-state having a lower standard free energy than the uncomplexed transition-state because of a higher entropy factor. It has been suggested that the neighbouring donor molecule reduces the extent of ordering of the solvent molecules in the transition state. Support for the proposal that these effects are the result of specific complex-formation rather than some general solvent effect comes from the observation that hexaethylbenzene has virtually no effect on the rate. It is well established that this molecule is a very poor donor because of the steric effect of the ethyl groups, which prevents the sufficiently close approach of an acceptor molecule (see Chapter 6). Colter and his co-workers have also studied [86] the rate enhancements of the acetolysis of some dinitro-9-fluorenyl and dinitrobenzhydryl-p-toluenesulphonates. These show a similar behaviour although, as expected, the effects are not so large as with the stronger acceptor 2,4,7-trinitro-9-fluorenyl-p-toluene sul-phonate.

Kinetics of the recombination of iodine atoms in various gases argue the formation of an intermediate complex between the iodine-atom and the gas molecule or ion (the so-called chaperon). The heat of formation is considerably greater than that expected for a van der Waals-type complex and it is suggested that, in fact, the intermediate complex is a charge-transfer complex.[87-90] A similar explanation has been given for iodine-atom recombination in liquid solution, and, in terms of chlorine-atom complexes, for the solvent effects in certain chlorinations.[91] Estimates of the degree of association of iodine atoms with o-xylene in the liquid phase have been made by flash photometry.[92] From kinetic studies of the inhibition of hydrocarbon oxidation by aromatic amines and phenols, the reversible formation of charge-transfer complexes between inhibitor and alkylperoxy radicals has been postulated,[93-95] although measurements indicate that, at least in the case of the cumyl-peroxy radical–pyridine complex, the equilibrium constant is small.[96] Estimates of the equilibrium constant of this radical with triethylamine and with tri-n-butylamine are considerably higher.[97]

The polymerization of N-vinylcarbazole by the addition of one of a variety of electron acceptors has been described.[98, 99] These acceptors include chloranil, o-chloranil, bromanil, 2,3-dichloro-5,6-dicyano-p-benzoquinone, 1,4,5,8-tetrachloro-anthraquinone, 7,7,8,8-tetracyano-quinodimethane, tetracyanoethylene, 1,3,5-trinitrobenzene, 2,4-dinitro-chlorobenzene, trichloroacetonitrile, pyromellitic dianhydride, iodanil, p-benzoquinone, trichloroethylene, tetrachloroethylene, tetranitro-methane, and maleic anhydride. It has been suggested[98] that a charge-transfer complex is formed, which undergoes electron transfer to produce a pair of radical ions; the carbazole cation radical then reacts successively with molecules of free N-vinylcarbazole to form the polymer.

$$M + A \; \rightleftharpoons \; \underset{\text{charge-transfer}}{(M.A)} \; \rightleftharpoons \; M^{\cdot +} + A^{\cdot -}$$

$$M^{\cdot +} + nM \; \longrightarrow \; M^{\cdot +}_{n+1} \; \underset{-A^{\cdot -}}{\overset{+A^{\cdot -}}{\rightleftharpoons}} \; M^{+}_{n+1} \, O^{\cdot -} \tag{11.11}$$

There is, as yet, no final vindication of this plausible mechanism. The polymerization of N-vinylcarbazole and of alkylvinyl ethers has also been initiated by the electron acceptors: tropylium, pyrylium, N-methyl-acridinium, N-methylphenazonium and xanthylium.[100, 101]

The electron acceptor 2,4,6-trinitrostyrene and the electron donor 4-vinylpyridine copolymerize exothermically at room temperature.[102] Products of molecular weight 6000–8000 are obtained in about 60% yield. The presence, or absence, of oxygen appears to make no difference. Again, it is reasonable to postulate a mechanism involving the formation of a charge-transfer complex followed by electron transfer.

The participation of a charge-transfer complex in the copolymerization of p-dioxene and maleic anhydride has been suggested by Iwatsuki and Yamashita.[103]

II.C. Photochemical Reactions which may proceed through Charge-Transfer Complexes

Whereas the ground states of weak charge-transfer complexes contain only small contributions from dative structures, in the excited states dative structures should be dominant. Thus, suitable irradiation should transform a weakly bonded complex, in which the component molecules are fairly widely separated, into an effectively ionic complex, in which the molecular components are closer together. Both these changes should be conducive to further irreversible chemical reaction in suitable systems.

The simplest reaction which could ensue is photo-ionization of the complex, that is, a splitting of the complex in the excited state. For a charge-transfer complex, derived from neutral singlet-state donor and acceptor species, the two fragments formed will be free radical ions. This ionization, which has been seen in an earlier section in this chapter to occur when there is a sufficient energy-gain through solvation, may also be achieved in suitable systems by irradiation within the charge-transfer band. In this way the free radical anions of chloranil, 1,3,5-trinitrobenzene, tetracyanoethylene and pyromellitic dianhydride have been identified by irradiation of solutions of these acceptors in tetra-hydrofuran.[3, 104, 105] The absence of an observed spin signal from the tetrahydrofuran cation radical may be explained in terms of the band broadening which will arise through the exchange of this positive hole with other molecules of tetrahydrofuran throughout the solution. Observations of photo-conductivity confirm the ionic nature of the products.[105] The recombination processes have also been studied. An example of photo-excitation from an ionic ground state to a non-ionic excited state to produce two *neutral* radicals has been demonstrated by Kosower and Linqvist[106] by the flash photolysis of pyridinium iodides.

Related to the photo-ionization of a charge-transfer complex are the interesting observations of Simons and Tatham.[107] A solution of carbon tetrabromide and an electron donor such as diphenylamine, N-methyl-diphenylamine or triphenylamine in an isopentane–methylcyclohexane solvent mixture is cooled to 77°K. If the resulting glass is irradiated within the charge-transfer band, at which energies the components are transparent, colour centres develop. There is good evidence that these centres are due to electrons trapped in potential wells, formed by aggregates of carbon tetrabromide in the frozen glass.

The photo-addition of maleic anhydride to benzene has been shown by two groups of workers[108, 109] to form the 2:1 adduct XXXVI. The observation that radiation of energy below *ca.* 35,700 cm^{-1} (~280 nm) is effective, provides good evidence that the reaction is proceeding via

XXXVI

the charge-transfer complex rather than by a photo-excited state of either maleic anhydride or benzene, since neither component has any significant absorption below 35,700 cm^{-1}, whereas the major part of the intermolecular charge-transfer band is at lower energies than this. Furthermore, although the product was observed when solutions of maleic anhydride in benzene, in the absence of a diluting solvent, were irradiated, no product was observed when solutions in cyclohexane were used.[108] In the latter case it was estimated that only 3% of the maleic anhydride was complexed, compared with 32% in the former case. The observation that molecular oxygen did not inhibit the addition significantly, suggests that the reaction is proceeding via an excited singlet, rather than a triplet level. In this and many related reactions, benzophenone can act as a photo-sensitizer:[108, 110, 111] here the participation of triplet states is probable. The behaviour of substituted benzenes with maleic anhydride on irradiation has also been studied.[112] It was found that photo-addition was only observed in systems where an intermolecular charge-transfer transition occurred. However, this in itself does not appear to be a sufficient condition. Thus, durene and naphthalene both show intermolecular charge-transfer transitions, but in neither case has an adduct been obtained.[112] In the same paper it was reported that hexamethylbenzene plus maleic anhydride in cyclohexane solution also showed a charge-transfer band but gave no adduct. However, more recently, a photochemical reaction of a methylcyclohexane solution of these two components, using suitably filtered radiation, has been reported.[113] The products included pentamethylbenzylsuccinic anhydride (XXXVII). The suggested mechanism of a proton-transfer reaction of the two components in the excited state of the charge-transfer complex (XXXVIII)[114] to yield the radicals XXXIX and XL appears reasonable. The alternative formation of a diradical (XLI) seems less probable since none of the final products from such an intermediate was detected. This contrasts with the photochemical reaction of maleic anhydride with cyclohexene, discussed below, where there is strong support for a diradical intermediate. When a solidified layer obtained from a molten mixture of hexamethylbenzene and maleic anhydride was irradiated no adduct was detected. The solid substrate contains no

excited charge-transfer
complex

XXXVIII   XLI

XXXVII   XXXIX   XL

(11.12)

charge-transfer complex,[114] which is consistent with the postulate that
the formation of XXXVII, where it occurs, is via the charge-transfer
complex, although comparisons between processes in the liquid and solid
phases should be treated with caution.

Barltrop *et al.*[115] have made an elegant study of the photochemical
reactions of cyclohexene with various olefinic acceptors. The experiments
were designed so that irradiation occurred only within the respective
charge-transfer band. The three volatile products from the fumaronitrile–cyclohexene interaction are cyclohex-2-enylsuccinonitrile (XLII),
cis-bicyclo[4,2,0]octane-*trans*-7,8-dinitrile (XLIII) and, as a minor
component, bicyclohex-2-enyl (XLIV). If the reaction proceeds via the

XLII   XLIII   XLIV

charge-transfer complex, then, for maximum overlap between the
π-orbitals of the donor and acceptor moieties of the complex, the most
favoured arrangement is for the two double bonds to lie parallel, one
vertically above the other (XLV). If binding via the excited state
(XLVI) occurs through the ion centres, which seems likely in a non-polar
solvent, then the diradicals XLVII ($R^1 = R^3 = H$, $R^2 = R^4 = CN$; and
$R^2 = R^4 = H$, $R^1 = R^3 = CN$) will be formed. (Barltrop and his co-
workers have pointed out that it is immaterial to the argument concerning the stereochemistry of the products whether the ion, or the radical

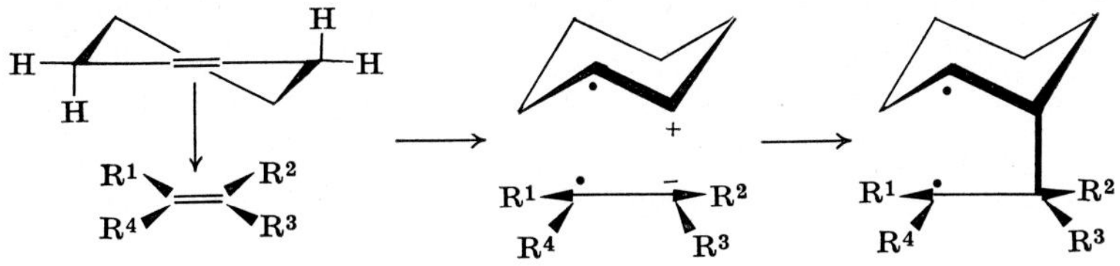

$$\tag{11.13}$$

charge-transfer complex	excited charge-transfer complex	
XLV	XLVI	XLVII

centres, in XLV combine first.) It is reasonable to suppose that the bond formed in XLVII is axial to the six-membered ring.[116] In the case of the fumaronitrile interaction, XLVII represents two structures: either $R^1 = R^3 = H, R^2 = R^4 = CN$, or $R^1 = R^3 = CN, R^2 = R^4 = H$. Secondary bond-formation in either case will give the *cis-trans* isomer XLIII, which is the only bicyclo-octane product observed.

In the maleonitrile–cyclohexene system the two products of primary bond-formation from the excited charge-transfer complex will be either the *endo*-diradical XLVII ($R^1 = R^2 = CN$, and $R^3 = R^4 = H$) or the *exo*-diradical XLVII ($R^1 = R^2 = H$, and $R^3 = R^4 = CN$). *Immediate* secondary bond-formation from the *endo*-diradical will give the *cis-cis-endo* isomer XLVIII and the *cis-cis-exo* isomer XLIX from the *exo*-radical. However,

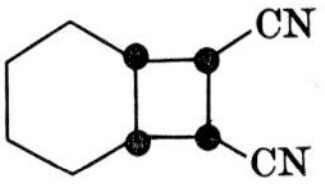

XLVIII

it is argued that, to minimize the dipole-dipole repulsion, there is rotation about the central carbon-carbon bond of the maleonitrile moiety of XLVII ($R^1 = R^2 = CN$, $R^3 = R^4 = H$) and XLVII ($R^1 = R^2 = H$, $R^3 = R^4 = CN$). This will lead to the formation of some XLIII in either case. This product is, in fact, formed along with XLVIII and XLIX as well as bicyclohex-2-enyl (XLIV).

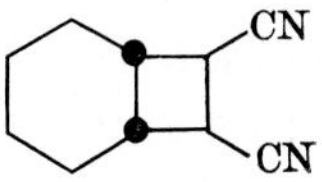

XLIX

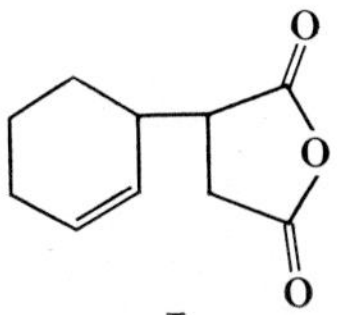

L

In the maleic anhydride–cyclohexene system, bicyclohex-2-enyl (XLIV) and cyclohex-2-enylsuccinic anhydride (L) were detected together with the *cis-cis-exo* (LI), *cis-cis-endo* (LII) and *cis-trans* (LIII) bicyclooctane derivatives. As in the maleonitrile system, there are *exo* and *endo* configurations of the charge-transfer complex. These yield the diradicals LIV and LV respectively, which could give rise via primary bond-formation followed directly by secondary bond-formation to two of the observed products, namely LI and LII. It is suggested that, in the *endo*-diradical, there will be a strong tendency for ring inversion to occur to give LVI, from which either LII or the third observed bicyclo-octane LIII may be formed.

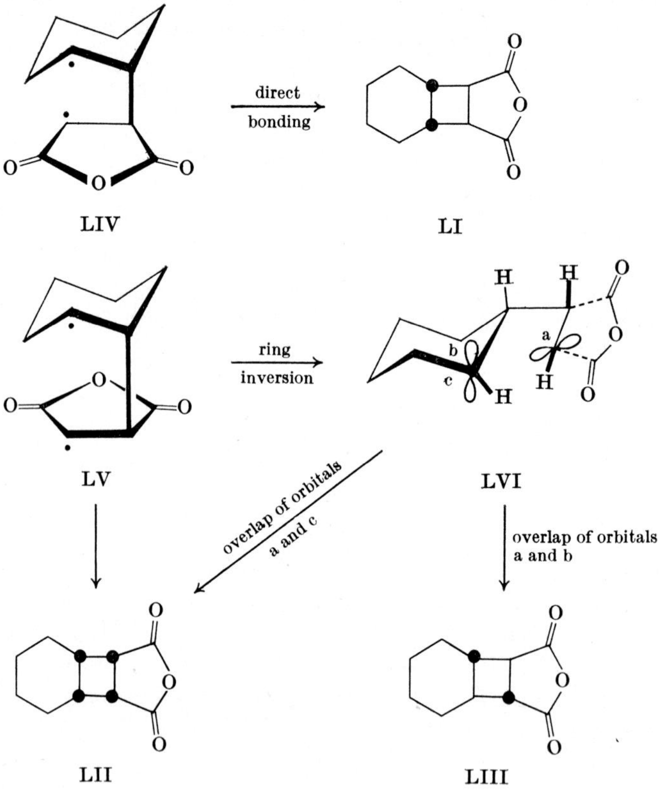

Corey and his co-workers[117] have suggested that charge-transfer complexes are formed between excited 2-cyclohexenone and various donor alkenes in order to account for the particular 1, 2-cycloaddition products. For example, in the reaction with isobutylene, both *trans-* and *cis*-7,7-dimethylbicyclo[4,2,0]octan-2-one (LVII) are obtained, the former predominating, together with 8,8-dimethylbicyclo[4,2,0]octan-2-one (LVIII). The other two products of this reaction are 2- and 3-(β-methylallyl)-cyclohexanone (LIX, LX).*

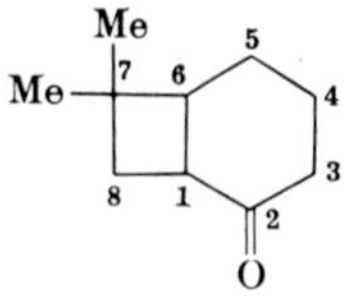

LVII 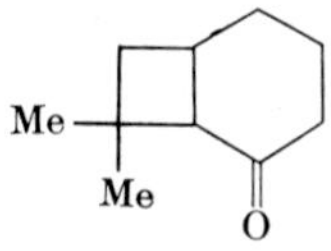   LVIII 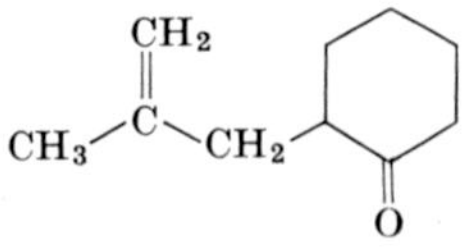   LIX

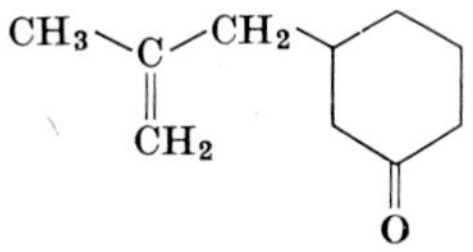

LX 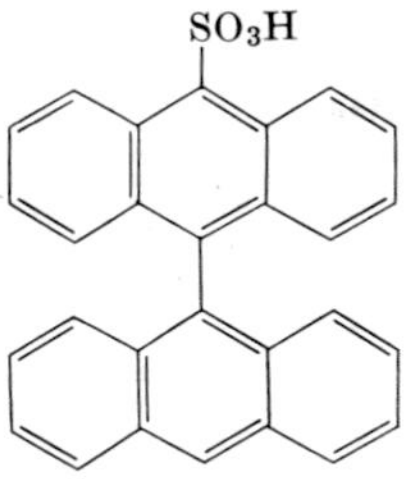   LXI

The formation of anthracene-9-sulphonic acid and of the acid (LXI) from irradiated solutions of anthracene and sulphur dioxide in carbon tetrachloride or in *n*-hexane may arise via the charge-transfer complex, which, through photo-excitation and subsequent aerial oxidation, could yield the observed products.[118]

It has been suggested [119] that the photo-oxidations of various hydrocarbons may occur through charge-transfer complex formation. The similarity of the dependences of the oxidation rate and of the charge-transfer absorption intensity upon the oxygen and substrate concentrations provides some support for this postulate, though in itself it is not conclusive.

Photochemical reactions of amines, particularly aliphatic amines, with halo-subsituted methanes may proceed via charge-transfer complex formation.[120] Such reactions include the formation of triethylammonium chloride from solutions of triethylamine in carbon tetrachloride.[121, 122] Kosower [1] has suggested a possible chain process for such reactions.

* In addition to the work cited above, the closely related dimethyl maleate–cyclohexene system has been studied [A. Cox, P. de Mayo and R. W. Yip, *J. Am. chem. Soc.* **88**, 1043 (1966) and references therein]. Whereas the maleic anhydride–cyclohexene charge-transfer spectrum is immediately obvious, the dimethyl maleate system shows no evidence of complex formation although the dimethyl maleate does show an enhancement of the $n \rightarrow \pi^*$ transition.

The photo-oxidation of diethyl ether by molecular oxygen in the absence of photo-sensitizers has been described.[123] Light of wavelength within the range 218–278 nm is effective. This is outside the absorption of either oxygen or ether, but it does include a large portion of the ether–oxygen charge-transfer band, which has a maximum at *ca.* 210 nm.

REFERENCES

1. E. M. Kosower, *in* "Progress in Physical Organic Chemistry," Vol. 3, eds. S. G. Cohen, A. Streitwieser, Jr. and R. N. Taft, Interscience, New York (1965), p. 81.
2. L. J. Andrews and R. M. Keefer, "Molecular Complexes in Organic Chemistry," Holden-Day, San Francisco (1964), p. 146.
3. R. L. Ward, *J. chem. Phys.* **39**, 852 (1963).
4. K. M. C. Davis and M. C. R. Symons, *J. chem. Soc.* 2079 (1965).
5. D. Bijl, H. Kainer and A. C. Rose-Innes, *Naturwissenschaften* **41**, 303 (1954).
6. H. Kainer and A. Überle, *Chem. Ber.* **88**, 1147 (1955).
7. J. W. Eastman, G. Engelsma and M. Calvin, *J. Am. chem. Soc.* **84**, 1339 (1962).
8. I. Isenberg and S. L. Baird, Jr., *J. Am. chem. Soc.* **84**, 3803 (1962).
9. R. Foster and T. J. Thomson, *Trans. Faraday Soc.* **58**, 860 (1962).
10. H. A. Torrey and W. H. Hunter, *J. Am. chem. Soc.* **34**, 702 (1912).
11. H. Kainer, D. Bijl and A. C. Rose-Innes, *Nature, Lond.* **178**, 1462 (1956).
12. G. T. Pott, "Molionic Lattices," Thesis, University of Groningen (1966).
13. G. T. Pott and J. Kommandeur, *Molec. Phys.* **13**, 373 (1967).
14. R. Foster and T. J. Thomson, *Trans. Faraday Soc.* **59**, 1059 (1963).
15. P. H. Emslie, R. Foster and T. J. Thomson, *Recl Trav. chim. Pays-Bas Belg.* **83**, 1311 (1964).
16. M. Rosenblum, R. W. Fish and C. Bennett, *J. Am. chem. Soc.* **86**, 5166 (1964).
17. O. W. Webster, W. Mahler and R. E. Benson, *J. Am. chem. Soc.* **84**, 3678 (1962).
18. W. Liptay, G. Briegleb and K. Schindler, *Z. Elektrochem.* **66**, 331 (1962).
19. F. E. Stewart, M. Eisner and W. R. Carper, *J. chem. Phys.* **44**, 2866 (1966).
20. R. Foster and J. W. Morris, unpublished work.
21. D. H. Anderson, R. M. Elofson, H. S. Gutowsky, S. Levine and R. B. Sandin, *J. Am. chem. Soc.* **83**, 3157 (1961); R. M. Elofson, D. H. Anderson, H. S. Gutowsky, S. Levine and K. F. Schulz, *J. Am. chem. Soc.* **85**, 2622 (1963).
22. R. B. Sandin, R. M. Elofson and K. F. Schulz, *J. org. Chem.* **30**, 1819 (1965).
23. S. Carter, J. N. Murrell, E. J. Rosch, N. Trinajstić and P. A. H. Wyatt, *J. chem. Soc.* (B), 477 (1967).
24. D. Buckley, S. Dunstan and H. B. Henbest, *J. chem. Soc.* 4880 (1957).
25. R. Foster, *Recl Trav. chim. Pays-Bas Belg.* **83**, 711 (1964).
26. E. A. C. Lucken, F. Ramirez, V. P. Catto, D. Rhum and S. Dershowitz, *Tetrahedron* **22**, 637 (1966).
27. Z. Rappoport, *J. chem. Soc.* 4498 (1963).
28. Z. Rappoport and A. Horowitz, *J. chem. Soc.* 1348 (1964).
29. P. G. Farrell, J. Newton and R. F. M. White, *J. chem. Soc.* (B), 637 (1967).
30. N. S. Isaacs, *J. chem. Soc.* (B), 1053 (1966).
31. R. Foster and P. Hanson, *Tetrahedron* **21**, 255 (1965).
32. G. N. Sausen, V. A. Engelhardt and W. J. Middleton, *J. Am. chem. Soc.* **80**, 2815 (1958).

33. J. K. Williams, D. W. Wiley and B. C. McKusick, *J. Am. chem. Soc.* **84**, 2210 (1962).
34. R. Foster and I. Horman, *J. chem. Soc.* (B), 1049 (1966).
35. E. A. Braude, L. M. Jackman, R. P. Linstead and G. Lowe, *J. chem. Soc.* 3123 (1960).
36. R. Foster and I. B. C. Matheson, unpublished work.
37. D. H. Reid, M. Fraser, B. B. Molloy, H. A. S. Payne and R. G. Sutherland, *Tetrahedron Lett.* 530 (1961).
38. I. Horman, unpublished work.
39. W. J. Middleton, R. E. Heckert, E. L. Little and C. G. Krespan, *J. Am. chem. Soc.* **80**, 2783 (1958).
40. P. Brown and R. C. Cookson, *Tetrahedron* **21**, 1977 (1965).
41. P. Brown and R. C. Cookson, *Tetrahedron* **21**, 1993 (1965).
42. S. Nagakura, *J. Am. chem. Soc.* **80**, 520 (1958).
43. H. Yada, J. Tanaka and S. Nagakura, *Bull. chem. Soc. Japan* **33**, 1660 (1960).
44. G. Kortüm and H. Wilski, *Z. phys. Chem.* **202**, 35 (1953).
45. J. Kleinberg, E. Colton, J. Sattizahn and C. A. VanderWerf, *J. Am. chem. Soc.* **75**, 442 (1953).
46. S. G. W. Ginn and J. L. Wood, *Trans. Faraday Soc.* **62**, 777 (1966).
47. C. Reid and R. S. Mulliken, *J. Am. chem. Soc.* **76**, 3869 (1954).
48. R. A. Zingaro and W. E. Tolberg, *J. Am. chem. Soc.* **81**, 1353 (1959).
49. A. I. Popov and R. T. Pflaum, *J. Am. chem. Soc.* **79**, 570 (1957).
50. I. Haque and J. L. Wood, *Spectrochim. Acta* **23A**, 959 (1967).
51. O. Hassel and H. Hope, *Acta chem. scand.* **15**, 407 (1961).
52. S. N. Bhat and C. N. R. Rao, *J. Am. chem. Soc.* **88**, 3216 (1966).
53. H. C. Brown and J. D. Brady, *J. Am. chem. Soc.* **74**, 3570 (1952).
54. G. A. Olah, S. J. Kuhn and S. H. Flood, *J. Am. chem. Soc.* **83**, 4571 (1961).
55. W. S. Tolgyesi, *Can. J. Chem.* **43**, 343 (1965).
56. M. J. S. Dewar, *Nature, Lond.* **156**, 784 (1945); *J. chem. Soc.* **406**, 777, (1946).
57. M. J. S. Dewar, "The Electronic Theory of Organic Chemistry," Clarendon Press, Oxford (1949).
58. M. J. S. Dewar and A. P. Marchand, *A. Rev. phys. Chem.* **16**, 321 (1965).
59. P. W. Robertson, *J. chem. Soc.* 1267 (1954).
60. H. C. Brown and K. L. Nelson, *J. Am. chem. Soc.* **75**, 6292 (1953).
61. R. M. Keefer, A. Ottenberg and L. J. Andrews, *J. Am. chem. Soc.* **78**, 255 (1956).
62. E. Berliner and M. C. Beckett, *J. Am. chem. Soc.* **79**, 1425 (1957).
63. L. J. Andrews and R. M. Keefer, *J. Am. chem. Soc.* **79**, 1412 (1957).
64. R. M. Keefer and L. J. Andrews, *J. Am. chem. Soc.* **78**, 5623 (1956).
65. B. Z. Senkowski and G. S. Panson, *J. org. Chem.* **26**, 943 (1961).
66. J.-E. Dubois and F. Garnier, *Tetrahedron Lett.* 3961 (1965).
67. J.-E. Dubois and F. Garnier, *Tetrahedron Lett.* 3047 (1966).
68. J.-E. Dubois and F. Garnier, *J. Chim. phys.* **63**, 351 (1966).
69. J.-E. Dubois and F. Garnier, *Spectrochim. Acta* **23A**, 2279 (1967).
70. C. R. Allen, A. J. Brook and E. F. Caldin, *J. chem. Soc.* 2171 (1961).
71. R. Foster, *J. chem. Soc.* 3508 (1959).
72. R. Foster and R. K. Mackie, *Tetrahedron* **16**, 119 (1961).
73. S. A. Penkett, Thesis, University of Leeds (1963).
74. R. Foster and C. A. Fyfe, *Rev. pure appl. Chem.* **16**, 61 (1966).
75. E. Buncel, A. R. Norris and K. E. Russell, *Quart. Rev.* (*London*) **22**, 123 (1968).
76. R. Foster and C. A. Fyfe, *J. chem. Soc.* (B), 53 (1966).
77. R. Foster and C. A. Fyfe, *Tetrahedron* **22**, 1831 (1966).

78. W. Liptay and N. Tamberg, *Z. Elektrochem.* **66**, 59 (1962).
79. G. Briegleb, W. Liptay and M. Canter, *Z. phys. Chem. Frankf. Ausg.* **26**, 55 (1960).
80. R. Foster and R. K. Mackie, *J. chem. Soc.* 3843 (1962).
81. S. D. Ross and I. Kuntz, *J. Am. chem. Soc.* **76**, 3000 (1954).
82. F. M. Menger and M. L. Bender, *J. Am. chem. Soc.* **88**, 131 (1966).
83. A. K. Colter and L. M. Clemens, *J. Am. chem. Soc.* **87**, 847 (1965).
84. A. K. Colter and S. S. Wang, *J. Am. chem. Soc.* **85**, 114 (1963).
85. A. K. Colter, G. H. Megerle and P. S. Ossip, *J. Am. chem. Soc.* **86**, 3106 (1964).
86. A. K. Colter, F. F. Guzik and S. H. Hui, *J. Am. chem. Soc.* **88**, 5754 (1966).
87. G. Porter and J. A. Smith, *Nature, Lond.* **184**, 446 (1959).
88. G. Porter and J. A. Smith, *Proc. R. Soc.* **A261**, 28 (1961).
89. G. Porter, *Discuss. Faraday Soc.* **33**, 198 (1962).
90. S. J. Rand and R. L. Strong, *J. Am. chem. Soc.* **82**, 5 (1960).
91. G. A. Russell, *J. Am. chem. Soc.* **80**, 4987 (1958).
92. R. L. Strong and J. Pérano, *J. Am. chem. Soc.* **89**, 2535 (1967).
93. C. E. Boozer and G. S. Hammond, *J. Am. chem. Soc.* **76**, 3861 (1954).
94. C. E. Boozer, G. S. Hammond, C. E. Hamilton and J. N. Sen, *J. Am. chem. Soc.* **77**, 3233 (1955).
95. J. R. Thomas and C. A. Tolman, *J. Am. chem. Soc.* **84**, 2930 (1962).
96. J. R. Thomas, *J. Am. chem. Soc.* **85**, 591 (1963).
97. J. R. Thomas, *J. Am. chem. Soc.* **85**, 593 (1963).
98. H. Scott, G. A. Miller and M. M. Labes, *Tetrahedron Lett.* 1073 (1963).
99. L. P. Ellinger, *Chemy Ind.* 1982 (1963).
100. C. E. H. Bawn, C. Fitzsimmons and A. Ledwith, *Proc. chem. Soc.* 391 (1964).
101. C. E. H. Bawn, R. Carruthers and A. Ledwith, *Chem. Commun.* 522 (1965).
102. N. C. Yang and Y. Gaoni, *J. Am. chem. Soc.* **86**, 5022 (1964).
103. S. Iwatsuki and Y. Yamashita, *Makromolek. Chem.* **89**, 205 (1965).
104. C. Lagercrantz and M. Yhland, *Acta chem. scand.* **16**, 1043 (1962).
105. D. F. Ilten and M. Calvin, *J. chem. Phys.* **42**, 3760 (1965).
106. E. M. Kosower and L. Linqvist, *Tetrahedron Lett.* 4481 (1965).
107. J. P. Simons and P. E. R. Tatham, *J. chem. Soc.* (A), 854 (1966).
108. D. Bryce-Smith and J. E. Lodge, *J. chem. Soc.* 2675 (1962).
109. J. S. Bradshaw, *J. org. Chem.* **31**, 3974 (1966).
110. G. O. Schenk and R. Steinmetz, *Tetrahedron Lett.* No. 21, 1 (1960).
111. D. Bryce-Smith and B. Vickery, *Chemy Ind.* 429 (1961).
112. D. Bryce-Smith and A. Gilbert, *J. chem. Soc.* 918 (1965).
113. Z. Raciszewski, *J. chem. Soc.* (B), 1147 (1966).
114. Z. Raciszewski, *J. chem. Soc.* (B), 1142 (1966).
115. J. A. Barltrop and R. Robson, *Tetrahedron Lett.* 597 (1963); R. Robson, P. W. Grubb and J. A. Barltrop, *J. chem. Soc.* 2153 (1964).
116. D. H. R. Barton and R. C. Cookson, *Quart. Rev. (London)* **10**, 44 (1956).
117. E. J. Corey, J. D. Bass, R. LeMahieu and R. B. Mitra, *J. Am. chem. Soc.* **86**, 5570 (1964).
118. T. Nagai, K. Terauchi and N. Tokura, *Bull. chem. Soc. Japan* **39**, 868 (1966).
119. J. C. W. Chien, *J. phys. Chem., Ithaca* **69**, 4317 (1965).
120. D. P. Stevenson and G. M. Coppinger, *J. Am. chem. Soc.* **84**, 149 (1962).
121. R. F. Collins, *Chemy Ind.* 704 (1957).
122. R. Foster, *Chemy Ind.* 1354 (1960).
123. V. I. Stenberg, R. D. Olson, C. T. Wang and N. Kulevsky, *J. org. Chem.* **32**, 3227 (1967).

Chapter 12

Biochemical Systems

12.A. Introduction

In 1952, in what has become a classical paper, Mulliken [1] suggested that charge-transfer complexes may play an important role in biological systems. Some possible implications have been discussed in a book by Szent-Györgyi. [2] Many workers have produced evidence which, it is claimed, supports this proposition. Various aspects of the problems involved have been reviewed. [3-5]

In a few cases, it has been suggested that intramolecular charge-transfer interactions across space may be involved in the *structures* of some biological macromolecules and various cellular systems.* The major concern, however, has been with the proposal that charge-transfer complexes are involved in biochemical *reactions*. There are two steps in exploring this possibility: firstly to show the existence of charge-transfer complex formation in the system; then to demonstrate that it is an essential part of the biochemical process. Although the capacity of many compounds of biochemical importance to form charge-transfer complexes may be readily demonstrated if suitable partners for complex formation are chosen, such donor–acceptor *pairs* may not *per se* be of particular biochemical interest. There are relatively few examples where both components are of biological interest. In Chapter 11 it has been seen that to demonstrate the essential participation of charge-transfer complexes

---

* See also Chapter 3.

in relatively simple non-biochemical reactions requires more than the detection of such a complex in the reaction mixture. The vastly more complicated nature of biochemical reactions has so far prevented any demonstration of the direct involvement of charge-transfer complexes in the mechanisms of such reactions. Even if a correlation between a type of biological activity and a property of the charge-transfer complex with some second component were to be observed, it would not of necessity imply a causal relationship between the two. Both properties could derive from some other, more basic, common feature.

In some instances, the existence of charge-transfer complex formation between components in a given biochemical system has been proposed on the most tenuous of evidence. The mere production of a colour when two biochemical (or other) reagents are mixed has often been given as the sole evidence for charge-transfer complex formation. Obviously such an observation by itself is insufficient to justify this conclusion. More positive evidence, such as the systematic variation of the energy of the absorption band for, say, complexes of a series of related donors with a given acceptor compared with other series of complexes with other acceptors or with the ionization potentials of the donors (see Chapter 3), is desirable before the attribution of a particular colouration to charge-transfer complex formation is made. Indeed, from the evidence given in Chapter 3 it is apparent that the visual observation of colour formation is not even a necessary condition for charge-transfer complex formation. In solution the dissociation of the complex may be so high that any charge-transfer absorption may be detected only with difficulty. The energy of such a transition may in any case be sufficiently high so that no absorption occurs in the visible region. Even careful spectrophotometric measurements in the ultraviolet–visible region may not reveal very significant intensity increases and even less likely is it that discrete new maxima will be observed. Virtually all the absorption due to a charge-transfer transition can be hidden underneath the absorptions characteristic of the component moieties of the complex, together with the absorptions of the free components which will be in equilibrium with the complex in solution. In some cases, optical absorptions which develop slowly have been assigned as arising from charge-transfer complex formation between reactants. There is in fact no evidence that charge-transfer complexes are formed slowly. It is likely that such absorptions result from products of normal covalent chemical reactions. This tendency to postulate charge-transfer interaction in systems containing biochemically active components, without adequate experimental evidence, has recently been remarked upon by Kosower[3] and independently by Pullman and Pullman.[5]

Recent discussion has mainly centred around particular groups of compounds, including model compounds containing what are thought to be essential structural features. These compounds include the oxidation–reduction coenzymes, particularly the pyridinium moiety of pyridinium nucleotides, the isoalloxazine moiety of flavin nucleic acids and nucleic acid bases, indoles, amino acids and proteins, carotenes, quinones and porphins. Amongst synthetic compounds with pronounced physiological activity, the carcinogenic hydrocarbons and the phenothiazine drugs have received particular attention. Simple Hückel molecular orbital calculations suggest that many of these compounds fall within the classification of good donors (low positive value of β for the highest-occupied molecular orbital), or good electron acceptors (see Table 12.1). The calculations of Pullman and Pullman [4] indicate that some compounds are both good donors and good acceptors; such compounds involve porphyrins, carotenes and retinenes. It has also been pointed out that those polybenzenoid aromatic hydrocarbons with filled orbitals of high energy will also have low-energy vacant orbitals so that the hydrocarbons which are good donors will also be good acceptors (see also Section 12.B.7).

Quite apart from the experimental difficulties of detecting a possible charge-transfer absorption band, there remains the fact that for most charge-transfer complexes, as defined in Chapter 1, charge-transfer forces are *not* the major contributing factor to the stability of the complex in the ground state. Consequently, correlations between electron-donating ability of the one component (or electron-accepting ability of the second component) and the association constant, or free energy, or enthalpy of formation, cannot *in general* be assumed to be a necessary feature of a group of charge-transfer complexes (see Chapter 7). This is especially important to remember when relatively large component molecules are involved; in such instances, steric effects and other features of the molecule may completely overwhelm any trend in thermodynamic quantities resulting from charge-transfer forces.

12.B. Possible Charge-Transfer Complexation

Since 1950, charge-transfer complexes have been postulated in a variety of biochemical systems. Some of the major groups of compounds involved are discussed in this section.

12.B.I. Pyridinium Ions

The pyridinium ion is of particular biological interest. Nicotinamide adenine dinucleotide* (NAD$^+$) (I, R = H) and nicotinamide adenine

* Alternatively called diphosphopyridine nucleotide (DPN), the reduced form being written as DPNH.

TABLE 12.1. Energy coefficients of molecular orbitals (in β units).*

Compound	Highest-filled molecular orbital	Lowest-empty molecular orbital
Purine	0·69	−0·74
Adenine	0·49	−0·87
Guanine	0·31	−1·05
Hypoxanthine	0·40	−0·88
Xanthine	0·44	−1·01
Uric acid	0·17	−1·19
Uracil	0·60	−0·96
Thymine	0·51	−0·96
Cytosine	0·60	−0·80
Barbituric acid	1·03	−1·30
Alloxan	1·03	−0·76
Phenylalanine	0·91	−0·99
Tyrosine	0·79	−1·00
Histidine	0·66	−1·16
Tryptophan	0·53	−0·86
Riboflavin	0·50	−0·34
Pteridine	0·86	−0·39
2-Amino-4-hydroxypteridine	0·49	−0·65
2,4-Diaminopteridine	0·54	−0·51
2,4-Dihydroxypteridine	0·65	−0·66
Folic acid	0·53	−0·65
Porphin	0·30	−0·24
2,3-Divinylporphin	0·29	−0·23
1-Vinyl-5-formylporphin	0·30	−0·21
α-Carotene	0·10	−0·19
β-Carotene	0·08	−0·18
Vitamin A_1	0·23	−0·31
Vitamin A_2	0·20	−0·26
Retinene	0·28	−0·26
p-Benzoquinone	1	−0·23
1,4-Naphthoquinone	1	−0·33
9,10-Anthraquinone	1	−0·44
Benzohydroquinone	0·63	−1
Naphthohydroquinone	0·41	−0·71
Anthrahydroquinone	0·23	−0·53
NAD^+	1·03	−0·36
NADH	0·30	−0·92
FMN	0·50	−0·34
$FMNH_2$	−0·11	−0·95

* Ref. 5.

dinucleotide phosphate* (NADP$^+$) (I, R $=$ OP(OH)$_2$) are typical pyridine nucleotide coenzymes. They are of importance in biochemical processes involving NAD$^+$ in oxidation-reduction in which the pyridinium ring of NAD$^+$ (II, R $=$ ribosyl adenine phosphate residue) is reduced by hydride addition in the 4-position to give NADH (III, R $=$ ribosyl

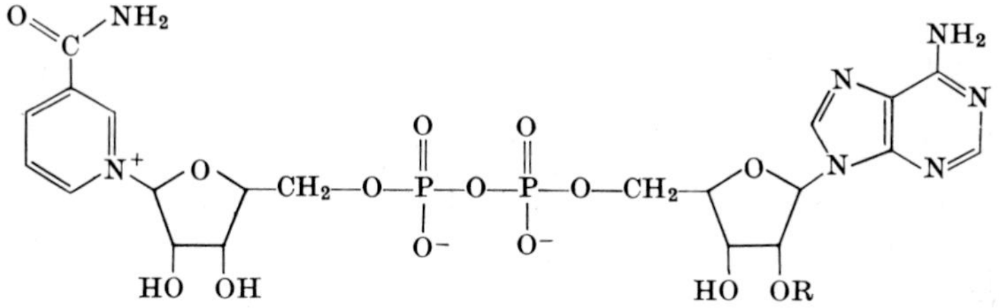

I

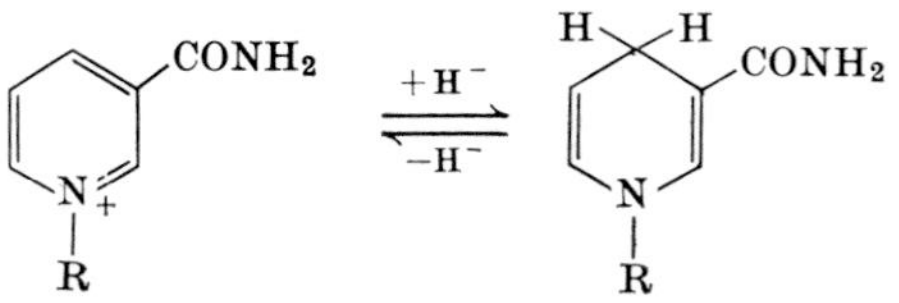

(12.1)

II   III

adenine phosphate residue). The coenzyme NADP$^+$ may be reduced in a similar way to NADPH. Biochemical interest in these coenzymes had led to a study of simple model compounds containing the pyridinium part of these structures, in particular 1-alkylpyridinium ions and various substituted derivatives including the 3-carbamido compounds. In his initial studies of the action of the iodide ion on pyridinium ions, Kosower[6] considered the possibility that the iodide became attached to the pyridinium ring by a covalent bond to yield a structure IV. Kosower

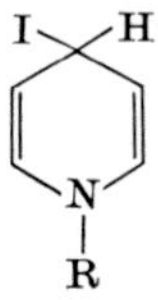

IV

and his co-workers[7-13] were later able to show that the products formed did not have such structures but were in fact intermolecular charge-transfer complexes in which the pyridinium ion acts as the electron-acceptor and the iodide ion as the electron-donor species. Such complexes

* Alternatively called triphosphopyridine nucleotide (TPN), the reduced form being written as TPNH.

are essentially ion-pairs in the ground state. Absorption of energy within the charge-transfer band causes excitation to a covalent excited state. The properties of this group of complexes have been discussed in Chapters 3 and 7. The extensive studies of Kosower and his co-workers leave no doubt as to the nature of these iodide–pyridinium interactions, and in particular to the ability of the pyridinium ion to act as an electron acceptor. However, it does not necessarily follow that charge-transfer complexing participates in such redox processes as the NAD^+–NADH equilibrium [equation (12.1)].

Kosower and Bauer [14] have suggested that a charge-transfer complex (V) is formed by the action of sodium dithionite on 1-alkyl-3-carbon-iodopyridinium ion. This reaction is of particular interest since the

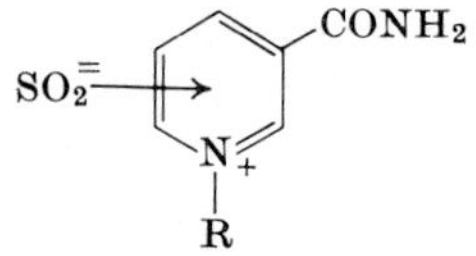

$$SO_2^= \longrightarrow \quad \text{(pyridinium ring)} \quad CONH_2$$
$$N^+$$
$$R$$
$$V$$

reduction of NAD^+ itself by sodium dithionite yields a product which is identical with the enzymatically produced NADH,[15] whereas some other methods of reduction, for example by electrolysis or with sodium borohydride, yield a reduced form of NAD^+, the solution quickly becoming yellow, the colour being more intense than that of the final product. Yarmolinsky and Colowick [16] have argued that this product is in fact the σ-complex (VI), which then decomposes to form NADH and sulphite ion. Recent measurements [17] of the n.m.r. spectra of the dithionite reduction of a model NAD^+ ion leave no doubt that the product formed is the σ-complex (VI) rather than the charge-transfer complex (V).

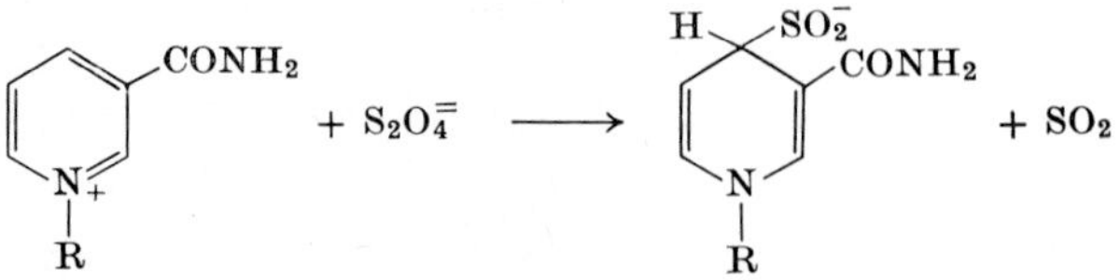

$$\text{(ring)} CONH_2 \quad + S_2O_4^= \longrightarrow \quad \text{(ring)} \begin{array}{c} H \quad SO_2^- \\ CONH_2 \end{array} \quad + SO_2 \qquad (12.2)$$
$$N^+ \qquad\qquad\qquad N$$
$$R \qquad\qquad\qquad R$$
$$(NAD^+) \qquad\qquad\qquad VI$$

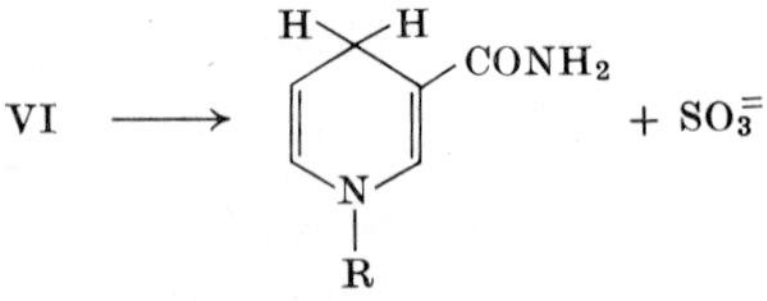

$$VI \longrightarrow \quad \text{(ring)} \begin{array}{c} H \quad H \\ CONH_2 \end{array} \quad + SO_3^= \qquad (12.3)$$
$$N$$
$$R$$
$$NADH$$

Charge-transfer complex formation appears to occur in simple analogues of the NAD^+–NADH couple; for example, with II, III (R = n-propyl)[18] and II, III (R = benzyl),[19] measurable association constants have been obtained. By contrast, the degree of complex formation between NAD^+ and NADH is at the most very small.[19] This reluctance to undergo intermolecular association may be the result of intramolecular charge-transfer interaction in NAD^+ between the pyridinium ring and the adenine moiety.[19] Estimates of the degree of association have been made from the increase in optical absorption in the visible region on mixing. Although the degree of association and the spectral absorption are typical of charge-transfer complexes, this evidence alone does not provide a strong argument for the nature of the complex. Mixtures of reduced flavin mononucleotide ($FMNH_2$) (VII)

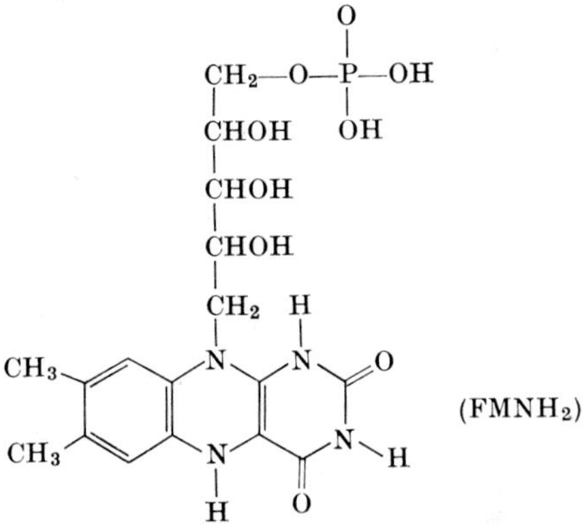

VII

with NAD^+ show a new absorption band. This is almost certainly due to an intermolecular charge-transfer transition. The maximum has been separately reported as occurring at 21,000 cm^{-1} (480 nm)[20] and at 22,000 cm^{-1} (455 nm).[21] Quantitative spectrophotometric measurements on solutions containing NAD^+ analogues with $FMNH_2$ provide support for the existence of charge-transfer complexes in such systems. For interaction between a series of substituted methylpyridinium chlorides and $FMNH_2$ the variation of optical absorption with concentration is consistent with 1:1 complex formation. Values of association constant and optical characteristics of the complexes, summarized in Table 12.2, are consistent with charge-transfer complex formation. A more vital argument for the nature of the interaction is that, apart from the ions which contain free carboxyl groupings and NAD^+ itself, the energy of the absorption band extra to the absorption of the compound

12

species is proportional to the energy of the lowest-empty molecular orbital of the particular NAD^+ analogue[21] (Table 12.2). The relatively high energy of the absorption band of the $FMNH_2$–NAD^+ complex may be the result of steric effects of the large group attached at the ring-nitrogen atom in NAD^+.

Massey and Palmer[20] have suggested that the green complexes they observed to be formed between the reduced form of the flavoprotein enzyme, lipoyl dehydrogenase, and either NAD^+ (14,000 cm^{-1}) (700 nm),

TABLE 12.2. Properties of the complexes $FMNH_2$–NAD^+ analogues and the lowest-empty (LE) and highest-occupied (HO) molecular orbital energy coefficients of the NAD^- analogues.*

R in N-methyl-pyridinium chloride	ν_{CT}/cm^{-1}	ϵ_{max}^{AD}	K_c^{AD}/l mol^{-1}	LE†/-β	HO†/-β
3-CONH$_2$	19,000	830	4·3	0·34 (0·34)	0·56 (0·56)
4-CONH$_2$	15,300	600	5·0	0·16 (0·17)	0·58 (0·58)
3-COOCH$_3$	18,500	1000	7·7	0·34 (0·34)	0·64 (0·64)
4-COOCH$_3$	14,100	1050	8·8	0·15 (0·16)	0·66 (0·66)
3-COOH	22,000	1200	2·5	0·34 (0·35)	0·77 (0·77)
4-COOH	22,000	2500	0·9	0·16 (0·17)	0·79 (0·78)
3-COCH$_3$	17,400	80	7·6	0·30 (0·31)	1·21 (1·21)
4-COCH$_3$	12,500	—‡	—‡	0·06 (0·08)	1·16 (1·16)
3-NH$_2$	18,900	300	4·1	0·36 (0·37)	0·36 (0·36)
4-NH$_2$	—§	—§	—§	0·48 (0·48)	0·44 (0·44)
H	18,500	600	1·1	0·36 (0·36)	1·16 (1·16)
ADPR‖	22,000	560	11·4	0·34 (0·34)	0·56 (0·56)

* Ref. 21.
† Calculated without (with) inclusion of hyperconjugation effect of the methyl groups.
‡ Gradually decomposes.
§ Cannot form complex with $FMNH_2$.
‖ Represents the adenine–phosphate–ribosyl moiety of NAD^+.

or the hypoxanthine derivative of NAD^+ (14,000 cm^{-1}) (700 nm), or oxidized thionicotinamide adenine dinucleotide (13,500 cm^{-1}) (740 nm) are charge-transfer complexes. The position of the absorption bands characteristic of the complexes are in the correct order with respect to the electron affinities of the acceptors as measured by their redox potentials. The intensities of the bands increase as the temperature is reduced. The effect is reversible. No complexing was observed by $NADP^+$ or oxidized nicotinamide mononucleotide.

Various workers[20-26] have shown that NAD^+, $NADP^+$ and various simple pyridinium analogues give broad featureless absorptions, extra to the absorption of the component species, when mixed with indole, certain indole derivatives including tryptophan, or other electron donors. The possibility that these are a result of charge-transfer complex formation is supported by Shifrin's[27] observation that 1-(β-indolylethyl)-3-carbamidopyridinium chloride (VIII) has an absorption in methanol

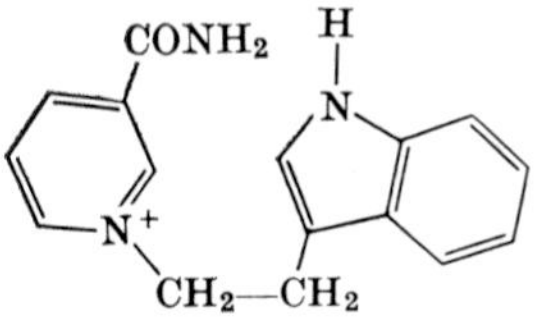

VIII

which, by comparison with an equimolar mixture of tryptamine hydrochloride and nicotinamide methochloride, is characterized by a broad absorption band tailing into the 22,000 cm^{-1} (450 nm) range of the visible spectrum, with an estimated maximum at 30,800 cm^{-1} (325 nm). Shifrin reasonably concludes that this low energy absorption is the result of a charge-transfer transition across space between the indole ring and the pyridinium ring in VI. This will be more intense than the measured absorption in mixtures of nicotinamide methochloride and tryptamine hydrochloride, where the charge-transfer absorption will only arise in the relatively few correctly orientated donor–acceptor pairs. Further confirmation of the nature of the interaction in VIII has also been obtained by Shifrin,[28] who has studied the behaviour of compounds containing both *para*-substituted phenyl moieties and the 3-carbamido-pyridinium system incorporated in the same molecule (IX). The absorption spectra of methanolic solutions of IX, where X is —NH$_2$, —OH,

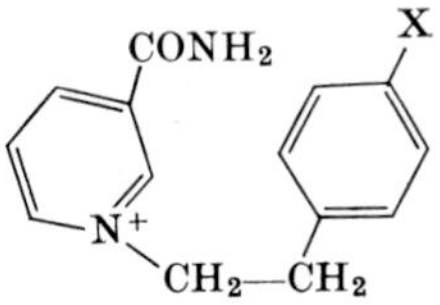

IX

—OCH$_3$, —CH$_3$, —Cl or H, were compared with the absorptions of equimolar mixtures of the corresponding 4-X-phenylethylamines and 1-methylnicotinamide perchlorate (1-methyl-3-carbamidopyridinium perchlorate) in methanol. The difference-spectra have maxima, the energies of which are proportional to the ionization potentials of the corresponding substituted benzenes C$_6$H$_5$X. This correlation, together

with the observation that the intensities of these transitions are con-
centration-independent, suggests that the absorptions are the result of
charge-transfer interactions across space between the pyridinium ring
and the benzene moiety in each of these compounds (see also
Chapter 3).

The broad absorption shown by solutions of glyceraldehyde-3-
phosphate dehydrogenase containing NAD^+ may be the result of a
charge-transfer interaction,[8, 25] again between the pyridinium ring acting
as the acceptor site and an indole moiety as the donor site.

Evidence for complex formation between pyridinium ions and neutral
aromatic hydrocarbons has been given recently.[29] These appear to be
charge-transfer complexes. In the complexes formed between porphyrins
and a variety of 1-substituted nicotinamides in aqueous solution, it is
claimed that the effect of charge transfer is small. This contrasts with the
interaction of porphyrins with some biologically inactive compounds
(see Section 12.B.9).

12.B.2. Flavins

Two coenzymes utilized in flavoproteins are flavin mononucleotide
(FMN) (X) and flavin adenine dinucleotide (FAD) (XI), both of which
are derivatives of riboflavin (XII). As in the case of the pyridine coenzymes, these compounds are involved in biochemical oxidation–
reduction processes in which hydrogen transfer at the isoalloxazine ring
occurs. The reduced forms, $FMNH_2$ and $FADH_2$, are represented by the
structures XIII, in which R corresponds to the side-chain in FMN and
FAD respectively (see also structure VII).

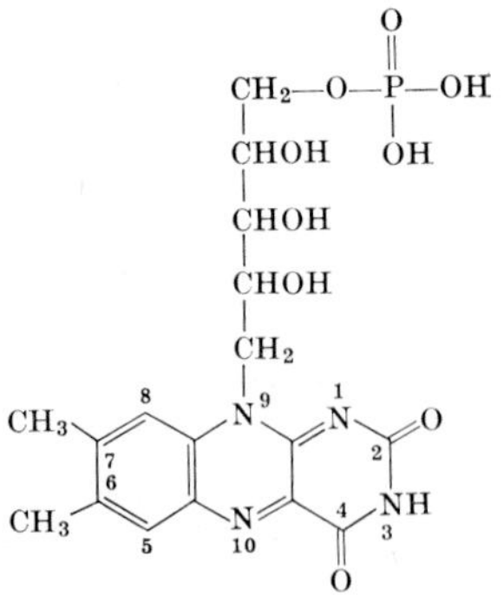

(FMN)
X

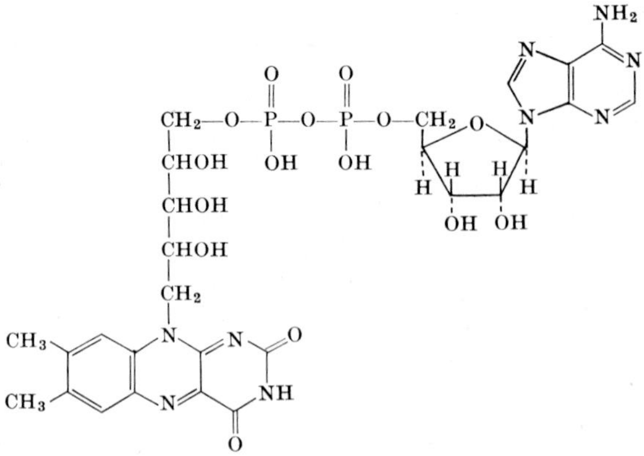

(FAD)
XI

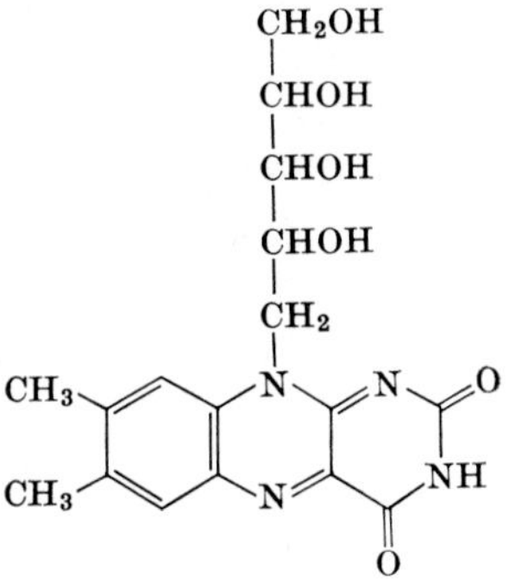

CH$_2$OH
|
CHOH
|
CHOH
|
CHOH
|
CH$_2$

XII

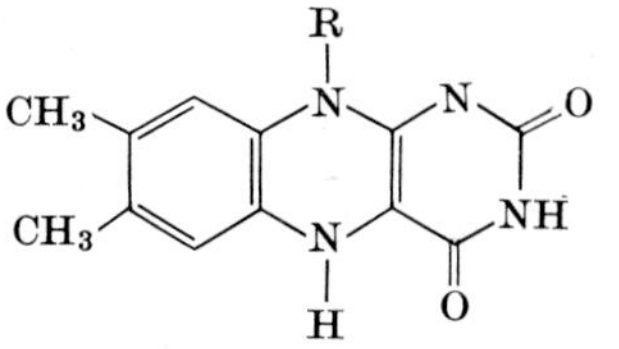

XIII

Interactions in solution between FMN, FAD or riboflavin and various donors including hydrocarbons,[30] indoles,[22, 31–36] NADH,[22, 35] NADPH,[22] purines and pyrimidines,[36, 37] as well as other compounds not having notable donor properties[36, 37] have been reported. In some of these systems there is little doubt that complete electron transfer occurs to form the flavin semiquinone.[31, 34, 35, 38, 39] Although the e.s.r. absorption, when observed, does not provide evidence of charge-transfer complex formation, such paramagnetic species may be formed via charge-transfer complexes (Chapter 2) and the radicals so formed may themselves be involved in charge-transfer complexes. Mixtures of FMNH$_2$ and FMN show a new broad absorption with a maximum at ~11,000 cm^{-1} (900 nm). Massey and Palmer[20] suggest that this is due to charge-transfer complex formation. There is strong evidence for intra-molecular complexing between the adenine and flavin moieties of FAD.[40–42]

Many of the claims for any type of complexing have been based on differences in the spectra of the mixed reagents compared with the sum of the absorptions of the unmixed components. In some cases, the main spectral feature is the decrease in the absorption of the flavin on the addition of the second component. Quenching of the flavin fluorescence-emission also often occurs. Enhancement of absorption at long wave-lengths, when observed, is usually slight. Correlations between association constants of a series of complexes of a common acceptor and the electron-donor properties of the donor component (e.g. in the interaction of riboflavin with certain purines and pyrimidines) do not necessarily imply that the complexes are charge-transfer complexes. Wilson[36] has pointed out significant differences in the properties of some flavin–indole complexes compared with flavin–purine complexes. In the latter group, complexes are formed not only by FMN but also by the fully reduced

$FMNH_2$. By contrast, various indoles and pyrrole only show significant spectral changes with FMN. Even for these indole–FMN complexes there is as yet no compelling argument that a significant charge-transfer contribution is involved. Evidence of charge-transfer transitions has been obtained in complexes between various phenols and FMN, riboflavin and related compounds, including those systems in strongly acid solution where the flavin exists mainly as a mono-protonated species.[38, 39] Thus, the complexes between 9-methylisoalloxazine hydrochloride and various naphthalene diols have absorption bands, the energies of which decrease as ionization potentials* of the component diols decrease.

It has been suggested that 1,3,7,9-tetramethyluric acid [43] and various amino acids [44] form charge-transfer complexes with the semiquinone derived from riboflavin.

The anaerobic photo-reduction of FMN by its own side-chain is inhibited by substances which are expected to be good donors towards FMN, e.g. DL-tryptophan. L-tyrosine, caffeine and serotonin.[45] Since the reduction of FMN by NADH in the dark is unaffected by serotonin, it is argued that the inhibition must be sought in some property of the flavin molecule which participates in the photochemical process. This has led Radda and Calvin [45] to suggest that it is in fact due to the formation of a charge-transfer complex between the triplet state of FMN acting as the acceptor, and the inhibitor acting as the donor.

12.B.3. Quinones

Very few charge-transfer studies appear to have been made on biologically active quinones. In general, it is anticipated that quinones will act as electron acceptors, although this property in a particular quinonoid system will be very dependent on the nature and extent of substitution (see, for example, Chapter 7).

The compound 2-methyl-1,4-naphthoquinone (XIV), also called

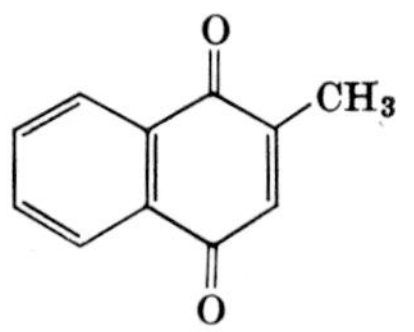

XIV

vitamin K_3, which was used at one time under the name menadione as a synthetic substitute for vitamin K_1, has been shown by Cilento and Sanioto [46] to complex with aromatic hydrocarbons. The general

* Calculated by a S.C.F. method.

behaviour of these systems suggests that charge-transfer complexation is involved although, because of the low stability of the complexes under the conditions used in these studies, their estimates of the thermo-dynamic constants may need revision. Some of the difficulties involved in such evaluations have been discussed in Chapter 6. The recent determin-ations of the stabilities of the complexes of 2-methyl-1,4-naphtha-quinone with a porphyrin and with chlorophyll by Williams and his co-workers [47] are referred to further in Section 12.B.9. Complex formation between the antioxidant N-phenyl-2-naphthylamine and vitamin K_3 has recently been described.[48]

It would be of interest to study the effect of the side-chain of vitamins K_1 (XV), of K_2 (XVI), of the ubiquinones (XVII, $n = 6, 7, 8, 9$ or 10) and of α-tocopherylquinone (XVIII) in their behaviour as electron acceptors

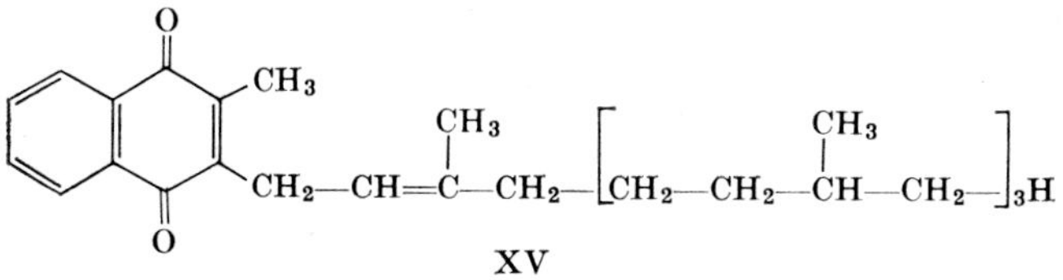

XV

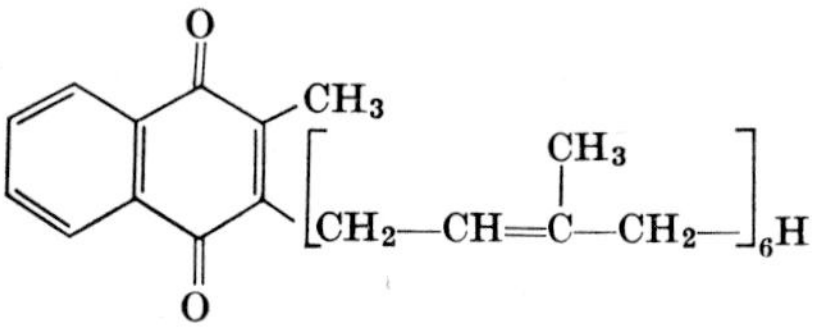

XVI

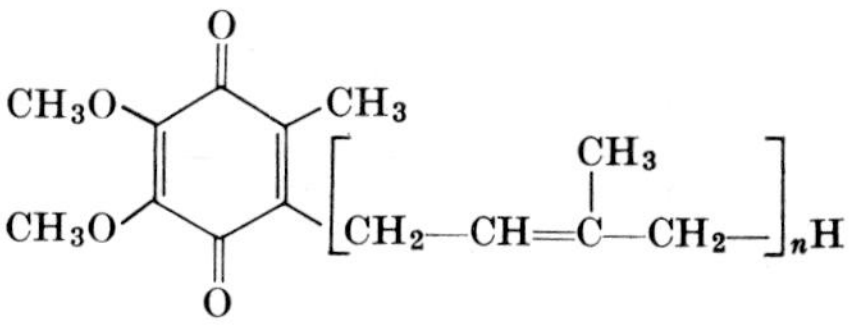

XVII

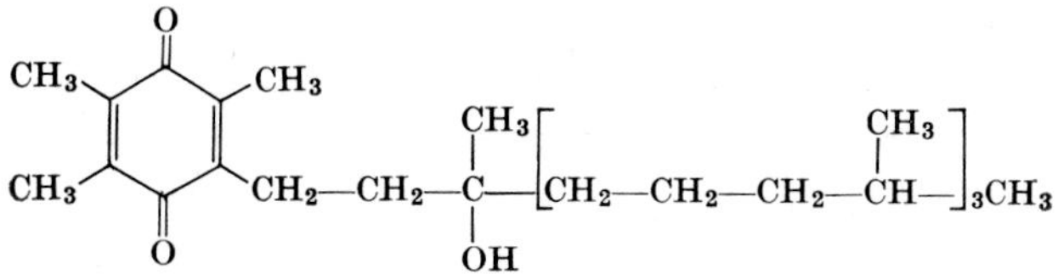

XVIII

in charge-transfer complex formation. Many other quinones of biological importance may show significant complexing ability with electron donors; for example, the ortho-quinone adrenochrome (XIX).

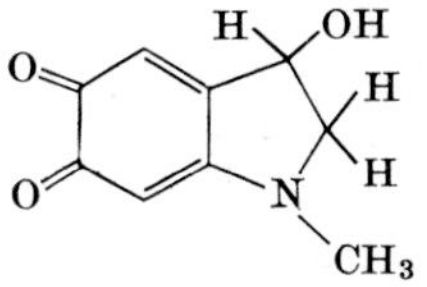

XIX

12.B.4. Indoles

Reference has already been made in Section 12.B.1 to the intra-molecular charge-transfer interaction across space of the indole moiety of 1-(β-indolylethyl)-3-carbamidopyridinium chloride (VIII) with the pyridine ring of this compound, and also to indole–NAD$^+$ and indole–FMN interactions. The particularly strong donor properties of indoles, which are also shown in the more usual systems in which the donor and acceptor sites are in different molecules, have been commented upon by Szent-Györgyi and his co-workers.[49, 50] They suggest that this is the result of a localization of the donor site at the 2–3 position in indole (XX).

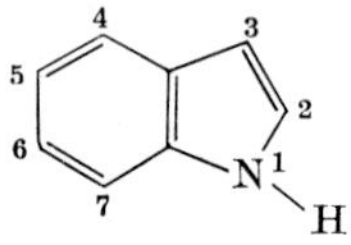

XX

Calculation of the formal negative charge distribution shows a high density in the 3-position.[51]

There is little doubt that indoles do form charge-transfer complexes with some electron-acceptor species. The absorption maxima observed when indole and substituted indoles are mixed with various acceptors vary in a manner similar to the maxima of the corresponding hexa-methylbenzene-acceptor complexes (see Chapter 3). The reversible dissociation of the complexes on dilution and on warming, together with the magnitude of the enthalpy of formation of typical complexes (e.g. $\Delta H^{\ominus c} = -1 \cdot 9$ kcal mol^{-1} for indole–chloranil in carbon tetrachloride), is added evidence for the charge-transfer complexing potentialities of indoles with acceptors.[52]

More recently the stabilities of complexes of some indoles and related compounds with the acceptors 1,3,5-trinitrobenzene and 1,4-dinitro-benzene have been measured,[53] using an n.m.r. method described in Chapter 6. The stabilities of the complexes are high; for example, the indole–1,3,5-trinitrobenzene complex is more than twice as stable as the corresponding N,N-dimethylaniline–1,3,5-trinitrobenzene complex in chloroform at 33·5°C. Furthermore, the trend of the association constant values and the changes in the chemical shifts of the various protons in indole on complex formation (see Chapter 5) support the conclusions of Szent-Györgyi and his co-workers,[48–50] namely, that electron donation appears to derive more from the 3-position than from any other part of the indole molecule. They had suggested, however, that this semi-localization of binding only occurs when interaction involves *localized* vacant orbitals of the acceptor, whereas the n.m.r. results suggest that the effect also occurs when the acceptor has available only delocalized vacant molecular orbitals. Recent molecular-orbital calculations[54] on indoles and substituted indoles using the frontier-electron principle,[55, 56] in conjunction with the experimentally determined charge-transfer transition energies referred to above, support this suggestion. Some evidence for a similar type of localization in the solid phase has been obtained from X-ray crystallography.[57] The interaction of various indoles with thiamine is discussed in Section 12.B.6 below.

12.B.5. Amino Acids and Proteins

The possible involvement of charge-transfer complexes in the binding of FMN to protein and of amino acid charge-transfer complexes in the inhibition of the photochemical reduction of FMN has already been mentioned (Section 12.B.2). The interaction of proteins with aromatic hydrocarbons is discussed in Section 12.B.7.

Most of the studies of the behaviour of amino acids and proteins have been concerned with the behaviour of one particular amino acid, tryptophan. The strong donor properties of the indole ring (Section 12.B.4) appear to make this amino acid somewhat exceptional, although Cilento and Tedeschi[25] have given some evidence for complexing between a pyridinium model compound of NAD^+ with tyrosine and with phenylalanine. Spectral evidence has been obtained for charge-transfer complex formation between both NAD^+ and model pyridinium compounds with chymotrypsinogen, a protein of high tryptophan content.[58]

Evidence of charge-transfer complex formation between proteins and chloranil leaves little doubt that proteins can participate in such complexes.[59] The relationship of such complexes to semi-conductivity in proteins and protein complexes has been studied, particularly by Eley and his co-workers,[60–62] and the subject has been discussed recently by Pullman and Pullman.[4]

In polar solvents, the interaction of amino acids and chloranil involves electron-transfer processes with the formation of the semiquinone ion. The reaction appears to be similar to the action of aliphatic amines on high-potential quinones.[63] There seems to be no evidence as yet to decide whether or not this type of reaction proceeds via a charge-transfer complex (see Chapter 11). With the weaker acceptor, iodine, in aqueous solutions, it is suggested that immediate complexing of amino acids occurs with I^+, rather than with molecular iodine.[63] The spectra of such solutions show optical absorptions corresponding to the triiodide ion (I_3^-).

12*

12.B.6. Nucleic Acid Bases and Related Compounds

The structure of nucleic acids is dominated by the hydrogen-bonded base-pairing in the double helix. Nevertheless, interest has been shown in the possibility of charge-transfer complexation in other aspects of the biochemical behaviour of nucleic acid and of its base components. Estimates of the ionization potentials of the nucleic acid bases, as potential components in such complexes, show that they should be reasonably effective π-donors particularly in the case of guanine.[64–70] For comparisons of donor ability amongst the purines and pyrimidines as a group, values of the resonance integral coefficients, obtained by Hückel LCMO calculations, are available (Table 12.1). Care must be taken in applying arguments concerning the behaviour of the nucleic acid bases to nucleic acids themselves. For example, there appear to be differences in the hydrogen-bonded interaction of DNA with actinomycin, compared with the purine–actinomycin interaction.[71]

Crystallographic studies of 8-azaguanine (XXI) indicate an intermolecular separation between successive planes, which is ~0·2 Å less than

$$
\begin{array}{c}
\text{O} \\
\text{H}-\overset{1}{\text{N}} \quad 6 \quad \overset{7}{\text{N}} \\
\quad \quad \quad \text{N} \, 8 \\
\text{H}_2\text{N} \quad \overset{2}{\text{N}} \quad \text{N} \, 9 \\
\quad \quad 3 \quad \quad \text{H}
\end{array}
$$

XXI

the normal van der Waals distance.[72] This has led to the suggestion[72] that charge-transfer forces are involved, although as yet no appropriate optical transition has been observed. However, calculations[70] suggest that the resonance energy for the intermolecular interaction is very small and consequently charge-transfer forces can play only a negligible role in such nucleic acid structures. 8-Azaguanine is of interest since it acts as a cell poison.[73] It is suggested that the incorporation of such a molecule into the RNA chain may cause a looping of the RNA molecule either by the interaction of two azaguanine residues or the interaction of one with a natural base residue. These loops will form obstructions and will prevent the movement of the ribosome along the chain and so inhibit the normal protein synthesis in the cell.

Attempts have been made to form charge-transfer complexes between nucleic acid bases or nucleosides and the simple strong acceptor chloranil, by dissolving the two components in dimethylsulphoxide solution.[74, 75] Although new absorption bands were observed, these developed only slowly at room temperature. It seems unlikely that reactions with significant activation energies can involve the formation of structures which we describe as charge-transfer complexes. The suggestion by Slifkin[76] that an electron transfer process occurs between the two solute species, which is then followed by some further chemical reaction, seems a more likely explanation of the colour formation, especially when the nature of the solvent is considered.

Evidence for some type of complexing between thiamine and various indoles has been obtained from n.m.r. spectroscopy.[77] In D_2O solutions in which thiamine appears as XXII, all the proton magnetic resonances

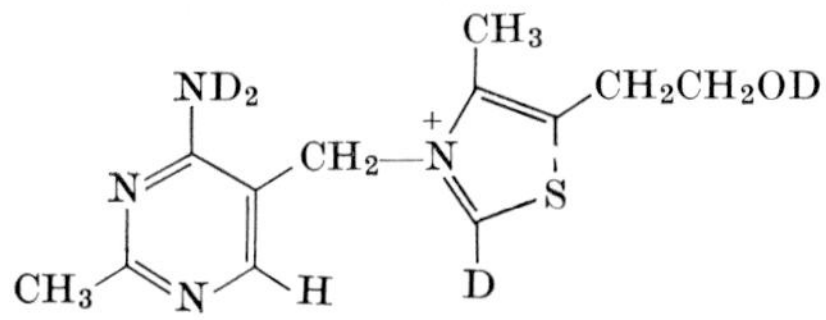

XXII

move upfield as the indole is added. The effect is very marked for the aromatic-ring proton, the methylene-bridge protons and the pyrimidine–methyl protons. A general discussion of this effect, in terms of complex formation, has been given in Chapter 5. The major perturbation being concentrated on the pyrimidine ring suggests that the interaction with the indole molecule occurs at this ring. These observations in themselves do not prove that charge-transfer forces make any significant contribution to the binding, although the nature of the two interacting species and the localization of interaction suggest that this is likely.

The question of charge-transfer interaction between aromatic hydrocarbons and nucleic acids or nucleic acid bases is discussed in the following section.

12.B.7. Aromatic Hydrocarbon Carcinogens

Cancer is not a single disease: it is a term which describes that whole range of conditions which involve cell malignancies. Chemical carcinogens include certain aromatic hydrocarbons, aromatic amines, urethanes, nitrogen mustards, carbon tetrachloride as well as a variety of inorganic substances.[78–81] The term "chemical carcinogenesis" is consequently a blanket expression which covers what is almost certainly a wide variety of biochemical mechanisms. In the case of such carcinogens as the alkylating agents, covalent bond formation is undoubtedly involved. The absence of *obviously* reactive chemical sites in aromatic hydrocarbon carcinogens (but see below) has marked off these compounds as a single group, although there is no evidence that they have a common mode of action.[82]

Many aromatic hydrocarbons form well-defined complexes with electron acceptors, both in the solid phase and in solution. The numerous measurements which have been made on such systems (see Chapters 3 and 7) leave no doubt that in most cases charge-transfer complexes are formed.[45, 83-88] Some highly carcinogenic aromatic hydrocarbons, such as dibenz[a,h]anthracene (1,2,5,6-dibenzanthracene (XXIII), 9,10-dimethylbenz[a]anthracene (9,10-dimethyl-1,2-benzanthracene) (XXIV), benzo[a]pyrene (3,4-benzopyrene) (XXV), and 20-methylcholanthrene (XXVI), have low ionization potentials. However, the claims

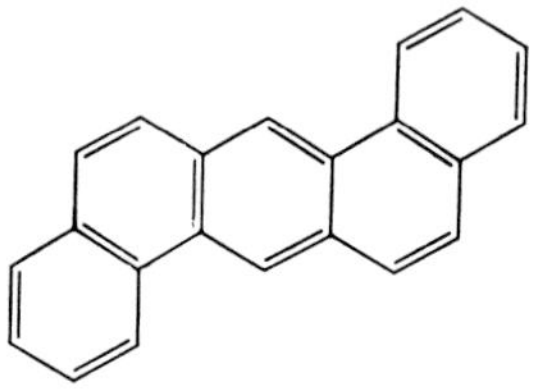

XXIII

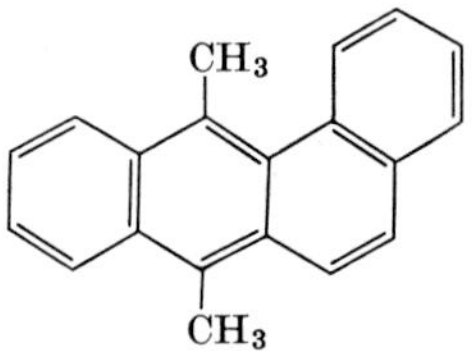

XXIV

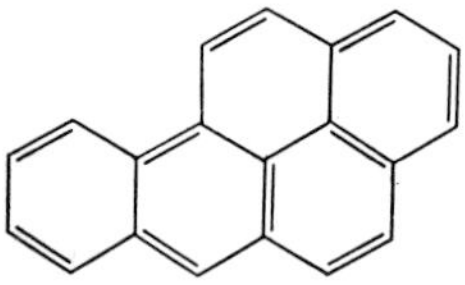

XXV

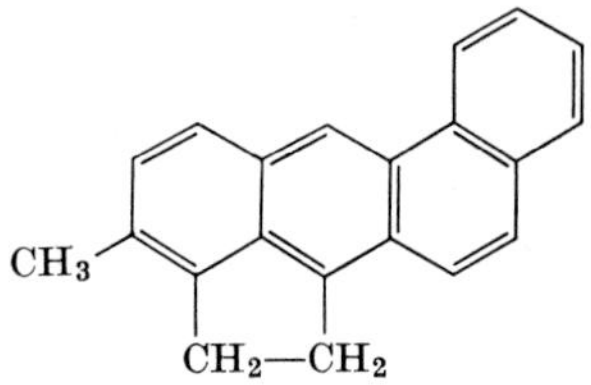

XXVI

have been varied for a correlation between carcinogenicity and ionization potential or some other parameter such as the calculated highest-filled molecular orbital, the experimentally determined energy of the charge-transfer band, including visual estimates of the colour,[46, 89-91] or the stability of the complex with a given acceptor in solution.[92, 93] Experimental difficulties in measuring small association constants (Chapter 6) may lead to incorrect values, so that some of the conclusions concerning the presence or absence of a correlation between experimental values of association constant and carcinogenicity may not be meaningful. Certainly, when a sufficiently large number of hydrocarbons, both biologically active and inactive, are compared, it becomes apparent[81, 91] that there is no simple correlation between carcinogenic activity and electron-donor properties of the hydrocarbons (Table 12.3). There is a similar lack of evidence[81, 94] to support the hypothesis of a direct

TABLE 12.3. Electron donor (or acceptor) properties of hydrocarbons as measured by the absolute value of the coefficient k of their highest-filled (or lowest-empty) molecular orbital.*†

Compound‡	k	Carcino-genicity
Benzene	1	−
Triphenylene	0·684	−
Naphthalene	0·618	−
Phenanthrene	0·605	−
Benzo[c]phenanthrene (3,4-benzphenanthrene)	0·566	+
Dibenzo[e,l]pyrene (1,2,6,7-dibenzpyrene)	0·555	−
Dibenzo[a,g]phenanthrene (1,2,5,6-dibenzphenanthrene)	0·550	+
Coronene	0·539	−
Dibenzo[c,g]phenanthrene (3,4,5,6-dibenzphenanthrene)	0·535	−
Dibenzo[a,c]phenanthrene (1,2,3,4-dibenzphenanthrene)	0·532	+
Chrysene	0·520	−
Picene	0·501	−
Dibenz[a,c]anthracene (1,2,3,4-dibenzanthracene)	0·499	−
Benzo[e]pyrene (1,2-benzpyrene)	0·497	−
Dibenz[a,j]anthracene (1,2,7,8-dibenzanthracene)	0·492	+
Dibenz[a,h]anthracene (1,2,5,6-dibenzanthracene)	0·473	++
Benz[a]anthracene (1,2-dibenzanthracene)	0·452	±
Pyrene	0·445	−
Benzo[a]perylene (1,2-benzperylene)	0·439	−
Pentaphene	0·437	−
Dibenzo[b,g]phenanthrene (2,3,5,6-dibenzphenanthrene)	0·419	−
Anthracene	0·414	−
Dibenzo[a,h]phenanthrene (1,2,6,7-dibenzphenanthrene)§	0·405	−
Dibenzo[a,l]pyrene (1,2,3,4-dibenzpyrene)	0·398	++
Dinaphtho[1,2-b; 1,2-k]chrysene (4,5,10,11-di-1′,2′-naphthochrysene)	0·383	−
Benzo[a]pyrene (3,4-benzpyrene)	0·371	+++
Dibenzo[a,l]naphthacene (1,2,9,10-dibenznaphthacene)	0·361	−
Anthra[1,2-a]anthracene (1′,2′-anthra-1,2-anthracene)	0·360	−
Dibenzo[a,j]naphthacene (1,2,7,8-dibenznaphthacene)	0·358	−
Dibenzo[a,c]naphthacene (1,2,3,4-dibenznaphthacene)	0·356	−
Dibenzo[b,k]perylene (2,3,8,9-dibenzperylene)	0·356	−
Anthra[2,1-a]anthracene (2′,1′-anthra-1,2-anthracene)	0·348	−
Perylene	0·347	−
Dibenzo[a.i]pyrene (3,4,9,10-dibenzpyrene)	0·342	+++
Benzo[a]naphthacene (1,2-benznaphthacene)	0·327	−
Dibenzo[a,h]pyrene (3,4,8,9-dibenzpyrene)	0·303	+++
Naphtho[2,3-a]pyrene (2′,3′-naphtho-3,4-pyrene)	0·303	−
Naphthacene	0·295	−
Anthanthrene	0·291	−
Dinaphtho[2,3-a; 2′,3′-i]pyrene (3,4,9,10-di(2′,3′-naphtho-pyrene)	0·273	−
Pentacene	0·220	−
Naphthodianthrene	0·177	−

* Ref. 103.
† The smaller k the better the donor (and acceptor) properties of the hydrocarbon.
‡ For alternative name see Compound Index.
§ Described as 2,3,7,8-dibenzphenanthrene in Ref. 103.

connection between carcinogenicity and electron affinity amongst the aromatic hydrocarbons.[95, 96] Arguments that the electron-accepting ability should fall within a certain range[97–100] have been criticized[101] for theoretical reasons; in any case, no strong correlation is observed if a sufficiently large number of aromatic hydrocarbons are taken. The suggestion that the hydrocarbon carcinogens must be simultaneously good electron donors and good electron acceptors has been made.[90, 102] This postulate has met with considerable criticism.[103] For example, it has been pointed out that for unsubstituted aromatic hydrocarbons there is inevitably a correlation between their electron-donating and electron-accepting properties. Again, when a large group of aromatic compounds is taken, any apparent correlation between such properties and their carcinogenicity shown by a smaller and somewhat selected group of compounds, disappears. Birks[104] has suggested that the biochemical action of the hydrocarbon carcinogens may be through an energy-transfer process from an excited state of the protein moiety of a hydro-carbon–protein complex to the hydrocarbon in an excited state. Some problems of this mechanism have been discussed by Pullman.[105]

In fact, the carcinogenic properties of certain aromatic hydrocarbons may be directly dependent on their chemical reactivity, which is far greater than was at one time believed. For example, covalent bonding to protein *in vivo* can occur. Various hydrocarbons, when painted on mouse skin, rapidly become attached to protein by strong chemical bonds.[106–108] There appears to be a correlation between this chemical reactivity and the ability of the hydrocarbon to form skin tumours.[108, 109] There are, however, some exceptions such as dibenz[a,c]anthracene (1,2,3,4-dibenzanthracene) (XXVII) and anthanthrene (XXVIII), which

XXVII   XXVIII

are strongly bound but which are non-carcinogenic. However, these exceptions have a marked difference in their electrophoretic behaviour compared with those strongly bound hydrocarbons which are bio-logically active.[110, 111] Degradative evidence whereby, for example, 2-phenylphenanthrene-3,2'-dicarboxylic acid (XXIX) is obtained from the dibenz[a,h]anthracene–protein adduct[112, 113] shows that the reactive centre of this hydrocarbon is in the 5,6-position, the so-called K region

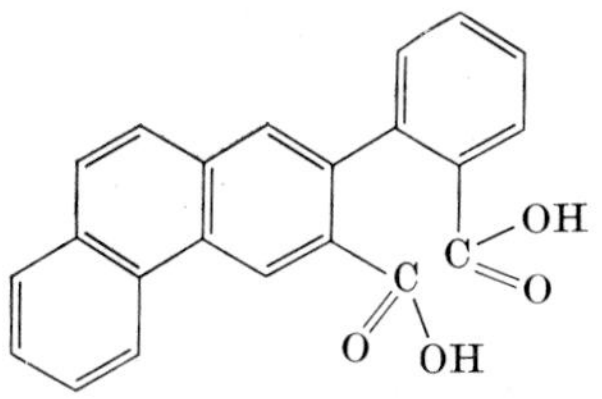

XXIX

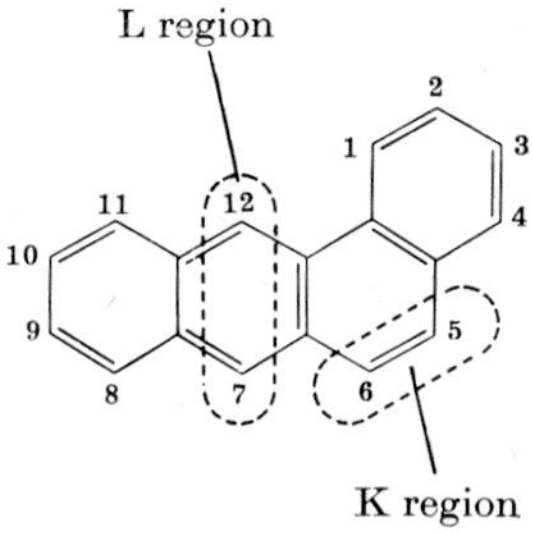

XXX

(see structure XXX).* This is convincing evidence in support of the proposals that the localized reactivity in this region of the benz[a]anthracenes is important for their carcinogenicity.[81, 105, 114–118] The estimates of bond order,[116] free valence[119] and localization energy[120, 121] indicate a high reactivity in this region.[122] Reaction with osmium tetroxide is known to occur across this bond.[123] In order to explain cases such as unsubstituted benz[a]anthracene which has a reactive K-region but is not carcinogenic, it is suggested that the reactivity in another region of the molecule, the L-region (see structure XXX) must be low for the compound to be biologically active.[105, 118, 124] Reasonable correlations have been obtained by making this assumption.[81]

The heterocyclic compound tricycloquinazoline (XXXI) is of interest in that it is carcinogenic for mouse skin[125, 126] although it has no region corresponding to a K-region and does not bind strongly with epidermal protein.[127] Evidence for the interaction of XXXI and of the analogue XXXII with DNA, which may involve charge transfer, has been given recently.[128]

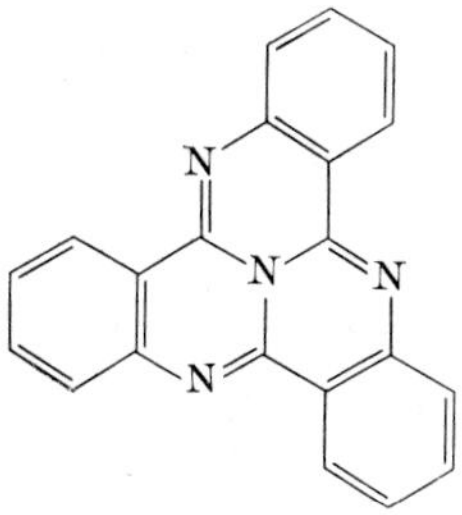

XXXI

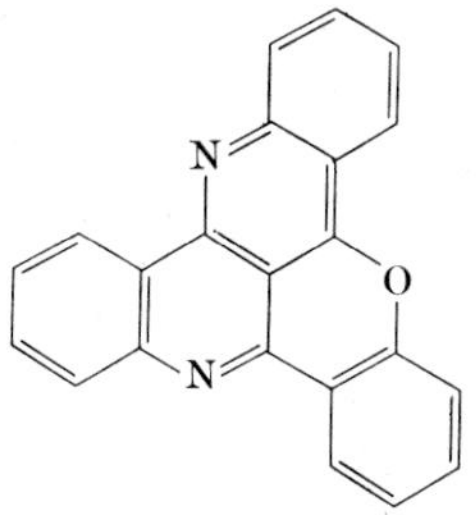

XXXII

Molecular shape is apparently important, as was noted in the early work by Cook.[129] Recently, the direct increase in carcinogenicity as the aromatic hydrocarbons become structurally similar to the steroids has been noted.[130] This suggests that such carcinogens may act at the same

* U.P.A.C. numbering.

sites as the steroid hormones. Contrary to earlier ideas, it is now established that many of these polycyclic aromatic hydrocarbons deviate considerably from planarity.[131]

However, the lack of carcinogenicity may be due to other biochemical modifications of the hydrocarbon; for example oxidation or hydroxylation to yield an inactive product (so-called detoxification mechanisms). Such biochemicals may, of course, produce the opposite effect, namely, to form active carcinogens from inactive substances. Although there is as yet no evidence of this amongst the aromatic hydrocarbon carcinogens, metabolic products which are active are known in other groups of chemical carcinogens; for example, the biochemical hydroxylation of 2-naphthylamine to form the active 1-hydroxy-2-naphthylamine.[132] There is evidence that simpler non-carcinogenic aromatic hydrocarbons such as benzene and naphthalene can affect proteins, causing, for example, enzyme inhibition.[133]

Although proteins are considered to be probably the most important site for interaction with carcinogens, another possible site of action is at a nucleic acid.[4, 134] The biochemical mechanism may have features common to the phenomenon of chemical mutagenesis which almost certainly involves nucleic acids.[134, 135] Many observations of chemical interaction between aromatic hydrocarbons and purines, pyrimidines or nucleic acids, refer to the solubilization of the hydrocarbon.[136–141] On the basis of model systems [142, 143] it appears that the mode of interaction is the insertion of the hydrocarbon molecules between adjacent nucleic acid basic pairs.[140, 141] This results in disordered sections in the helical structure of the DNA molecule. It has been observed [140] that in RNA, where the helical structure is not so dominant, the effect of the addition of hydrocarbons is small. The solubilizing power towards a given hydrocarbon increases with the electron-donating ability, for a series of purines.[137, 144, 145] Although such a correlation might be expected if charge-transfer complexation were involved, this evidence by itself does not justify such an assumption. Indeed, there is no correlation between the electron affinities of the hydrocarbons and their solubilization by 1,3,7,9-tetramethyluric acid,[146] which would be expected if charge transfer was important. Furthermore, no charge-transfer bands have been reported for the purine–hydrocarbon systems although, for the reasons given at the beginning of this chapter, this may be accounted for by the relatively poor electron-accepting properties of the hydrocarbons. There is now a growing opinion that the dominant force of attraction in these complexes is not charge transfer but rather dipole–induced-dipole interaction between the polar purine and the polarizable hydrocarbon,[141, 147] or possibly dispersion forces.[148]

Nevertheless, it still remains uncertain as to whether such nucleic acid complexes or corresponding protein–hydrocarbon complexes have any charge-transfer character and whether, whatever their nature, they have a significant part to play in the processes of carcinogenesis. The lack of correlation between solubilization and carcinogenicity would argue against the latter proposition.[105]

Although the evidence for the involvement of charge-transfer complexing between the hydrocarbon and biological substrate is at present more negative than positive, it is very apparent that our knowledge of the processes involved is so tenuous that it is impossible to do little more than speculate. It is in some ways surprising that attempts have been made to relate this type of biological activity with charge-transfer complex formation when there are so many uncertainties concerning the initial step in the biochemical reaction. It is partly a reflection of our ignorance that such correlations have been hopefully sought.

12.B.8. Drug Interactions

Little progress has yet been made in the general problem of drug action at the molecular level.[149] Some substances such as the anaesthetics and other lipotropic drugs often show a low specificity and may well act by an essentially "physical" mechanism. However, such compounds are atypical in that the majority of physiologically active compounds foreign to the biochemical substrate show a high degree of specificity. This particular characteristic is even more prominent in the case of natural chemical substances.

The use of the term "receptor"[150, 151] to denote the site of action of a compound is in itself an indication of our ignorance concerning the molecular action of the drug. Added to this are the problems of penetration, absorption, transportation and excretion, as well as other hazards of the various enzymatically controlled biochemical reactions which the drug may encounter. The basic process of stimulation at the molecular level has given rise to two modifications of the receptor theory. Some workers consider that a drug acts so long as it is attached to the receptor but that different drugs have different efficacies when they occupy the receptor site.[152, 153] An alternative explanation of the differences shown by two drugs at the same receptor is that the activation of the receptor depends on the rate of encounter with a drug at the receptor (Paton's kinetic hypothesis).[154]

Two drugs which have a similar gross effect on the tested tissue may nevertheless be acting at different receptor sites. On occasion, this may in fact be demonstrated. The immediate effect of the attached drug may be due to the perturbed receptor, or it may be caused by the drug-moiety of the complex affecting other processes. These may not readily be determined, particularly since the electron distribution and conformation of both parts of the drug-receptor complex may be very different from those of either part in isolation.

Such problems have caused pharmacologists to turn their attention to the relationship between the chemical structure of drugs and their physiological action. These studies, which can generally be made at least semi-quantitative, may lead to a better understanding of the molecular nature of drug-receptor interactions.

In many cases, it would appear that the binding of the drug to the receptor is of relatively low energy, certainly considerably less than that involved in normal covalent bonding. Thus, in typical pharmacological assays, a drug can be washed out of a preparation and replaced by another drug which can in turn be replaced by the original drug. This drug may produce the same, or nearly the same, response for a given dose as initially. The whole process may be repeated many times. Incidentally, Paton's rate theory of stimulation [154] demands that the drug-receptor complex shall be capable of dissociation, since on this theory a high stimulant activity demands, in general, not only a high rate of association but also a high rate of dissociation in order to maintain the rate of association.

Chemically, this reversible behaviour could occur by ionic association, through hydrogen bonding, by other weaker forces which might include charge-transfer forces, or, what is more likely, a combination of several of these types of forces.

As a first step to determine whether charge-transfer forces are in any way involved, the ability of drugs and related compounds to form charge-transfer complexes with well-characterized electron acceptors or electron donors, mainly in non-aqueous solvents, has been studied. Evidence for complexing by indoles with electron acceptors has already been mentioned (Section B.4).

Considerable interest has been shown in the phenothiazine drugs. Various workers have noted the low ionization potentials of these compounds.[155, 156] Calculations of the highest-occupied molecular orbital of phenothiazine (XXXIII) and of chlorpromazine (XXXIV)

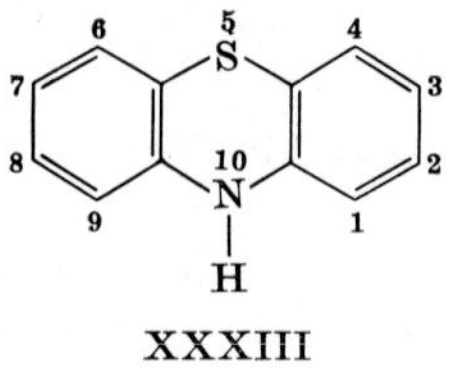

XXXIII

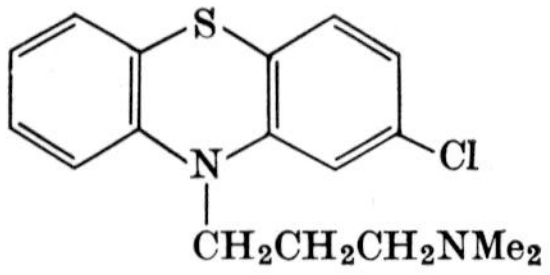

XXXIV

also indicate that these molecules should be strong electron donors,[157, 158] though the estimate is somewhat modified when the sulphur d-orbitals are taken into account.[159] Various phenothiazines and the structurally related phenoxazine (**XXXV**) and thianthrene (**XXXVI**) show electronic

XXXV   XXXVI

absorption band(s) extra to the absorptions of the components when mixed with one of a variety of acceptors in acetonitrile.[160] When the energies of the maxima of these bands are plotted against energy maxima for the corresponding hexamethylbenzene–electron-acceptor complexes,

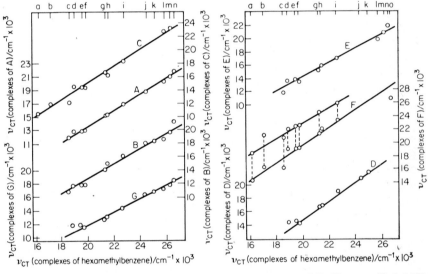

[R. Foster and P. Hanson, Ref. 160]

Fig. 12.1. Plots of ν_{CT} for complexes of: (A) phenothiazine; (B) 3,7-dimethyl-phenothiazine; (C) 10-methylphenothiazine; (D) chlorpromazine; (E) phenoxazine; (F) thianthrene; (G) N,N,N′,N′-tetramethyl-p-phenylenediamine, with the acceptors: (a) 2,3-dichloro-5,6-dicyano-p-benzoquinone; (b) 2,3-dicyano-p-benzoquinone; (c) tetracyanoethylene; (d) bromanil; (e) chloranil; (f) trichloro-p-benzoquinone; (g) 2,6-dichloro-p-benzoquinone; (h) 2,5-dichloro-p-benzoquinone; (i) chloro-p-benzoquinone; (j) p-benzoquinone; (k) methyl-p-benzoquinone; (l) 1,3,5-trinitrobenzene; (m) 1,4-dinitrobenzene; (n) 2,4,6-trinitrotoluene; (o) iodine; all against ν_{CT} for the corresponding hexamethylbenzenes. (The plots are separated into two groups for clarity.)

straight lines are obtained (Fig. 12.1), which are typical of charge-transfer complexes (see Chapter 3). The equilibria have been measured for various phenothiazine drugs, including promethazine (XXXVII), ethopropazine (XXXVIII), diethazine (XXXIX), trimeprazine (XL) and promazine (XLI), with the electron acceptor 1,4-dinitrobenzene in

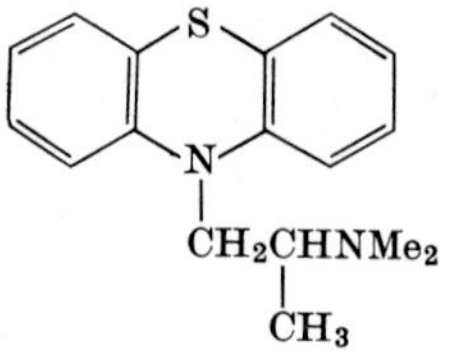

XXXVII

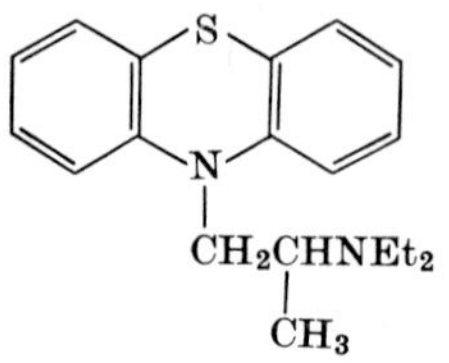

XXXVIII

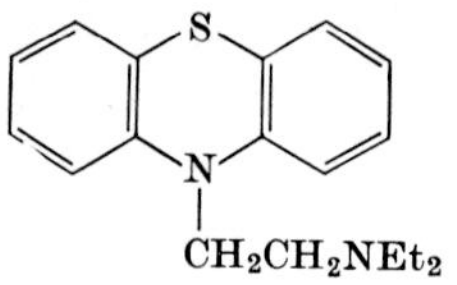

XXXIX

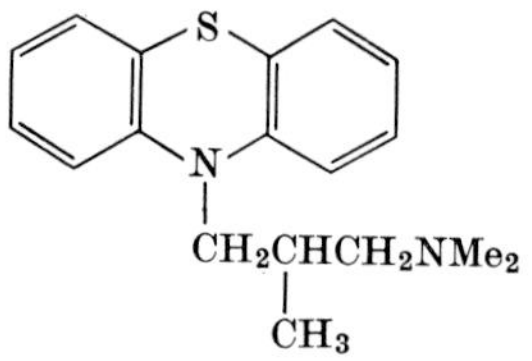

XL

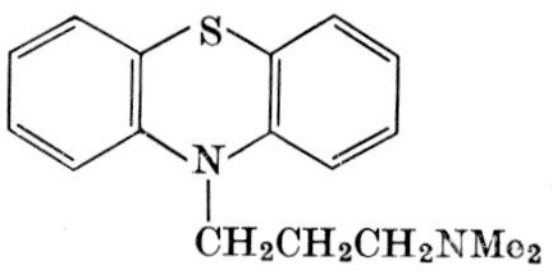

XLI

chloroform and in carbon tetrachloride solution.[161] The n.m.r. chemical shift method described in Chapter 6 was used. The results (Table 12.4) show that, relative to N,N-dimethylaniline, these compounds are good electron donors and form 1:1 complexes with electron acceptors. Biochemically inactive 10-methylphenothiazine has a higher association constant than any of the drug molecules measured, all of which have a large side-chain in the 10-position. This suggests that complex formation may be reduced by steric interference from this side-chain in these particular complexes. By contrast, a double role by the phenothiazine drug molecules at a biological site of action has been suggested,[162] whereby the drug acts as a π-electron donor in a charge-transfer interaction with one receptor and as a hydrogen-bonding agent via a protonated aliphatic amino nitrogen atom in the side-chain at a second receptor. These two suggestions are not necessarily conflicting. They do illustrate, however, the care needed in drawing conclusions about possible biochemical behaviour based on observations on model systems, particularly those not intended to represent the drug-receptor complex.

Electrical properties of the phenothiazine and chlorpromazine complexes with iodine have been measured,[163] both in the solid phase and in solution. The results suggest that these complexes may occur with stoichiometries other than 1:1.

TABLE 12.4. Association constants (K_r^{AD}) and chemical shifts (Δ_0) for protons in the acceptor moiety for complexes of various phenothiazine donor molecules with 1,4-dinitrobenzene in $CHCl_3$ solution and in CCl_4 solution at $33 \cdot 5°C$.*

		CHCl$_3$ solution		CCl$_4$ solution	
	Donor	K_r^{AD}/kg mol	Δ_0 †/Hz	K_r^{AD}/kg mol	Δ_0 †/Hz
XXXIV	Chlorpromazine	$0 \cdot 20_5$	143	$1 \cdot 19$	$58 \cdot 6$
XXXVII	Promethazine	$0 \cdot 23_8$	153	$1 \cdot 32$	$72 \cdot 2$
XXXIX	Diethazine	$0 \cdot 28_8$	132	$1 \cdot 28$	$63 \cdot 0$
XXXVIII	Ethopropazine	$0 \cdot 23_0$	167	$1 \cdot 25$	$76 \cdot 0$
XLI	Promazine	$0 \cdot 29_3$	135	$1 \cdot 40$	$71 \cdot 3$
XL	Trimeprazine	$0 \cdot 27_4$	146	$1 \cdot 24$	$77 \cdot 0$
	10-Methylpheno-thiazine	$0 \cdot 35_9$	124	—‡	—‡
	Dimethylaniline	$0 \cdot 14_3$	138	$0 \cdot 798$	$68 \cdot 7$

* Ref. 161.
† Proton chemical shifts measured at $60 \cdot 004$ MHz.
‡ Too insoluble to be measured.

N.m.r. studies have been made[164] on a group of local anaesthetics, including benzocaine (XLII), procaine (XLIII), lignocaine (xylocaine) (XLIV), and prilocaine (XLV), complexed with 1,3,5-trinitrobenzene and

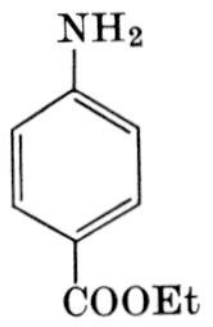

XLII

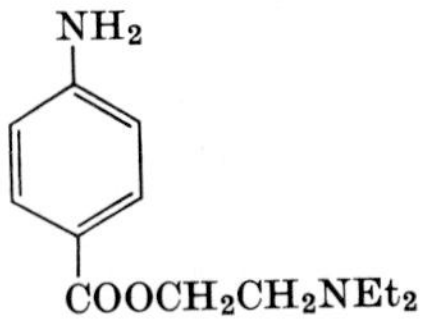

XLIII

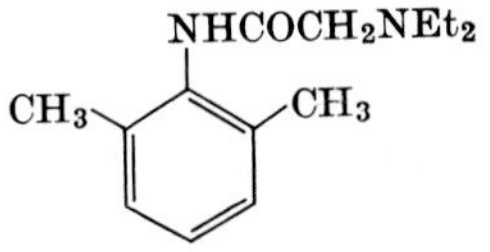

XLIV

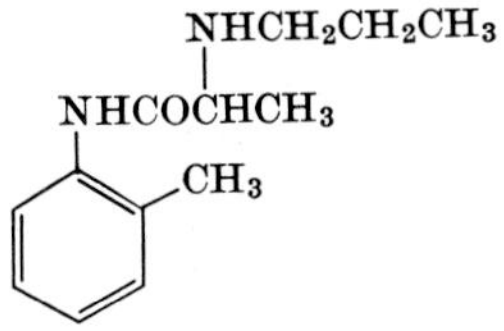

XLV

with 1,4-dicyanotetrafluorobenzene (Table 12.5). From a comparison of benzocaine or procaine with aniline, the inclusion of a carbonyl substituent with its —M effect does not appear to reduce the donor property as it appears in the stability of the complex. In the case of procaine, the

TABLE 12.5. Association constants (K_r^{AD}) and relative chemical shift values for nuclear magnetic resonance of nuclei in the acceptor moiety of the complex for various electron-acceptor–local anaesthetic drug complexes and related complexes at 33·5°C.*

Acceptor		1,3,5-Trinitrobenzene		2,3,5,6-Tetrafluoro-1,4-dicyanobenzene	
Donor	Solvent	K_r^{AD}/kg mol^{-1}	Δ_0 †/Hz	K_r^{AD}/kg mol^{-1}	Δ_0 ‡/Hz
Benzocaine	CHCl$_3$	0·5$_1$	57	0·5$_4$	104
Procaine	CCl$_4$	3·4	39	6·2$_7$	49
Procaine	CHCl$_3$	0·5$_6$	63	0·6$_4$	84
Lignocaine	CCl$_4$	9·6	43	11·1	88
Lignocaine	CHCl$_3$	0·7$_0$	58		
Prilocaine	CCl$_4$	8·0	37	7·5$_3$	78
Prilocaine	CHCl$_3$	0·7$_9$	51	0·5$_9$	133
Aniline	CCl$_4$	2·3$_8$	53	4·1$_8$	76
Aniline	CHCl$_3$	0·5$_4$	66	0·5$_6$	122
2,6-Dimethylaniline	CCl$_4$	3·5	61	4·3$_5$	99
2,6-Dimethylaniline	CHCl$_3$	0·8$_2$	70	0·8$_0$	141

* Ref. 164.
† Relative shift of ^{1}H in the acceptor measured at 60·004 MHz.
‡ Relative shift of ^{19}F in the acceptor measured at 56·462 MHz.

association is measurably greater than for aniline. Similar increases in association are observed in lignocaine and in procaine, compared with 2,6-dimethylaniline (Table 12.5). Although the increases are small when chloroform solutions are used, the differences are considerable for carbon tetrachloride solutions. This increase in association may be the result of the cooperative effect of n-electron donation by the aliphatic amino group, which could interact with the opposite side of the acceptor from the interaction with the π-electron of the donor (XLVI). Competition by the solvent for the n-donor site in particular could account

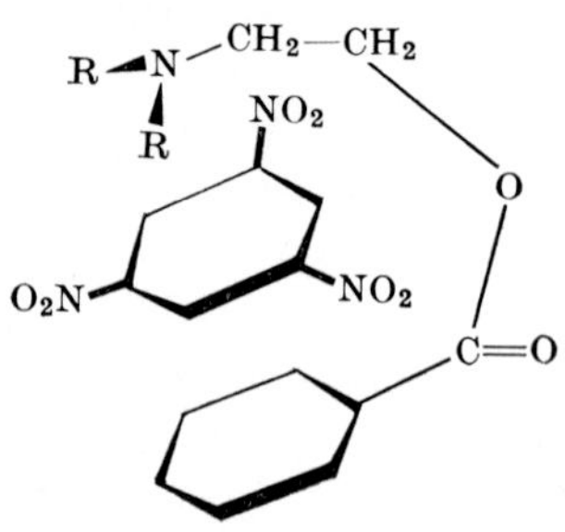

XLVI

for the differences caused by these solvents. Although the experimental method cannot differentiate between such a doubly bound complex and a mixture of two types of complex involving either π- or n-bonding from the donor, experiments with mixtures of π- and n-donors would seem to argue against the presence of two types of complex. The difference between the potentiating effect of the side chain in these compounds, and the adverse effect on association shown by the side chains in the phenothiazine drugs, is accounted for by the stereochemistry of the ring system in this latter group, which has a non-planar folded structure.

Although there is reasonable evidence for charge-transfer complex formation when 1,3,5-trinitrobenzene is mixed with various local anaesthetics of the procaine–benzocaine group, the colours observed[165] may not be wholly due to this type of interaction. Other products having high molar absorptivities in the visible region, when aliphatic amines are mixed with 1,3,5-trinitrobenzene and similar compounds, have been reported.[166] However, the n.m.r. determinations described above indicate that for these particular solutions the formation of products other than charge-transfer complexes is negligibly small, although the visual appearance can be deceptive. The relatively sharp absorption bands reported[167–169] for the difference spectra of mixtures of procaine hydrochloride with compounds such as thiamine, caffeine or dimethyluracil appear to be artifacts of the method.[170]

Despite some misleading observations, there now appears to be sufficient evidence to assert that many compounds used as drugs can participate in charge-transfer complex formation, usually as the donor species with simple acceptor molecules. A possible practical line of progress would be to take series of compounds closely related in structure and to test whether there is any relation between the charge-transfer complexes formed with a given second component and the physiological activities of the compounds. This would be analogous to the pharmacologist's quest for structure–action relationships. No extensive experiments of this type have yet appeared.

Another development is the study of interactions which are anticipated to be more akin to the drug-receptor complex. A knowledge of the behaviour of the interaction of drugs with amides, including simple polypeptides, would be of value. Agin[170] has reported that a transient magenta colour develops when procaine hydrochloride is mixed with RNA. He suggests that this is the result of charge-transfer complex formation. The band appears to be at low energy (18,200 cm^{-1}, 550 nm) for a charge-transfer complex between these particular components. Without the comparison of related electron-donor–electron-acceptor systems, it is difficult to be confident of such an assignment. Further-more, the colouration does not appear immediately, thus arguing against this type of complex.

The suggestion has been made[157] that the mechanism of action of chlorpromazine and other psychotropic drugs may in some way be due to the electron-donating ability of these compounds. This idea has been extended by Snyder and Merril,[171] who have claimed that there is a correlation between the highest-occupied molecular orbital for various tryptamines, amphetamines, phenylethylamines and lysergic acid diethylamide, and their hallucinogenic potency. It would be of interest to know something of the ability of these compounds to form charge-transfer complexes with various electron acceptors. Other workers have suggested that the psychotropic activity of chlorpromazine is due in fact to the positive ion radical,[172] formed by loss of a single electron. Evidence in support of its intercalation in the DNA helix has been given.[173]

12.B.9. Photo-synthesis

In photo-synthesis, light quanta are absorbed by chlorophyll, which somehow become the driving force for the overall photo-synthetic process, namely, the conversion of carbon dioxide and water into carbohydrates and oxygen.[174–176] A central problem is how this relatively high energy which is absorbed can be trapped and used in such a process rather than be immediately reradiated or dissipated as heat. The model experiments of Calvin et al.[177] with laminar phthalocyanine–electron acceptor systems (Chapter 9) have led to the suggestion that a similar process is occurring at the biological site of photo-chemical energy transfer, called the chloroplast, which also has a laminar structure.[178] In this scheme chlorophyll (XLVII) with its conjugated dihydro-porphyrin ring structure acts as the donor in forming a charge-transfer complex with plastoquinone, a quinone specific to chloroplasts. Light absorption by the chlorophyll causes electron transfer to the plasto-quinone. This is followed by hole-migration through neighbouring chlorophyll molecules so that the back-reaction is prevented. The migrating hole eventually reaches a cytochrome system at which electron-transfer from the iron of the cytochrome, or some other electron donor, restores the chlorophyll to its initial state.

Some evidence for the formation of charge-transfer complexes by chlorophyll and other porphyrin-like molecules has recently been obtained.[47, 179–183] Some of these porphyrins are involved in the related electron-transport chain in respiration. Addition of an electron acceptor to a solution of a porphyrin causes a decrease in intensity and a broaden-ing of the band structure of the porphyrin. The spectra show isosbestic

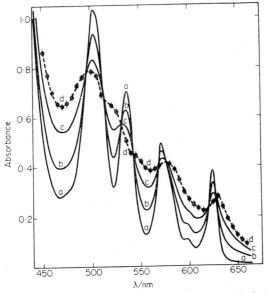

CH=CH₂ X
α
CH₃— I   II —C₂H₅
N   N
δ   Mg   β
N   N
CH₃ IV   III —CH₃
H
CH₂   H
| |
CH₂   H—C—C=O
| |
COOC₂₀H₃₉   COOCH₃

phytyl

XLVII

chlorophyll a: X = CH₃
chlorophyll b: X = CHO

points, which may be taken as evidence of 1:1 complex formation (Fig. 12.2). Only little evidence has been obtained for the presence of an intermolecular charge-transfer transition. There is some indication of an

[M. Gouterman and P. E. Stevenson, Ref. 179]

FIG. 12.2. Absorption spectra of aetioporphyrin I (concentration 7.0×10^{-5} mol l⁻¹) in nitrobenzene solvent: (a) no 1,3,5-trinitrobenzene (TNB); (b) TNB 0.04 mol l⁻¹; (c) TNB 0.20 mol l⁻¹; (d) extrapolated pure complex absorption.

extra band hidden under the band system of the porphyrin in some cases. For example, in the 1,3,5-trinitrobenzene complexes of aetioporphyrin I (XLVIII) and of tetraphenylporphyrin, a charge-transfer absorption may be present in the region of 20,000 cm^{-1}.[79] (See Fig. 12.2.) A similar behaviour is shown by zinc phthalocyanine (XLIX) in the presence of an

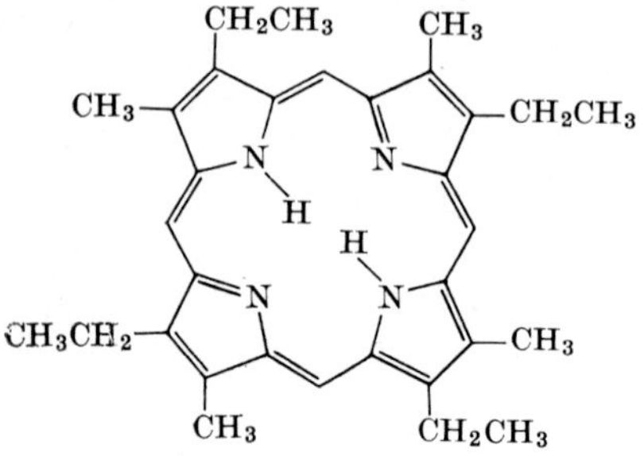

XLVIII 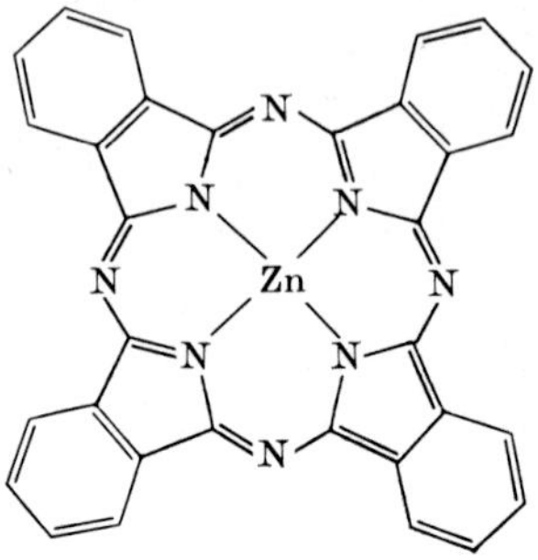   XLIX

electron acceptor.[184] However, as Mauzerall[185] points out, the direct spectroscopic evidence for intermolecular charge-transfer absorption is not large.

TABLE 12.6. Association constants (K_c^{AD}) for complexes of various electron acceptors with some porphyrin-type molecules* at ca. 23°C.

Acceptor	Donor†	Solvent	λ/nm‡	K_c^{AD}/l mol^{-1}
2,4,7-Trinitrofluorenone	Meso p	CHCl$_3$	497	2006 ± 340
2,4,7-Trinitrofluorenone	Aetio p	CHCl$_3$	499	2356 ± 355
2,4,7-Trinitrofluorenone	Co(II) meso p	CHCl$_3$	522	3038 ± 402
2,4,7-Trinitrofluorenone	Ag(II) meso p	CHCl$_3$	561	2261 ± 256
2,4,7-Trinitrofluorenone	Co(II) meso p	CH$_2$Cl$_2$	552	970 ± 124
1,3,5-Trinitrobenzene	Meso p	CHCl$_3$	497	173 ± 28
1,3,5-Trinitrobenzene	Aetio p	CHCl$_3$	499	334 ± 64
1,3,5-Trinitrobenzene	Co(II) meso p	CHCl$_3$	552	102 ± 11.5
1,3,5-Trinitrobenzene	Ag(II) meso p	CHCl$_3$	561	134 ± 13
1,3,5-Trinitrobenzene	Mn(III) meso p	CHCl$_3$	520	~30
1,3,5-Trinitrobenzene	Co(II) meso p	CH$_2$Cl$_2$	552	102 ± 11·5
Benzotrifuroxan	Co(II) meso p	CH$_2$Cl$_2$	552	674 ± 128
4,6-Dinitrobenzofuroxan	Co(II) meso p	CH$_2$Cl$_2$	552	155 ± 27
3,5-Dinitrobenzonitrile	Co(II) meso p	CH$_2$Cl$_2$	552	39 ± 9
2,4,6-Trinitrotoluene	Co(II) meso p	CH$_2$Cl$_2$	552	10 ± 3·7
Vitamin K$_3$	Co(II) meso p	CH$_2$Cl$_2$	552	3·8 ± 4·5

* Ref. 182.
† Meso p ≡ mesoporphyrin IX dimethyl ester, aetio p ≡ aetioporphyrin I.
‡ Wavelength at which optical measurements were made.

Association constants and other thermodynamic constants have been determined for several porphyrin–acceptor systems using optical,[179–182] n.m.r.[180, 183] and osmometric[180] techniques (Table 12.6). The comparable values for the association constants of a given acceptor with cobalt(II)mesoporphyrin IX dimethyl ester (L), with the corresponding

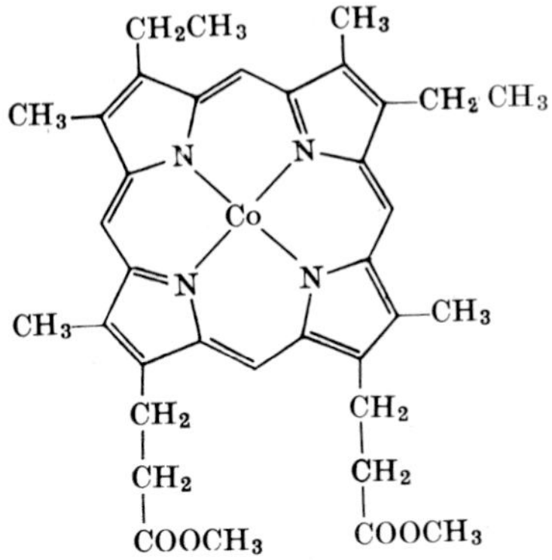

L

Ag(II) derivative and with the metal-free porphyrin (Table 12.6) suggest that it is the electrons of the porphyrin ring and not those of the metal atom which are primarily involved in complex formation.[182] The general trend of larger association constants for complexes of a given donor with expectedly stronger electron acceptors may afford some support* for the suggestion that there is a significant contribution from charge-transfer forces stabilizing the complexes in one or two cases. There is some evidence of porphyrins containing trivalent metals behaving as an electron acceptor.[182] The reported stabilities of the complexes of haematoporphyrin (LI) with aliphatic amine donors seem remarkably

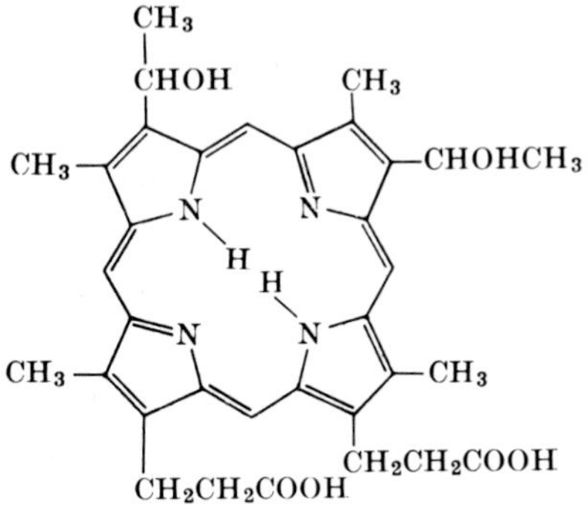

LI

* Care must be taken with such evidence; it can be misleading, see Chapter 7.

large.[186] In the case of metal-containing porphyrin-like molecules, binding between a nitrogen base and the central metal atoms can occur.[187-189]

Paraquat (N,N'-dimethyl-4-4'-dipyridyliumdichloride, LII) is an inhibitor of photo-synthesis.[190-192] The mechanism of this action is not

LII

known, but it may involve an electron-transfer process in which the paraquat di-cation is reduced to the free radical mono-cation. The observation[182] that paraquat will complex with porphyrin-like molecules is therefore of particular interest. Other examples of complexing of the paraquat di-cation with electron donors are given in Chapter 10.

REFERENCES

1. R. S. Mulliken, *J. Am. chem. Soc.* **74**, 811 (1952).
2. A. Szent-Györgyi, "Introduction to a Submolecular Biology," Academic Press, New York and London (1960).
3. E. M. Kosower, *in* "Progress in Physical Organic Chemistry," Vol. 3, eds. S. G. Cohen, A. Streitweiser and R. W. Taft, Interscience, New York and London (1965), p. 141.
4. B. Pullman and A. Pullman, "Quantum Biochemistry," Interscience, New York and London (1963).
5. A. Pullman and B. Pullman, *in* "Quantum Theory of Atoms, Molecules and the Solid State," ed. P. O. Löwdin, Academic Press, New York and London (1966), p. 345.
6. E. M. Kosower, *J. Am. chem. Soc.* **77**, 3883 (1955).
7. E. M. Kosower and P. E. Klinedinst, Jr., *J. Am. chem. Soc.* **78**, 3493 (1956).
8. E. M. Kosower, *J. Am. chem. Soc.* **78**, 3497 (1956).
9. E. M. Kosower and J. C. Burbach, *J. Am. chem. Soc.* **78**, 5838 (1956).
10. E. M. Kosower, *J. Am. chem. Soc.* **80**, 3253 (1958).
11. E. M. Kosower, J. A. Skorcz, W. M. Schwarz, Jr. and J. W. Patton, *J. Am. chem. Soc.* **82**, 2188 (1960).
12. E. M. Kosower and J. A. Skorcz, *J. Am. chem. Soc.* **82**, 2195 (1960).
13. E. M. Kosower, "Molecular Biochemistry," McGraw-Hill, New York (1962), p. 180.
14. E. M. Kosower and S. W. Bauer, *J. Am. chem. Soc.* **82**, 2191 (1960).
15. O. Warburg, W. Christian and A. Griese, *Biochem. Z.* **282**, 157 (1935).
16. M. B. Yarmolinsky and S. P. Colowick, *Biochim. biophys. Acta* **20**, 177 (1956).
17. W. S. Caughey and K. A. Schellenberg, *J. org. Chem.* **31**, 1978 (1966).
18. R. Unzelman, J. Ludowieg and L. A. Strait, *Experientia* **20**, 506 (1964).
19. G. Cilento and S. Schreier, *Archs. Biochem. Biophys.* **107**, 102 (1964).
20. V. Massey and G. Palmer, *J. biol. Chem.* **237**, 2347 (1962).

21. T. Sakurai and H. Hosoya, *Biochim. biophys. Acta* **112**, 459 (1966).
22. I. Isenberg and A. Szent-Györgyi, *Proc. natn. Acad. Sci. U.S.A.* **45**, 1229 (1959).
23. G. Cilento and P. Giusti, *J. Am. chem. Soc.* **81**, 3801 (1959).
24. S. G. A. Alivisatos, F. Ungar, A. Jibril and J. A. Mourkides, *Biochim. biophys. Acta* **51**, 361 (1961).
25. G. Cilento and P. Tedeschi, *J. biol. Chem.* **236**, 907 (1961).
26. A. Fulton and L. E. Lyons, *Aust. J. Chem.* **20**, 2267 (1967).
27. S. Shifrin, *Biochim. biophys. Acta* **81**, 205 (1964).
28. S. Shifrin, *Biochim. biophys. Acta* **96**, 173 (1965).
29. G. Cilento and D. L. Sanioto, *Archs. Biochem. Biophys.* **110**, 133 (1965).
30. D. B. McCormick, Heng-Chun Li and R. E. Mackenzie, *Spectrochim. Acta* **23A**, 2353 (1967).
31. I. Isenberg, A. Szent-Györgyi and S. L. Baird, Jr., *Proc. natn. Acad. Sci. U.S.A.* **46**, 1307 (1960).
32. H. A. Harbury and K. A. Foley, *Proc. natn. Acad. Sci. U.S.A.* **44**, 662 (1958).
33. H. A. Harbury, K. F. LaNoue, P. A. Loach and R. M. Amick, *Proc. natn. Acad. Sci. U.S.A.* **45**, 1708 (1959).
34. I. Isenberg and A. Szent-Györgyi, *Proc. natn. Acad. Sci. U.S.A.* **44**, 857 (1958).
35. I. Isenberg, S. L. Baird, Jr. and A. Szent-Györgyi, *Proc. natn. Acad. Sci. U.S.A.* **47**, 245 (1961).
36. J. E. Wilson, *Biochemistry, Wash.* **5**, 1351 (1966).
37. L. D. Wright and D. B. McCormick, *Experientia* **20**, 501 (1964).
38. D. E. Fleischman and G. Tollin, *Biochim. biophys. Acta* **94**, 248 (1965).
39. D. E. Fleischman and G. Tollin, *Proc. natn. Acad. Sci. U.S.A.* **53**, 38 (1965).
40. J. C. M. Tsibris, D. B. McCormick and L. D. Wright, *Biochemistry, Wash.* **4**, 504 (1965).
41. B. M. Chassy and D. B. McCormick, *Biochemistry, N.Y.* **4**, 2612 (1965).
42. See Ref. 32 and references cited therein.
43. M. A. Slifkin, *Biochim. biophys. Acta* **109**, 617 (1965).
44. M. A. Slifkin, *Nature, Lond.* **197**, 275 (1963).
45. G. K. Radda and M. Calvin, *Nature, Lond.* **200**, 464 (1963).
46. G. Cilento and D. L. Sanioto, *Ber. Bunsenges. physik. Chem.* **67**, 426 (1963).
47. H. A. O. Hill, A. J. Macfarlane, B. E. Mann and R. J. P. Williams, *Chem. Commun.* 123 (1968).
48. A. Ledwith and D. H. Iles, *Chem. in Britain* **4**, 266 (1968).
49. A. Szent-Györgyi and I. Isenberg, *Proc. natn. Acad. Sci. U.S.A.* **46**, 1334 (1960); A. Szent-Györgyi, I. Isenberg and J. McLaughlin, *Proc. natn. Acad. Sci. U.S.A.* **47**, 1089 (1961).
50. B. Smaller, I. Isenberg and S. L. Baird, Jr., *Nature, Lond.* **191**, 168 (1961).
51. B. Pullman and A. Pullman, "Les théories électroniques de la chimie organique," Masson, Paris (1952).
52. R. Foster and P. Hanson, *Trans. Faraday Soc.* **60**, 2189 (1964).
53. R. Foster and C. A. Fyfe, *J. chem. Soc.* (B), 926 (1966).
54. J. P. Green and J. P. Malrieu, *Proc. natn. Acad. Sci. U.S.A.* **54**, 659 (1965).
55. K. Fukui, T. Yonezawa, C. Nagata and H. Shingu, *J. chem. Phys.* **22**, 1433 (1954).
56. K. Fukui, T. Yonezawa and C. Nagata, *J. chem. Phys.* **26**, 831 (1957).
57. A. W. Hanson, *Acta Crystallogr.* **17**, 559 (1964).
58. P. E. Wilcox, E. Cohen and Wen Tan, *J. biol. Chem.* **228**, 999 (1957).
59. K. M. C. Davis, D. D. Eley and R. S. Snart, *Nature, Lond.* **188**, 724 (1960).

60. D. D. Eley, G. D. Parfitt, M. J. Perry and D. H. Taysum, *Trans. Faraday Soc.* **49**, 79 (1953).
61. M. H. Cardew and D. D. Eley, *Discuss. Faraday Soc.* **27**, 115 (1959).
62. D. D. Eley and D. I. Spivey, *Trans. Faraday Soc.* **56**, 1432 (1960).
63. M. A. Slifkin, *Spectrochim. Acta* **20**, 1543 (1964).
64. A. Pullman and B. Pullman, *Bull. Soc. chim. Fr.* 766 (1958); *ibid.*, 594 (1959).
65. B. Pullman, *J. chem. Soc.* 1621 (1959).
66. A. Veillard and B. Pullman, *Compt. Rend.* **253**, 2277 (1961).
67. A. Veillard and B. Pullman, *J. Theoret. Biol.* **4**, 37 (1963).
68. Ref. 4, pp. 217, 220, 224.
69. A. Pullman and M. Rossi, *Biochim. biophys. Acta* **88**, 211 (1964).
70. M.-J. Mantione and B. Pullman, *Compt. Rend.* **262**, [D] 1492 (1966).
71. E. Reich, *Science, N.Y.* **143**, 684 (1964).
72. W. M. Macintyre, *Science, N.Y.* **147**, 507 (1965).
73. R. W. Brockman, C. Sparks, D. J. Hutchison and H. E. Skipper, *Cancer Res.* **19**, 177 (1959).
74. P. Machmer and J. Duchesne, *Nature, Lond.* **206**, 618 (1965).
75. J. Duchesne, P. Machmer and M. Read, *Compt. Rend.* **260**, 2081 (1965).
76. M. A. Slifkin, *Biochim. biophys. Acta* **103**, 365 (1965).
77. H. Z. Sable and J. E. Biaglow, *Proc. natn. Acad. Sci. U.S.A.* **54**, 808 (1965).
78. R. Schoental, *in* "Polycyclic Hydrocarbons," Vol. I, ed. E. Clar, Academic Press, New York and London (1964), Ch. 18.
79. D. B. Clayson, "Chemical Carcinogenesis," Churchill, London (1962).
80. P. Emmelot, *in* "Molecular Pharmacology," Vol. II, ed. E. J. Ariëns, Academic Press, New York and London (1964), p. 53.
81. A. Pullman and B. Pullman "Cancérisation par les Substances Chimiques et Structure Moléculaire," Masson, Paris (1955).
82. G. M. Badger, *Adv. Cancer Res.* **2**, 73 (1954).
83. R. Bhattacharya and S. Basu, *Trans. Faraday Soc.* **54**, 1286 (1958).
84. M. J. S. Dewar and A. R. Lepley, *J. Am. chem. Soc.* **83**, 4560 (1961).
85. M. J. S. Dewar and H. Rogers, *J. Am. chem. Soc.* **84**, 395 (1962).
86. A. R. Lepley, *J. Am. chem. Soc.* **84**, 3577 (1962).
87. R. Beukers and A. Szent-Györgyi, *Recl Trav. chim. Pays-Bas Belg.* **81**, 255 (1962).
88. M. Nepraš and R. Zahradník, *Colln. Czech. chem. Commun. Engl. Edn* **29**, 1545 (1964).
89. A. Szent-Györgyi, I. Isenberg and S. L. Baird, Jr., *Proc. natn. Acad. Sci. U.S.A.* **46**, 1444 (1960).
90. A. C. Allison and T. Nash, *Nature, Lond.* **197**, 758 (1963).
91. S. S. Epstein, I. Bulon, J. Koplan, M. Small and N. Mantel, *Nature, Lond.* **204**, 750 (1964).
92. K. H. Takemura, M. D. Cameron and M. S. Newman, *J. Am. chem. Soc.* **75**, 3280 (1953).
93. R. E. Kofahl and H. J. Lucas, *J. Am. chem. Soc.* **76**, 3931 (1954).
94. A. C. Allison, M. E. Peover and T. A. Gough, *Nature, Lond.* **197**, 764 (1963).
95. J. E. Lovelock, A. Zlatkis and R. S. Becker, *Nature, Lond.* **193**, 540 (1962).
96. A. C. Allison and J. W. Lightbown, *Nature, Lond.* **189**, 892 (1961).
97. R. Mason, *Nature, Lond.* **181**, 820 (1958).
98. R. Mason, *Br. J. Cancer* **12**, 469 (1958).
99. R. Mason, *Discuss. Faraday Soc.* **27**, 129 (1959).
100. R. Mason, *Radiat. Res.* Supp. **2**, 452 (1960).

101. A. Pullman and B. Pullman, *Nature, Lond.* **196**, 228 (1962).

102. A. C. Allison and T. Nash, *Nature, Lond.* **199**, 469 (1963).

103. B. Pullman and A. Pullman, *Nature, Lond.* **199**, 467 (1963); cf. A. C. Allison and T. Nash, Ref. 102.

104. J. B. Birks, *Nature, Lond.* **190**, 232 (1961).

105. A. Pullman, *Biopolymers, Symp.* **1**, 47 (1964).

106. E. C. Miller, *Cancer Res.* **11**, 100 (1951).

107. W. G. Wiest and C. Heidelberger, *Cancer Res.* **13**, 246, 250, 255 (1953).

108. C. Heidelberger and M. G. Moldenhauer, *Cancer Res.* **16**, 442 (1956).

109. V. T. Oliverio and C. Heidelberger, *Cancer Res.* **18**, 1094 (1958).

110. G. R. Davenport, C. W. Abell and C. Heidelberger, *Cancer Res.* **21**, 599 (1961).

111. C. W. Abell and C. Heidelberger, *Cancer Res.* **22**, 931 (1962).

112. P. M. Bhargava, H. I. Hadler and C. Heidelberger, *J. Am. chem. Soc.* **77**, 2877 (1955).

113. P. M. Bhargava and C. Heidelberger, *J. Am. chem. Soc.* **78**, 3671 (1956).

114. A. Pullman, *Compt. Rend.* **221**, 140 (1945).

115. A. Pullman, *Compt. Rend. Soc. Biol.* **139**, 1056 (1945).

116. A. Pullman, *Annls. Chim.* [12] **2**, 5 (1947).

117. P. Daudel and R. Daudel, *Bull. Soc. Chim. biol.* **31**, 353 (1949).

118. A. Pullman and B. Pullman, *Adv. Cancer Res.* **3**, 117 (1955).

119. R. Daudel, *Compt. Rend. Soc. Biol.* **142**, 5 (1948).

120. A. Pullman and J. Baudet, *Compt. Rend.* **237**, 986 (1953).

121. R. L. Flurry, Jr., *J. Med. Chem.* **7**, 668 (1964).

122. C. A. Coulson, *Adv. Cancer Res.* **1**, 1 (1953).

123. G. M. Badger, *J. chem. Soc.* 456 (1949); *ibid.*, 1809 (1950).

124. A. Pullman, *Compt. Rend.* **236**, 2318 (1953).

125. R. W. Baldwin, G. J. Cunningham and M. W. Partridge, *Br. J. Cancer* **13**, 94 (1959).

126. R. W. Baldwin, G. J. Cunningham, M. W. Partridge and H. J. Vipond, *Br. J. Cancer* **16**, 275 (1962).

127. R. W. Baldwin, H. C. Palmer and M. W. Partridge, *Br. J. Cancer* **16**, 740 (1962).

128. C. Nagata, M. Kodama, A. Imamura and Y. Tagashira, *Gann* **57**, 75 (1966).

129. J. W. Cook, "Chemistry and Cancer," Institute of Chemistry, London (1943).

130. N. C. Yang, A. J. Castro, M. Lewis and T.-W. Wong, *Science, N.Y.* **134**, 386 (1961).

131. J. Iball, *Nature, Lond.* **201**, 916 (1964).

132. G. M. Bonser, D. B. Clayson and J. W. Jull, *Br. med. Bull.* **14**, 146 (1958).

133. W. H. Vogel, R. Snyder and M. P. Schulman, *J. Pharmac. exp. Ther.* **146**, 66 (1964).

134. B. Pullman, *Biopolymers, Symp.* **1**, 141 (1964).

135. J. D. Watson and F. H. C. Crick, *Nature, Lond.* **171**, 964 (1953).

136. N. Brock, H. Druckrey and H. Hamperl, *Arch. exp. Path. Pharmak.* **189**, 709 (1938).

137. H. Weil-Malherbe, *Biochem. J.* **40**, 351, 363 (1946).

138. E. Boyland and B. Green, *Br. J. Cancer* **16**, 347 (1962).

139. E. Boyland and B. Green, *Biochem. J.* **84**, 54P (1962).

140. E. Boyland and B. Green, *Br. J. Cancer* **16**, 507 (1962).

141. A. M. Liquori, B. De Lerma, P. Ascoli, C. Botré and M. Trasciatti, *J. molec. Biol.* **5**, 521 (1962).

142. P. De Santis, E. Giglio and A. M. Liquori, *Nature, Lond.* **188**, 47 (1960).

143. P. De Santis, E. Giglio, A. M. Liquori and A. Ripamonti, *Nature, Lond.* **191**, 900 (1961).
144. A. Pullman and B. Pullman, *Biochim. biophys. Acta* **36**, 343 (1959).
145. B. Pullman and A. Pullman, *Rev. mod. Phys.* **32**, 428 (1960).
146. B. L. Van Duuren, *J. phys. Chem., Ithaca* **68**, 2544 (1964).
147. B. L. Van Duuren, *Nature, Lond.* **210**, 622 (1966).
148. B. Pullman, P. Claverie and J. Caillet, *Science, N.Y.* **147**, 1305 (1965).
149. "Molecular Pharmacology," Vol. I, ed. E. J. Ariëns, Academic Press, New York and London (1964).
150. A. J. Clark, "The Mode of Action of Drugs on Cells," Williams and Wilkins, Baltimore, Md. (1937).
151. J. H. Gaddum, *J. Physiol., Lond.* **61**, 141 (1926); *ibid.,* **89**, 7P (1937).
152. E. J. Ariëns, J. M. van Rossum and A. M. Simonis, *Pharmac. Rev.* **9**, 218 (1957).
153. R. P. Stephenson, *Br. J. Pharmac. Chemother.* **11**, 379 (1956).
154. W. D. M. Paton, *Proc. R. Soc.* **B154**, 21 (1961).
155. D. R. Kearns and M. Calvin, *J. phys. Chem., Ithaca* **34**, 2026 (1961).
156. L. E. Lyons and J. C. Mackie, *Nature, Lond.* **197**, 589 (1963).
157. G. Karreman, I. Isenberg and A. Szent-Györgyi, *Science, N.Y.* **130**, 1191 (1959).
158. Ref. 4, p. 804.
159. M. K. Orloff and D. D. Fitts, *Biochim. biophys. Acta* **47**, 596 (1961).
160. R. Foster and P. Hanson, *Biochim. biophys. Acta* **112**, 482 (1966).
161. R. Foster and C. A. Fyfe, *Biochim. biophys. Acta* **112**, 490 (1966).
162. T. Nash and A. C. Allison, *Biochem. Pharmac.* **12**, 601 (1963).
163. F. Gutmann and H. Keyzer, *J. chem. Phys.* **46**, 1969 (1967).
164. R. Foster and C. A. Fyfe, unpublished work.
165. W. A. Strickland, Jr. and L. Robertson, *J. Pharm. Sci.* **54**, 452 (1965).
166. R. Foster and C. A. Fyfe, *Rev. pure Appl. Chem.* **16**, 61 (1966) and references therein.
167. T. Eckert, *Arch. Pharm., Berl.* **295**, 233 (1962).
168. T. Eckert, *Naturwissenschaften* **49**, 18 (1962).
169. T. Eckert, *Arzneimittel-Forsch.* **12**, 8 (1962).
170. D. Agin, *Nature, Lond.* **205**, 805 (1965).
171. S. H. Snyder and C. R. Merril, *Proc. natn. Acad. Sci. U.S.A.* **54**, 258 (1965).
172. L. H. Piette, G. Bulow and I. Yamazaki, *Biochim. biophys. Acta* **88**, 120 (1964).
173. S. Ohnishi and H. M. McConnell, *J. Am. chem. Soc.* **87**, 2293 (1965).
174. M. Calvin, *J. theoret. Biol.* **1**, 258 (1961).
175. "Light and Life," eds. W. D. McElroy and B. Glass, Johns Hopkins Press, Baltimore (1961).
176. D. I. Arnon, M. Losada, M. Nozaki and K. Tagawa, *Nature, Lond.* **190**, 601 (1961).
177. D. R. Kearns, G. Tollin and M. Calvin, *J. chem. Phys.* **32**, 1020 (1960).
178. R. B. Park and N. G. Pon, *J. molec. Biol.* **3**, 1 (1961).
179. M. Gouterman and P. E. Stevenson, *J. chem. Phys.* **37**, 2266 (1962).
180. J. R. Larry, *Diss. Abstr.* **27**, 2316B (1967).
181. H. A. O. Hill, A. J. Macfarlane and R. J. P. Williams, *Chem. Commun.* 905 (1967).
182. A. J. Macfarlane, Thesis, Oxford (1968).
183. H. A. O. Hill, B. E. Mann and R. J. P. Williams, *Chem. Commun.* 906 (1967).
184. P. J. McCartin, *J. Am. chem. Soc.* **85**, 2021 (1963).

185. D. Mauzerall, *Biochemistry, Wash.* **4**, 1801 (1965).
186. J. G. Heathcote, G. J. Hill, P. Rothwell and M. A. Slifkin, *Biochim. biophys. Acta* **153**, 13 (1968).
187. P. E. Wei, A. H. Corwin and R. R. Arellano, *J. org. Chem.* **27**, 3344 (1962).
188. C. B. Storm, A. H. Corwin, R. R. Arellano, M. Martz and R. Weintraub, *J. Am. chem. Soc.* **88**, 2525 (1966).
189. G. R. Seeley, *J. phys. Chem., Ithaca* **71**, 2091 (1967).
190. A. T. Jagendorf and M. Avron, *J. biol. Chem.* **231**, 277 (1958).
191. R. Hill and D. A. Walker, *Pl. Physiol., Lancaster* **34**, 240 (1959).
192. W. R. Boon, *Chemy Ind.* **782** (1965).

Chapter 13

Applications of Charge-Transfer Complexes in Organic Chemistry

13.A. Formation of Derivatives for the Purposes of Identification and Separation

Over the years a large number of complexes, which might now be described as charge-transfer complexes, have been prepared either as a means of separating substances or for the purposes of identification.[1] Many, though by no means all, of these complexes are those obtained by the interaction of aromatic hydrocarbons with electron acceptors. The best known of these are the hydrocarbon picrates.* Although picric acid and the related styphnic acid (2,4,6-trinitroresorcinol) are often the electron acceptors used, many other electron acceptors may be more suitable in specific cases. 1,3,5-Trinitrobenzene yields stable solid complexes with a variety of electron donors. In the past two decades 2,4,7-trinitrofluorenone (TNF) has also become a popular reagent in this role.[2] Tetracyanoethylene (TCNE) forms complexes with a variety of electron donors.[3] However, the solid complexes are often difficult to separate despite the large association constants for many of these complexes in solution. Furthermore, the high chemical reactivity of this reagent (see Chapter 11) limits its usefulness as a complexing agent. Less notice has

---

* The charge-transfer hydrocarbon picrates should not be confused with the amine picrates, which are substituted ammonium picrate salts formed by proton transfer.

been given to other reagents such as benzotrifuroxan [4, 5] and 4,6-dinitro-benzofuroxan,[4] which form remarkably stable solid complexes with a wide range of aromatic hydrocarbons. These and some other less usual electron acceptors are referred to in Chapter 1. The melting points of a selection of solid charge-transfer complexes are given in Table 13.1. Such complexes have been used to separate components in mixtures utilizing the differences in relative stabilities and/or solubilities of the complexes in solution. Regeneration of the parent compound from the isolated complex is usually a simple process, since the binding energy of the complex is small. Dissolution in a solvent of relatively high polarity is often sufficient to cause complete decomposition. Alternatively, one component may be removed by column chromatography.[2, 6-8] For example, in many electron-donor complexes with polynitro aromatic acceptors, the second component is retained when a benzene solution of the complex is passed down an alumina column. A chemical reaction, such as acidification in the case of amine charge-transfer complexes, may also be used to separate the components.

13.B. The Possible Use of Solid Charge-Transfer Complexes in X-ray Structural Determinations

A possible application of solid charge-transfer complexes is to provide a crystal lattice containing a heavy atom in order to facilitate the phasing and consequent structural determination from X-ray diffraction measurements. Thus, by determining the structure of a total complex of, say, an electron donor with an acceptor containing a heavy atom, the structure of the donor moiety alone is obtained. Newman and his co-workers [9] have synthesized such acceptor molecules containing heavy atoms, namely 4-bromo- and 4-iodo-2,5,7-trinitrofluorenone (I, R = Br

I

and R = I) although they appear to have been used primarily for the determination of the total complex, rather than because of interest in the donor component. For the weaker π-π complexes, there appears to be little distortion of the component molecules in the complex (see Chapter 8).

TABLE 13.1. Some so▮

Acceptor	Benzotrifuroxan			Benzotrifurazan			2,4,7-Trinitrofluorenone		
Donor	Colour	M.p.*	Ref.†	Colour	M.p.*	Ref.†	Colour	M.p.*	Ref
Benzene	Colourless	d	a						
Durene				Colourless	149–52	c			
Hexamethyl-benzene	Pale yellow	207d	a	Pale yellow	226	c			
Biphenyl	Pale yellow	164–6	a	Colourless	115–16	c			
Naphthalene	Greenish-yellow	250–2d	a	Cream	172	c	Yellow	151·2–154	g
2-Phenyl-naphthalene	Bright yellow	158	a				Orange-yellow	169·5–170·5	j
Anthracene							Red	193·8–194·0	j
Phenanthrene				Cream	204–6	c	Brilliant yellow	196·4–197·2	j
Chrysene	Orange	256–8	m				Dull yellow	247·8–249·0	j
Pyrene	Yellow	280–2	m	Yellow	256	a	Deep red	242–3	j
Perylene				Red	267–70	c	Black	270–1	j
Benz[a]-anthracene							Red	223·6–224·0	j
Azulene	Black	140–5	m	Dark green	168–70	c			
trans-Stilbene				Yellow	153	c	Bright red	148·4–148·8	j
Indole	Pale yellow	180	a	Yellow	200	c			
Carbazole							Brownish-red	173·4–174·4	j
Aniline	Orange	133–4d	m						
N,N-Dimethyl-aniline	Crimson	118–20d	m						

* d—reported decomposition. In fact the majority probably decompose on melting.
† Refs.: [a] A. S. Bailey and J. R. Case, Tetrahedron 3, 113 (1958). [b] "Beilstein's Hanbuch der Organisch▮ Chemie," Springer-Verlag, Berlin. [c] A. S. Bailey and J. M. Evans, J. chem. Soc. (C), 2105 (1967). [d] A. S. Bail▮ B. R. Henn and J. M. Langdon, Tetrahedron 19, 161 (1963). [e] R. C. Fuson and B. C. McKusick, J. Am. che▮ Soc. 65, 60 (1943). [f] "Organic Reagents for Organic Analysis," Hopkins and Williams, London (195▮ [g] M. Orchin, L. Reggel and E. O. Woolflock, J. Am. chem. Soc. 69, 1225 (1947). [h] A. S. Bailey, R. J. Williams and J. D. Wright, J. chem. Soc. 2579 (1965). [i] R. A. Friedel, M. Orchin and L. Reggel, J. A▮ chem. Soc. 70, 199 (1948). [j] M. Orchin and E. O. Woolflock, J. Am. chem. Soc. 68, 1727 (1946). [k] R. C. Jo▮

13.C. Molecular Weight Determinations

Consider a charge-transfer complex, derived from a donor or acceptor of unknown molecular weight and a second component of known identity, which can be obtained as a stoichiometric solid. When a small quantity of the solid is dissolved in a polar solvent, normally dissociation will be effectively complete. If the concentration of the second component can be determined in this solution, for example by direct spectrophotometry,[10] or indirectly by using a colourimetric reaction such as arge-transfer complexes.

1,2,4,5-Tetracyanobenzene			1,3,5-Trinitrobenzene			Picric acid		
Colour	M.p.*	Ref.†	Colour	M.p.*	Ref.†	Colour	M.p.*	Ref.†
						Bright yellow	84·3	b
llow	267–70	d	Greenish-yellow	100–1	e			
			Yellow	174	b	Orange-yellow	170	f
le yellow	211–14	h						
llow	257–63	h	Colourless	152	b	Yellow	150	f
			Yellow	113·5–115·2	i			
ange	277–80	d	Orange-yellow	164	f	Ruby-red	138	f
llow	248–51	h	Orange-yellow	164	f	Golden-yellow	143	f
ange	292–5d	h		188·5–189·5	k	Orange-red	174–5	l
ange	268–70	d	Yellow	245	n	Dark red	222	o
			Brick-red	248–9	p	Dark	223–224·5	q
			Orange	159·8–160·2	r	Dark red	141·5–142·5	s
ark	d	h	Brown	166·5–167·5	t	Brown-violet	144–8	b
				120	f		94–5	u
			Yellow	187	f			
			Orange-red	200	f			
ark red	224–77d	h	Orange-red	124–125·5	v			
eep purple	266–8	d	Dark violet	110–11	v			

d M. B. Neuworth, J. Am. chem. Soc. 66, 1497 (1944). l L. F. Fieser, M. Fieser and E. B. Hershberg Am. chem. Soc. 58, 1463 (1936). m A. S. Bailey, J. chem. Soc. 4710 (1960). n E. Hertel, Liebigs Ann., 451, 79 (1927). o E. Hintz, Ber. dt. chem. Ges. 10, 2141 (1877). p M. Orchin and R. A. Friedel, J. Am. chem. Soc 8, 573 (1946). q G. T. Morgan and J. G. Mitchell, J. chem. Soc. 536 (1934). r M. S. Newman and R. T. Hart Am. chem. Soc. 69, 298 (1947). s J. W. Cook, J. chem. Soc. 2524 (1931). t Pl. A. Plattner and A. St. Pfau elv. chim. Acta 20, 224 (1937). u Heilbron's "Dictionary of Organic Compounds," Eyre and Spottis- oode, London (1965). v S. D. Ross and M. M. Labes, J. Am. chem. Soc. 79, 76 (1957).

that obtained by the addition of base to 1,3,5-trinitrobenzene,[11] or by titration,[12] then, from a knowledge of the stoichiometry of the complex and the weight of the solid complex used to make up the solution, the molecular weight of the first component can be calculated. Alternatively, if the complex is of known stoichiometry, a weight, which bears a simple ratio to the molecular weight, will be obtained. The method works well for hydrocarbon picrates.[10, 12] Here the free picric acid has an absorption in the region of 380 nm and can usually be estimated photometrically,[10] providing that the hydrocarbon component does not absorb in this region. Inaccuracies due to incomplete dissociation may in fact be negligible, since, for many systems, the optical transitions of the components are only slightly perturbed in the complex (but see Chapter 3) and do not coincide with the intermolecular charge-transfer band. Obviously, in principle, the method may be applied to other types of complex, such as the amine picrate salts. Some examples are given in Table 13.2.

TABLE 13.2. Molecular weights of charge-transfer complexes estimated from the concentration of the acceptor when the complex is dissociated in solution (acceptor = 1,3,5-trinitrobenzene).*

Donor	M.p.† complex/°C	Obs. m.w.	Calc. m.w.
Acenaphthene	167·0–169·0	369·0	367·31
Acridine	115·0–115·5	392·2	392·32
Anthracene	116·3–164·0	389·8	391·33
Anthranilic acid	191·5–192·7	350·5	350·24
Carbazole	200·3–200·8	379·8	380·31
N,N-Dimethylaniline	107·5–109·5	319·1	334·29
Diphenylene oxide	98·0–99·0	379·9	381·29
Naphthalene	153·5–154·5	340·8	341·27
1-Naphthoic acid‡	190·0–192·5	271·9	278·7
1-Naphthol	175·0–176·0	360·4	357·27
1-Naphthylamine	218·2–218·6	357·3	356·29
Phenanthrene	162·7–163·7	390·4	391·33
Pyrene	250·0–250·5	415·0	415·35
trans-Stilbene§	122·0–123·0	610·2	606·46

* From Ref. 11.
† M.p.'s in sealed capillaries.
‡ Stoichiometry AD_2.
§ Stoichiometry A_2D.

13.D. Charge-Transfer Complex Formation in Chromatographic Separations

The selective absorption of different donor species on chromatographic stationary phases containing electron acceptors has been discussed in Chapter 6, where it was shown how the effect could be used to estimate the degree of association of the acceptor with a given donor; the acceptors include iodine,[13, 14] TNF[15] and 1,3,5-trinitrobenzene.[16] The same principle may be used to separate mixtures of donors. The converse procedure of separating electron acceptors on a stationary electron-donor phase has been used.[17]

Norman[18] has described the use of TNF, in the form of a liquid coated on firebrick, as a stationary phase up to temperatures of 200°C in gas–liquid chromatography.

Thin-layer chromatography (TLC), using silica gel impregnated with TNF, has been employed to separate various hydroaromatic compounds derived from phenanthrene, anthracene and chrysene.[19] Other separations of aromatic hydrocarbons have been made using thin layers of silica gel containing 1,3,5-trinitrobenzene,[17] and picric acid.[20] The behaviour of aromatic amines on thin layers containing 1,3,5-trinitrobenzene or picryl chloride has also been investigated.[21]

Column chromatography was used by Godlewicz[22] to separate hydrocarbon mixtures. Silica gel was impregnated with 1,3,5-trinitrobenzene, with the intention that the latter compound should act as a colour-indicator for the various hydrocarbons. It is probable, certainly in the light of more recent observations, that the 1,3,5-trinitrobenzene in fact plays an important role in the actual separation of the hydrocarbon mixture, as suggested by Harvey and Halonen.[19] Columns of silicic acid containing picric acid or TNF,[23, 24] and columns of tetrachlorophthalic anhydride and of tetrachlorophthalimide[25] have also been used. It is of interest that, whereas silica gel alone is relatively ineffective in separating mixtures of closely related aromatic hydrocarbons, silica gel containing 5% TNF provides a clean separation.[19] Amongst the homogeneous materials which have been used as the stationary phase in column chromatography is the polynitrostyrene referred to as nitrobenzylpolystyrene (II).[26]

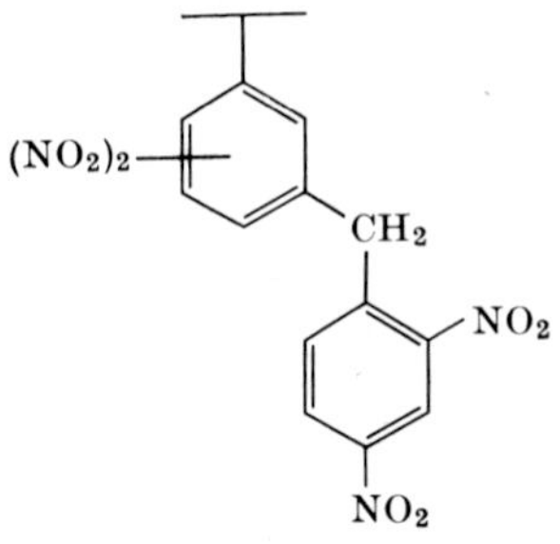

II

13.E. Resolution of Enantiomers

The possibility of resolving racemic mixtures of optical enantiomorphs through the formation of diastereoisomeric charge-transfer complexes was entertained by Brown and Hammick.[27] They investigated the complex formation between dimethyl 4,4',6,6'-tetranitro-2,2'-diphenate (III) and various asymmetric substituted naphthalene and acenaphthene

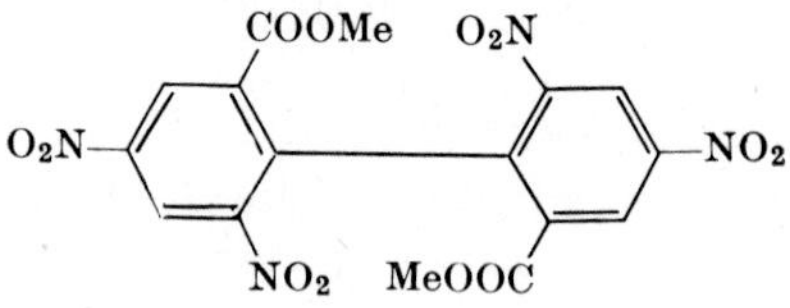

III

hydrocarbons in an unsuccessful attempt to resolve the hydrocarbons. The acid corresponding to III was also used.

More recently, the same principle has been applied by Newman and his co-workers [28, 29] to resolve the "overcrowded" molecule, hexahelicene (phenanthro[3,4-c]phenanthrene) (IV), which exists as left-handed and

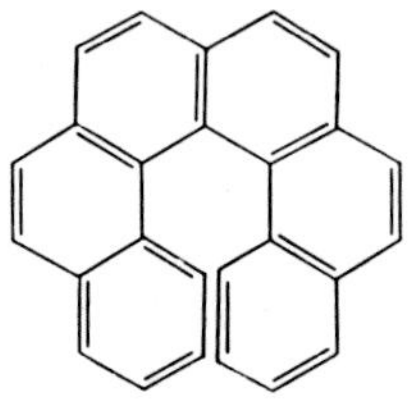

IV

right-handed spiral molecules. The racemic mixture was resolved by complex formation with the (−) isomer of 2-(2,4,5,7-tetranitro-9-fluorenylidineaminoöxy)-propionic acid [30] (V). This compound acts as an

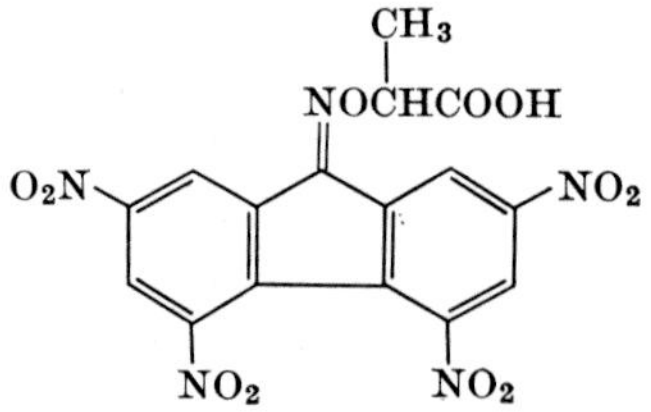

V

electron acceptor to form complexes of different stability with (+)- and with (−)-hexahelicene. Using the same principle, Klemm and Reed [31] have shown that a racemic mixture of α-naphthyl sec-butyl ether may be partially resolved using V absorbed on silicic acid. The partial resolution of 9-see-butylphenanthrene has been achieved in a similar manner.[32]

13.F. Elucidation of Nuclear Magnetic Resonance Spectra

Charge-transfer complex formation can have a differential effect on the various nuclei in a molecular component of a complex. Some simple cases, such as the greater change in the chemical shift of the α-protons in naphthalene compared with the β-protons when large quantities of 1,3,5-trinitrobenzene are added to solutions of naphthalene,[33] have already been discussed in Chapter 5. If the geometry of the complex is known, or alternatively, if sufficient correlations exist for spectra in which the line assignments are known, such relative changes in shift can be used to make assignments in the spectra of other molecules. The principle involved is similar to that used by Bhacca *et al.* from observations of the changes in chemical shifts of solute molecules when the solvent is changed from an isotropic to a non-isotropic medium (see Chapter 5). If the spectral assignments are known, the changes in chemical shift of different nuclei on complex formation may provide information about the geometry of the complex or about the structure of the component molecule.

A more complicated system which may involve charge-transfer complexing* has recently been described.[34] The assignment of the stero-conformation of 9,10-dipropylanthracene-9,10-diol (VI) has been made by making use of the anisotropic effect of the aromatic ring on the alkyl groups. When VI alone is dissolved in deutero-chloroform, the

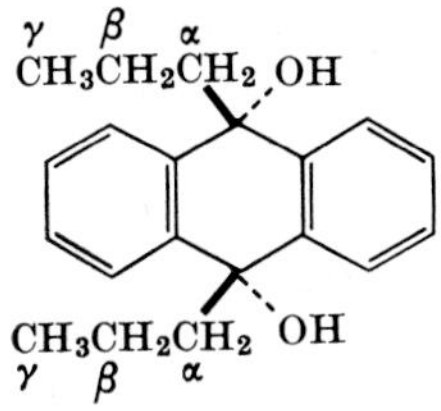

VI

pseudo-axial configuration of the alkyl groups causes the β-methylene and γ-methyl protons of the propyl groups to be equivalent. Addition of tetranitromethane to the system is claimed to reduce the anisotropy of the benzene rings and hence permit normal coupling within the propyl groups, as is observed.

13.G. Estimation of Solvent Polarity

Mention has already been made in Chapter 3 of the sensitivity of the energy in the intermolecular charge-transfer transition of pyridinium

---

* Although the authors' general interpretation is not questioned, there is some doubt as to whether these complexes of tetranitromethane, described in this paper as π-complexes, are in fact charge-transfer complexes (see Chapter 10, p. 283).

13*

iodides to changes in the nature of the solvent. This is a result of the degree of solvation of the ground state of the complex which is effectively an ion-pair. Kosower[35-39] has defined a "Z-value" scale of polarities of solvents such that the Z-value of a given solvent is the energy of the intermolecular charge-transfer band in kcal mol^{-1} of 1-ethyl-4-carbomethoxypridinium iodide (VII) when dissolved in that

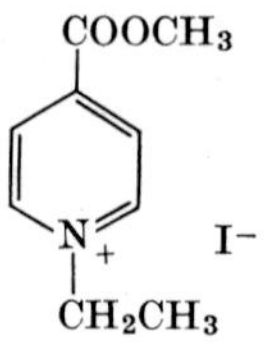

$$COOCH_3$$

VII

solvent (see Table 13.3). Such values can obviously be obtained rapidly. The main limitation to the range of solvents that may be measured is set by the actual solubility of VII in the particular solvent; this is chiefly restricted to solvents with very low solvent polarity (and therefore low Z-value).

TABLE 13.3. E_T* and Z values (measures of solvent polarity).

Solvent	E_T*/kcal mol^{-1} (25°C)	Z/kcal mol^{-1} (25°C)	Ref.†
Water	63·1	94·6	36
Formamide	56·6	83·3	36
Ethylene glycol	56·3	85·1	36
Methanol	55·5	83·6	36
Methylformamide	54·1		
Ethanol/water 80:20	53·6	84·8	36
2-Methoxyethanol	52·3		
Ethanol	51·9	79·6	36
Acetic acid	51·9‡	79·2	36
Benzyl alcohol	50·8	78·4	a
Propan-1-ol	50·7	78·3	36
Butan-1-ol	50·2	77·7	36
Propan-2-ol	48·6	76·3	36
3-Methylbutan-1-ol	47·0		
Decan-1-ol		73·3	b
4-Methyl-1,3-dioxol-2-one (propylene carbonate)	46·6		
Nitromethane	46·3		
Methyl cyanide (acetonitrile)	46·0	71·3	36
Dimethyl sulphoxide	45·0	71·1	36
Aniline	44·3		
Sulfolane	44·0		
2-Methyl-propan-2-ol (t-butyl alcohol)	43·9§	71·3	36
Dimethylformamide	43·8	68·5	36

TABLE 13.3.—*continued*

Solvent	E_T*/kcal mol^{-1} (25°C)	Z/kcal mol^{-1} (25°C)	Ref.†
Acetone	42·2	65·7	36
Nitrobenzene	42·0		
Benzonitrile	42·0	65·0	c
1,1,2,2-Tetrachloroethane		64·3	b
1,2-Dichloroethane	41·9	63·4	b
Acetophenone	41·3		
Dichloromethane	41·1	64·2	36
Tetramethyl urea	41·0		
Pyridine	40·2	64·0	36
Quinoline	39·4		
cis-1,2-Dichloroethene		63·9	c
Chloroform	39·1	63·2	36
α-Picoline	38·3		
1,1-Dichloroethane		62·1	b
1,2-Dimethoxyethane (diglyme)	38·2	62·1	d
Ethyl acetate	38·1	59·4	b
Fluorobenzene		60·2	c
o-Dichlorobenzene		60·0	c
Bromobenzene	37·5	59·2	c
Chlorobenzene	37·5	58·0	c
Tetrahydrofuran	37·4		
Anisole	37·2		
2,6-Lutidine	36·7		
Dioxane	36·0		
Diphenyl ether	35·3§		
Diethyl ether	34·6		
Benzene	34·5		
Toluene	33·9		
Carbon bisulphide	32·6		
Carbon tetrachloride	32·5		
Hexane	30·9		
2,2,4-Trimethylpentane (iso-octane)		60·1	36

* E_T values are energies, in kcal/mol^{-1} of the peak maximum of the intramolecular

charge-transfer absorption band of the phenolbetaine

in the various solvents [C. Reichardt, *Angew. Chem.* **77**, 30 (1965); *Angew. Chem. (Int. Edn.)* **4**, 29 (1965)] (see Chapter 3, p. 63).

† Refs. (to z-values): *a* R. Foster, unpublished work. *b* P. H. Emslie and R. Foster, *Recl Trav. chim. Pays-Bas Belg.* **84**, 255 (1965). *c* C. Walling and P. J. Wagner, *J. Am. chem. Soc.* **86**, 3368 (1964). *d* J. A. Berson, Z. Hamlet and W. A. Mueller, *J. Am. chem. Soc.* **84**, 297 (1962).

‡ Interpolated value.

§ Measured at 30°C.

TABLE 13.4. Ionization potentials of electron donors from energies of charge-transfer bands of complexes with various electron acceptors (A)

| | Ionization potential/eV | | | | | | | | | | | |
| | From intermolecular charge-transfer band of complex with acceptor A | | | | | | | | | | From other expt. methods | |
Donors	A = TNB	Ref.*	A = chloranil	Ref.*	A = I$_2$	Ref.*	A = bromanil	Ref.*	A = TCNE	Ref.*		Ref.*
Durene	8·4 / 8·3	a / c	8·3 / 8·0	a / c	8·4	a					8·025	b
Hexamethyl-benzene	7·9 / 7·8	a / c	7·95 / 7·8	a / c	8·0	a					7·85	d
Naphthalene	8·20	e			8·15	e	8·08	e, f	8·11		8·26 / 8·12 / 8·24	g / b / h
Anthracene	7·4 / 7·4	a / c	7·4 / 7·4	a / c	7·4	a	7·4	e, f			7·55	g
Naphthacene	6·94	a	7·0	a							7·15 / 6·95	i / j
Phenanthrene	8·15 / 8·15	a / e	8·1	a	7·85	a	8·22	e, f	8·24	e	8·03	g
Pyrene	7·51	e	7·55	a	7·5	a	7·48	e, f	7·45	e	7·72	j
Chrysene	7·62	e	7·8	a	7·75	a	7·68	e, f			8·01	j
Perylene	7·10	e	7·15	a			7·06	f	7·19	e		
Triphenylene			8·1	a	7·8	a	8·08	f			8·19	j
Benz[a]anthracene	7·48	e	7·6	a			7·56	e, f				
Benzo[a]pyrene	7·33	k	7·30	k					7·37	k		
Anthanthrene	6·95	e					7·01	e, f	7·22	e		

Compound					
Azulene	7·4 (43)	7·4 (43)	7·5 (43)		7·43, 7·72, 7·76 (44, l, h)
Biphenyl	8·35 (a)	8·45 (a)	8·4 (a)	8·64 (f)	8·22, 8·27 (b, h)
Biphenylene	7·42, 7·95, 7·85 (m)	7·48 (m)		7·68 (f)	7·95, 7·70 (h, b)
Aniline	7·6 (a)			7·37, 7·76 (n, o)	7·84 (h)
N-Methylaniline	7·3 (a)	7·3 (c)		7·58 (f)	7·58 (o)
N.N-Dimethyl-aniline	7·1 (c)	6·7 (c)		7·31 (f)	7·44 (o)
N,N,N′,N′-Tetra-methyl-p-phenylenediamine	6·6 (a), 6·35 (a)				
o-Phenylene-diamine	6·7 (c), 7·45 (a)			7·36 (f)	
Diphenylamine	7·4 (a)			7·31 (f)	7·59 (p)
o-Toluidine	7·75 (a)			7·69 (f)	7·68 (p)
m-Toluidine	7·75 (a)				7·57 (p)
p-Toluidine	7·65 (a)			7·58 (f)	7·57 (p)

* Refs.: a G. Briegleb and J. Czekalla, Z. Elektrochem. 63, 6 (1959). b K. Watanabe, T. Nakayama and J. R. Mottl, "Final Report on Ionisation Potentials of Molecules by a Photoionisation Method," Dept. Army 5B99-01-004 Ordnance R and D-TB2-0001 OOR-1624, University of Hawaii (1959). c R. Foster, Nature, Lond. 183, 1253 (1959). d R. Bralsford, P. V. Harris and W. C. Price, Proc. R. Soc. A258, 459 (1960). e H. Kuroda, Nature, Lond. 201, 1214 (1964). f M. Kinoshita, Bull. chem. Soc. Japan 35, 1609 (1962). g M. E. Wacks and V. H. Dibeler, J. chem. Phys. 31, 1557 (1959). h J. H. D. Eland, P. J. Shepherd and C. J. Danby, Z. Naturf. 21a, 1580 (1966). i J. G. Angus and G. C. Morris, J. molec. Spectrosc. 21, 310 (1966). j M. E. Wacks, J. chem. Phys. 41, 1661 (1964). k R. S. Becker and E. Chen, J. chem. Phys. 45, 2403 (1966). l R. J. van Brunt and M. E. Wacks, J. chem. Phys. 41, 3195 (1964). m P. H. Emslie, R. Foster and R. Pickles, Can. J. Chem. 44, 9 (1966). n D. G. Farnum, E. R. Atkinson and W. C. Lothrop, J. org. Chem. 26, 3204 (1961). o P. G. Farrell and J. Newton, J. phys. Chem., Ithaca 69, 3506 (1965). p P. G. Farrell and J. Newton, Tetrahedron Lett. 5517 (1966).

13.H. Estimation of Ionization Potentials, Electron Affinities and Hückel Coefficients of Molecules

From the discussion in Chapter 3, it is apparent that empirical relationships exist between ionization potentials of donors and the energy of the intermolecular charge-transfer absorption (ν_{CT}) for the corresponding complexes with a given acceptor, the values of ν_{CT} being in general only slightly dependent on the particular solvent for complexes formed from uncharged species. It cannot be too strongly emphasized that such correlations, particularly *linear* correlations, are theoretically unexpected [40] and are the result of relative values of certain terms in relationships such as equation (3.1) described in Chapter 3. In the same chapter it was also pointed out that, for strong electron-donor electron-acceptor interactions, the theoretical relationship described by equation (3.1) is no longer tenable and is replaced by a different relationship between ν_{CT} and ionization potential of the donor, namely equation (3.2). This difference is well illustrated by the donor ionization potential versus ν_{CT} plots for iodine complexes of n-donors and of the weaker π-donors [41] (Fig. 3.6, p. 48). Nevertheless, within these limitations, estimates of ionization potentials of donors may be obtained from such relationships. Mulliken [42] has pointed out that the ionization potentials used, and therefore those interpolated, should be the vertical, rather than adiabatic values. This point is discussed in Chapter 3.

The technique is attractive in that determinations of ν_{CT} can be obtained rapidly using a standard spectrophotometer, provided the charge-transfer band is resolved. It provides values of ionization potential of molecules for which the standard methods may be experimentally difficult. This applies particularly to molecules of low volatility, which limits the use of measurements in the gaseous phase using photoionization, Rydberg series determinations, electron impact, and photoelectron spectroscopy. Some empirical relationships between ionization potential of donors and ν_{CT} for series of complexes with various acceptors are given in Chapter 3. In cases where more than one intermolecular charge-transfer band is observed, the multiplicity may be due to transitions from more than one level in the donor molecule (thereby possibly giving information about other ionization potentials besides the first). Alternatively, the multiple bands may be due to the participation of more than one energy level in the acceptor (see Chapter 3). If the precautions indicated above are taken, reasonable values of ionization potential can be obtained. For example, Finch [43] estimated the ionization potential of azulene to be 7·4–7·5 eV, based on ν_{CT} values from various complexes in solution. After this work was published, Clark [44] determined the ionization potential from Rydberg series measurements in the vapour phase and obtained a value of $7 \cdot 431 \pm 0 \cdot 006$ eV. Some values of ionization potential interpolated from ν_{CT} data are given in Table 13.4. With less certainty, electron affinities of electron-acceptor molecules (E^A) may be estimated from corresponding empirical correlations of E^A and ν_{CT} for series of complexes with a common donor species (Table 13.5). A

TABLE 13.5. Electron affinities of various electron acceptors estimated from charge-transfer spectra of their complexes with electron donors.*

Acceptor	E^A/eV
p-Benzoquinone	$0 \cdot 7_7$
Nitro-p-benzoquinone	$1 \cdot 45$
Cyano-p-benzoquinone	$1 \cdot 04$
Fluoro-p-benzoquinone	$0 \cdot 91$
Chloro-p-benzoquinone	$0 \cdot 97$
Bromo-p-benzoquinone	$0 \cdot 98$
Iodo-p-benzoquinone	$0 \cdot 95$
Phenyl-p-benzoquinone	$0 \cdot 87$
Methyl-p-benzoquinone	$0 \cdot 75$
Methoxy-p-benzoquinone	$0 \cdot 65$
2,3-Dichloro-p-benzoquinone	$1 \cdot 1$
2,5-Dichloro-p-benzoquinone	$1 \cdot 1_5$
2,6-Dichloro-p-benzoquinone	$1 \cdot 2$
2,3-Dicyano-p-benzoquinone	$1 \cdot 7$
Trichloro-p-benzoquinone	$1 \cdot 3$
2,3-Dichloro-5,6-dicyano-p-benzoquinone	$1 \cdot 9_5$
Chloranil	$1 \cdot 37$
Bromanil	$1 \cdot 37$
Iodanil	$1 \cdot 3_6$
o-Chloranil	$1 \cdot 5_5$
o-Bromanil	$1 \cdot 6$
1,2-Dinitrobenzene	0
1,3-Dinitrobenzene	$0 \cdot 3$
1,4-Dinitrobenzene	$0 \cdot 7$
1,3,5-Trinitrobenzene	$0 \cdot 7$
2,4,7-Trinitrofluorenone	$1 \cdot 1$
2,4,6-Trinitrotoluene	$0 \cdot 6$
1,4-Dicyanobenzene	0
1,3,5-Tricyanobenzene	$0 \cdot 1$
1,2,4,5-Tetracyanobenzene	$0 \cdot 4$
Pyromellitic dianhydride	$0 \cdot 8_5$
Tetrachlorophthalic anhydride	$0 \cdot 5_6$
Maleic anhydride	$0 \cdot 11$
Dichloromaleic anhydride	$0 \cdot 53$
Tetracyanoethylene	$2 \cdot 2$
7,7,8,8-Tetracyanoquinodimethane	$1 \cdot 7$

* Based on a compilation by G. Briegleb [*Angew. Chem.* **76**, 326 (1964); *Angew Chem.* (*Int. Edn.*) **3**, 617 (1964)], to which the reader is referred for a more comprehensive list. The values of E^A are calculated from equation (3.12) relative to E^A (chloranil) $= 1 \cdot 37$ eV. Some additions and amendments have been made to the originally quoted values.

major difficulty in this case is the paucity of E^A values. The problem is referred to in Chapter 3.

By using corresponding correlations in the molecular orbital descriptions [45-52] of the charge-transfer transition (Chapter 3), estimates of the donor highest-occupied orbital energy level coefficient (χ_i) may be obtained if sufficient values of χ_i for other donors are available to establish an initial correlation.[45] Here again, attention must be paid to the limitation of such relationships as emphasized by Mulliken[40] and referred to at the beginning of this section. From the plots of χ_i (donor) against ν_{CT}, extrapolation to the value of ν_{CT} corresponding to $\chi_i = 0$ should yield the energy of the lowest-unoccupied orbital of the acceptor. A selection of such values is given in Table 13.6.

TABLE 13.6. Values of the Hückel coefficient (χ_i) and the resonance integral (β) for various acceptors derived from plots of equation (3.3).*

Acceptor	χ_i	β	Ref.†
7,7,8,8-Tetracyanoquinodimethane	−0·038	−3·50	52
Bromanil	−0·072	−3·97	52
Tetracyanoethylene	−0·11	−3·06	45
Chloranil	−0·12	−3·74	52
Iodanil	−0·13	−3·66	52
1,4,5,8-Naphthalene tetracarboxylic dianhydride	−0·14	−3·81	52
9-Dicyanomethylene-2,4,7-trinitrofluorenone	−0·21	−3·13	a
Mellitic trianhydride	−0·22	−3·21	52
2,4,7-Trinitrofluorenone	−0·30	−3·12	47
3-Nitro-1,8-naphthalic anhydride	−0·32	−3·58	52
1,2,4,5-Tetracyanobenzene	−0·335	−3·21	b
Pyromellitic dianhydride	−0·41	−2·87	52
1,3,5-Trinitrobenzene	−0·46	−3·00	45
Iodine (I$_2$)	−0·85	−2·29	52

* Mainly quoted in Ref. 52.

† Refs.: ^{a} T. K. Mukherjee and L. A. Levasseur, *J. org. Chem.* **30**, 644 (1965). ^{b} A. Zweig, J. E. Lehnsen, W. G. Hodgson and W. H. Jura, *J. Am. chem. Soc.* **85**, 3937 (1963).

REFERENCES

1. P. Pfeiffer, "Organische Molekülverbindungen," 2nd edition, Ferdinand Enke, Stuttgart (1927).
2. M. Orchin and E. O. Woolfolk, *J. Am. chem. Soc.* **68**, 1727 (1946).
3. R. E. Merrifield and W. D. Phillips, *J. Am. chem. Soc.* **80**, 2778 (1958).
4. A. S. Bailey and J. R. Case, *Proc. chem. Soc.* 176, 211 (1957); *Tetrahedron* **3**, 113 (1958).
5. A. S. Bailey, *J. chem. Soc.* 4710 (1960).
6. Pl. A. Plattner and A. St. Pfau, *Helv. chim. Acta* **20**, 224 (1937).
7. E. Lederer and M. Lederer, "Chromatography," Elsevier, Amsterdam (1957), p. 65.

8. A. U. Rahman and O. L. Tombesi, *J. Chromat.* **23**, 312 (1966).
9. M. S. Newman and J. Blum, *J. Am. chem. Soc.* **86**, 5600 (1964).
10. K. G. Cunningham, W. Dawson and F. S. Spring, *J. chem. Soc.* 2305 (1951).
11. J. C. Godfrey, *Analyt. Chem.* **31**, 1087 (1959).
12. E. K. Andersen, *Acta chem. scand.* **8**, 157 (1954).
13. R. J. Cvetanović, F. J. Duncan, W. E. Falconer and W. A. Sunder, *J. Am. chem. Soc.* **88**, 1602 (1966).
14. W. E. Falconer and R. J. Cvetanović, *J. Chromat.* **27**, 20 (1967).
15. A. R. Cooper, C. W. P. Crowne and P. G. Farrell, *Trans. Faraday Soc.* **62**, 2725 (1966); *ibid.*, **63**, 447 (1967).
16. A. R. Cooper, C. W. P. Crowne and P. G. Farrell, *J. Chromat.* **27**, 362 (1967).
17. M. Frank-Neumann and P. Jössang, *J. Chromat.* **14**, 280 (1964).
18. R. O. C. Norman, *Proc. chem. Soc.* 151 (1958).
19. R. G. Harvey and M. Halonen, *J. Chromat.* **25**, 294 (1966).
20. H. Kessler and E. Müller, *J. Chromat.* **24**, 469 (1966).
21. D. B. Parihar, S. P. Sharma and K. K. Verma, *J. Chromat.* **31**, 120 (1967).
22. M. Godlewicz, *Nature, Lond.* **164**, 1132 (1949).
23. A. Berg and J. Lam, *J. Chromat.* **16**, 157 (1964).
24. L. H. Klemm, D. Reed and C. D. Lind, *J. org. Chem.* **22**, 739 (1957).
25. N. P. Buu-Hoï and P. Jacquignon, *Experientia* **13**, 375 (1957).
26. J. T. Ayres and C. K. Mann, *Analyt. Chem.* **36**, 2185 (1964).
27. B. R. Brown and D. Ll. Hammick, *J. chem. Soc.* 1395 (1948).
28. M. S. Newman, W. B. Lutz and D. Lednicer, *J. Am. chem. Soc.* **77**, 3420 (1955).
29. M. S. Newman and D. Lednicer, *J. Am. chem. Soc.* **78**, 4765 (1956).
30. M. S. Newman and W. B. Lutz, *J. Am. chem. Soc.* **78**, 2469 (1956).
31. L. H. Klemm and D. Reed, *J. Chromat.* **3**, 364 (1960).
32. L. H. Klemm, K. B. Desai and J. R. Spooner, Jr., *J. Chromat.* **14**, 300 (1964).
33. R. Foster and C. A. Fyfe, *J. chem. Soc.* (B), 926 (1966).
34. C. R. M. Butt, D. Cohen, L. Hewitt and I. T. Millar, *Chem. Commun.* 309 (1967).
35. E. M. Kosower, *J. Am. chem. Soc.* **80**, 3261 (1958).
36. E. M. Kosower, *J. Am. chem. Soc.* **80**, 3253 (1958).
37. E. M. Kosower and G.-S. Wu, *J. Am. chem. Soc.* **83**, 3142 (1961).
38. E. M. Kosower, G.-S. Wu and T. S. Sorensen, *J. Am. chem. Soc.* **83**, 3147 (1961).
39. E. M. Kosower, W. D. Closson, H. L. Goering and J. R. Grass, *J. Am. chem. Soc.* **83**, 2013 (1961).
40. R. S. Mulliken, *Proc. Int. Conf. on Co-ordination Compounds*, Amsterdam (1955), p. 371; *Recl. Trav. chim. Pays-Bas Belg.* **75**, 845 (1956).
41. R. S. Mulliken and W. B. Person, *A. Rev. phys. Chem.* **13**, 107 (1962).
42. R. S. Mulliken, *J. Am. chem. Soc.* **74**, 811 (1952).
43. A. C. M. Finch, *J. chem. Soc.* 2272 (1964).
44. L. B. Clark, *J. chem. Phys.* **43**, 2566 (1965).
45. M. J. S. Dewar and A. R. Lepley, *J. Am. chem. Soc.* **83**, 4560 (1961).
46. M. J. S. Dewar and H. Rogers, *J. Am. chem. Soc.* **84**, 395 (1962).
47. A. R. Lepley, *J. Am. chem. Soc.* **84**, 3577 (1962).
48. A. R. Lepley, *J. Am. chem. Soc.* **86**, 2545 (1964).
49. A. R. Lepley and C. C. Thompson, Jr., *J. Am. chem. Soc.* **89**, 5523 (1967).
50. M. Nepraš and R. Zahradník, *Tetrahedron Lett.* 57 (1963).
51. M. Nepraš and R. Zahradník, *Colln. Czech. chem. Commun. Engl. Edn* **29**, 1545, 1555 (1964).
52. S. A. Berger, *Spectrochim. Acta* **23A**, 2213 (1967).

Appendix Table I

Semiconductive properties of some charge-transfer complexes

Donor	Acceptor*	State†	Ratio D:A (other than 1:1)	Resistivity (room temp.)/ohm cm	$Ea\ddagger$/eV	Ref.§
Perylene	Fluoranil	SC		$6{\cdot}6 \times 10^{13}$||	1·46||	66
Perylene	Fluoranil			2×10^{14}¶	1·46¶	66
N,N,N',N'-Tetramethylbenzidine	Fluoranil			$3{\cdot}4 \times 10^{12}$		72
p-Aminodiphenylamine	Fluoranil			$2{\cdot}8 \times 10^{12}$		72
p-Ansidine	Chloranil			10^{10}		a
	Chloranil			10^{11}		a
Diaminodurene	Chloranil			$2{\cdot}8 \times 10^{3}$	0·50	a
Diaminodurene	Chloranil	SC		$7{\cdot}0 \times 10^{4}$**	0·52	53
				$6{\cdot}9 \times 10^{4}$††		53
Diaminodurene	Chloranil			$8{\cdot}4 \times 10^{4}$‡‡		53
				$3{\cdot}4 \times 10^{4}$	0·58	53
1,5-Diaminonaphthalene	Chloranil	SC		$1{\cdot}3 \times 10^{9}$**	1·2, 1·58	53
				$6{\cdot}3 \times 10^{11}$††	1·48	53
				$2{\cdot}0 \times 10^{11}$‡‡	1·48	53
1,5-Diaminonaphthalene	Chloranil			$7{\cdot}2 \times 10^{11}$	1·30	a
1,5-Diaminonaphthalene	Chloranil			10^{11}		72
1,5-Diaminonaphthalene	Chloranil			$6{\cdot}1 \times 10^{10}$		63
1,5-Diaminonaphthalene	Chloranil			$1{\cdot}2 \times 10^{11}$	1·16	a
1,8-Diaminonaphthalene	Chloranil			10^{11}		53
1,6-Diaminopyrene	Chloranil	SC		10^{9}**	0·38	53
				10^{6}††	0·38	53
				10^{9}‡‡	0·38	53
1,6-Diaminopyrene	Chloranil			4×10^{3}	0·38	53
1,6-Diaminopyrene	Chloranil			10^{4}		a
1,6-Diaminopyrene	Chloranil			$1{\cdot}2 \times 10^{4}$		72
N,N-Dimethylaniline	Chloranil			10^{10}		a
N,N-Dimethylaniline	Chloranil			5×10^{7}§§	0·47	102
	Chloranil			$8{\cdot}1 \times 10^{8}$||||		102

Compound	Acceptor	Ratio/Note	Resistivity	E_a	Ref
Hexamethylbenzene	Chloranil		10^{11}		*a*
Perylene	Chloranil		10^{8}		*a*
Perylene	Chloranil		2.8×10^{11}		72
p-Phenylenediamine	Chloranil		10^{7}	0·86	*a*
p-Phenylenediamine	Chloranil		$\sim 2 \times 10^{7}$	0·8	48 *b*
p-Phenylenediamine	Chloranil		4.3×10^{6}		72
p-Phenylenediamine	Chloranil		1.5×10^{7}	0·92	63
Pyrene	Chloranil		10^{11}		*a*
N,N,N′,N′-Tetramethylbenzidine	Chloranil		2.3×10^{7}		72
N,N,N′,N′-Tetramethyl-p-phenylenediamine	Chloranil		10^{9}		*a*
N,N,N′,N′-Tetramethyl-p-phenylenediamine	Chloranil		1.3×10^{4}§§ ; 2.0×10^{4}‖	0·53	102 ; 102
o-Tolidine	Chloranil		3.8×10^{9}	1·06	63
Phthalocyanine	Chloranil	DF	~ 100	~0·4	67
Tetrathiotetracene	o-Chloranil	3:1	2–4	0·2	62
Violanthrene	o-Chloranil		10^{3}	0·4	68
1,6-Diaminopyrene	o-Chloranil				*a*
N,N-Dimethylaniline	Bromanil		9×10^{7}§§ ; 1.5×10^{9}‖	0·45	102 ; 102
p-Phenylenediamine	Bromanil		10^{10}		*a*
N,N,N′,N′-Tetramethyl-p-phenylenediamine	Bromanil		4.2×10^{4}§§ ; 1.3×10^{5}‖	0·56	102 ; 102
Tetrathiotetracene	o-Bromanil	3:1	6–8	0·2	62
1,6-Diaminopyrene	Iodanil		10^{6}		*a*
N,N-Dimethylaniline	Iodanil		3×10^{7}§§ ; 1.7×10^{8}‖	0·43	102 ; 102
p-Phenylenediamine	Iodanil		10^{11}		*a*
N,N,N′,N′-Tetramethyl-p-phenylenediamine	Iodanil		1.1×10^{5}§§ ; 1.5×10^{6}‖	0·59	102 ; 102
Dibenzo[c,d]phenothiazine	DDQ	2:1	20·7	0·42	75
Dibenzo[c,d]phenothiazine	DDQ	2:1	17	0·18–0·2	49
Dibenzo[c,d]phenothiazine	DDQ	1:1	5×10^{3}		49
Ferrocene	DDQ		3×10^{10}	1·4	*c*
Perylene	DDQ		3×10^{6}	0·90	26
p-Phenylenediamine	DDQ		10^{6}	0·74	26

Appendix Table I—continued

Donor	Acceptor*	State†	Ratio D:A (other than 1:1)	Resistivity (room temp.)/ ohm cm	$Ea\ddagger$/eV	Ref.§
Pyrene	DDQ			10^{13}	1·8	26
Dibenzo[c,d]phenothiazine	DBNQ		1:1	$\sim 10^{8}$	0·26	49
Dibenzo[c,d]phenothiazine	DBNQ		3:2	240	1·55	49
Benzo[e]pyrene	TCNQ			$5 \cdot 3 \times 10^{11}$	1·22	93
Benzo[a]pyrene	TCNQ			$4 \cdot 3 \times 10^{10}$	1·1	93
Carbazole	TCNQ			$7 \cdot 0 \times 10^{10}$	0·28	45
1,6-Diaminopyrene	TCNQ			0·5	1·1	47
N-Ethylcarbazole	TCNQ			$1 \cdot 8 \times 10^{13}$	1·5	45
Poly(N-vinylcarbazole)	TCNQ		31:1	$1 \cdot 1 \times 10^{16}$	1·4	45
Poly(N-vinylcarbazole)	TCNQ		34:1	$5 \cdot 0 \times 10^{15}$	1·3	45
Poly(N-vinylcarbazole)	TCNQ		35:1	$2 \cdot 4 \times 10^{14}$	1·4	45
Poly(N-vinylcarbazole)	TCNQ		38:1	$1 \cdot 2 \times 10^{14}$	1·1	45
Poly(N-vinylcarbazole)	TCNQ		57:1	$1 \cdot 4 \times 10^{14}$	2·54	45
Aniline	TNB			10^{17}	1·72	60
Anthanthrene	TNB			10^{18}	1·8	60
Anthracene	TNB			9×10^{17}	2·72	60
p-Chloroaniline	TNB			5×10^{13}	2·40	60
Chrysene	TNB			7×10^{18}		60
Coronene	TNB			10^{13}	2·08	a
N,N-Dimethylaniline	TNB			10^{11}	2·36	a
N,N-Dimethylaniline	TNB			10^{16}	1·66	60
1-Naphthylamine	TNB			3×10^{7}	2·48	60
Perylene	TNB			10^{19}	2·04	60
Phenanthrene	TNB			7×10^{18}	2·20	60
p-Phenylenediamine	TNB			8×10^{16}	1·62	60
Pyrene	TNB			10^{20}	1·14	60
N,N,N′,N′-Tetramethylbenzidine	TNB			10^{17}		60
Violanthrene	TNB			5×10^{13}		60
Coronene	Picric acid			10^{12}		a
Coronene	TNF			10^{12}		a

Compound	Acceptor	Ratio	K	ΔH	Ref.
Acenaphthene	TCNE		$5\cdot3 \times 10^{13}$	$2\cdot04$	61
Anthanthrene	TCNE		$1\cdot1 \times 10^{10}$	$1\cdot34$	61
Azulene	TCNE		$4\cdot7 \times 10^{10}$	$0\cdot75$	61
Benzidine	TCNE		1×10^{14}	$1\cdot42$	63
Benzo[a]pyrene	TCNE		4×10^{13}	$1\cdot15$–$1\cdot35$	94
Benzo[e]pyrene	TCNE		3×10^{14}	$1\cdot77$	94
p,p′-Diaminodiphenylmethane	TCNE		10^{6}		63
1,5-Diaminonaphthalene	TCNE		10^{7}	$0\cdot80$	63
Dimethoxybenzene	TCNE		10^{11}		a
Hexamethylbenzene	TCNE		10^{11}		a
Hexamethylbenzene	TCNE		$4\cdot1 \times 10^{13}$	$1\cdot16$	61
Naphthalene	TCNE		$3\cdot2 \times 10^{15}$	$2\cdot48$	61
Pentamethylbenzene	TCNE		$4\cdot4 \times 10^{13}$	$1\cdot11$	61
Perylene	TCNE		10^{11}		a
Perylene	TCNE		$2\cdot4 \times 10^{12}$	$1\cdot44$	61
Phenanthrene	TCNE		$2\cdot2 \times 10^{12}$	$1\cdot52$	61
o-Phenylenediamine	TCNE		10^{16}		63
m-Phenylenediamine	TCNE		6×10^{14}	$1\cdot86$	63
p-Phenylenediamine	TCNE		2×10^{15}	$1\cdot56$	63
Pyrene	TCNE		$4\cdot5 \times 10^{15}$	$1\cdot65$	61
Pyrene	TCNE		10^{10}		62
Tetrathiotetracene	TCNE	3:2	15	$0\cdot20$	63
o-Tolidine	TCNE		10^{16}		63
Violanthrene	TCNE		$5\cdot2 \times 10^{8}$	$0\cdot35$	61
Benzidine	I_2	1:0.78	$1\cdot6 \times 10^{5}$	$0\cdot68$	43
Benzidine	I_2	1:0.94	$6\cdot2 \times 10^{2}$	$0\cdot38$	43
Benzidine	I_2	1:1.26	$2\cdot2$	$0\cdot38$	43
Benzidine	I_2	1:1.49	12	$0\cdot38$	43
Benzo[a]phenothiazine	I_2		20	$0\cdot4$	d
Benzo[c]phenothiazine	I_2		20	$0\cdot28$–$0\cdot4$	d
Acridine	I_2		10^{13}		a
Chlorpromazine	I_2	2:3 DF	420	$0\cdot5$	a
Coronene	I_2		$2\cdot8 \times 10^{8}$		64
Coronene	I_2		10^{9}		a
N-Methylphenothiazine	I_2		$1\cdot4$	$0\cdot28$	d
1,5-Diaminonaphthalene	I_2	1:0.70	$6\cdot1 \times 10^{7}$	$1\cdot32$	43
1,5-Diaminonaphthalene	I_2	1:0.86	$1\cdot6 \times 10^{6}$	$0\cdot84$	43

Appendix Table I—continued

Donor	Acceptor*	State†	Ratio D:A (other than 1:1)	Resistivity (room temp.)/ ohm cm	$Ea\ddagger$/eV	Ref.§
1,5-Diaminonaphthalene	I_2		1:0·90	$1·0 \times 10^6$	0·80	43
1,5-Diaminonaphthalene	I_2		1:1·01	$3·1 \times 10^5$	0·64	43
Perylene	I_2		1:3	2–3	0·06	57
Perylene	I_2		1:3	9–10		40
Perylene	I_2		1:3	6·3	0·038	72
Perylene	I_2		2:3	8		57
Perylene	I_2		2:3	3·0	0·06	70
Perylene	I_2		2:3	9–10		40
Perylene	I_2		2:3	10		a
Perylene	I_2		1:0·23	6×10^3	0·09	40
Perylene	I_2		1:0·86	74	0·06	40
Perylene	I_2		1:1·08	27	0·06	40
Perylene	I_2		1:1·56	16	0·06	40
Perylene	I_2		1:1·80	19	0·06	40
Perylene	I_2		1:2·42	19	0·05	40
Perylene	I_2		1:2·96	20	0·06	40
Perylene	I_2		1:5·8	22	0·07	40
Phenothiazine	I_2		2:3	350		e
Phenothiazine	I_2			20	0·34	d
p-Phenylenediamine	I_2		0·45:1	$5·9 \times 10^9$	1·6	42
p-Phenylenediamine	I_2		0·67:1	$3·5 \times 10^7$	1·22	42
p-Phenylenediamine	I_2		0·82:1	$1·7 \times 10^5$	0·82	42
p-Phenylenediamine	I_2		1·03:1	$2·7 \times 10^5$	0·88	42
p-Phenylenediamine	I_2		1·36:1	$1·04 \times 10^6$	0·96	42
p-Phenylenediamine	I_2		1·62:1	$5·4 \times 10^6$	1·18	42
Poly(4-vinylpyridine)	I_2		1:2	10^4	1·3	f
Poly(4-vinylpyridine)	I_2		1:0·3	10^7		f
Pyranthrene	I_2			17	0·09	g
Pyrene	I_2		1:2	75	0·28***	57

o-Tolidine	I_2		1:0·94	$3·5 \times 10^3$	0·54	43
o-Tolidine	I_2		1:1·02	290	0·48	43
o-Tolidine	I_2		1:1·29	29	0·36	43
o-Tolidine	I_2		1:1·53	91	0·14, 0·15	43
Violanthrene	I_2		1:2	45	0·25	44
Violanthrene	I_2		1:3·17	127	0·14	41
Violanthrene	I_2		1:1·90	18·0	0·18	41
Violanthrene	I_2		1:1·31	24·0	0·18	41
Violanthrene	I_2		1:0·118	220	0·45	41
Violanthrene	I_2		$1:10^{-2}$	$3·1 \times 10^5$	0·45	41
Violanthrene	I_2		$1:3·6 \times 10^{-3}$	$2·8 \times 10^7$		41
Violanthrene	I_2		$1:8 \times 10^{-4}$	6×10^8	0·44	41
Violanthrene	I_2		$1:5 \times 10^{-4}$	$1·4 \times 10^9$	0·5	41
Violanthrene	I_2	DF			0·13	68
Perylene	Br_2		1:2·2	7·8	0·20	44
Pyranthrene	Br_2		1:1·6	220	0·20	44
Violanthrene	Br_2		1:2·2	66		44
Acridine	ICl			10^{13}		a

* Acceptors: DDQ = 2,3-dichloro-5,6-dicyano-p-benzoquinone; DBNQ = 2,3-dibromo-5,6-dicyano-p-benzoquinone; TCNQ = 7,7,8,8-tetracyanoquinodimethane; TNB = 1,3,5-trinitrobenzene; TNF = 2,4,7-trinitrofluorenone; TCNE = tetracyanoethylene.

† Microcrystalline, unless otherwise stated. SC = single crystal. DF = deposited film.

‡ Ea defined by the expression $\rho = \rho_0 Ea/2RT$.

§ Refs.: (numbers refer to references in Chapter 9): [a] M. M. Labes, R. Sehr and M. Bose, $J. chem. Phys.$ **33**, 868 (1960). [b] K. Pignoń and K. Lorenz, $Rocz. Chem. Ann. Soc. Chim. Polon.$ **40**, 699 (1966). [c] R. L. Brandon, J. H. Osiecki and A. Ottenberg, $J. org. Chem.$ **31**, 1214 (1966). [a] Y. Matsunaga, $Helv. phys. Acta$ **36**, 800 (1963). [e] F. Gutmann and H. Keyzer, $J. chem. Phys.$ **46**, 1969 (1967). [f] S. B. Mainthia, P. L. Kronick and M. M. Labes, $J. chem. Phys.$ **41**, 2206 (1964). [g] H. Akamatu and H. Inokuchi, in "Symposium on Electrical Conductivity in Organic Solids," eds. H. Kallmann and M. Silver, Interscience, New York (1961), p. 279.

∥ Parallel to c-axis.

¶ Perpendicular to c-axis.

** Along x-axis.

†† Along y-axis.

‡‡ Along z-axis.

§§ ac measurement.

∥∥ dc measurement.

*** 0·14 eV at low temperature.

Appendix Table 2

References to determinations of association constant and of other thermodynamic constants for some organic electron-donor–electron-acceptor systems.

Acceptor	Donor	Method*	Constants determined	Reference
Iodine (I_2)	N,N-Dimethylacetamide	Spectro, calorimetry	$K, \Delta H^\ominus, \Delta S^\ominus$	R. S. Drago, T. F. Bolles and R. J. Niedzielski, J. Am. chem. Soc. **88**, 2717 (1966)
	N,N-Dimethylpropionamide	Spectro	$K, \Delta H^\ominus, \Delta S^\ominus$	R. S. Drago, D. A. Wenz and R. L. Carlson, J. Am. chem. Soc. **84**, 1106 (1962)
	N,N-Dimethylacetamide	Spectro	$K, \Delta H^\ominus, \Delta S^\ominus$	R. S. Drago, R. L. Carlson, N. J. Rose and D. A. Wenz, J. Am. chem. Soc. **83**, 3572 (1961)
	N,N-Dimethylacetamide	Infrared	$K, \Delta H^\ominus, \Delta S^\ominus$	C. D. Schmulbach and R. S. Drago, J. Am. chem. Soc. **82**, 4484 (1960)
	N,N-Dimethylacetamide, N,N-dimethylformamides	Spectro	$K, \Delta H^\ominus, \Delta S^\ominus$	R. S. Drago, D. A. Wenz and R. L. Carlson, J. Am. chem. Soc. **85**, 1106 (1962)
	N,N-Dimethylbenzamides	Spectro	$K, \Delta H^\ominus, \Delta S^\ominus$	R. L. Carlson and R. S. Drago, J. Am. chem. Soc. **85**, 505 (1963)
	N,N-Dimethylbenzamide	Spectro	$K, \Delta H^\ominus, \Delta S^\ominus$	R. L. Carlson and R. S. Drago, J. Am. chem. Soc. **84**, 2320 (1962)
	Sulphoxides, diethyl sulphite, tetramethylene sulphone	Spectro	$K, \Delta H^\ominus$	R. S. Drago, B. B. Wayland and R. L. Carlson, J. Am. chem. Soc. **85**, 3125 (1963)
	Acetonitrile	Spectro	K	A. I. Popov and W. A. Deskin, J. Am. chem. Soc. **80**, 2976 (1958)
	Acetonitrile, chloroacetonitriles	Spectro	$K, \Delta H^\ominus, \Delta S^\ominus$	W. B. Person, W. C. Golton and A. I. Popov, J. Am. chem. Soc. **85**, 891 (1963)
	Propionitrile	Spectro	K	P. Klaeboe, J. Am. chem. Soc. **85**, 871 (1963)
	Valeronitrile	Spectro	$K, \Delta H^\ominus, \Delta S^\ominus$	J. A. Maguire, A. Bramley and J. J. Banewicz, Inorg. Chem. **6**, 1752 (1967)

Benzonitrile	Spectro	K	P. Klaeboe, J. Am. chem. Soc. **84**, 3458 (1962)
Dioxane	Spectro	K	R. S. Drago and N. J. Rose, J. Am. chem. Soc. **81**, 6141 (1959).
Dioxane	Spectro	$K, \Delta H^{\ominus}$	J. A. A. Ketelaar, C. van de Stolpe and H. R. Gersmann, Recl Trav. chim. Pays-Bas Belg. **70**, 499 (1951)
Dioxane	Spectro	$K, \Delta H^{\ominus}$	J. A. A. Ketelaar, C. van de Stolpe, A. Goudsmit and W. Dzcubas, Recl Trav. chim. Pays-Bas Belg. **71**, 1104 (1952)
Diethers	Spectro	$K, \Delta H^{\ominus}, \Delta S^{\ominus}$	A. F. Garito and B. B. Wayland, J. phys. Chem., Ithaca **71**, 4061 (1967)
Diethyl ether, benzene	Spectro	$K, \Delta H^{\ominus -}$	F. T. Lang and R. L. Strong, J. Am. chem. Soc. **87**, 2345 (1965)
Diethyl ether	Spectro	$\Delta H^{\ominus}$	P. A. D. de Maine and P. Carapellucci, J. molec. Spectrosc. **7**, 83 (1961)
Diethyl ether, 2-methyl-propan-2-ol	Spectro	K	J. S. Ham, J. chem. Phys. **20**, 1170 (1952)
Diethyl ether, trimethylene oxide, 2-methyltetrahydro-furan, tetrahydrofuran, tetrahydropyran	Spectro	$K, \Delta H^{\ominus}, \Delta S^{\ominus}$	M. Brandon, M. Tamres and S. Searles, Jr., J. Am. chem. Soc. **82**, 2129 (1960)
Diethyl ether, trimethylene oxide, 2-Methyltetrahydro-furan, tetrahydrofuran, tetrahydropyran	Spectro	K	M. Tamres and M. Brandon, J. Am. chem. Soc. **82**, 2134 (1960)
Diethyl ether, methanol, ethanol	Spectro	$K, \Delta H^{\ominus}$	P. A. D. de Maine, J. chem. Phys. **26**, 1192 (1957)
7-Oxabicyclo[2.2.1]heptane	Spectro	$K, \Delta H^{\ominus}, \Delta S^{\ominus}$	M. Tamres, S. Searles, Jr. and J. M. Goodenow, J. Am. chem. Soc. **86**, 3934 (1964)
Dioxane, and seleno- and thio- analogues of dioxane	Spectro	$K, \Delta H^{\ominus}, \Delta S^{\ominus}$	J. D. McCullough and I. C. Zimmermann, J. phys. Chem., Ithaca **65**, 888 (1961)
Acetone, cyclohexanone, methyl acetate, 3-phenyl-sydnone	Infrared	K	H. Yamada and K. Kozima, J. Am. chem. Soc. **82**, 1543 (1966)

Appendix Table 2.—continued

Acceptor	Donor	Method*	Constants determined	Reference
Iodine (I₂)	Carbonyl and thiocarbonyl compounds	Spectro	$K, \Delta H^{\ominus}$	K. R. Bhaskar, S. N. Blat, A. S. N. Murthy and C. N. R. Rao, *Trans. Faraday Soc.* **62**, 788 (1966)
	Acetone, cyclic ketones, cyclohexanol	Spectro	$K, \Delta H^{\ominus}, \Delta S^{\ominus}$	D. Wobschall and D. A. Norton, *J. Am. chem. Soc.* **87**, 3559 (1965)
	Dimethylcyanamide	Spectro	$K, \Delta H^{\ominus}, \Delta S^{\ominus}$	E. Augdahl and P. Klaeboe, *Acta chem. scand.* **19**, 807 (1965)
	Substituted diphenyl selenides	Spectro	K	J. D. McCullough and B. A. Eckerson, *J. Am. chem. Soc.* **73**, 2954 (1951)
	Aldehydes	Infrared	K	E. Augdahl and P. Klaeboe, *Acta chem. scand.* **16**, 1637 (1962)
	Alcohols, *n*-butyl ether, ethanol, ethyl acetate	Spectro	K	R. M. Keefer and L. J. Andrews, *J. Am. chem. Soc.* **75**, 3561 (1953)
			K	J. H. Hildebrand and B. L. Glascock, *J. Am. chem. Soc.* **31**, 26 (1909)
	Ethanol	Calorimetry	$\Delta H^{\ominus}$	K. Hartley and H. A. Skinner, *Trans. Faraday Soc.* **46**, 621 (1950)
	Acetic acid	Spectro	K	R. E. Buckles and J. F. Mills, *J. Am. chem. Soc.* **75**, 552 (1953)
	Carbonyl compounds	Spectro	$K, \Delta H^{\ominus}$	R. L. Middaugh, R. S. Drago and R. J. Niedzielski, *J. Am. chem. Soc.* **86**, 388 (1964)
	Tetramethylurea, tetra-methylthiourea	Spectro	$K, \Delta H^{\ominus}, \Delta S^{\ominus}$	R. P. Lang, *J. phys. Chem.*, *Ithaca* **72**, 2129 (1968)
	Isothiocyanate esters	Spectro	$K, \Delta H^{\ominus}, \Delta S^{\ominus}$	E. Plahte, J. Grundnes and P. Klaeboe, *Acta chem. scand.* **19**, 1897 (1965)
	Thiourea, thioacetamide	Spectro	$K, \Delta H^{\ominus}, \Delta S^{\ominus}$	R. P. Lang, *J. Am. chem. Soc.* **84**, 1185 (1962)
	Thioureas, thiocarbanilides	Spectro	$K, \Delta H^{\ominus}, \Delta S^{\ominus}$	K. R. Bhaskar, R. K. Gosavi and C. N. R. Rao, *Trans. Faraday Soc.* **62**, 29 (1966)

Thiodimethylacetamide, thioanisole, tetramethyl-thiourea	Spectro	K, $\Delta H^{\ominus}$	R. J. Niedzielski, R. S. Drago and R. L. Middaugh, *J. Am. chem. Soc.* **86**, 1694 (1964)
Thiacycloalkanes	Spectro	K	J. D. McCullough and D. Mulvey, *J. Am. chem. Soc.* **81**, 1291 (1959)
Dialkylsulphides, tetrahydro-thiophene	Spectro	K, $\Delta H^{\ominus}$, $\Delta S^{\ominus}$	M. Kroll, *J. Am. chem. Soc.* **90**, 1097 (1968)
Dimethyl sulphide, dimethyl selenide	Spectro	K	N. W. Tideswell and J. D. McCullough, *J. Am. chem. Soc.* **79**, 1031 (1957)
Diphenyl selenide	Spectro	K	J. D. McCullough, *J. Am. chem. Soc.* **64**, 2672 (1942)
Disulphides, 1,2-dithiane	Calorimetry, spectro	K, $\Delta H^{\ominus}$, $\Delta S^{\ominus}$	B. Nelander, *Acta chem. scand.* **20**, 2289 (1966)
Diethyl sulphide, *trans*-2,3-butylene sulphide thia- cycloalkanes	Spectro	K, $\Delta H^{\ominus}$, $\Delta S^{\ominus}$	M. Tamres and S. Searles, Jr., *J. phys. Chem.*, *Ithaca* **66**, 1099 (1962)
Diethyl sulphide, diethyl disulphide, amides	Spectro	K, $\Delta H^{\ominus}$, $\Delta S^{\ominus}$	H. Tsubomura and R. P. Lang, *J. Am. chem. Soc.* **83**, 2085 (1961)
Diethyl sulphide	Spectro	K, $\Delta H^{\ominus}$, $\Delta S^{\ominus}$	S. P. McGlynn, *Radiat. Res. Supp.* **2**, 300 (1960)
Diethyl sulphide	Spectro	K, $\Delta H^{\ominus}$, $\Delta S^{\ominus}$	J. M. Goodenow and M. Tamres, *J. chem. Phys.* **43**, 3393 (1965); M. Tamres and J. M. Goodenow, *J. phys. Chem.*, *Ithaca* **71**, 1982 (1967)
Diethyl disulphide, ethyl mercaptan, diethyl sulphide	Spectro	K, $\Delta H^{\ominus}$, $\Delta S^{\ominus}$	M. Good, A. Major, J. N. Chaudhuri and S. P. McGlynn, *J. Am. chem. Soc.* **83**, 4329 (1961)
Sulphides, thiacycloalkanes	N.m.r.	K	E. T. Strom, W. L. Orr, B. S. Snowden, Jr. and D. E. Woessner, *J. phys. Chem.*, *Ithaca* **71**, 4017 (1967)
Thiacycloalkanes	Spectro	K, $\Delta H^{\ominus}$, $\Delta S^{\ominus}$	J. D. McCullough and I. C. Zimmermann, *J. phys. Chem.*, *Ithaca* **66**, 1198 (1962)
Thioanisoles	Spectro	K, $\Delta H^{\ominus}$, $\Delta S^{\ominus}$	J. van der Veen and W. Stevens, *Recl Trav. chim. Pays-Bas Belg.* **82**, 287 (1963).
Thioanisole, tolyl methyl sulphides	Spectro	K, $\Delta H^{\ominus}$	V. Ramakrishnan and P. A. D. de Maine, *J. Miss. Acad. Sci.* **10**, 82 (1964)

Appendix Table 2—*continued*

Acceptor	Donor	Method*	Constants determined	Reference
Iodine (I_2)	Dimethylsulphoxide	Spectro	K, $\Delta H^\ominus$, $\Delta S^\ominus$	P. Klaeboe, *Acta chem. scand.* **18**, 27 (1964)
	Sulphoxides	Spectro	K, $\Delta H^\ominus$, $\Delta S^\ominus$	P. Klaeboe, *Acta chem. scand.* **18**, 999 (1964)
	Sulphoxides	Spectro	K, $\Delta H^\ominus$, $\Delta S^\ominus$	J. Grundnes and P. Klaeboe, *Trans. Faraday Soc.* **60**, 1991 (1964)
	Sulphur (S_8)	Spectro	K	J. Jander and G. Türk, *Chem. Ber.* **97**, 25 (1964)
	Hydrogen sulphide	Spectro	K, $\Delta H^\ominus$	J. Jander and G. Türk, *Chem. Ber.* **98**, 894 (1965)
	Diphenyl selenides, dimethyl selenide	Spectro	K, $\Delta H^\ominus$, $\Delta S^\ominus$	J. D. McCullough and I. C. Zimmermann, *J. phys. Chem., Ithaca* **64**, 1084 (1960)
	Diphenyl selenide	Spectro	K	J. D. McCullough, *J. Am. chem. Soc.* **64**, 2672 (1942)
	Selenocyclopentane, selenocyclohexane	Spectro	K, $\Delta H^\ominus$, $\Delta S^\ominus$	J. D. McCullough and A. Brunner, *Inorg. Chem.* **6**, 1251 (1967)
	Diphenyl selenium oxide	Spectro	K, $\Delta H^\ominus$, $\Delta S^\ominus$	J. Grundnes and P. Klaeboe, *Acta chem. scand.* **18**, 2022 (1964)
	Trimethylamine	Spectro	K	R. S. Drago and N. J. Rose, *J. Am. chem. Soc.* **81**, 6141 (1959)
	iso-Propylamine, diethylamine	Spectro	K	S. Kobinata and S. Nagakura, *J. Am. chem. Soc.* **88**, 3905 (1966)
	Ammonia, aliphatic amines	Spectro	K, $\Delta H^\ominus$, $\Delta S^\ominus$	H. Yada, J. Tanaka and S. Nagakura, *Bull. chem. Soc. Japan* **33**, 1660 (1960)
	Triethylamine	Spectro	K, $\Delta H^\ominus$, $\Delta S^\ominus$	S. Nagakura, *J. Am. chem. Soc.* **80**, 520 (1958)
	Triethylamine	Calorimetry	K, $\Delta H^\ominus$	C. D. Schmulbach and D. M. Hart, *J. Am. chem. Soc.* **86**, 2347 (1964)
	Aminoboranes	Spectro	K, $\Delta H^\ominus$, $\Delta S^\ominus$	I. D. Eubanks and J. J. Lagowski, *J. Am. chem. Soc.* **88**, 2425 (1966)
	Phosphine oxides, phosphine sulphides and phosphine selenides	Spectro	K	R. A. Zingaro, R. E. McGlothlin and E. A. Meyers, *J. phys. Chem., Ithaca* **66**, 2579 (1962)

Pyridine, methylpyridines	Spectro	K	A. I. Popov and R. H. Rygg, J. Am. chem. Soc. 79, 4622 (1957)
Styrylpyridines	Spectro	$K, \Delta H^\ominus, \Delta S^\ominus$	G. Aloisi, G. Cauzzo, G. Giacometti and U. Mazzucato, Trans. Faraday Soc. 61, 1406 (1965)
Styrylpyridines	Spectro	K	U. Mazzucato, G. Aloisi and G. Cauzzo, Trans. Faraday Soc. 62, 2685 (1966)
Pyridine	Optical	$K, \Delta H^\ominus, \Delta S^\ominus$	C. Reid and R. S. Mulliken, J. Am. chem. Soc. 76, 3869 (1954).
Pyridine	Infrared	K	A. G. Maki and E. K. Plyler, J. phys. Chem., Ithaca 66, 766 (1962)
Pyridine	Spectro	K	G. Aloisi, G. Cauzzo and U. Mazzucato, Trans. Faraday Soc. 63, 1858 (1967)
Pyridine	Far infra-red	K	R. F. Lake and H. W. Thompson, Proc. R. Soc. A297, 440 (1967)
Pyridines	Spectro	K	W. J. McKinney, M. K. Wong and A. I. Popov, Inorg. Chem. 7, 1001 (1968)
Pyridines	Spectro	K	A. I. Popov and R. H. Rygg, J. Am. chem. Soc. 79, 4622 (1957)
Aza-aromatics	Spectro	K	J. N. Chaudhuri and S. Basu, Trans. Faraday Soc. 55, 898 (1959)
Pyridines	Spectro	K	K. R. Bhaskar and S. Singh, Spectrochim. Acta 23A, 1155 (1967)
Pyridines	Spectro	K	H. D. Bist and W. B. Person, J. phys. Chem., Ithaca 71, 2750 (1967)
Pyridines	Spectro	K	R. D. Srivastava and G. Prasad, Spectrochim. Acta 22, 825 (1966)
Pyridines, quinolines	Spectro	$K, \Delta H^\ominus, \Delta S^\ominus$	V. G. Krishna and B. B. Bhowmik, J. Am. chem. Soc. 90, 1700 (1968)
N-Oxides	Spectro	$K, \Delta H^\ominus, \Delta S^\ominus$	T. Kubota J. Am. chem. Soc. 87, 458 (1958)
N-Oxides	Spectro	$K, \Delta H^\ominus, \Delta S^\ominus$	T. Kubota, M. Yamakawa, M. Takasuka, K. Iwatani, H. Akazawa and I. Tanaka, J. phys. Chem., Ithaca 71, 3597 (1967)
Dipyridylethylenes	Spectro	K	G. Giacometti, U. Mazzucato and S. Parolini, Tetrahedron Lett. 3733 (1964)

Appendix Table 2—continued

Acceptor	Donor	Method*	Constants determined	Reference
Iodine (I_2)	Azanaphthalenes	Spectro	K	I. Ilmet and M. Krasij, *J. phys. Chem., Ithaca* **70**, 3755 (1966)
	Thiophene, 2-methylfuran, furan, N-methylpyrrole	Spectro	$K, \Delta H^{\ominus}, \Delta S^{\ominus}$	R. P. Lang, *J. Am. chem. Soc.* **84**, 4438 (1962)
	Furan	Spectro	$\Delta H^{\ominus}$	E. I. Ginns and R. L. Strong, *J. phys. Chem., Ithaca* **71**, 3059 (1967)
	Pentamethylene tetrazole	Spectro	K	A. I. Popov, C. C. Bisi and M. Craft, *J. Am. chem. Soc.* **80**, 6513 (1958)
	Cyclohexene, vinyl halides	Spectro	K	L. J. Andrews and R. M. Keefer, *J. Am. chem. Soc.* **74**, 458 (1952)
	Alkyl halides	Spectro	K	L. J. Andrews and R. M. Keefer, *J. Am. chem. Soc.* **74**, 1891 (1952)
	Olefins, chloro-olefins	Spectro	K	J. A. A. Ketelaar and C. van de Stolpe, *Recl Trav. chim. Pays-Bas Belg.* **71**, 805 (1952)
	Tri-n-butyl phosphate	Spectro	$K, \Delta H^{\ominus}, \Delta S^{\ominus}$	H. Tsubomura and J. M. Kliegman, *J. Am. chem. Soc.* **82**, 1314 (1960)
	Triphenylarsine	Spectro	$K, \Delta H^{\ominus}, \Delta S^{\ominus}$	E. Augdahl, J. Grundnes and P. Klaeboe, *Inorg. Chem.* **4**, 1475 (1965)
	Iodine	Spectro	$\Delta H^{\ominus}$	P. A. D. de Maine, *J. chem. Phys.* **24**, 1091 (1956)
	Iodine	Spectro	$\Delta H^{\ominus}$	R. M. Keefer and T. L. Allen, *J. chem. Phys.* **25**, 1059 (1956)
	Iodine	Spectro	$\Delta H^{\ominus}$	M. M. de Maine, P. A. D. de Maine and G. E. McAlonie, *J. molec. Spectrosc.* **4**, 271 (1960)
	Iodine	Spectro	$\Delta H^{\ominus}$	M. Tamres, W. K. Duerksen and J. M. Goodenow, *J. phys. Chem., Ithaca* **72**, 966 (1968)
	Thianthrene, phenoxathiin, diphenylene oxide	Spectro	$K, \Delta H^{\ominus}, \Delta S^{\ominus}$	A. Kuboyama, *J. Am. chem. Soc.* **86**, 164 (1964)
	Aromatic amines	Spectro	$K, \Delta H^{\ominus}, \Delta S^{\ominus}$	H. Tsubomura, *J. Am. chem. Soc.* **82**, 40 (1960)

Aromatic amines	K	Spectro	A. K. Chandra and D. C. Mukherjee, *Trans. Faraday Soc.* **60**, 62 (1964)
Triphenylamine	K	Redox	W. H. Bruning, R. F. Nelson, L. S. Marcoux and R. N. Adams, *J. phys. Chem., Ithaca* **11**, 3055 (1967)
Aromatic hydrocarbons	K	Spectro	R. Bhattacharya and S. Basu, *Trans. Faraday Soc.* **54**, 1286 (1958)
Benzene	$K, \Delta H^{\ominus}$	Spectro	T. M. Cromwell and R. L. Scott, *J. Am. chem. Soc.* **72**, 3825 (1950)
Aromatic hydrocarbons, chlorobenzenes, dioxane, olefins, chloro-olefins	$K, \Delta H^{\ominus}, \Delta S^{\ominus}$	Spectro	J. A. A. Ketelaar, *J. Phys. Radium, Paris* **15**, 197 (1954)
Aromatic hydrocarbons	K	Spectro	R. M. Keefer and L. J. Andrews, *J. Am. chem. Soc.* **74**, 4500 (1952)
Anthracene, phenanthrene	K	Spectro	J. Peters and W. B. Person, *J. Am. chem. Soc.* **86**, 10 (1964)
Alkylbenzenes, 2-methyl-propan-2-ol	$K, \Delta H^{\ominus}, \Delta S^{\ominus}$	Spectro	R. M. Keefer and L. J. Andrews, *J. Am. chem. Soc.* **77**, 2164 (1955)
Aromatic sterically-hindered hydrocarbons	$K, \Delta H^{\ominus}, \Delta S^{\ominus}$	Spectro	R. E. Lovins, L. J. Andrews and R. M. Keefer, *J. phys. Chem., Ithaca* **68**, 2553 (1964)
Stilbenes	$K, \Delta H^{\ominus}, \Delta S^{\ominus}$	Spectro	S. Yamashita, *Bull. chem. Soc. Japan* **32**, 1212 (1959)
Naphthalenes	$\Delta H^{\ominus}$	Spectro	P. A. D. de Maine and J. Peone, Jr., *J. molec. Spectrosc.* **4**, 262 (1960)
Fluorobenzenes and fluorotoluenes	K	Spectro	M. Tamres, *J. phys. Chem., Ithaca* **68**, 2621 (1964)
Benzene, mesitylene	K	Spectro	H. A. Benesi and J. H. Hildebrand, *J. Am. chem. Soc.* **71**, 2703 (1949)
Naphthalene	K	Spectro	N. W. Blake, H. Winston and J. A. Patterson, *J. Am. chem. Soc.* **73**, 4437 (1951)
Naphthalene	K	Cryo, spectro	R. D. Srivastava and P. A. D. de Maine, *J. Miss. Acad. Sci.* **10**, 51 (1964)
Aromatic hydrocarbons	K	Spectro	J. S. Ham, *J. Am. chem. Soc.* **76**, 3881 (1954)

Appendix Table 2—continued

Acceptor	Donor	Method*	Constants determined	Reference
Iodine (I$_2$)	Bromo derivatives of aromatic hydrocarbons	Spectro	K	T. Mch. Spotswood, *Aust. J. Chem.* **15**, 278 (1962)
	Mesitylene, durene, hexamethylbenzene	Infrared	K	J. Morcillo and E. Gallego, *An. R. Soc. esp. de Fís. Quím.* **56(B)**, 263 (1960)
	Aromatic hydrocarbons	Spectro	K	R. S. Drago and N. J. Rose, *J. Am. chem. Soc.* **81**, 6141 (1959)
	Aromatic ethers, biphenyl	Spectro	K	P. A. D. de Maine, *J. chem. Phys.* **26**, 1189 (1957)
	Alkylbenzenes	Spectro	K	M. Tamres, D. R. Virzi and S. Searles, *J. Am. chem. Soc.* **75**, 4358 (1953)
	Aromatic hydrocarbons	Spectro	K	M. Chowdhury and S. Basu, *J. chem. Phys.* **32**, 1450 (1960)
	Olefins	Spectro	$K, \Delta H^{\ominus}, \Delta S^{\ominus}$	J. G. Traynham and J. R. Olechowski, *J. Am. chem. Soc.* **81**, 571 (1959)
	Olefins	GLC	K(rel)	R. J. Cvetanović, F. J. Duncan, W. E. Falconer and W. A. Sunder, *J. Am. chem. Soc.* **88**, 1602 (1966)
	Olefins	GLC	K(rel)	W. E. Falconer and R. J. Cvetanović, *J. Chromat.* **27**, 20 (1967)
Iodine (atomic)	Hexamethylbenzene	Spectro	K	R. L. Strong, *J. phys. Chem., Ithaca* **66**, 2423 (1962)
	Hexamethylbenzene	Spectro	K	R. L. Strong and J. Pérano, *J. Am. chem. Soc.* **83**, 2843 (1961)
	o-Xylene	Spectro	$K, \Delta H^{\ominus}, \Delta S^{\ominus}$	R. L. Strong and J. Pérano, *J. Am. chem. Soc.* **89**, 2535 (1967)

Iodine mono-chloride 14	Acetic acid	Spectro	K	R. E. Buckles and J. F. Mills, J. Am. chem. Soc. 75, 552 (1953)
	Acetonitrile	Spectro	K	A. I. Popov and W. A. Deskin, J. Am. chem. Soc. 80, 2976 (1958)
	Acetonitriles chloroaceto-nitriles	Spectro	$K, \Delta H^{\ominus}, \Delta S^{\ominus}$	W. B. Person, W. C. Golton and A. I. Popov, J. Am. chem. Soc. 85, 891 (1963)
	Benzonitrile	Spectro	K	P. Klaeboe, J. Am. chem. Soc. 84, 3458 (1962)
	Propionitrile	Spectro	$K, \Delta H^{\ominus}, \Delta S^{\ominus}$	P. Klaeboe, J. Am. chem. Soc. 85, 871 (1963)
	Propionitrile	Spectro	K	R. E. Buckles and J. F. Mills, J. Am. chem. Soc. 76, 4845 (1954)
	Nitriles	Spectro	K	F. Shah-Malak and J. H. P. Utley, Chem. Commun. 69 (1967)
	Nitriles	Infrared	K	E. Augdahl and P. Klaeboe, Spectrochim. Acta 19, 1665 (1963)
	N,N-Dimethylactamide	Spectro	$K, \Delta H^{\ominus}, \Delta S^{\ominus}$	R. S. Drago and D. A. Wenz, J. Am. chem. Soc. 84, 526 (1962)
	Dimethylcyanamide	Spectro	$K, \Delta H^{\ominus}, \Delta S^{\ominus}$	E. Augdahl and P. Klaeboe, Acta chem. scand. 19, 807 (1965)
	Aldehydes	Infrared	K	E. Augdahl and P. Klaeboe, Acta chem. scand. 16, 1647 (1962)
	2-Iodopropane	Spectro, kinetic	K	L. J. Andrews and R. M. Keefer, J. Am. chem. Soc. 75, 543 (1953)
	Pentamethylene tetrazole	Spectro	K	A. I. Popov, C. C. Bisi and M. Craft, J. Am. chem. Soc. 80, 6513 (1958)
	Dioxane	Spectro	K	A. I. Popov, C. C. Bisi and W. B. Person, J. phys. Chem., Ithaca 64, 691 (1960)
	Pyridines	Spectro	K	A. I. Popov and R. H. Rygg, J. Am. chem. Soc. 79, 4622 (1957)
	Aromatic amines	Spectro	K	A. K. Chandra and D. C. Mukherjee, Trans. Faraday Soc. 60, 62 (1964)
	Alkylbenzenes, chloro-toluenes, 4-methoxytoluene	Spectro	$K, \Delta H^{\ominus}, \Delta S^{\ominus}$	N. Ogimachi, L. J. Andrews and R. M. Keefer, J. Am. chem. Soc. 77, 4202 (1955)
	Aromatic hydrocarbons	Spectro	K	R. M. Keefer and L. J. Andrews, J. Am. chem. Soc. 74, 4500 (1952)

Appendix Table 2—continued

Acceptor	Donor	Method*	Constants determined	Reference
Iodine mono-chloride	Alkylbenzenes, chloro-benzene, bromobenzene	Spectro	K	R. M. Keefer and L. J. Andrews, J. Am. chem. Soc. **72**, 5170 (1950)
	Benzene	Infrared	K	A. I. Popov, R. E. Humphrey and W. B. Person, J. Am. chem. Soc. **82**, 1850 (1960)
Iodine mono-bromide	Acetonitrile	Spectro	K	A. I. Popov and W. A. Deskin, J. Am. chem. Soc. **80**, 2976 (1958)
	Acetonitrile, chloroacetonitriles	Spectro	$K, \Delta H^{\ominus}, \Delta S^{\ominus}$	W. B. Person, W. C. Golton and A. I. Popov, J. Am. chem. Soc. **85**, 891 (1963)
	Benzonitrile	Spectro	K	P. Klaeboe, J. Am. chem. Soc. **84**, 3458 (1962)
	Propionitrile	Spectro	K	P. Klaeboe, J. Am. chem. Soc. **85**, 871 (1963)
	Aldehydes	Infrared	K	E. Augdahl and P. Klaeboe, Acta chem. scand. **16**, 1655 (1962)
	Dimethylcyanamide	Spectro	$K, \Delta H^{\ominus}, \Delta S^{\ominus}$	E. Augdahl and P. Klaeboe, Acta chem. scand. **19**, 807 (1965)
	Pyridines	Spectro	K	A. I. Popov and R. H. Rygg, J. Am. chem. Soc. **79**, 4622 (1957)
	Pentamethylene tetrazole	Spectro	K	A. I. Popov, C. C. Bisi and M. Craft, J. Am. chem. Soc. **80**, 6513 (1958)
	Aromatic hydrocarbons	Spectro	K	R. D. Whitaker and H. H. Sisler, J. phys. Chem., Ithaca **67**, 523 (1963)
Iodine cyanide (cyanogen iodide)	Dioxane, metrazole, pyridine	Infrared	K	A. I. Popov, R. E. Humphrey and W. B. Person, J. Am. chem. Soc. **82**, 1850 (1960)
	Sulphoxides	Infrared	K	E. Augdahl and P. Klaeboe, Acta chem. scand. **18**, 18 (1964)
	Diethyl ether, ethanol, tetra-hydrofuran, dioxane, pyridine	Spectro	$K, \Delta H^{\ominus}, \Delta S^{\ominus}$	H. D. Bist and W. B. Person, J. phys. Chem., Ithaca **71**, 3288 (1967)

Bromine	Proprionitrile	Spectro	K	P. Klaeboe, J. Am. chem. Soc. **85**, 871 (1963)
	Benzonitrile	Spectro	K	P. Klaeboe, J. Am. chem. Soc. **84**, 3458 (1962)
	N,N-Dimethylacetamide	Spectro	K, $\Delta H^\ominus$, $\Delta S^\ominus$	R. S. Drago and D. A. Wenz, J. Am. chem. Soc. **84**, 526 (1962)
	Thiocarbanilides	Spectro	K	K. R. Bhaskar, R. K. Gosavi and C. N. R. Rao, Trans. Faraday Soc. **62**, 29 (1966)
	Diphenylselenide	Spectro	K	N. W. Tideswell and J. D. McCullough, J. Am. chem. Soc. **79**, 1031 (1957)
	Substituted diphenyl selenides	Spectro	K, $\Delta H^\ominus$, $\Delta S^\ominus$	J. D. McCullough and M. K. Barsh, J. Am. chem Soc. **71**, 3029 (1949)
	Naphthalene	Spectro	K	N. W. Blake, H. Winston and J. A. Patterson, J. Am. chem. Soc. **73**, 4437 (1951)
	Benzene, methylbenzenes, halobenzenes	Spectro	K	R. M. Keefer and L. J. Andrews, J. Am. chem. Soc. **72**, 4677 (1950)
	Methyl iodide	Spectro	K	L. J. Andrews and R. M. Keefer, J. Am. chem. Soc. **74**, 1891 (1952)
	Quinolines	Spectro	K	J. J. Eisch and B. Jaselskis, J. org. Chem. **28**, 2865 (1963)
	Mesitylene	Kinetic, spectro	K	R. M. Keefer, J. H. Blake, III, and L. J. Andrews, J. Am. chem. Soc. **76**, 3062 (1954)
Chlorine	Benzene, m-xylene	Spectro	K	L. J. Andrews and R. M. Keefer, J. Am. chem. Soc. **73**, 462 (1951)
Hydrogen chloride	Benzene, alkylbenzenes	Solubility	K, $\Delta H^\ominus$, $\Delta S^\ominus$	H. C. Brown and J. J. Melchiore, J. Am. chem. Soc. **87**, 5269 (1965)
	Benzene, alkylbenzenes	Solubility	$K(\text{rel})$	H. C. Brown and J. D. Brady, J. Am. chem. Soc. **74**, 3570 (1952)
Hydrogen bromide	Benzene, alkylbenzenes	Solubility	K, $\Delta H^\ominus$, $\Delta S^\ominus$	H. C. Brown and J. J. Melchiore, J. Am. chem. Soc. **87**, 5269 (1965)

Appendix Table 2—continued

Acceptor	Donor	Method*	Constants determined	Reference
Carbon tetrachloride	Hexamethylbenzene	Spectro	K	R. F. Weimer and J. M. Prausnitz, *J. chem. Phys.* **42**, 3643 (1965)
	Benzene, mesitylene	Spectro	K	R. Anderson and J. M. Prausnitz, *J. chem. Phys.* **39**, 1225 (1963)
	Hexamethylbenzene	Spectro	$K, \Delta H^{\ominus}$	F. Dörr and G. Buttgereit, *Ber. Bunsenges. physik. Chem.* **67**, 867 (1963)
	Aromatic amines	Spectro	$K, \Delta H^{\ominus}$	K. M. C. Davis and M. F. Farmer, *J. chem. Soc.* (B) 28 (1967)
	Tetracene	Spectro	K	K. M. C. Davis and M. F. Farmer, *J. chem. Soc.* (B) 859 (1968)
	Triethylamine	Spectro	K	D. P. Stevenson and G. M. Coppinger, *J. Am. chem. Soc.* **84**, 149 (1962)
	Benzene, methylbenzenes	Spectro	K	H. Kellawi and D. R. Rosseinsky, Chemical Society Anniversary Meeting, Exeter (1967); *J. chem. Soc.* (A), 1207 (1969).
Fluorotrichloro-methane	Triethylamine	Spectro	K	D. P. Stevenson and G. M. Coppinger, *J. Am. chem. Soc.* **84**, 149 (1962)
Bromotrichloro-methane	Triethylamine	Spectro	K	D. P. Stevenson and G. M. Coppinger, *J. Am. chem. Soc.* **84**, 149 (1962)
tert-Butylhalides	Benzene, monohalobenzenes	N.m.r.	$K, \Delta H^{\ominus}, \Delta S^{\ominus}$	R. C. Fort, Jr. and T. R. Lindstrom, *Tetrahedron* **23**, 3227 (1967)
Adamantyl halides	Benzene, monohalobenzenes	N.m.r.	$K, \Delta H^{\ominus}, \Delta S^{\ominus}$	R. C. Fort, Jr. and T. R. Lindstrom, *Tetrahedron* **23**, 3227 (1967)

Donor	Acceptor system	Method	Property	Reference
Sulphur dioxide	N,N-Dimethylacetamide	Spectro	$K, \Delta H^{\ominus}, \Delta S^{\ominus}$	R. S. Drago and D. A. Wenz, *J. Am. chem. Soc.* **84**, 526 (1962)
	Alcohols, benzene, hydroquinone	Spectro	$K, \Delta H^{\ominus}, \Delta S^{\ominus}$	P. A. D. de Maine, *J. chem. Phys.* **26**, 1036, 1042, 1049 (1957)
	Olefins, benzene, methylbenzenes	Spectro	$K, \Delta H^{\ominus}, \Delta S^{\ominus}$	D. Booth, F. S. Dainton and K. J. Ivin, *Trans. Faraday Soc.* **55**, 1293 (1959)
	Benzene, alkylbenzenes, chlorobenzene	Spectro	K	L. J. Andrews and R. M. Keefer, *J. Am. chem. Soc.* **78**, 4169 (1951)
	Triethylamine	Spectro	$K, \Delta H^{\ominus}, \Delta S^{\ominus}$	S. D. Christian and J. Grundnes, *Nature, Lond.* **214**, 1111 (1967); J. Grundnes and S. D. Christian, *J. Am. chem. Soc.* **90**, 2239 (1968)
	Anthracene	Spectro	$K, \Delta H^{\ominus}, \Delta S^{\ominus}$	T. Nagai, K. Terauchi and N. Tokura, *Bull. chem. Soc. Japan* **39**, 868 (1966)
	Aromatic tertiary amines	Pressure	$K, \Delta H^{\ominus}, \Delta S^{\ominus}$	W. E. Byrd, *Inorg. Chem.* **1**, 762 (1962)
Carbonyl cyanide	Diethyl ether, benzene	Spectro	K	J. Prochorow, *J. chem. Phys.* **43**, 3394 (1965)
	Diethyl ether, benzene	Spectro	K	J. Prochorow and A. Tramer, *J. chem. Phys.* **44**, 4545 (1966)
Nitrobenzene	N-Alkylanilines	Spectro	K	B. Dale, R. Foster and D. Ll. Hammick, *J. chem. Soc.* 3986 (1954)
	N-Methylacetamide	Spectro	K	O. D. Bonner and G. B. Woolsey, *Tetrahedron* **24**, 3625 (1968)
	Tetrakis(dimethylamine)ethylene	Spectro	K	P. R. Hammond and R. H. Knipe, *J. Am. chem. Soc.* **89**, 6063 (1967)
	Hexamethylbenzene	Spectro	K†	N. B. Jurinski and P. A. D. de Maine, *J. Am. chem. Soc.* **86**, 3217 (1964)
p-Chloronitrobenzene	Tetrakis(dimethylamino)ethylene	Spectro	K	P. R. Hammond and R. H. Knipe, *J. Am. chem. Soc.* **89**, 6063 (1967)
p-Nitroanisole	Tetrakis(dimethylamino)ethylene	Spectro	K	P. R. Hammond and R. H. Knipe, *J. Am. chem. Soc.* **89**, 6063 (1967)

Appendix Table 2—continued

Acceptor	Donor	Method*	Constants determined	Reference
o-Dinitro-benzene	N,N,N′,N′-Tetramethyl-p-phenylenediamine	Spectro	K	R. Foster and T. J. Thomson, *Trans. Faraday Soc.* **59**, 2287 (1963)
	N-Alkylanilines	Spectro	K	B. Dale, R. Foster and D. Ll. Hammick, *J. chem. Soc.* 3986 (1954)
m-Dinitro-benzene	Hexamethylbenzene	Spectro	K†	N. B. Jurinski and P. A. D. de Maine, *J. Am. chem. Soc.* **86**, 3217 (1964)
	Aniline	Spectro	K	J. Landauer and H. M. McConnell, *J. Am. chem. Soc.* **74**, 1221 1952)
	Substituted anilines	Spectro	K	B. Dale, R. Foster and D. Ll. Hammick, *J. chem. Soc.* 3986 (1954)
	N,N,N′,N′-Tetramethyl-p-phenylenediamine	Spectro	K	R. Foster and T. J. Thomson, *Trans. Faraday Soc.* **59**, 2287 (1963)
	Ferrocene	Spectro	K	B. Hetnarski, *Bull. Acad. pol. Sci. Sér. Sci. chim.* **13**, 523 (1965)
	N-Methylacetamide	Spectro	K	O. D. Bonner and G. B. Woolsey, *Tetrahedron* **24**, 3625 (1968)
p-Dinitro-benzene	Hexamethylbenzene	Spectro	K†	N. B. Jurinski and P. A. D. de Maine, *J. Am. chem. Soc.* **86**, 3217 (1964)
	Aniline	Spectro	K	J. Landauer and H. M. McConnell, *J. Am. chem. Soc.* **74**, 1221 (1952)
	Aromatic hydrocarbons, substituted anilines	N.m.r.	K	R. Foster and C. A. Fyfe, *Trans. Faraday Soc.* **61**, 1626 (1965)
	Phenothiazines	N.m.r.	K	R. Foster and C. A. Fyfe, *Biochim. biophys. Acta* **112**, 490 (1966)
	Indoles, benzothiophene, thiophene, pyrrole	N.m.r.	K	R. Foster and C. A. Fyfe, *J. chem. Soc.* (B), 926 (1966)

	N,N,N',N'-tetramethyl-p-phenylenediamine	Spectro	$K, \Delta H^\ominus, \Delta S^\ominus$	R. Foster and T. J. Thomson, *Trans. Faraday Soc.* **59**, 2287 (1963)
	Ferrocene	Spectro	K	B. Hetnarski, *Bull. Acad. pol. Sci. Sér. Sci. chim.* **13**, 523 (1965)
	N-Alkylanilines	Spectro	K	B. Dale, R. Foster and D.Ll. Hammick, *J. chem. Soc.* 3986 (1954)
2,4-Dinitro-chlorobenzene	Substituted anilines	Spectro	$K, \Delta H^\ominus, \Delta S^\ominus$	S. D. Ross, M. Bassin and I. Kuntz, *J. Am. chem. Soc.* **76**, 4176 (1954)
	Aniline	Spectro, kinetic	$K, \Delta H^\ominus, \Delta S^\ominus$	S. D. Ross and I. Kuntz, *J. Am. chem. Soc.* **76**, 3000 (1954)
	Ferrocene	Spectro	K	B. Hetnarski, *Bull. Acad. pol. Sci. Sér. Sci. chim.* **13**, 523 (1965)
1,2,3-Trinitro-benzene	Hexamethylbenzene	Spectro	$K, \Delta H^\ominus$	C. E. Castro, L. J. Andrews and R. M. Keefer, *J. Am. chem. Soc.* **80**, 2322 (1958)
	Hexamethylbenzene	Spectro	K	R. Foster, *J. chem. Soc.* 1075 (1960)
1,2,4-Trinitro-benzene	Hexamethylbenzene	Spectro	K	R. Foster, *J. chem. Soc.* 1075 (1960)
1,3,5-Trinitro-benzene	Olefins	GLC	$K(rel)$	R. J. Cvetanović, F. J. Duncan and W. E. Falconer, *Can. J. Chem.* **42**, 2410 (1964)
	Diphenylpolyenes	Spectro	$K, \Delta H^\ominus, \Delta S^\ominus$	G. Briegleb, J. Czekalla and A. Hauser, *Z. phys. Chem. Frankf. Ausg.* **21**, 114 (1959)
	Aromatic hydrocarbons, ethylenes	Spectro	$K, \Delta H^\ominus, \Delta S^\ominus$	G. Briegleb and J. Czekalla, *Z. Elektrochem.* **59**, 184 (1955)
	Aromatic hydrocarbons	Spectro	$K†$	N. B. Jurinski and P. A. D. de Maine, *J. Am. chem. Soc.* **86**, 3217 (1964)
	Hexamethylbenzene	Spectro	$K, \Delta H^\ominus, \Delta S^\ominus$	G. Briegleb and J. Czekalla, *Naturwissenschaften* **41**, 448 (1954)
	Benzene	Spectro	K	D. M. G. Lawrey and H. M. McConnell, *J. Am. chem. Soc.* **74**, 6175 (1952)

Appendix Table 2—continued

Acceptor	Donor	Method*	Constants determined	Reference
1,3,5-Trinitrobenzene	Benzene	Spectro	K	J. M. Corkill, R. Foster and D. Ll. Hammick, J. chem. Soc. 1202 (1955)
	Hexamethylbenzene	Spectro	K	R. Foster, J. chem. Soc. 1075 (1960)
	Aromatic hydrocarbons, anilines	N.m.r.	K	R. Foster and C. A. Fyfe, Trans. Faraday Soc. 61, 1626 (1965)
	Benzene, benzene-d_6	N.m.r.	K	P. H. Emslie and R. Foster, Tetrahedron 21, 2851 (1965)
	Naphthalene	Spectro	$K, \Delta H^{\ominus}$	C. C. Thompson, Jr. and P. A. D. de Maine, J. Am. chem. Soc. 85, 3096 (1963)
	Naphthalene	Spectro	$K, \Delta H^{\ominus}, \Delta S^{\ominus}$	N. Christodouleas and S. P. McGlynn, J. chem. Phys. 40, 166 (1964)
	Naphthalene	Spectro	K	A. S. Bailey and J. R. Case, Tetrahedron 3, 113 (1958)
	Hexamethylbenzene	Spectro	K	J. Czekalla, G. Briegleb, W. Herre and R. Glier, Z. Elektrochem. 61, 537 (1957)
	Hexamethylbenzene	Spectro	K	S. D. Ross, M. M. Labes and M. Schwarz, J. Am. chem. Soc. 78, 343 (1956)
	Aromatic hydrocarbons	Spectro	K	A. H. Ewald, Trans. Faraday Soc. 64, 733 (1968)
	Aromatic hydrocarbons	Spectro	$K, \Delta H^{\ominus}$	C. C. Thompson, Jr. and P. A. D. de Maine, J. phys. Chem., Ithaca 69, 2766 (1965)
	Methylbenzenes	N.m.r.	$K, \Delta H^{\ominus}, \Delta S^{\ominus}$	R. Foster, C. A. Fyfe and M. I. Foreman, Chem. Commun. 913 (1967)
	Biphenyls	Spectro	$K, \Delta H^{\ominus}$	C. E. Castro and L. J. Andrews, J. Am. chem. Soc. 77, 5189 (1955); C. E. Castro, L. J. Andrews and R. M. Keefer, 80, 2322 (1958)
	Substituted anilines, anthracene	Spectro	K	S. D. Ross, M. Bassin and I. Kuntz, J. Am. chem. Soc. 76, 4176 (1954)
	Aromatic hydrocarbons, aromatic amines	Spectro	$K, \Delta H^{\ominus}, \Delta S^{\ominus}$	A. Bier, Recl. Trav. chim. Pays-Bas Belg. 75, 866 (1956)

Anilines, naphthalene	Spectro	K, $\Delta H^\ominus$, $\Delta S^\ominus$	S. D. Ross and M. M. Labes, J. Am. chem. Soc. 77, 4916 (1955)
Pyrene, N,N-dimethylaniline	Spectro	K	A. S. Bailey, B. R. Henn and J. M. Langdon, Tetrahedron 19, 161 (1963)
Aniline	Spectro	K	J. Landauer and H. M. McConnell, J. Am. Chem. Soc. 74, 1221 (1952)
Substituted anilines	Spectro	K	R. Foster and D. Ll. Hammick, J. chem. Soc. 2685 (1954)
N,N-dimethylaniline	Spectro	K	S. D. Ross and M. M. Labes, J. Am. chem. Soc. 79, 76 (1957)
N,N,N′,N′-Tetramethyl-p-phenylenediamine	Spectro	K, $\Delta H^\ominus$, $\Delta S^\ominus$	R. Foster and T. J. Thomson, Trans. Faraday Soc. 59, 2287 (1963)
N,N,N′,N′-Tetramethyl-p-phenylenediamine p-dimethoxybenzene	Spectro	K, $\Delta H^\ominus$, $\Delta S^\ominus$	S. Iwata, H. Tsubomura and S. Nagakura, Bull. chem. Soc. Japan 37, 1506 (1964)
Diphenylamine	Spectro	K	R. Foster, D. Ll. Hammick and A. A. Wardley, J. chem. Soc. 3817 (1953)
Aromatic bromo-compounds	Spectro	K	T. Mch. Spotswood, Aust. J. Chem. 15, 278 (1962)
N-Methylacetamide	Spectro	K	O. D. Bonner and G. B. Woolsey, Tetrahedron 24, 3625 (1968)
Piperidine	Spectro	K, $\Delta H^\ominus$	W. Liptay and N. Tamberg, Z. Elektrochem. 66, 59 (1962)
Pentamethylenete trazole	Spectro, n.m.r.	K	T. C. Wehman and A. I. Popov, J. chem. Soc. 70, 3688 (1966)
Indoles, benzothiophene, thiophene, pyrrole	N.m.r.	K	R. Foster and C. A. Fyfe, J. chem. Soc. 926 (1966)
N-iso-Propylcarbazole	Spectro	K, $\Delta H^\ominus$, $\Delta S^\ominus$	J. H. Sharp, J. phys. Chem., Ithaca 70, 584 (1966)
Dithienyls, cyclopendadithiophenes	Spectro	K, $\Delta H^\ominus$, $\Delta S^\ominus$	A. Kraak and H. Wynberg, Tetrahedron 24, 3881 (1968)
Ferrocene	Spectro	K, $\Delta H^\ominus$, $\Delta S^\ominus$	B. Hetnarski, Bull. Acad. pol. Sci. Sér. Sci. chim. 13, 515 (1965)
Substituted ferrocenes	Spectro	K	B. Hetnarski, Bull. Acad. pol. Sci. Sér. Sci. chim. 13, 557 (1965)

Appendix Table 2—continued

Acceptor	Donor	Method*	Constants determined	Reference
1,3,5-Trinitro-benzene	Chlorophylls and related compounds	Spectro	K	J. R. Larry, *Diss. Abstr.* **27**, 2316-B (1967)
	Haematoporphyrin	Spectro	$K, \Delta H^{\ominus}$	J. G. Heathcote, G. J. Hill, P. Rothwell and M. A. Slifkin, *Biochim. biophys. Acta* **153**, 13 (1968)
	Porphyrins	Spectro	$K, \Delta H^{\ominus}, \Delta S^{\ominus}$	M. Gouterman and P. E. Stevenson, *J. chem. Phys.* **37**, 2266 (1962)
	Cobalt(II)mesoporphyrin IX dimethyl ester	Spectro	K	H. A. O. Hill, A. J. Macfarlane and R. J. P. Williams, *Chem. Commun.* 905 (1967)
	Cobalt(II)mesoporphyrin IX dimethyl ester	Spectro, n.m.r.	K	H. A. O. Hill, A. J. Macfarlane, B. E. Mann and R. J. P. Williams, *Chem. Commun.* 123 (1968)
	Zinc phthalocyanine	Spectro	K	P. J. McCartin, *J. Am. chem. Soc.* **85**, 2021 (1963)
2,4,6-Trinitro-toluene	Anthracene	Spectro	K	S. D. Ross, M. Bassin and I. Kuntz, *J. Am. chem. Soc.* **76**, 4176 (1954)
	Hexamethylbenzene	Spectro	$K^{\dagger}$	N. B. Jurinski and P. A. D. de Maine, *J. Am. chem. Soc.* **86**, 3217 (1964)
	Hexamethylbenzene	Spectro	K	J. Czekalla, G. Briegleb, W. Herre and R. Glier, *Z. Elektrochem.* **61**, 537 (1957)
	N-Alkylanilines	Spectro	K	B. Dale, R. Foster and D. Ll. Hammick, *J. chem. Soc.* 3986 (1954)
	Zinc phthalocyanine	Spectro	K	P. J. McCartin, *J. Am. chem. Soc.* **85**, 2021 (1963)
	Cobalt(II)mesoporphyrin IX dimethyl ester	Spectro	K	H. A. O. Hill, A. J. Macfarlane and R. J. P. Williams, *Chem. Commun.* 905 (1967)
	Cobalt(II)mesoporphyrin IX dimethyl ester	Spectro, n.m.r.	K	H. A. O. Hill, A. J. Macfarlane, B. E. Mann and R. J. P. Williams, *Chem. Commun.* 123 (1968)

		Method	Measured	Reference
2,4,6-Trinitro-m-xylene 14**	N-Alkylanilines	Spectro	K	B. Dale, R. Foster and D. Ll. Hammick, J. chem. Soc. 3986 (1954)
Picryl chloride (1-chloro-2,4,6-trinitrobenzene)	Hexamethylbenzene	Spectro, kinetic	K	S. D. Ross, M. Bassin, M. Finkelstein and W. A. Leach, J. Am. chem. Soc. 76, 69 (1954)
	Hexamethylbenzene	Spectro	K	S. D. Ross, M. M. Labes and M. Schwarz, J. Am. chem. Soc. 78, 343 (1956)
	Hexamethylbenzene	Spectro	K	J. Czekalla, G. Briegleb, W. Herre and R. Glier, Z. Elektrochem. 61, 537 (1957)
	Aromatic hydrocarbons	Spectro	K	A. H. Ewald, Trans. Faraday Soc. 64, 733 (1968)
	Anthracene	Spectro	K	S. D. Ross, M. Bassin and I. Kuntz, J. Am. chem. Soc. 76, 4176 (1954)
	N-iso-Propylcarbazole	Spectro	$K, \Delta H^{\ominus}, \Delta S^{\ominus}$	J. H. Sharp, J. phys. Chem., Ithaca 70, 584 (1966)
	Ferrocene	Spectro	K	B. Hetnarski, Bull. Acad. pol. Sci. Sér. Sci. chim. 13, 515 (1965)
	Ethylferrocene	Spectro	K	B. Hetnarski, Bull. Acad. pol. Sci. Sér. Sci. chim. 13, 557 (1965)
Picric acid	Naphthalene, m-dinitrobenzene, 1,3,5-trinitrobenzene	Spectro	$K, \Delta H^{\ominus}, \Delta S^{\ominus}$	S. D. Ross and I. Kuntz, J. Am. chem. Soc. 76, 74 (1954)
	Aromatic hydrocarbons	Spectro	$K, \Delta H^{\ominus}, \Delta S^{\ominus}$	G. Briegleb, J. Czekalla and A. Hauser, Z. phys. Chem. Frankf. Ausg. 21, 99 (1959)
	Substituted naphthalenes	Spectro	K	P. D. Gardner, R. L. Brandon, N. J. Nix and I. Y. Chang, J. Am. chem. Soc. 81, 3413 (1959)
	Substituted naphthalenes	Partition	K	P. D. Gardner, W. E. Stump, J. Am. chem. Soc. 79, 2759 (1957)
	Alkylbenzenes	Partition	K	H. D. Anderson and D. Ll. Hammick, J. chem. Soc. 1089 (1950)
	Aromatic hydrocarbons	Partition	K	R. Foster, D. Ll. Hammick and S. F. Pearce, J. chem. Soc. 244 (1959)
	Anthracene	Spectro	K	S. D. Ross, M. Bassin and I. Kuntz, J. Am. chem. Soc. 76, 4176 (1954)

Appendix Table 2—continued

Acceptor	Donor	Method*	Constants determined	Reference
Picric acid	Ferrocene	Spectro	K	B. Hetnarski, *Bull. Acad. pol. Sci. Sér. Sci. chim.* **13**, 515 (1965)
Picryl iodide	Anthracene	Spectro	K	S. D. Ross, M. Bassin and I. Kuntz, *J. Am. chem. Soc.* **76**, 4176 (1954)
2,4,6-Trinitro-anisole	Anthracene	Spectro	K	S. D. Ross, M. Bassin and I. Kuntz, *J. Am. chem. Soc.* **76**, 4176 (1954)
1-X-2,4,6-Trinitrobenzene	Hexamethylbenzene	Spectro	K	J. Czekalla, G. Briegleb, W. Herre and R. Glier, *Z. Elektrochem.* **61**, 537 (1957)
1,3-Difluoro-2,4,6-trinitro-benzene	Hexamethylbenzene	N.m.r.	K	R. Foster and C. A. Fyfe, *Chem. Commun.* 642 (1965)
1,2,3,5-Tetra-nitrobenzene	Hexamethylbenzene Hexamethylbenzene	Spectro N.m.r.	K K	R. Foster, *J. chem. Soc.* 1075 (1960) R. Foster and C. A. Fyfe, *Trans. Faraday Soc.* **62**, 1400 (1966)
Bipicryl	*p*-Toluidine, hexamethylbenzene	Spectro	$K, \Delta H^{\ominus}$	C. E. Castro, L. J. Andrews and R. M. Keefer, *J. Am. chem. Soc.* **80**, 2322 (1958)
Dinitro-naphthalenes	N,N,N',N'-Tetramethyl-*p*-phenylenediamine	Spectro	K	P. H. Emslie and R. Foster, *Tetrahedron* **20**, 1489 (1964)

2,4,7-Trinitro-fluorenone	Naphthalene	Spectro	K	A. S. Bailey and J. R. Case, *Tetrahedron* **3**, 113 (1958)
	Naphthalene	Spectro	$K, \Delta H^{\ominus}, \Delta S^{\ominus}$	N. Christodouleas and S. P. McGlynn, *J. chem. Phys.* **40**, 166 (1964)
	Substituted naphthalenes	Spectro	K	L. H. Klemm and J. N. Sprague, *J. org. Chem.* **19**, 1464 (1954)
	Naphthalene, substituted naphthalenes	Spectro	K	L. H. Klemm, J. N. Sprague and H. Ziffer, *J. org. Chem.* **20**, 200 (1955)
	Benzanthracenes, benzo[c]-phenanthrenes	Spectro	K	K. H. Takemura, M. D. Cameron and M. S. Newman, *J. Am. chem. Soc.* **75**, 3280 (1953)
	Aromatic bromo-compounds	Spectro	K	T. Mch. Spotswood, *Aust. J. Chem.* **15**, 278 (1962)
	Substituted anilines	Spectro, GLC	$K, \Delta H^{\ominus}, \Delta S^{\ominus}$	A. R. Cooper, C. W. P. Crowne and P. G. Farrell, *Trans. Faraday Soc.* **62**, 2725 (1966)
	Heterocyclics	GLC	K	A. R. Cooper, C. W. P. Crowne and P. G. Farrell, *Trans. Faraday Soc.* **63**, 447 (1967)
	N-*iso*-Propylcarbazole	Spectro	$K, \Delta H^{\ominus}, \Delta S^{\ominus}$	J. H. Sharp, *J. Phys. Chem., Ithaca* **70**, 584 (1966)
	Pentamethylenetetrazole	Spectro	K	T. C. Wehman and A. I. Popov, *J. phys. Chem., Ithaca* **70**, 2688 (1966)
	Zinc phthalocyanine	Spectro	K	P. J. McCartin, *J. Am. chem. Soc.* **85**, 2021 (1963)
	Cobalt(II)mesoporphyrin IX dimethyl ester	Spectro	K	H. A. O. Hill, A. J. Macfarlane and R. J. P. Williams, *Chem. Commun.* 905 (1967)
	Cobalt(II)mesoporphyrin IX dimethyl ester, chlorophyll *a*	Spectro, n.m.r.	K	H. A. O. Hill, A. J. Macfarlane, B. E. Mann and R. J. P. Williams, *Chem. Commun.* 123 (1968)
Nitro-*p*-terphenyls	N,N-Dimethyl-*p*-toluidine	Spectro	$K, \Delta H^{\ominus}, \Delta S^{\ominus}$	R. L. Hansen, *J. phys. Chem., Ithaca* **70**, 1646 (1966)
5,6-Dinitro-benzofuroxan	Naphthalene	Spectro	K	A. S. Bailey and J. R. Case, *Tetrahedron* **3**, 113 (1958)

Appendix Table 2—continued

Acceptor	Donor	Method*	Constants determined	Reference
4,6-Dinitro-benzofuroxan	Naphthalene	Spectro	K	A. S. Bailey and J. R. Case, *Proc. chem. Soc.* 176 (1957); *Tetrahedron* **3**, 113 (1958)
	Cobalt(II)mesoporphyrin IX dimethyl ester	Spectro, n.m.r.	K	H. A. O. Hill, A. J. Macfarlane, B. E. Mann and R. J. P. Williams, *Chem. Commun.* **123** (1968)
6-Nitrobenzo-difuroxan	Naphthalene	Spectro	K	A. S. Bailey and J. R. Case, *Proc. chem. Soc.* 176 (1957); *Tetrahedron* **3**, 113 (1958)
Benzotrifuroxan	Naphthalene	Spectro	K	A. S. Bailey and J. R. Case, *Proc. chem. Soc.* 176 (1957); *Tetrahedron* **3**, 113 (1958)
	N,N-Dimethylaniline	Spectro	K	A. S. Bailey and J. M. Evans, *Chemy. Ind.* 1424 (1964)
	Cobalt(II)mesoporphyrin IX dimethyl ester	Spectro, n.m.r.	K	H. A. O. Hill, A. J. Macfarlane, B. E. Mann and R. J. P. Williams, *Chem. Commun.* **123** (1968)
Benzotrifurazan	N,N-Dimethylaniline, pyrene	Spectro	K	A. S. Bailey and J. M. Evans, *Chemy. Ind.* 1424 (1964)
Tetrachloroph-thalic anhydride	Pyrene	Spectro	K	M. E. Peover, *Trans. Faraday Soc.* **60**, 417 (1964)
	Pyrene	Spectro	K	A. S. Bailey, B. R. Henn and J. M. Langdon, *Tetrahedron* **19**, 161 (1963)
	Naphthalene	Optical	$K, \Delta H^{\ominus}, \Delta S^{\ominus}$	N. Christodouleas and S. P. McGlynn, *J. chem. Phys.* **40**, 166 (1964)
	Hexamethylbenzene	Spectro	K	J. Czekalla, G. Briegleb, W. Herre and R. Glier, *Z. Elektrochem.* **61**, 537 (1957)

Acceptor	Donor	Method	Measurement	Reference
	Hexamethylbenzene	Spectro	K, $\Delta H^{\ominus}$, $\Delta S^{\ominus}$	J. Czekalla and K.-O. Meyer, *Z. phys. Chem. Frankf. Ausg.* **27**, 185 (1961)
	Bromo-substituted aromatic hydrocarbons	Spectro	K	T. Mch. Spotswood, *Aust. J. Chem.* **15**, 278 (1962)
	Aromatic hydrocarbons and azahydrocarbons	Spectro	K	M. Chowdhury and S. Basu, *Trans. Faraday Soc.* **56**, 335 (1960)
	Quinolines	Spectro	K	M. Chowdhury, *J. phys. Chem., Ithaca* **65**, 1899 (1966)
Tetrabromophthalic anhydride	Aromatic hydrocarbons	Spectro	K	R. D. Srivastava and P. D. Gupta, *Spectrochim. Acta* **24A**, 373 (1968)
	Naphthalene	Spectro	K, $\Delta H^{\ominus}$, $\Delta S^{\ominus}$	N. Christodouleas and S. P. McGlynn, *J. chem. Phys.* **40**, 166 (1964)
Tetraiodophthalic anhydride	Naphthalene	Spectro	K, $\Delta H^{\ominus}$, $\Delta S^{\ominus}$	N. Christodouleas and S. P. McGlynn, *J. chem. Phys.* **40**, 166 (1964)
Pyromellitic dianhydride	Benzene, alkylbenzenes, anisole, chlorobenzene	Spectro	K	L. L. Ferstandig, W. G. Toland and C. D. Heaton, *J. Am. chem. Soc.* **83**, 1151 (1961)
	Alkylbenzenes	Spectro	K	Y. Nakayama, Y. Ichikawa and T. Matsuo, *Bull. chem. Soc. Japan* **38**, 1674 (1965)
	Condensed aromatic hydrocarbons	Spectro	K	I. Ilmet and P. M. Rashba, *J. phys. Chem., Ithaca* **71**, 1140 (1967)
3-Nitro-1,8-naphthalic anhydride	Condensed aromatic hydrocarbons	Spectro	K	I. Ilmet and S. A. Berger, *J. phys. Chem., Ithaca* **71**, 1534 (1967)
Dichloromaleic anhydride	Methylbenzenes	Spectro	K†	C. H. J. Wells, *Tetrahedron* **22**, 1985 (1966)
	Benzene, anisole	Spectro	K	L. J. Andrews and R. M. Keefer, *J. Am. chem. Soc.* **75**, 3776 (1953)

Appendix Table 2—continued

Acceptor	Donor	Method*	Constants determined	Reference
Maleic anhydride	Mesitylene, styrene	Spectro	K, $\Delta H^{\ominus}$	W. G. Barb, *Trans. Faraday Soc.* **49**, 143 (1953)
	Hexamethylbenzene	Spectro	K, $\Delta H^{\ominus}$	Z. Raciszewski, *J. chem. Soc.* (B), 1142 (1966)
	Aromatic hydrocarbons, anisole	Spectro	K	L. J. Andrews and R. M. Keefer, *J. Am. chem. Soc.* **75**, 3776 (1953)
p-Benzoquinone	Benzene	Spectro	K	L. J. Andrews and R. M. Keefer, *J. Am. chem. Soc.* **75**, 3776 (1953)
	Methylbenzenes	N.m.r.	K	R. Foster and C. A. Fyfe, *Trans. Faraday Soc.* **62**, 1400 (1966)
	Aromatic hydrocarbons	Spectro	K	M. Chowdhury, *Trans. Faraday Soc.* **57**, 1482 (1961)
	Aromatic hydrocarbons, hydroquinone, dimethyl ether, phenol, anisole	Spectro	$\Delta H^{\ominus}$	A. Kuboyama and S. Nagakura, *J. Am. chem. Soc.* **77**, 2644 (1955)
	N,N-Dimethylaniline	Spectro	K, $\Delta H^{\ominus}$, $\Delta S^{\ominus}$	G. Briegleb and J. Czekalla, *Z. Elektrochem.* **58**, 249 (1954)
	Hydroquinone	Solubility	K, $\Delta H^{\ominus}$	A. Bertoud and S. Kunz, *Helv. chim. Acta* **21**, 17 (1938)
	Hydroquinone	Spectro	K	R. E. Moser, H. G. Cassidy, *J. Am. chem. Soc.* **87**, 3463 (1965)
	Hydroquinone dimethyl ether, phenol	Spectro	$\Delta H^{\ominus}$	H. Tsubomura, *Bull. chem. Soc. Japan* **26**, 304 (1953)
2,3-Dicyano-p-benzoquinone	Alkylbenzenes	Spectro	K	P. R. Hammond, *J. chem. Soc.* 3113 (1963)
	Hexamethylbenzene	N.m.r.	K	R. Foster and C. A. Fyfe, *Trans. Faraday Soc.* **62**, 1400 (1966)

Acceptor	Donor	Method	Measurement	Reference
2,3-Dichloro-5,6-dicyano-p-benzoquinone	Alkylbenzenes	Spectro	K	P. R. Hammond, J. chem. Soc. 3113 (1963)
	Aromatic hydrocarbons	Spectro	$K\dagger$	R. Foster and I. Horman, J. chem. Soc. (B), 171 (1966)
	Aromatic hydrocarbons	Spectro	K	R. D. Srivastava and G. Prasad, Spectrochim. Acta 22, 1869 (1966)
	Hexamethylbenzene	Polarography	K	M. E. Peover, Trans. Faraday Soc. 60, 417 (1964)
Fluoranil	Benzene, alkylbenzenes	N.m.r.	K	N. M. D. Brown, R. Foster and C. A. Fyfe, J. chem. Soc. (B), 406 (1967)
	Benzene, alkylbenzenes	N.m.r.	$K, \Delta H^{\ominus}, \Delta S^{\ominus}$	R. Foster, C. A. Fyfe and M. I. Foreman, Chem. Commun. 913 (1967)
Chloranil	Aromatic hydrocarbons	Spectro	$K, \Delta H^{\ominus}, \Delta S^{\ominus}$	G. Briegleb, J. Czekalla and G. Reuss, Z. phys. Chem. Frankf. Ausg. 30, 333 (1961)
	Alkylbenzenes	Spectro	K	E. A. Halevi and M. Nussim, J. chem. Soc. 876 (1963)
	Hexamethylbenzene, pyrene	Polarography	K	M. E. Peover, Trans. Faraday Soc. 60, 417 (1964)
	Aromatic hydrocarbons	Spectro	K	S. K. Chakrabarti and S. Basu, Trans. Faraday Soc. 60, 465 (1964)
	Benzene	Spectro	K	J. M. Corkill, R. Foster and D. Ll. Hammick, J. chem. Soc. 1202 (1955)
	Alkylbenzenes	Spectro	K	R. Foster, D. Ll. Hammick and B. N. Parsons, J. chem. Soc. 555 (1956)
	Hexamethylbenzene	Spectro	K	R. Foster, D. Ll. Hammick and P. J. Placito, J. chem. Soc. 3881 (1956)
	Naphthalene	Spectro	K	P. A. D. de Maine and R. D. Srivastava, J. Miss. Acad. Sci. 10, 67 (1964)
	Naphthalene	Spectro	$K, \Delta H^{\ominus}, \Delta S^{\ominus}$	N. Christodouleas and S. P. McGlynn, J. chem. Phys. 40, 166 (1964)
	Aromatic hydrocarbons	Spectro	K	A. H. Ewald, Trans. Faraday Soc. 64, 733 (1968)
	Aniline and halo-substituted anilines	Spectro	K	W. R. Carper, R. M. Hedges, H. N. Simpson, J. phys. Chem., Ithaca 69 1707 (1965)

Appendix Table 2—continued

Acceptor	Donor	Method*	Constants determined	Reference
Chloranil	N,N-Dimethylaniline, hexa-methylbenzene	Spectro	$K, \Delta H^{\ominus}, \Delta S^{\ominus}$	G. Briegleb and J. Czekalla, Z. Elektrochem. 58, 249 (1954)
	Phloroglucinol	Spectro	K	S. Carter, J. chem. Soc. (A), 404 (1968)
	Aromatic hydrocarbons, naphthols, naphthylamines	Spectro	K	D. Nespoulous, J. Salvinien and P. Viallet, Compt. Rend. (Ser. C), 264, 941 (1967)
	Haematoporphyrin	Spectro	$K, \Delta H^{\ominus}$	J. G. Heathcote, G. J. Hill, P. Rothwell and M. A. Slifkin, Biochim biophys. Acta 153, 13 (1968)
	Indole	Spectro	$K, \Delta H^{\ominus}, \Delta S^{\ominus}$	R. Foster and P. Hanson, Trans. Faraday Soc. 60, 2189 (1964)
Bromanil	Aromatic hydrocarbons	Spectro	K	S. K. Chakrabarti and S. Basu, Trans. Faraday Soc. 60, 465 (1964)
	Hexamethylbenzene	Spectro	K	R. Foster, D. Ll. Hammick and P. J. Placito, J. chem. Soc. 3881 (1956)
	Aromatic hydrocarbons	Spectro	K	S. K. Chakrabarti and S. Basu, Trans. Faraday Soc. 60, 465 (1964)
	Hexamethylbenzene	Spectro	K	R. Foster, D. Ll. Hammick and P. J. Placito, J. chem. Soc. 3881 (1956)
Iodanil	Hexamethylbenzene	Spectro	K	R. Foster, D. Ll. Hammick and P. J. Placito, J. chem. Soc. 3881 (1956)
	Aromatic hydrocarbons	Spectro	K	S. K. Chakrabarti and S. Basu, Trans. Faraday Soc. 60, 465 (1964)
Other substituted p-benzoquinones	Hexamethylbenzene	Spectro	$K, \Delta H^{\ominus}$	P. R. Hammond, J. chem. Soc. 471 (1964)
	Hexamethylbenzene	Spectro	K	R. Foster, D. Ll. Hammick and P. J. Placito, J. chem. Soc. 3881 (1956)

Acceptor	Donor	Technique	Method	Reference
	Hexamethylbenzene	N.m.r.	K	R. Foster and C. A. Fyfe, *Trans. Faraday Soc.* **62**, 1400 (1966)
	Hexamethylbenzene	Spectro	K	J. Czekalla, G. Briegleb, W. Herre and R. Glier, *Z. Elektrochem.* **61**, 537 (1957)
2-Methyl-1,4-naphthoquinone (menadione)	7-Methylbenz[a]anthracene	Spectro	K	G. Cilento and D. L. Sanioto, *Z. phys. Chem.* **223**, 333 (1963)
	Aromatic hydrocarbons	Spectro	K, $\Delta H^\ominus$, $\Delta S^\ominus$	G. Cilento and D. L. Sanioto, *Ber. Bunsenges. physik. Chem.* **67**, 426 (1963)
Riboflavin	Aromatic hydrocarbons	Fluorescence quenching	K	D. B. McCormick, Heng-Chun Li and R. E. MacKenzie, *Spectrochim. Acta* **23A**, 2353 (1967)
Tetracyanoquinodimethane	Benzene, alkylbenzenes	N.m.r.	K	M. W. Hanna and A. L. Ashbaugh, *J. phys. Chem., Ithaca* **68**, 811 (1964)
	Aromatic hydrocarbons	Spectro	K	L. R. Melby, R. J. Harder, W. R. Hertler, W. Mahler, R. E. Benson and W. E. Mochel, *J. Am. chem. Soc.* **84**, 3374 (1962)
	Methylbenzenes	Polarography	K	R. D. Holm, W. R. Carper and J. A. Blancher, *J. Phys. Chem., Ithaca* **71**, 3960 (1967)
	Hexamethylbenzene	N.m.r.	K	R. Foster and C. A. Fyfe, *Trans. Faraday Soc.* **62**, 1400 (1966)
	Hexamethylbenzene, pyrene	Polarography	K	M. E. Peover, *Trans. Faraday Soc.* **60**, 417 (1964)
	Aminoazobenzenes	Spectro	K	W. Damerau, *Z. Naturf.* **21B**, 937 (1966)
	Pentamethylenetetrazole	Spectro	K	T. C. Wehman and A. I. Popov, *J. phys. Chem., Ithaca* **70**, 3688 (1966)
3,5-Dinitrobenzonitrile	Cobalt(II)mesoporphyrin IX dimethyl ester	Spectro	K	H. A. O. Hill, A. J. Macfarlane and R. J. P. Williams, *Chem. Commun.* 905 (1967)
	Cobalt(II)mesoporphyrin IX dimethyl ester	Spectro, n.m.r.	K	H. A. O. Hill, A. J. Macfarlane, B. E. Mann and R. J. P. Williams, *Chem. Commun.* 123 (1968)

Appendix Table 2—continued

Acceptor	Donor	Method*	Constants determined	Reference
Dicyanobenzenes	N,N,N′,N′-Tetramethyl-p-phenylenediamine	Spectro	K	R. Foster and T. J. Thomson, Trans. Faraday Soc. **59**, 2287 (1963)
1,4-Dicyano-2,3,5,6-tetra-fluorobenzene	Methylbenzenes	N.m.r.	K	N. M. D. Brown, R. Foster and C. A. Fyfe, J. chem. Soc. (B), 406 (1967)
1,3,5-Tricyano-benzene	N,N-Dimethylaniline	Spectro	K	A. S. Bailey, B. R. Henn and J. M. Langdon, Tetrahedron **19**, 161 (1963)
	N,N,N′,N′-Tetramethyl-p-phenylene diamine	Spectro	$K, \Delta H^{\ominus}, \Delta S^{\ominus}$	R. Foster and T. J. Thomson, Trans. Faraday Soc. **59**, 2287 (1963)
	Hexamethylbenzene	N.m.r.	K	R. Foster and C. A. Fyfe, Trans. Faraday Soc. **62**, 1400 (1966)
1,2,4,5-Tetracyanobenzene (pyromellitonitrile)	Benzene, hexamethylbenzene	Spectro	$K, \Delta H^{\ominus}, \Delta S^{\ominus}$	S. Iwata, J. Tanaka and S. Nagakura, J. Am. chem. Soc. **88**, 894 (1966)
	Hexamethylbenzene	N.m.r.	K	R. Foster and C. A. Fyfe, Trans. Faraday Soc. **62**, 1400 (1966)
	Hexamethylbenzene, pyrene	Polarography	K	M. E. Peover, Trans. Faraday Soc. **60**, 417 (1964)
	Pyrene, N,N-dimethylaniline, durene	Spectro	K	A. S. Bailey, B. R. Henn and J. M. Langdon, Tetrahedron **19**, 161 (1963)
	N,N,N′,N′-Tetramethyl-p-phenylene diamine	Spectro	$K, \Delta H^{\ominus}, \Delta S^{\ominus}$	R. Foster and T. J. Thomson, Trans. Faraday Soc. **59**, 2287 (1963)
Tetrachlorophthalonitrile (tetrachloro-1,2-dicyanobenzene)	Pyrene, N,N-dimethylaniline	Spectro	K	A. S. Bailey, B. R. Henn and J. M. Langdon, Tetrahedron **19**, 161 (1963)

2,4,6-Tricyano-s-triazine	Pyrene	Spectro	K	A. S. Bailey, B. R. Henn and J. M. Langdon, Tetrahedron, **19**, 161 (1963)
Tetracyano-ethylene	Aromatic hydrocarbons	Spectro	K, $\Delta H^{\ominus}$, $\Delta S^{\ominus}$	G. Briegleb, J. Czekalla and G. Reuss, Z. phys. Chem. Frankf. Ausg. **30**, 333 (1961)
	Aromatic hydrocarbons	Spectro	K	M. J. S. Dewar and C. C. Thompson, Jr., Tetrahedron supp. **7**, 97 (1966)
	Aromatic hydrocarbons	Spectro	K	A. H. Ewald, Trans. Faraday Soc. **64**, 733 (1968)
	Benzene	Spectro	K	J. R. Gott, W. G. Maisch, J. chem. Phys. **39**, 2229 (1963)
	Benzene	Spectro	K, $\Delta H^{\ominus}$, $\Delta S^{\ominus}$	P. H. Emslie and R. Foster, Tetrahedron **21**, 2851 (1965)
	Naphthalene	Spectro	K	G. D. Johnson and R. E. Bowen, J. Am. chem. Soc. **87**, 1655 (1965)
	p-Xylene	Spectro	K, $\Delta H^{\ominus}$, $\Delta S^{\ominus}$	M. Kroll and M. L. Ginter, J. phys. Chem., Ithaca **69**, 3671 (1965)
	Hexamethylbenzene, pyrene	Polarography	K	M. E. Peover, Trans. Faraday Soc. **60**, 417 (1964)
	Hexamethylbenzene	Spectro	K, $\Delta H^{\ominus}$, $\Delta S^{\ominus}$	R. Foster and I. B. C. Matheson, Spectrochim. Acta **23A**, 2037 (1967)
	Methylbenzenes	Spectro	K, $\Delta H^{\ominus}$, $\Delta S^{\ominus}$	M. Kroll, J. Am. chem. Soc. **90**, 1097 (1968)
	Pyrene	Spectro	K	R. Foster and I. Horman, J. chem. Soc. (B), 171 (1966)
	Pyrene, naphthalene	Spectro	K, $\Delta H^{\ominus}$, $\Delta S^{\ominus}$	H. Kuroda, T. Amano, I. Ikemoto and H. Akamatu, J. Am. chem. Soc. **89**, 6056 (1967)
	Biphenylenes	Spectro	K, $\Delta H^{\ominus}$, $\Delta S^{\ominus}$	D. G. Farnum, E. R. Atkinson and W. C. Lothrop, J. org. Chem. **26**, 3204 (1961)
	Biphenylene	Spectro	K, $\Delta H^{\ominus}$, $\Delta S^{\ominus}$	P. H. Emslie, R. Foster and R. Pickles, Can. J. Chem. **44**, 9 (1966)
	Stilbenes	Spectro	K	W. H. Laarhoven and R. J. F. Nivard, Recl Trav. chim. Pays-Bas Belg. **84**, 1478 (1965)
	Nitro-p-terphenyls, p-terphenyl, biphenyl, benzene	Spectro	K, $\Delta H^{\ominus}$, $\Delta S^{\ominus}$	R. L. Hansen, J. phys. Chem., Ithaca **70**, 1646 (1966)

Appendix Table 2—continued

Acceptor	Donor	Method*	Constants determined	Reference
Tetracyano-ethylene	Cyclohexene, aromatic hydro-carbons, pyridine, halo-benzenes, anisole	Spectro	K	R. E. Merrifield and W. D. Phillips, *J. Am. chem. Soc.* **80**, 2778 (1958)
	Paracyclophanes	Spectro	K	L. A. Singer and D. J. Cram, *J. Am. chem. Soc.* **85**, 1080 (1963)
	Aromatic ethers	Spectro	K	T. Matsuo and O. Higuchi, *Bull. chem. Soc. Japan* **41**, 518 (1968)
	N,N-Dimethylaniline	Spectro	K, $\Delta H^{\ominus}$	Z. Rappoport, *J. chem. Soc.* 4498 (1963)
	N-Methylaniline	Spectro	K, $\Delta H^{\ominus}$, $\Delta S^{\ominus}$	Z. Rappoport and A. Horowitz, *J. chem. Soc.* 1348 (1964)
	Aromatic amines	Spectro	K	N. S. Isaacs, *J. chem. Soc.* (B), 1053 (1966)
	N,N,N′,N′-Tetramethyl-*p*-phenylenediamine	Spectro	K	W. Liptay, G. Briegleb and K. Schindler, *Z. Elektrochem.* **66**, 331 (1962)
	Ferrocenes	Spectro	K, $\Delta H^{\ominus}$, $\Delta S^{\ominus}$	E. Adman, M. Rosenblum, S. Sullivan and T. N. Margulis, *J. Am. chem. Soc.* **89**, 4540 (1967)
	Ferrocene	Spectro	K	M. Rosenblum, R. W. Fish and C. Bennett, *J. Am. chem. Soc.* **86**, 5166 (1964)
	Phenylborazines	Spectro	K	K. A. Muszkar and B. Kirson, *Israel J. Chem.* **2**, 57 (1964)
	Hexamethylborazole	Spectro	K	N. G. S. Champion, R. Foster and R. K. Mackie, *J. chem. Soc.* 5060 (1961)
	Tetrahydrofuran, tetra-hydropyran, dioxane	Spectro	K, $\Delta H^{\ominus}$, $\Delta S^{\ominus}$	R. Vars, L. A. Tripp and L. W. Pickett, *J. phys. Chem.*, *Ithaca* **66**, 1754 (1962)
	N-*iso*-Propylcarbazole	Spectro	K, $\Delta H^{\ominus}$, $\Delta S^{\ominus}$	J. H. Sharp, *J. phys. Chem.*, *Ithaca* **70**, 584 (1966)
	Tetrazoles	Spectro	K	T. C. Wehman and A. I. Popov, *J. phys. Chem.*, *Ithaca* **70**, 3688 (1966)

Compound		Method	Quantities	Reference
Indole		Spectro	K, $\Delta H^{\ominus}$, $\Delta S^{\ominus}$	Farrell, Trans. Faraday Soc. 62, 18 (1961)
Haematoporphyrin		Spectro	K, $\Delta H^{\ominus}$	R. Foster and P. Hanson, Tetrahedron 21, 255 (1965)
				J. G. Heathcote, G. J. Hill, P. Rothwell and M. A. Slifkin, Biochim. biophys. Acta 153, 13 (1968)
Camphor, substituted camphors, fenchone, 3-methyl-cyclohexanone		Spectro	K, $\Delta H^{\ominus}$, $\Delta S^{\ominus}$	H. G. Kuball and K. Henschel, Z. phys. Chem. Frankf. Ausg. 50, 60 (1966)
Disulphides, dithioctic acid		Spectro	K, $\Delta H^{\ominus}$	W. M. Moreau and K. Weiss, Nature, Lond. 208, 1203 (1965); J. Am. chem. Soc. 88, 204 (1966)
Disulphides, thioctic acid		Spectro	K	W. M. Moreau and K. Weiss, Nature, Lond. 208, 1203 (1965)
Dimethylsulphoxide		Spectro	K, $\Delta H^{\ominus}$, $\Delta S^{\ominus}$	F. E. Stewart, M. Eisner and W. R. Carper, J. chem. Phys. 44, 2866 (1966)
1,1-Dicyano-2,2-bis(trifluoromethyl)ethene	Aromatic hydrocarbons	Spectro	K	W. J. Middleton, J. org. Chem. 30, 1402 (1965)
Dicyano-methylenehexa-fluorocyclobutane	Aromatic hydrocarbons	Spectro	K	W. J. Middleton, J. org. Chem. 30, 1402 (1965)
Pyridinium ions	Iodide ion	Spectro	K	E. M. Kosower and J. C. Burbach, J. Am. chem. Soc. 78, 5838 (1956)
Tropylium ion	Mesitylene	Spectro	K	M. Feldman and S. Winstein, J. Am. chem. Soc. 83, 3338 (1961)
Phosphoro-nitrilic chloride trimer	Naphthalene	Spectro	K	P. A. D. de Maine and R. D. Srivastava, J. Miss. Acad. Sci. 10, 67 (1964)

Appendix Table 2—continued

Acceptor	Donor	Method*	Constants determined	Reference
Phosphoronitrilic chloride tetramer	Naphthalene	Spectro	K	P. A. D. de Maine and R. D. Srivastava, *J. Miss. Acad. Sci.* **10**, 67 (1964)
Di-*n*-propyl tetrachlorophthalate	Benzene, methylbenzenes	GLC	K	D. F. Cadogan and J. H. Purnell, *J. chem. Soc.* (A), 2133 (1968)
Acridine	Tetrakis(dimethylamino)-ethylene	Spectro	K	P. R. Hammond and R. H. Knipe, *J. Am. chem. Soc.* **89**, 6063 (1967)
Phenazine	Tetrakis(dimethylamino)-ethylene	Spectro	K	P. R. Hammond and R. H. Knipe, *J. Am. chem. Soc.* **89**, 6063 (1967)
Polycyclic aromatic hydrocarbons	Methylbenzenes‡	Spectro	K	W. E. Wentworth and E. Chen, *J. phys. Chem., Ithaca* **67**, 2201 (1963)
Hexafluorobenzene	Aromatic amines	Spectro, n.m.r.	$K, \Delta H\ominus$	T. G. Beaumont and K. M. C. Davis, *J. chem. Soc.* (B), 1131 (1967)
Polyfluorobenzenes polyfluoro-chloro-benzenes	Aromatic amines	Spectro	$K, \Delta H\ominus, \Delta S\ominus$	P. R. Hammond, *J. chem. Soc.* (A), 145 (1968)

* Spectro = spectrophotometry in ultraviolet–visible region.
† Measured at two or more temperatures.
‡ Commented upon by P. R. Hammond and R. H. Knipe [*J. Am. chem. Soc.* **89**, 6063 (1967)].

AUTHOR INDEX

Numbers in parentheses are reference numbers and are included to assist in locating references. Numbers in *italics* refer to the page on which the reference is listed in full. Numbers prefixed with the letter A refer to the pages in Appendix Table I in which Chapter 9 references are cited.

A

Abell, C. W., 354(110, 111) *371*

Abidi, M. S. F. A., 64(112) *92*

Abraham, R. J., 113(100) *124*, 174(151) *178*

Abrahams, S. C., 230(42, 43) *250*

Acker, D. S., *9*, 256(27) *273*

Adams, R. N., 269(91) *275, 403*

Adman, E., 284(79) 286(79) 288(79) *301, 426*

Aggarwal, S. L., 255(15) *273*

Agin, D., 363(170) *372*

Aiga, H., 34(6) 35(6) *38*, 67(6) 70(6) 71, 72(6) *89*

Akamatu, H., 34(15, 24) 35(15) 36, 37(15, 24) *39*, 67(24) 69(24) 70(24) *89*, 160(129) 169(129) *178*, 234(80) 239(80) 245(80) *251*, 257(39, 40, 41, 44, 46, 51) 258(39, 40, 41, 44, 46) 259(60) 260(60) 261, 263(41, 79) 264, 265(79) 269(89) 272(89) A392(60) A394(40) A395(41, 44) *274, 275, 395, 425*

Akazawa, H., *401*

Alewaeters, R., 96(11) *102*, 231(49) *250*

Alivisatos, S. G. A., 342(24) *369*

Al-Joboury, M. I., 42(46, 47, 49) *90*

Allen, C. R., 322(70) *333*

Allen, T. L., 297(148) 298(148) *302, 402*

Allerhand, A., 106(18) *122*

Alley, Jr., S. K., 108(65) *123*

Allison, A. C., 352(90, 94) 354(90, 96, 102) 360(162) *370, 371, 372*

Allred, A. L., 113, *124*, 141(46) *176*

Aloisi, G., *401*

Amano, T., 34(15) 35(14) 37(15) *89*, 160(129) 169(129) *178, 425*

Amick, R. M., 345(33) *369*

Andersen, E. K., 377(12) *389*

Andersen, T. N., 262(76) 263(76) *275*

Anderson, D. H., 311(21) *332*

Anderson, H. D., 147(66) *176*, 198, *214*, *415*

Anderson, J. E., 108(52) 121(118) *123, 124*, 253(8) *273*

Anderson, J. S., 14(25) *16*

Anderson, R., 187(25, 26) *214*, 278(16) 279(16) *299, 408*

Andrews, L. J., 16(36, 38, 48) *17, 40, 74*, 94(5) *102*, 148(69, 70, 71) 152(102) 154(102) *176, 177*, 180(72) 184(54, 76) 190(7) 195(54, 55) 196(55, 57) 198(54, 55) 204(72) 206, 207, *213, 214, 215, 280*, 288(89) *301*, 303, 321(2, 61, 63, 64) *332, 333, 398, 402, 403, 405, 406, 407, 409, 411, 412, 416, 419, 420*

Androes, G. M., 271(92) *275*

Anex, B. G., 34(27) *89*, 256(25) *273*

Angus, J. G., *385*

Arellano, R. R., 368(187,188) *373*

Ariëns, E. J., 357(152) *372*

Arnon, D. I., 364(176) *372*

Arthur, Jr., P., 256(31) *273*

Ascoli, P., 356(141) *372*

Ashbaugh, A. L., 112, 117(78) *123*, 142(54) *176, 423*

Asmus, E., 170(144) *178*

Atack, D., 182(16) 184(16) *213*

Atkinson, E. R., 203, *385, 425*

Augdahl, E., 100(29, 31, 33) *103, 398, 402, 405, 406*

Aust, R. B., 67(134) *92*, 262(73) *275*

Avron, M., 368(199) *373*

Ayres, J. T., 297(127) *302*, 379(26) *389*

K

N

S

Turner, A. B., 107(45) *123*

Turner, D. W., 42(46, 47, 48, 49, 50, 51) *90*

Turner, S., 113(100) *124*

Twiselton, D. R., 109(68) 111(88) 113 (88) 114(88) 117(88) 119(88) *123*, 145(61) 150(86) 160(61) *176, 177*, 181(50) 193(50) 194(50) 195(50) 196(50) 201(50) *214*

U

Überle, A., 34(7) 35(7) *89*, 256(21) *273*, 304(6) *332*

Uchida, T., *219*, 257(40, 41) 258(40, 41) 263(41) A394(40) A395(41) *274*

Ulrich, P., 293(111) *302*

Ungar, F., 342(24) *369*

Unzelman, R., 341(18) *368*

Ur, H., 260(64) A393(64) *274*

Utley, J. H. P., *405*

V

Vahlensieck, H. J., 82(192) 85(192) *93*

van Brunt, R. J., *385*

van Dam, J., 152(197) *177*

van de Stolpe, C., 132(22) 133(22) 140(22) 173(150) *175, 178, 397, 402*

van der Veen, J., 192(45) *214, 399*

VanderWerf, C. A., 319(45) *333*

Van Duuren, B. L., 356(146, 147) *372*

van Rossum, J. M., 357(152) *372*

Vars, R., *426*

Vasudevan, K., *11*

Vedeneyev, V. I., 42(52) 87(52) *90*

Veillard, A., 350(66, 67) *370*

Venkatasubramanian, N. K., 107(41) *122*

Verhoeven, J. W., 78(173) 79(173) *93*

Verma, K. K., 379(21) *389*

Viallet, P., *422*

Vickery, B., 327(111) *334*

Vipond, H. J., 355(125) *371*

Virzi, D. R., *185*, 281(49) *300, 404*

Vogel, W. H., 356(133) *371*

Voigt, E. M., *38*, 41(153) *44*, 65(121, 122) 67(143, 153) 68, 69(143) 70(143, 153) *92, 93*

Vos, A., 233(73) 241(73) *251*, 256(36) 257(36) 269(36) *274*

Vosburgh, W. C., 170(140) *178*

W

Wacks, M. E., *385*

Waegell, B., 107(50) 110(50) *123*

Wagner, P. J., *383*

Wain, R. L., 13(20) *16*

Walker, D. A., 368(191) *373*

Wallenfels, K., *9*, *10*, 61(95) *91*, 201(68, 69) *215*, 292(96) *301*

Walling, C., *383*

Wallwork, S. C., 233(66, 68, 70, 72, 74) 234(55, 56, 57, 67, 71, 72) 235, 236(55, 56) 237(56, 57) 239(66, 67, 68, 70, 71, 72, 83) 241(72, 74) 242 (72, 74) 243, 245(66, 72, 83) 248, 249. *250, 251*, 256(35) *274*

Walz, H., 252(3) 254(3) *273*, 277(8) *299*

Wang, C. T., 332(123) *334*

Wang, J. H., 292(101) 293(101) *301*

Wang, S. S., 152(103, 104) 153(103, 104) 154(104) *177*, 324(84) *334*

Warburg, O., 340(15) *368*

Ward, R. L., 304(3) 326(3) *332*

Wardley, A. A., 131(20) *175, 413*

Watanabe, K., 42(38, 39) 48(39) *90, 385*

Watari, F., 96(44) 100(44) 101(44) *103*

Watson, J. D., 356(135) *371*

Waugh, J. S., 115(105) *124*

Wayland, B. B., 190(42) 212, *214, 215, 396, 397*

Webster, O. W., 284(84) 285(84) 287 (84) *301*, 309(17) *332*

Wehman, T. C., *413, 417, 423, 426*

Wei, P. E., 368(187) *373*

Weil-Malherbe, H., 356(137) *371*

Weimer, R. F., 187(27) *214*, 278(17) 279(17, 32) *299, 300, 408*

Weintraub, R., 368(188) *373*

Weiss, F. T., 149(74) *176*

Weiss, J. J., 14, *16*, 253, *273*

Weiss, K., *38*, 190(39) *203, 214, 427*

Wells, C. H. J., *419*

SUBJECT INDEX

COMPOUND INDEX

See also Appendix Table 1 and Appendix Table 2 on pages 390 and 396 respectively. This index does not include solvents.

A

Acenaphthene, 47, 65, 69, 208, 260, 261, 378
Acenaphthylene, 50, 57–59, 203
Acepleiadylene, 233, 239, 245
Acetone, 21, 22, 219, 222
Acetonitrile, substituted acetonitriles, 95, 96, 192, 219, 222, 223
Acetyl-*p*-benzoquinone, 52, 54, 60
Acid anhydrides, 108
Acridine, 49, 378
Acridinium, 293, 325
Adenine, 338
Adrenochrome, 347
Aetioporphyrin I, and Co(II), Ag(II), Mn(III) derivatives, 365, 366
Alcohols, 4, 206, 219, 221, 222
Alkenes, 149, 283, 322, 330
Alkylbenzenes—*see* benzenes, alkyl; *also* individual compounds
Alloxan, 338
Amides, thioamides, 190
Amines, aliphatic, including ammonia, 4, 26, 47, 48, 100–102, 108, 136, 181, 185, 191, 207, 218, 220, 254, 255, 279, 282, 312, 319, 325, 331, 363
Amino acids, 338, 346, 348
2-Amino-1,4-naphthoquinone, 52
4-Amino-4′-nitro (*cis* and *trans* – 1,2-diphenylcyclopentane), 79
4-Amino-4′-nitrodiphenylmethane, 79
4-Amino-4′-nitro(1,2-diphenylethane), 79
4-Amino-4′-nitro(1,3-diphenylpropane) 79
α-(4-Aminophenyl)-ω-(4-nitrophenyl) alkanes, 79, 298
Ammonia—*see* amines, aliphatic
Aniline, 40, 41, 45, 47, 61, 154, 155, 181, 197, 200, 208, 261, 295, 361, 362, 376, 377, 385

Anilines, C-alkyl (*see also* individual compounds), 200, 201, 208, 362, 385
Anilines, N-alkyl (*see also* individual compounds), 200, 201, 282
Anisole, 40, 41, 45, 47, 61, 69, 197
Anthanthrene, 69, 260, 261, 353, 354, 384
Anthra[1,2-*a*]anthracene (1′,2′-anthra-1,2-anthracene), 353
Anthra[2,1-*a*]anthracene (2′,1′-anthra-1,2-anthracene; dibenzo[*bk*]chrysene), 353
Anthracene, 34, 37, 40, 41, 45, 47, 50, 58, 59, 68, 69, 77, 82, 83, 85, 117, 118, 208, 233, 234, 239, 245, 261, 262, 282, 294, 318, 323, 331, 353, 376–379, 384
Anthracene-9 sulphonic acid, 331
Anthranilic acid, 378
1,4-Anthraquinone, 52
9,10-Anthraquinone, 52
Aromatic hydrocarbons (*see also* individual compounds), 31, 58, 77, 147, 148
Aza-aromatics (*see also* individual compounds), 49
8-Azaguanine, 350
3-Azapyrylium, 293
Azulene, 40, 41, 45, 50, 57–59, 203, 233, 239, 245, 260, 261, 376, 377, 385–387

B

Barbituric acid, 338
Benz[*a*]anthracene, 40, 41, 47, 50, 57, 59, 69, 83, 353, 355, 376, 377, 384
Benzene, 3, 4, 14, 21, 22, 28, 40, 41, 45, 47, 57, 58, 74, 84, 86, 87, 95, 96, 98, 106, 107, 109, 111, 114, 118–121, 143, 181, 185, 188, 189, 194, 195, 197, 202, 206, 208–211, 216, 230, 231, 234, 254, 278–281, 294, 326, 327, 353, 356, 376, 377

3-Chloropyridine, 100

Chlorpromazine, 358–361, 363, 364

Chrysene, 40, 41, 47, 50, 57, 59, 69, 353, 376, 377, 379, 384

Cobaltocene, 284, 287, 288

Collidine—*see* pyridines, methyl

Copper(II) - 8 - hydroxyquinolinate—*see* bis-8- hydroxyquinolinatopalladium-II

Coronene, 50, 57, 59, 353

Coumarins, 108

Cumylperoxy radical, 325

Cyanine dyes, 108

Cyanobenzene—*see* benzonitrile

Cyano-*p*-benzoquinone, 52, 54, 60, 387

Cyanuric chloride, 229

Cycloheptatriene, 318

Cyclohexa-1,4-dione, 218, 227

Cyclohexane, 280

Cyclohexene, 208, 327–331

2-Cyclohexenone, 330

Cyclohex-2-enylsuccinic anhydride, 330

Cyclohex-2-enylsuccinonitrile, 328

Cytosine, 338

D

Decacyclene (diacenaphtho[1,2-*j*:1′,2′-*l*]-fluoranthene), 59

Diaminodurene, 272

9,9′-Dianthryl-10-sulphonic acid, 331

Dibenz[*a,c*]anthracene (1,2,3,4-dibenz-anthracene), 50, 57, 59, 353, 354

Dibenz[*a,h*]anthracene (1,2,5,6-dibenz-anthracene), 50, 57, 59, 69, 352–354

Dibenz[*a,j*]anthracene (1,2,7,8-dibenz-anthracene), 57, 353

Dibenz[*g,p*]chrysene, 59

Dibenzofuran, *see* diphenylene oxide

Dibenzo[*a,c*]naphthacene (1,2,3,4-di-benznaphthacene), 353

Dibenzo[*a,j*]naphthacene (1,2,7,8-di-benznaphthacene), 353

Dibenzo[*a,l*]naphthacene (1,2,9,10-di-benznaphthacene), 353

Dibenzo[*b,k*]perylene (2,3,8,9-dibenz-perylene; dibenzo[*fg,qr*]pentacene), 353

Dibenzo[*a,c*]phenanthrene (1,2,3,4-di-benzphenanthrene; benzo[*g*]chry-sene), 353

Dibenzo[*a,g*]phenanthrene (1,2,5,6-di-benzphenanthrene, benzo[*c*]chry-sene), 353

Dibenzo[*a,h*]phenanthrene (1,2,6,7-di-benzphenanthrene; benzo[*b*]chry-sene), 353

Dibenzo[*bg*]phenanthrene (2,3,5,6-di-benzphenanthrene), 353

Dibenzo[*c,g*]phenanthrene (3,4,5,6-di-benzphenanthrene), 353

Dibenzo[*a,e*]pyrene (Naphtho[1,2,3,4-*def*]chrysene), 50, 57, 59

Dibenzo[*a,h*]pyrene (3,4,8,9-dibenz-pyrene, dibenzo[*b, def*]chrysene), 353

Dibenzo[*a,i*]pyrene (3,4,9,10-dibenz-pyrene, benzo[*rst*]pentaphene), 353

Dibenzo[*a,l*]pyrene (1,2,3,4-dibenz-pyrene; dibenzo[*def, p*]chrysene), 59, 353

Dibenzo[*b,e*]pyrene (Dibenzo[*b, def*] chrysene), 50, 57

Dibenzo[*e,l*]pyrene (1,2,6,7-dibenz-pyrene; dibenzo[*fg, op*]naphtha-cene), 353

Dibenzyl (1,2-diphenylethane), 199, 200

2,6-Dibromo-*p*-benzoquinone, 306

Dibromopyromellitic dianhdyride, 52, 54

2,3-Dichloro-*p*-benzoquinone, 306, 387

2,5-Dichloro-*p*-benzoquinone, 52, 54, 83, 146, 147, 204–206, 359, 387

2,6-Dichloro-*p*-benzoquinone, 52, 54, 55, 307, 359, 387

2,3 - Dichloro - 5,6 - dicyano - *p* - benzo - quinone (DDQ), 8, 41, 52, 54, 153, 155, 201, 204, 205, 258, 260, 287, 293, 294, 296, 308, 317, 318, 325, 359, 387

Di-*p*-chlorodiphenyltellurium diiodide, 230

Dichloromaleic anhydride, 387

Dichloromethane, 187

1,4-Dicyanobenzene, 387

2,3-Dicyano-*p*-benzoquinone, 41, 114, 204, 205, 359, 387

1,2 - Dicyano - 1,2 - *bis*(trifluoromethyl) - ethene, 11